Bioinspired Materials for Fog Harvesting

Bioinspired Materials for Fog Harvesting

Principles, Design, Preparation, and Applications

Zhiguang Guo and Weimin Liu

WILEY-VCH

Authors

Prof. Zhiguang Guo
Hubei University
430062 Wuhan
China

Prof. Weimin Liu
Lanzhou Institute of Chemical Physics
Chinese Academy of Sciences
730000 Lanzhou
China

Cover Design: Wiley
Cover Images: © Zhiguang Guo,
© Weimin Liu

Typesetting: Straive, Chennai, India
Printing: CPI Group (UK) Ltd,
Croydon CR0 4YY

Library of Congress Card No.: applied for

British Library Cataloguing-in-Publication Data A catalogue record for this book is available from the British Library.

Bibliographic information published by the Deutsche Nationalbibliothek
The Deutsche Nationalbibliothek lists this publication in the Deutsche Nationalbibliografie; detailed bibliographic data are available on the Internet at <http://dnb.d-nb.de>.

Print ISBN: 978-3-527-35400-9
ePDF ISBN: 978-3-527-84745-7
ePub ISBN: 978-3-527-84744-0
oBook ISBN: 978-3-527-84746-4

C9783527354009_011025

Contents

Preface

This is my third time writing a preface for one of my upcoming books. Looking back, I began my journey in scientific research over twenty years ago as a PhD student, which makes us realize that time flies. Over the past two decades, a lot has changed both in my personal life and in the world. The only thing that remains constant is my love for and dedication to scientific research.

The field of fog harvesting has become a hot topic of research in the past decade, as freshwater resources are becoming increasingly scarce around the world. When a new field emerges and grows quickly, as judged by the number of publications, presentations, and patents. In the past decade, more than 8000 papers have been published. It is well known that the fog harvesting of a surface depends mainly on three factors: fog capture, fog droplet coalescence, and droplet transport. Discovering the relationship among them in enhancing the efficiency of fog harvesting is very important for us.

To this end, the idea for this book, titled *Bioinspired Materials for Fog Harvesting* (BMFH), dates back to 2014, when I published the first paper on fog harvesting, and my research group had just exceeded twenty members. More scientists paid their attention to this field, and the growing number of papers, preprints, presentations, and word-of-mouth discussions, the more important this field will be. Till now, I have published more than 100 papers in this field. It is time for me to summarize the progress and challenges in this field.

I am pleased to co-author this book with Prof Weimin Liu, a full professor at the Lanzhou Institute of Chemical Physics and an Academician of the Chinese Academy of Sciences. He also was my PhD supervisor during 2004–2007. *BEMFH* presents the recent progress, ideas, and research from my group and many of the world's foremost scientists working in the field of fog-harvesting materials. We discuss the theory, fundamentals, fabrications, properties, applications, and uses of these very unique materials inspired by nature.

BMFH covers 13 aspects of the fog-harvesting story, as follows:

1) Chapter 1 analyzes the current state of global water resources and discusses the necessity of fog harvesting and the development of innovative fog harvesting materials. It also emphasizes the bionic design principles inspired by natural organisms and outlines the methods for evaluating the performance of these engineering materials.

2) Chapter 2 identifies natural prototypes for fog collection, detailing the collection strategies employed by these biomimetic models, such as cacti, iron orchids, pitcher plants, desert beetles, spider silk, and desert lizards.
3) Chapter 3 introduces the fundamental principles of fog harvesting, including the nucleation mechanism of water molecules, droplet growth theory, and droplet transport mechanisms. It also presents a theoretical model for atmospheric water collection based on adsorption.
4) Chapter 4 provides a comprehensive summary of the factors influencing the fog-harvesting process, discussing the relationship between each factor's impact and its efficiency.
5) Chapter 5 reviews a range of microscopic materials utilized in fog harvesting, in which how variations in microstructure lead to differing wettability were emphasized.
6) Chapter 6 categorizes the fog-harvesting materials introduced in the previous chapters based on their microstructural characteristics.
7) Chapter 7 explores how the mechanical properties of fog-harvesting materials influence their efficiency.
8) Chapter 8 outlines various surface preparation methods and highlights how they yield different micro- and nano-scale surface characteristics.
9) Chapters 9 and 10, present two types of fog-harvesting systems (non-phase change and phase change).
10) Chapter 11 assesses the feasibility and potential applications of fog collection in current and future texts.
11) Chapters 12 and 13 summarize the challenges in the field of fog harvesting and propose future research directions.

The book is aimed at the following readers:

- Advanced students and instructors in the fields of science and engineering.
- Professional scientists and engineers, who may be trained in more traditional disciplines but who need to learn about this emerging area.
- Policymakers and management experts looking for an understanding of scientific challenges, prospective uses, and emerging markets for fog harvesting.

The overall goal is to capture the multidisciplinary and multifunctional flavor of biomimetic engineering materials for fog harvesting, while providing in-depth discussions of select areas.

Importantly, I would like to express my sincere appreciations to the contributors for their effort and dedication in preparing the chapters, to Hubei University, and to my colleagues and my PhD students: Dr Shangzhen Xie, Dr Fuchao Yang, Dr Xia Li, Dr Xian Zhang, Dr Ke Pei, Dr Shulun Ai, Mr Yuxuan He, Mr Guangyi Tian, Mr Pengyu Zhang, Miss Huayang Zhang, Miss Danyan Zhan, Miss Huimin Zhang, Miss Yidan Zhang, and Miss Anhui Yu for their continuous assistance and hard work in drawing the figures, in compliling the chapters, and for their corrected proof for this book. I also thank Dr Lifen Yang of Wiley for giving me the idea of starting a book at the beginning of this year, and her publication team for their continuous help and support. Finally, I am indebted to my wife, Yan Xin, and my two daughters, Zhehan and Zheyuan, for their kind and endless enthusiasm and support since we have experienced the difficult moment of the COVID-19 outbreak in Wuhan from 2020 to 2022, when I completed the first draft of this book with their help.

This work was supported by the National Natural Science Foundation of China (Grant No. 51735013).

December 2024

Zhiguang Guo
Wuhan

1 Introduction

Water is the source of terrestrial life and the necessary condition for reproduction and the continuation of life. Despite oceans cover 65% of the Earth's surface and contain 97.2% of global water, large portions are saline and undrinkable. Human use is limited to fresh water from freshwater lakes, rivers, and groundwater systems, accounting for only 2.53% of the world's water. Of this, about 70% is locked in polar and alpine glaciers, which are difficult to extract. Currently, our primary access to water includes rivers, freshwater lakes, and shallow groundwater – about 0.3% of global fresh water and 0.0007% of total world water. Although the ocean accounts for a significant proportion of global water resources (mostly saline), most available water comes from ice caps, glaciers, groundwater, inland seas, and rivers: 1.8%, 0.9%, 0.02%, and 0.001%, respectively. Regionally, nine countries, including Brazil, Russia, Canada, China, the United States, Indonesia, India, Colombia, and Congo, hold more than 60% of the world's fresh water, but at the same time, nearly a fourth (about 80 countries) of the global population faces severe shortages. Globally, about 1.5 billion people, or one sixth of the planet, suffer from inadequate supplies of water, with an additional third of countries amount to a billion people experiencing this crisis.

Harvesting water from ambient air holds potential for freshwater provision. Given all indicators, 3.8 billion people experienced varying levels of water scarcity in 2005 [1]. This is expected to persist until population growth peaks in 2050 [2]. Surface rivers constitute approximately 40% of precipitation across global land [3]. Nevertheless, WHO's report forecasts a 20–30% surge in demand for water by 2050 [4]. Urbanization and climate change compound this issue, with half of the urban populace projected to encounter water shortages by 2050 (Figure 1.1) [5].

Researchers analyze the scarcity of water resources across two dimensions, supply vs. consumption, while also appraising the quality of the supplied water in correlation to factors such as demographics, agricultural productivity, and topographical layout. Of the 304 cities examined, an astonishing 273 (about 89.8%) were confronted with water security risks in 2015. Among these, 262 municipalities (or roughly 86.1%) faced quality risk, another 85 entities (about 27.9%) ventured into scarce risk, and an additional 74 municipal governments (about 24.3%) encountered

Bioinspired Materials for Fog Harvesting: Principles, Design, Preparation, and Applications,
First Edition. Zhiguang Guo and Weimin Liu.

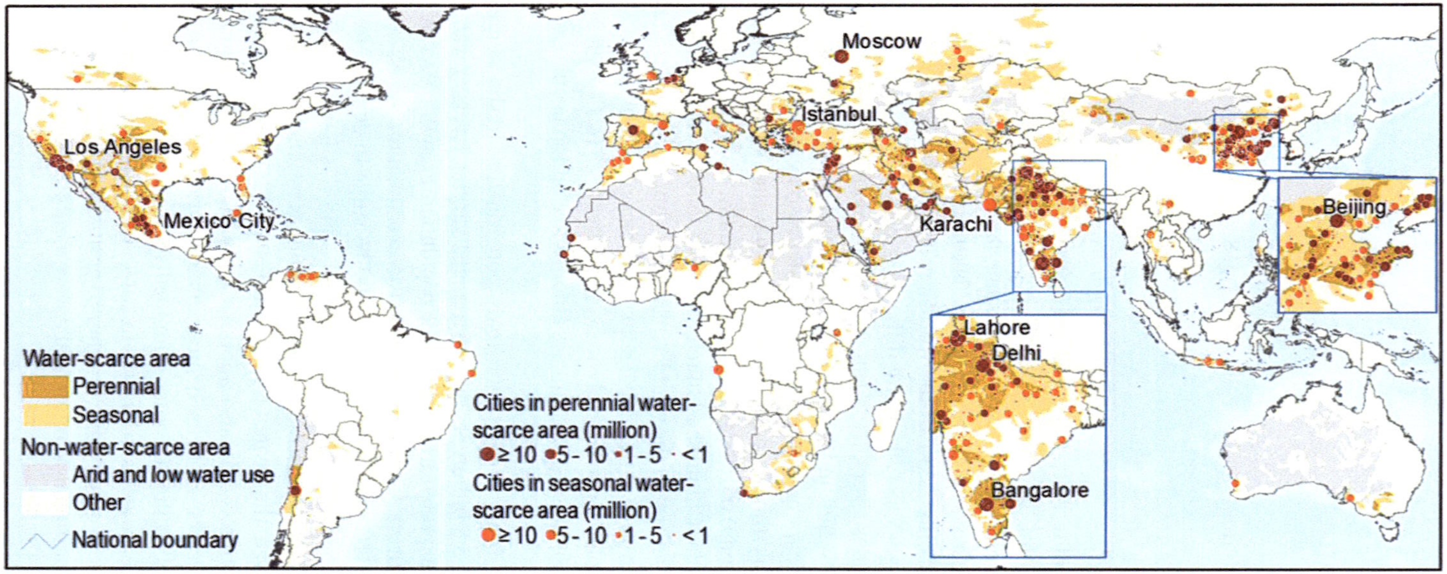

Figure 1.1 Global scarcity of water. Distribution of large cities in hydric stress regions (capitals exceeding 10 million inhabitants in 2016). Source: [5] / Springer Nature / CC BY 4.0.

dual risks (Figure 1.1). Moreover, 69 cities (around 22.7%) were marked as severely exposed to extreme water quality risk, and another 48 (or almost 15.8%) being severely encumbered by severe shortage of water scarcity risk, while 7 (about 2.3%) were grappling with both extremely high-water quality risk and severe water scarcity risk.

Strangely enough, among the evaluated eighteen megacities (i.e. those with population exceeding 10 million), an overwhelming 25 (around 92.6%) were affected by water security risks. Among these, 23 cities (about 85.2%) grappled with quality risks, an additional 10 (about 37.0%) were involved in scarce risk, and 8 (or around 29.6%) had to contend with dual risks. Notably, eleven of these major cities including São Paulo, Mumbai, and Dhaka were severely under threat of highly risky water quality, whereas five major cities (Delhi, Beijing, Los Angeles, Moscow, and Bangalore) were severely burdened by severe water scarcity risk, with Bangalore bearing the brunt of dual risks on both ends [6].

In terms of technology, desalination is recognized as an effective method to alleviate freshwater scarcity crisis. Broadly, desalination can be categorized into thermally driven (distillation) and membrane-based processes. Thermal seawater desalination remains the predominant method in the Middle East region [7]. However, membrane desalination methods exemplified by reverse osmosis (RO) have rapidly grown since the 1960s, and more seawater desalination plants opt for RO technology [8]. Through analysis and comparison, it can be found that one of the paramount challenges associated with RO technology is high energy consumption, which may escalate greenhouse gas emissions [9]. Despite extensive work researchers undertaking in integrating seawater desalination with renewable energy to address energy consumption issues, seawater desalination to obtain freshwater is impractical in inland regions distant from the ocean. Furthermore, even at small scales, seawater desalination technology not only necessitates accessible brackish water sources but also competent operators and maintenance personnel, which significantly restricts the application scope of seawater desalination. However, the geographical environment and climatic conditions of remote or inland areas, as well as issues concerning cost, energy consumption, secondary pollution, etc., impose limitations on the development of the technology [10]. Consequently, there is an urgent need to develop a promising technology to tackle situations in remote areas or areas where economic levels do not advocate centralized treatment and distribution networks [11].

As an immense renewable reservoir, the atmosphere contains approximately 12 900 km^3 of water [12], sufficient to meet part of the water needs of households, agriculture, and industry. Even in remote regions and arid deserts, abundant water vapor is present; the water in the atmosphere is deemed an enormous renewable reservoir. Capturing water vapor from air not only saves significant labor and energy costs but is universally viable across vast regions of Earth. The water vapor in the air is relatively free of impurities and bacteria, and the water harvested from air can reach potable water standards without elaborate purification and sterilization processes. Unlike seawater desalination, when water is extracted from air, it hardly disrupts the water cycle and does not have any detrimental impact on nearby crucial water resources after transportation. Even in the driest desert regions, water vapor is widely dispersed in the form of molecules in the atmosphere. Atmospheric water harvesting (AWH) produces fresh water through the collection of moisture in the air, enabling sustainable water transport without substantial infrastructure and geography or hydrology restrictions [13]. A viable water heating technology should meet the following primary criteria: robust water harvesting capabilities, low energy consumption, affordability, stability, and negligible constraints imposed by the environment and climate [14]. Therefore, the development of efficient AWH technology paves the way to harness this portion of the atmospheric water resource and holds promise to address the freshwater shortage issue, particularly in geographically distant or rural regions where water transportation is costly [15].

1.1 Overview of Research on Atmospheric Water Collection

Traditional fog collectors are a simple yet sustainable system, implemented when a gauze material is exposed to cloud banks. Some droplets collide and accumulate on the gauze material, aggregating to form larger fog droplets, which are subsequently attracted by gravity and channeled into drainage pipes, ultimately reaching water tanks or distribution systems. They can be categorized as standard fog collectors (SFCs) and large fog collectors (LFCs) . SFCs typically find use in exploratory research for assessing the potential quantity of fog accumulated under specific conditions. LFCs primarily perform actual fog collection operations. If the performance of an SFC surpasses a certain threshold, it transitions seamlessly into an LFC. Since the mid-twentieth century, conventional fog collecting initiatives have enjoyed considerable success in Chile, subsequently being deployed in several parts across the world. These locales feature climates and geographies conducive to natural fog gathering predominantly arid tropical and subtropical regions with high altitude. In most nations, the prevalent fog collector preferred involves placing a Raschel grid vertically between two poles to harness moisture from fog [16]. The grid features a trihedral pattern, comprising fine filaments and pores at the millimeter level. Woven polyolefin Raschel nets are a favored capturing medium, treated with UV protection and possessing a shading coefficient of 35%. Eiffel, an innovative 3D fog collector, comprises two layers of Raschel meshes and ten meshes. It optimizes the collection of windblown fog parallel to a heat exchanger, achieving a collection efficiency of 281.2 L day^{-1} – approximately 10 times that of standard full-sized heat exchangers. However, this ideal scenario for collecting fog is subject to strict constraints. Consequently, the emergence of "Harp" and "Diagonal Harp" fog processors has made significant strides beyond these limitations. Comprising a series of closely spaced vertical stainless steel wires, close to a fog net lacking horizontal lines, they capture fog originating from all directions, enhancing the precipitation rate of liquid droplets, averting blockages, thereby enhancing the performance of the collector [17].

However, some of traditional fog collectors merely harvest a portion of fog droplets, exhibiting a relatively modest collection efficiency. A portion of droplets bypass the mesh without colliding with its fibers, while some may bounce off the resilient fibers back into airflow. Furthermore, much of the current efforts geared toward refining such systems focus primarily on investigating the mesh topology of conventional fog collectors. In comparison, there remains insufficient investment in functionally modifying the mesh surfaces and synthesizing enduring functional surfaces given the constraints imposed by wind velocity, atmospheric humidity, droplet size, and other geographical environmental factors [18].

In the comprehensive process of studying the design concept of a biomimetic fog water collector, we have innovated and advanced the fog water collector by thoroughly absorbing the abundant inspiration from biological structures. Using the natural biological water collection mechanism as a guide, we meticulously explored this concept and have consequently designed a series of efficient fog water collectors and systems. Currently, insects such as desert coleopteran, spider silk, cacti, pitcher plants, and toothedge mosses that possess potentiality for fog water gathering have been widely recognized. An integrated water collection process is principally composed of three segments: fog water capture, fog water distribution, and water resource transportation. The most critical aspect is the capture of fog, and the structural attributes and moist degree of the capture surface significantly influence its efficiency. While hydrophilic surfaces have the advantage of swiftness in capturing moisture in fog, their formidable water drop binding force can easily lead to decreased heat transfer efficiency and reduced fog water capture positions when larger droplets or films rapidly form on the surface. Conversely, hydrophobic surfaces are incapable of immediately collecting surrounding fog drops; when minute droplets vertically collide with the surface of hydrophobic materials, a small amount of liquid is repelled, posing an impediment to an efficient fog water capture process.

Simultaneously, the removal of collected droplets facilitates the initial coalescence and further deposition of new droplets in the capture position, thereby forming an efficient and continuous collection cycle. The smooth operation of this cycle relies on the surface possessing superior moisture transportation capabilities. Therefore, how to establish an optimal capture interface for achieving the highest efficient fog water collection efficiency has emerged as a significant challenge. By seeking inspiration from nature and conducting thorough research into organisms with exceptional fog collection abilities, we observed that these organisms developed a sophisticated water flow system for the collection, accumulation, and removal of minute droplets after a long evolutionary journey. For example, natural organisms such as spider silk and cacti also have their unique and intricate fog water capturing, transporting, and collection systems. The periodic spindle and joint structure formed by spider silk in humid air not only can effectively capture a large number of droplets, but its elaborate curvature gradient and surface energy gradient also lay the groundwork for droplet directional transmission. Through the analysis of the structure of cacti, for the purpose of efficiently gathering fog, its needle-like substructure progressively diverges, forming multilayered and functionalized forms.

This specialized form enables the centralized direction of captured, merged liquids and ultimately finalizes the work of end collection in the root hair body of the spines. In stark contrast, desert beetles living in the desert, also utilizing a hydrophilic back ridge to construct a hydrophobic periphery, accomplish the task of capturing fog and separating it effectively.

In addition to the previously discussed sophisticated fog collection mechanism consisting of the typical organisms such as desert beetles, spider silk, and cacti, Nepenthes has consistently demonstrated exceptional capabilities in continual water supply and precise positioning delivery. In a humidity conducive environment, with its excellent absorption of moisture-laden nectar and ingenious microstructure, Nepenthes is capable of creating a smooth and stable thin film of water molecules on the leaf surface, thereby significantly reducing fluid resistance within this region. This feature is cleverly harnessed to successfully capture and utilize insects as a food source. Through rigorous research analysis of Nepenthes' unique structure, human beings have progressively developed numerous irrigation surfaces inspired by water-skid animals, which, due to their advantages of lower surface energy and higher nucleation density, can dramatically stimulate the swift coalescence of fog drops if integrated into the implementation process of fog capture. Alternatively, a layer of lubricating oil is applied to further enhance the fog liquefaction effect, transforming the entire surface into a durable flat surface that facilitates more easy agglomeration of water droplets. The resultant water droplets exhibit minimal contact angle retention phenomenon during descent and strong fluidity, which not only expedites water drop removal but also creates ample space for new droplet nucleation. With reference to the behavior pattern of drops in nature, we note that drops often follow the path with minimum surface energy when inscribed on an Araucaria surface, which has served as an inspiration for our fabricated simulation of Araucaria tree leaf surfaces achieving directed transfer of liquid drops in the direction opposite to gravity. There is no doubt that the forthcoming research process will provide us with additional insights from the rich ecological systems of nature, whether it is regarding fog collection, growth provisioning, or fluid transmission mechanisms and methods, all of which serve as invaluable resources to enhance fog collection efficiency and alleviate the universal issue of insufficient freshwater resources.

To tackle the critical problem of constructing an appropriate fog collection system structure and adjusting optimal chemical composition for designing efficient fog collection facilities, it is imperative that we fully explore and draw lessons from biological water accumulation strategies in nature and conduct innovative scientific research based on these. Owing to the tireless efforts of numerous scientists, significant and breakthrough achievements have been made in contemporary novel fog collection designs inspired by biology. Therefore, in subsequent sections, we will delve into and summarize the innovative ideas of bionic fog collection. With the rapid advancement of nanotechnology, various high-performance bionic fog collection systems are emerging like mushrooms, and the detailed content will be thoroughly explained in subsequent chapters. We are confident that they will provide valuable enlightenment and technical reference for further production of bionic fog collectors. In the evolution of fog collection equipment, wettability theory holds an indispensable role, providing us with the essential knowledge for a more accurate understanding of the fog collection process and the design of more efficient collection devices. To provide a more comprehensive portrayal of the overall picture and development trend of the fog collection field, we have decided to provide relevant theoretical support in a chapter on the evolution process of incorporating the concept of bionic fog collection.

Upon collection of fog, this process can be effectively undertaken when the unique weather conditions that surround moisture-laden, fogy areas (i.e. relative humidity [RH] approaching saturation) are present. However, this ideal environment is not present universally across Earth's regions, particularly in mountainous terrains that are steep, arid zones, and coastal locales. While these locations may typically yield air flow with high humidity, due to a lack of abundant water resources, implementing fog collection becomes an exceedingly challenging task. From an outlook of biomimetics, the diverse biological structures within nature provide substantial insights, offering us fertile resources for constructing corresponding equipment or systems. Nevertheless, current biomimetic fog collection technology remains in the experimental phase in laboratory high humidity environments (RH exceeding 60%). To genuinely apply this technology to daily life, significant hurdles remain. Consider wind direction shifts as an example; the fog collector may fail to exhibit its anticipated effects at such times. To address this issue, we require the unique ability of the fog collector to cope with any potential complex environmental changes. Crucially, the entire fog collection process must be conducted outdoors, where it may be exposed to high wind speeds, which can inflict irreversible damage on the equipment. Therefore, enhancing the durability and stability of fog collection equipment is crucial. Furthermore, a vast majority of biomimetic fog collection devices employ a sophisticated framework with precision reaching nanoscale levels. Any damage to these sophisticated structures could result in a significant reduction, if not complete loss, of fog collection efficiency. Also, most collection apparatus rapidly expel captured water into the atmosphere, which exacerbates the rapid evaporation and loss of water. To address this concern, designing and refining a closed loop system for fog collection become imperative tasks, including circumventing prolonged contact of droplets between the collector and atmosphere for optimal sealing and storage effects, etc., these are crucial problems that demand our immediate attention. Finally, the safety of harvested drinking water after collection demands undivided attention. Prolonged wetting of the sample surface can harbor extensive bacterial growth; similarly, functional coatings adhering to the droplet surface may delaminate during wear, posing a potential health risk in terms of water quality.

In the operational execution of fog collection, we are practically delving into and exercising a fundamental physical phenomenon of physics that clearly exhibits its physics principle essence – phase transition process, which lucidly illustrates this foundational principle. Specifically, in the dew drop formation system, we meticulously manage the ambient temperature, progressively imposing air humidity on saturated conditions, thereby elevating the water vapor pressure to the saturated steam pressure of water. This ensuing situation propels cohesion between water molecules, leading to the final crystallization of dew. Concerning the execution details of low-temperature operations, we can either resort to passive cooling methodologies (such as the effective conversion of solar energy into the primary energy reservoir, or radiant cooling) or more aggressively employ active technological tactics (like heat pumps and chilled water units). However, it is crucial to highlight the previous studies predominantly focused on passive radiative condensers not dependent on additional energy support, whereas for those low-humidity areas requiring greater quantities of dew collection, externally sourced active energy-assisted cooling technologies are inherently needed [18]. Moreover, it is essential to acknowledge that under certain unique thermal and moisture conditions, although active cooling facilities are employed, still no feasible moisture retrieval can be attained due to excessively low dew point temperature. Similarly, considering all the aspects discussed, a meticulous analysis and contrast of conventional desalination technologies (such as RO) reveal that active refrigeration-driven dew collection indeed necessitates extensive consumption of energy, with an average energy expenditure generated in this procedure estimated at about 3.5–4.5 $kWh\,m^{-2}$. Nevertheless, given the procurement costs, maintenance costs, and scarce availability of high-quality saline resources, the practical feasibility of distributed water sources can be assessed and anticipated. Even though dew collection is constrained by specific climatic and geographical conditions, it distinctly outperforms fog collection. Thus, a comprehensive evaluation of various factors may lead us toward proactively engaging people in the action of dew collection as a novel yet practical solution to alleviate freshwater scarcity issues.

Concerning the severe climatic conditions and geographic challenges faced when harvesting fog and condensate, a highly promising method of atmospheric moisture collection is attracting increasing attention from researchers. This approach utilizes an adsorbent to process the intricate steps of water molecules in the air, thus effectively capturing moisture. Liquid droplets accumulate progressively within the absorbing material, ultimately forming liquid droplets, facilitating further collection. This is essentially the common approach to decrease the surface energy structure and absorb the aggregation of water molecules, and commonly utilized adsorbents include traditional adsorption materials, metal–organic framework (MOF) materials, polymer gel-based materials, and porous carbon, four categories of adsorbents. In the forthcoming fifth chapter, we will thoroughly analyze the mechanism of water vapor collection in the atmospheric environment, covering key aspects such as fundamental theory, material design, collection and release of phase change condensate, and future development directions. Despite the high value and rarity of freshwater resources available globally, current operational atmospheric condensate systems still remain scarce. Therefore, the commercial operation of atmospheric condensate technology must adhere to the following five core criteria: efficiency, resource conservation, adjustable size, exceptional adaptability, and long-term stability to ensure continuous smooth operation or at least maintain the normal function for an entire monsoon cycle annually. While existing commercial atmospheric water generators have been introduced, there are currently no products capable of comprehensively meeting these ideals.

Presently, various types of specialized atmospheric condensate collectors are being researched for various atmospheric condensate technologies, with the expectation of gradually mass production and market them. Such special collection equipment not only coordinates closely with the various reaction bodies that need water supply, but once an atmospheric environmental condensate system based on the adsorption principle is implemented, it can purposefully acquire water resources from the atmosphere, holding the promise to become the guarantor for providing stable, reliable, and superior quality water sources for miniature reaction systems such as power generation and chemical reactions. Therefore, in the sixth chapter of this book, we will delve into some actual cases that are currently underway for development and application, demonstrating the tremendous potential and considerable role of water collection in miniature reaction systems.

1.2 Development of Biomimetic Fog Collection

The earlier roots of superwetting bionic surfaces can be traced to studies on the lotus leaf. In recent years, research has advanced tremendously from considering complex structures like desert beetles' exoskeletons, pitcher plants' pitchers, cactus spines, and spider silk. Fundamental wettability theory is critically instrumental in guiding the evolution of bioinspired fog collectors during their development process. Thus, our discourse will commence with wettability theory by commencing with a theoretical dissection of static wetting models and dynamic motion theory models followed by an extensive exploration of the evolution of bioinspired fog collectors.

The static wetting model originates from the smooth interface of an ideal surface. Given the asymmetric attractions surrounding molecules or atoms on a perfect surface, particles at the outermost interface (such as atoms and molecules) have increased energy levels. When a droplet attaches to the surface, it tends to exist in its lowest energetic state, which forms the contact angle, i.e. the angle between the external boundary of the droplet and the solid surface. As it can be seen in Figure 1.2, the contact angle serves as the primary indicator for assessing the wettability of a rigid surface, and it can be vividly depicted through Young's equation that defines static contact angle as follows:

$$\cos\theta = \frac{(\gamma_{SV} - \gamma_{SL})}{\gamma_{LV}}. \tag{1.1}$$

In particular, the components represent the surface tension for the solid–gas, solid–liquid, and liquid–gas interfaces, respectively. θ denotes the apparent contact angle where a hydrophilic interface is identified if θ is below 90°, an intermediate hydrophobic one ($90° < \theta < 150°$), and finally a superhydrophobic state when θ surpasses 150°. It should be highlighted that Young's equation primarily serves to depict ideal smooth surfaces, rendering it inapplicable to rough interfaces. Consequently, in actual scenarios, describing static contact angles necessitates the account of nonideal rough surfaces, further evolving the science behind surface studies.

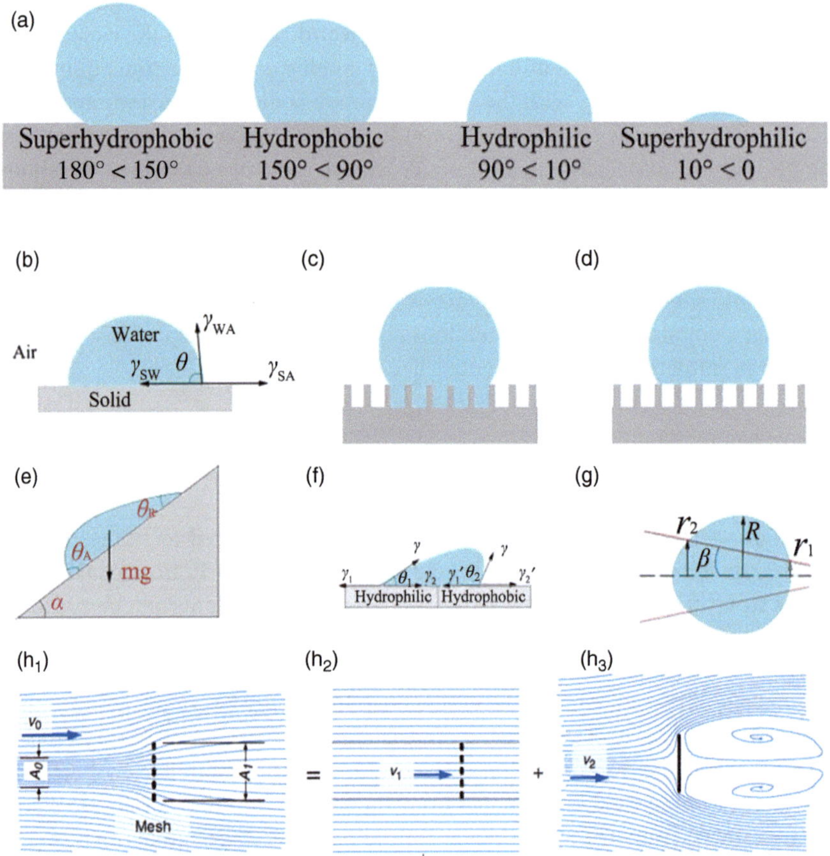

Figure 1.2 (a) Surface wettability. (b) Surface tension. (c) Wenzel model. (d) Cassie model. (e) Contact angle hysteresis. (f) Surface energy gradient. (g) Laplace pressure. Source: [19] / The Royal Society of Chemistry. (h) Superposition scheme; (h_1) represents the real case; (h_2) represents a mesh where all the flow is forced to pass through it, and (h_3) represents a non-permeable screen or plate. Source: [20] / with permission of Elsevier B.V.

In recognition of the presence of nonideal rough surfaces, researchers have developed two excellent theoretical models to elucidate how surface texture influences the contact angle of liquid droplets: Wenzel's model and Cassie–Baxter's model. In 1936, Wenzel introduced roughness into static wetting theory for the first time, fostering a revolutionary theoretical framework, as depicted in Figure 1.2c [21]. Wenzel's model may be expressed as follows: When the droplet is in full contact with the interface without interruption, it will be influenced by the surface's irregularities; subsequently, this influence will cause the droplet to completely cover the area where these features are present [22]:

$$\cos\theta_{\mathrm{w}} = r\cos\theta. \tag{1.2}$$

Wenzel introduced the concept of roughness (r, $r > 1$) into Young's equation. This ratio signifying the disparity between the actual contact area of an aqueous droplet with a solid surface and its relative surface area has been unequivocally defined as roughness. Hence, when the contact angle θ escalates beyond 90°, the roughness amplifies; conversely, if θ diminishes below 90°, the roughness diminishes. However, the limitation of Wenzel's model lies in its inability to depict interfaces composed of liquid–solid–gas under conditions that deviate from uniform wetting. When drops fail to permeate into concavities on a rough surface, they form a state of "locked" gas on the solid surface, thereby leading to a three-phase wetting condition comprising liquid–solid–gas, which can be delineated through Cassie–Baxter models,

$$\cos\theta_{CB} = f_{SL}\cos\theta_1 + f_{LV}\cos\theta_2. \tag{1.3}$$

As shown in Figure 1.2d, these models can be classified into two distinct scenarios: (i) liquid–solid phase interface and (ii) liquid–gas phase interface. Consequently, Eq. (1.3) could be further simplified,

$$\cos\theta_{CB} = f_{SL}(1 + \cos\theta_w) - 1. \tag{1.4}$$

Actually, both the Wenzel model and Cassie–Baxter model effectively describe two distinct wetting scenarios. Consequently, Marmur et al. have proposed a more realistic hybrid model where certain droplets infiltrate into the recesses of the rough structure, while some gas molecules are "imprisoned" therein [23],

$$\cos\theta^* = r_f f_{SL}\cos\theta + f_{SL} - 1. \tag{1.5}$$

Among them, r_f represents the proportion of the contact surface area between the liquid and solid to the total surface area.

Overall, droplets under the Wenzel model show a highly adhesive nature, while those under the Cassie–Baxter model show a tendency to roll off. Under external disturbances such as vibration and shock waves, the Cassie–Baxter state can effortlessly transition to the Wenzel state.

The theoretical models of dynamic movement are encapsulated in gradient wetting displacement, rolling, sliding, superlubricity, and hydrophilic and hydrophobic structure dynamics, as well as hydrophilic conduit motion. Droplets suspended on a surface with graded wettability exhibit an inclination toward migration from hydrophobic to hydrophilic zones, as illustrated in Figure 1.2e. The use of gradient wetting for directing droplet transport is a primary method used to achieve directional control of droplets on horizontal surfaces. For inclined surfaces, due to the influence of gravity, different transport modes may be selected by the fluid. For example, on a wet surface dominated by the Wenzel model, the predominant mode of motion is the sliding mode; however, on a wet surface governed by the Cassie–Baxter model, due to the larger static contact angle and weaker droplet adhesion force, droplets migrate through the rolling mode. On injection surfaces, the difference in tension between the droplet and injected fluid results in unique structures such as wetting ridges, which facilitate an ultra-sliding motion pattern characterized by low resistance and ease of slippage (Figure 1.3).

Figure 1.3 Various superwetting surfaces found in nature. Source: Reproduced with permission from [24] / Elsevier.

The inception of trees and shrubs accumulating substantial volumes of fog spans back several centuries, or perhaps as far back as human habitation among fogy mountain summits [25]. Nonetheless, in arid landscapes bereft of greenery, fog dissipates into dry interior regions, uncollected. Early literature citations and localized experimental trials involving fog accumulation exist; however, it wasn't until 1987 that a large-scale initiative was embarked upon, intended to provide sustainable water supply to a coastal village in Chile, home to approximately 330 residents. At Chile's elevated altitudes (780 m), the average droplet diameter falls within the vicinity of 8–12 μm, and the droplet density typically fluctuates between 100 and 400 cm^{-3}. Fog droplets are captured by a straightforward impaction method. A vast polypropylene net is strategically positioned along the trajectory of the droplets. As the droplets neared the surface, some circumvented the object, while others impacted the surface. The initial triumph of the Chilean fog collection endeavor, the subsequent field collections during the summer monsoon seasons in 1989 and 1990 in the Hofar region of Oman, and a pragmatic scheme implemented in Peru in 1990, collectively indicate that fog collection could potentially be universally applicable across regions or nations with distinct climatic and geographic attributes [26]. Following the recognition of the water-acquisition capabilities of flora and fauna in their natural habitats, various biomimetic fog collectors were devised. We shall now explore the progression of biomimetic fog collection, commencing with organisms such as the Namib Desert beetle and spider silk exhibiting water-acquisition capabilities.

In 2001, Parker and Lawrence unveiled the survival mechanism of the Namib Desert beetle [27]. Drops from the foggy breeze accumulate on hydrophilic protuberances until they attain critical dimensions, subsequently rolling down the hydrophobic valleys toward the beetle's maw [28]. Since this revelation, surfaces possessing both hydrophilic and hydrophobic characteristics have attracted considerable interest, culminating in the creation of diverse water-collection materials derived from substrates such as fabrics [29], cellulose films [30], composite nets [31], copper nets/sheets [31], and stainless steel nets [32]. To date, engineered hybrid surfaces of the Namib Desert beetle have been regulated in terms of both surface structure design and surface chefogry, yielding a plethora of invaluable research findings. It has been confirmed that surfaces featuring periodic peak concavities exhibit superior water collection efficacy due to the mitigation of the capture and aggregation processes of water, thereby diminishing the reevaporation of microdroplets. By scrutinizing the water-collection proficiency of nanostructured surfaces with varying patterns and morphologies, encompassing pure superhydrophilic surfaces, superhydrophobic surfaces, circular patterns, and star patterns, it was discovered that pure superhydrophilic and superhydrophobic surfaces displayed relatively modest water-collection efficiencies, while the star pattern exhibited the greatest collection potential [33]. Additional evidence indicates that the pentagram star pattern surpasses other point numbers of star patterns in terms of water-collection efficiency. The collection efficiency also appears to be profoundly influenced by the actual dimension of the pattern. When the star shape and total hydrophilic area remain consistent, reducing the pattern size amplifies the effect considerably. Nevertheless, attaining a high fog collection efficiency may necessitate sophisticated processing of the surface pattern, which invariably results in an augmentation in production time and scale constraints.

The spiderwebs are ubiquitous in our daily encounters, functioning as tools for arachnids to secure their prey. On fogy mornings, immediately preceding sunrise, one may witness spiderwebs abundant with water droplets. These complex constructs efficiently absorb humidity from ambient air. The fibrous strands consist of periodically spaced spindles constructed of disordered nanoparticles and junctions emitting nanofibers organized in a semi-ordered fashion [34]. Driven by the difference in roughness between spindles and junctions, not just a spatial structure gradient but also a surface energy gradient exists between them. This permits water droplets to spontaneously transition from the junction to the spindle due to these dual influences. Motivated by the water retention strategy of spider silk, scientists have engineered artificial spider silk. Currently, four methodologies can be utilized to fabricate biomimetic fog collectors: electrokinetics, impregnation, microfluidics, and flow coating [34].

Species belonging to the Cactaceae family indigenous to arid deserts display remarkable drought tolerance due to the cooperative interaction of conical structures on cactus thorns and wettability gradients. The synergy of surface energy gradients and Laplace forces underpins the capacity of cacti to exploit fog collection abilities [35]. Taking cues from cacti, several artificial conical structures have been devised via methods like magnetic particle-assisted molding [36], electrospinning, 3D printing [37], etc. To maximize the efficacy of fog collection, the apex angle and surface wettability of the conical structure have been scrutinized and refined by diverse research groups. Studies indicate that an optimal apex angle of approximately 10° facilitates the directional movement of trapped droplets on the hydrophobic layer-coated cone surface, thereby augmenting the water collection efficiency.

Given that conical spikes exhibit proficient fog capturing capability and directed transport to the base of the cone-shaped structure, the accumulation and preservation of these droplets become a pivotal concern. Consequently, a comprehensive fog capture system encompassing fog capture, coalescence, directional transportation, and collection necessitates development. Presented proposals incorporate Janus and porous membranes (comprising a hydrophobic surface and a hydrophilic surface), facilitating unidirectional droplet penetration prior to gravity causing droplets to descend into the collection receptacle. Alternatively, superhydrophilic sponges can be assembled at the base of the cone-shaped spine for continual water storage [38].

Unidirectional fluid transportation devoid of energy expenditure is an indispensable attribute of water capture, applicable for spontaneous water accumulation at precise locations. In sustained water capture procedures, prompt transit accelerates the process of water acquisition and reduces moisture volatilization. Chen et al.'s research established that the pitcher plant (*Nepenthes alata*) exhibits the ability to unilaterally transport superficial moisture through multiple groove micronetworks on its superior leaf margin. Moreover, Chen's group also disclosed the extremely rapid water capture prowess and transport procedure of Sarracenia trichomes [39]. Post droplet initiation by capillary forces to form a liquid layer on the hierarchical microchannels of the trichome surface, subsequent seized droplets attain ultrafast liquid transfer, with a velocity exceeding three orders of magnitude swifter than droplet transport instigated by Laplace pressure in the cactus spine. This expeditious droplet transport results from the shift from standard capillary forces to non-liquid–solid contact-induced superlubricity after the formation of the liquid layer. The ultrahydrophobic surface not only inhibits water molecule aggregation but also damages gas entrapped within its nanoscale architecture during operation, culminating in an irreversible transition from the Cassie state to the "sticky" Wenzel state. Conversely, the slippery perfusion surface derived from the pitcher plant surface, owing to the presence of slip fluid, forms a molecularly smooth surface, substantially augmenting the efficacy of droplet elimination and circumventing the problem of damage to the hydrophobic attributes of the ultrahydrophobic surface. Drawing inspiration from the distinctive structures of pitcher plants and rice leaves, Wong et al. proposed a hydrophilic directional rough surface, which not only boasts high droplet nucleation density but can also promptly eliminate droplets. This surface incorporates nano-textured directional microgrooves, wherein the nano-texture individually imparts hydrophilic liquid lubricants [40]. Through molecular dynamic simulations, they discerned that effective droplet nucleation was ascribed to hydrophilic surface functional groups, whereas the rapid removal of droplets was due to a substantial decrease in droplet adherence to the directed surface structure and slip interface. They further validated that the fabricated hydrophilic directional slip rough surface, endowed with a vast surface area, hydrophilic slip interface, and directed drive fluidity, surpasses traditional drive fluid surfaces in water collection applications.

The moss *Syntrichia caninervis*, one of the most prevalent mosses in desert locales, flourishes in severe aridity with scarce water resources. Truscott and coworkers identified in 2016 that *S. caninervis* employs hairy minuscule pili at the apices of each leaf to gather diverse sizes and shapes of water, encompassing clusters of water molecules, fog droplets, and rain drops [41]. Apart from conventional plants transporting water from the substrate, this reversed water collection system permits direct water capture from the atmosphere. Initially, clusters of water molecules in the air are apprehended and nucleated on the pili. Subsequently, small fog droplets amass in zones with elevated densities of small barbs, morphing into larger droplets. Then, the droplets swiftly transport directionally along the pili to the leaf base. On rainy days, pliant leaves and dense clusters of pili dissipate most impact forces and curtail moisture loss triggered by splashing, optimizing the retention of precipitating droplets.

Harnessing the distinctive microarchitecture displayed on termite wings, multiple species within this insect group have the capacity to fly in inclement weather conditions, exemplified by Nasutitermes sp. and Microcerotermes sp. The conventional configuration of a termite wing comprises hydrophobic, hair-like columns and star-shaped microlattice networks. Upon impact of macrodroplets on the wing surface, they bounce back due to repulsion forces exerted by hair-like columns. In contrast, when minute droplets come into contact with the wing, star-shaped microlattices provide anchorage for the droplet. As the droplet expands through microscale aggregation, it may eventually be propelled away. Drawing inspiration from the termite wing, Uchida and coworkers employed ultraviolet radiation-induced crystal growth methodology to mimic the dual wettability of the termite wing, culminating in two distinctively shaped needle-like crystals of varying dimensions [42]. Small needle-like crystals trigger droplet repulsion, while larger ones demonstrate adhesive properties to water droplets. This surface architecture displays exceptional water collection potential.

Through replication of natural models, scientists can ingeniously integrate two or more diverse water collection strategies, thereby generating optimally designed water collection systems that substantially augment their water collection efficiencies. For example, Park and Aizenber et al. were inspired by the nanofibrous desert beetle exhibiting hydrophilic protuberances in a hydrophobic environment, cacti thorns demonstrating geometric structural gradient morphologies, and pitcher plants with slippery internal surfaces, successfully crafting an asymmetric smooth surface whose moisture collection efficacy was appreciably higher than other surfaces. The synergistic fusion of vertex geometries, analogous to cactus thorns, guides droplets downslope by establishing a free energy profile, thereby facilitating directed droplet transportation; the smooth surface impregnated with fluid (derived from pitcher plants) further boosts transport performance, and the protuberance concept borrowed from the nanofibrous desert beetle [27]. Based on the three fundamental processes of water collection, rational design of water collection systems can seize and transport droplets even under gravitational influence. Indeed, the design of contemporary biomimetic fog collection apparatus or systems has progressed from single biomimicry to multi-biomimicry, and we shall explore additional biomimicry principles in the ensuing section.

Advances in water collection materials capable of harnessing diurnally fluctuations in ambient temperature and RH are an essential component in the evolution of atmospheric water collection apparatuses. Particularly, it pertains to an air-borne water harvesting system that operates through adsorption. Typically, water is procured during chilling nocturnal periods with heightened RH, subsequently desorbing during daytime as temperatures rise. Preferably, the energy necessitated for desorption is harnessed from solar radiation, obviating the necessity for supplementary energy input. Xin et al. fabricated a poly(n-isopropylacrylamide) sponge-like cotton fabric featuring temperature-responsive wettability alterations, capable of autonomously accumulating and discharging moisture from humid air. Microscopically, temperature fluctuations instigate structural transformations in thermoresponsive polymers, synergistically with augmented surface texture, allowing sponge-like material to reversibly alternate between superhydrophobic and superhydrophilic states [43]. This permits water to be harvested during nocturnal hours, followed by water desorption during daytime temperature ascension. Subsequently, investigators concentrated on devising alternative absorbent materials from a viewpoint of reduced energy expenditure and facile processing [44], encompassing hydrogels, three-dimensional porous structures [45], fabrics, nanoscale powders [46], and silica fibers [47]. For instance, Kim et al. synthesized a Ni-IRMOF74-III metal–organic framework architecture, which is adorned with azopyridine molecules. This MOF can undergo photochemically triggered cis/trans transitions, demonstrating superior water collection proficiency compared to other MOF-based water collectors, even under minimal temperature oscillation conditions.

Hydrogels, owing to their high hydrophilic nature and remarkable water retention capacity, are emerging as promising water collection and storage materials [48]. For instance, Wang et al. have devised a flexible hybrid light-thermal adsorbent comprising deliquescent salt and hydrogel, competent of capturing fog in the atmosphere [49]. Distinct from other hydrogels, this salt gel incorporates hygroscopic, nontoxic, environmentally friendly $CaCl_2$, preserving outstanding water absorption capacity even in low humidity settings. Due to the incorporation of hydrogel structure, the saltwater gel retains its solid form after substantial water absorption. The saltwater gel further incorporates carbon nanotubes and displays typical photothermal effects, swiftly discharging water upon routine sunlight exposure. With a judicious material design, the hydrogel can accumulate and discharge water intermittently at dusk and dawn.

Currently, atmospheric water collection systems employing adsorbent methods perform a unique daily water collection cycle, yet freshwater procured through this methodology still markedly lags behind in meeting escalating potable water requirements. Despite endeavors to implement techniques that curtail the adsorption–desorption cycle, collection efficacy remains restricted by the protracted cycle duration. Additionally, the absorbed water must be discharged in sunny conditions, thus impeding the ultimate production of fresh water on cloudy days. In addition, given climatic discrepancies across various nations and regions, and climate fluctuations across seasons, the adsorbent water harvesting capacity may be significantly reduced or even rendered ineffective. Hence, it is imperative to consider the potential impact of these objective factors during the material design and fabrication process. In arid regions, where the water absorption capacity of materials diminishes in low humidity environments, the water collection performance of most water collection materials is substandard. Consequently, researchers are innovating novel composite materials to permit their utilization in low humidity conditions, such as encapsulating hygroscopic salts within the adsorbent matrix [50].

1.3 Biomimetic Inspiration for Collecting Fog

The Namibian Desert, located within Africa, is recognized as one of the earliest and driest global deserts, serving also as the single coastal desert globally. Consequently, this desert presents a massive terrain of sand dunes influenced by fog. The ocean fog constitutes a primary source of nourishment for the vegetation, snakes, spiders, beetles, and lizards that inhabit this area. Over millennia of evolution, numerous creatures have adapted diverse tactics to survive while utilizing fog. For instance, Parker and Lawrence documented a fog collection strategy employed by the Namibian desert beetle Stenocara [27]. Studies indicate that the hydrophobic wax layer solely envelops the elevated regions of the insect's exoskeleton. Conversely, these elevated sites (ranging in diameter from 0.5 to 1.5 mm and positioned at intervals of 0.5–1.5 mm) lack the wax layer, demonstrating hydrophilic characteristics. It is postulated that this alternation between hydrophobicity and hydrophilicity contributes significantly to the beetle's proficient fog collection capability. The hydrophilic regions on its back efficiently trap minute droplets, facilitating their swift coalescence into larger ones. With sufficient aggregation, the large droplets exploit the hydrophobic nature of the adjacent region to swiftly roll toward the head where they are gathered and utilized. Motivated by this distinctive architecture of alternating hydrophilic projections and hydrophobic surroundings, several biomimetic materials have been engineered and exhibit superior fog collection abilities. Among them, superhydrophobic/superhydrophilic patterned surfaces with micro/nanoscale textures have been successfully produced through physical and chemical methodologies, resulting in a relatively advanced fabrication technique. The principal techniques encompass five categories: photolithography [51], mechanical micromilling [52], composite technology employing distinct wettability materials [53], stimulus-responsive fog capture technology [54], and inkjet printing technology [55] (Figurc 1.4).

In the production of water fog collectors using artificial desert beetle scales as a model, we primarily manipulate surface wettability and devise patterned surfaces to influence collection efficiency. In controlling surface wettability, we utilize selective chemical hydrophobic modifications, regulating the content of iron and cobalt particles on fiber fabrics to generate surfaces with varying degrees of hydrophobicity. Our experiments demonstrated that escalating the proportion of hydrophobic particles shortens the duration of the first drop of water rolling off the fiber surface. Conversely, augmenting hydrophilic particles prolongs this time. These superhydrophobic–superhydrophilic surfaces offer cost-effective and convenient advantages, but the distribution of hydrophobic and hydrophilic particles is disordered. We discovered in subsequent studies that the arrangement pattern of different hydrophilic and hydrophobic patterns significantly impacts water fog

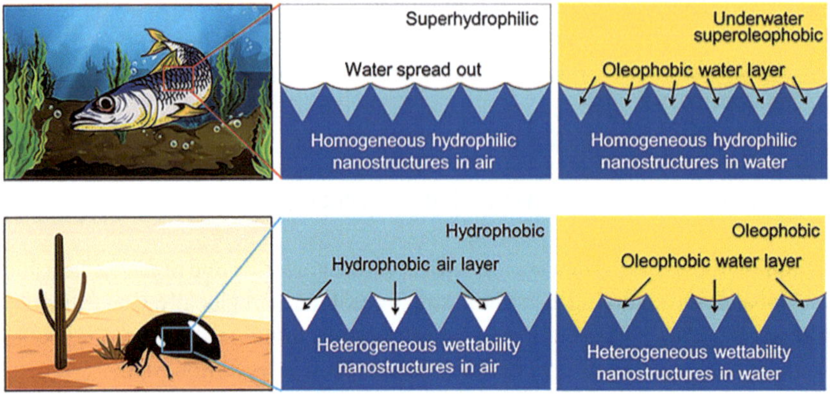

Figure 1.4 Surface wettability of fish scales and surface wettability of desert beetles. Source: [56] / American Chemical Society.

collection [57]. Indeed, from a morphological perspective, the differences between fabricated bionic surfaces are significant at the micro/nanoscale, but negligible at the macroscopic scale. The presence of hydrophilic regions primarily facilitates the rapid capture of liquid droplets. After experiencing the water fog supply phase, larger droplets can rapidly detach for the hydrophilic region to initiate another round of water fog capture. Therefore, it is crucial to investigate whether geometric structure design can expedite droplet detachment to strategically enhance water fog collection efficiency. We compared the process of water fog collection on two patterned surfaces: hemispherical hydrophilic projections and linear hydrophilic projections. It was found that over the same period, droplets could be captured more quickly and aggregated into a critical size on the surface arranged linearly. Moreover, the anisotropy of linear projections results in droplets picking up more droplets along their descent path, hence exhibiting superior water fog efficiency. Furthermore, considering the fluid characteristics of the water fog, diverse surface geometries will interfere with the movement of droplets in the fog. For instance, Aizenberg et al. observed in their study of condensation of water vapor on wet slippery projections that discontinuities in the surface can amplify the diffusion flux of water vapor, thereby enhancing the water supply rate. Thus, the impact of the geometric morphology of bionic surfaces on the overall water fog collection should not be overlooked, particularly its effect on the water fog supply process warrants further research. To this end, we produced a series of superhydrophobic projections with varying arrangements and quantities, revealing distinct effects on the enhancement of droplet growth rates. Using Fluent software, computational fluid dynamics (CFD) clearly illustrated that these projections can modify the dynamic pressure distribution over the surface and concentrate fluid (composed of microdroplets and air) around the projection. Droplets surrounding the projection exhibit higher velocity than those adjacent to the plane, thus these high-energy droplets can be captured more quickly by the projection surface. Consequently, these projections capture more microdroplets, and the growth speed of droplets on them can be substantially accelerated. The various patterned surfaces of the artificial desert beetle rely on gravity to induce droplet detachment. So how can we optimize the water transport process and improve water fog collection efficiency? The following sections will provide additional inspiration and strategies from spider silk and cacti, where the driving force can also be provided by surface energy gradients and Laplace pressure. How to optimize the design of three fundamental processes to realize an efficient water fog collection system has been our ongoing endeavor in both the past and present.

In 2010, Zheng and coworkers unveiled the concept of spider silk fog collection for the first time. They observed that under moist conditions, spider silk fabricates a structure consisting of alternating coarse conical spindle knots and fine junction points. This distinctive structure plays an integral role in directed water collection [58]. The spindle knot region is formed by highly randomized nanofibers, exhibiting high roughness; conversely, the junction region comprises orderly nanofibers arranged with lower roughness. Moreover, each repeating unit comprising spindle knots and junction points can be viewed as a spindly structure whose diameter varies gradually. Hence, spider silk in humid environments possesses unique surface energy and geometric structure gradients, which serve as driving forces for directional droplet transport. The surface energy gradient prompts droplets to move from the junction to the more hydrophilic spindle knot, while the geometric structure gradient generates a Laplace pressure difference, propelling droplets from the junction (high curvature area) to the conical spindle knot (low curvature area). Consequently, the mystery behind the efficient, continuous process of spider silk capturing, condensing, transporting, and accumulating fog has been unraveled, offering novel strategies for scientists to develop biomimetic fog collectors.

Undeniably, the size and surface morphology of artificial spindle knots, the spacing between spindle knots, etc., significantly affect the efficiency of fog collection when mimicking spider silk. For instance, Zheng et al. fabricated heterogeneous spindle knot microfibers using electrospinning. This technique effectively modulates the morphology of spindle knots by altering the concentration of calcium chloride in the fluid [58]. Experiments show that the morphology of spindle knots directly impacts the efficiency of fog collection. Greater roughness implies a higher gradient of surface energy, resulting in superior fog collection efficiency. Furthermore, this technology allows precise control of the distance between spindle knots and junction points by adjusting the flow rate. They discovered that when the interval between two spindle knots increases, it becomes challenging for droplets on adjacent spindle knots to coalesce, leading to prolonged maximum droplet formation time, extended water collection period, and reduced water collection efficiency. In addition, factors influencing the critical droplet detachment volume were examined, revealing that the geometry of the prepared spider silk fiber significantly influences the size of suspended droplets [59]. This is attributed to the hump (spindle knot), which enhances the stability of the triple phase contact line through the combined effects of slope and curvature, thereby providing sufficient capillary adhesion for suspended droplets. Therefore, incorporating small-sized spindle knots between two larger spindle knots into multi-scale, multi-gradient biomimetic fibers was proven to be an effective strategy to enhance the fiber's ability to suspend droplets and achieve higher water collection efficiency [60].

The cactus can be seen as a sophisticated combination of architecture and functionality within a dew collection system. Its spines, composed of three distinct structural features, namely, pointed barbs at the tip, gradient grooves in the middle, and banded hairlike structures at the base, synergistically facilitate the capture and directional transportation of droplets. Like spider silk's mechanism for fog collection and transport, the initial droplet is captured and grows simultaneously on both the barb and the main stem of the spine. Subsequently, the large droplet on the barb migrates toward the root of the barb and merges with the droplet on the main stem. Given that the entire spine is a tapered geometric structure, there exists a difference in the Laplace pressure from the apex to the root. Moreover, the spine is adorned with a hydrophobic plant wax layer, and the groove presents an anisotropic distribution; the density of the spine tip is higher than that of the root, thus creating a gradient of surface energy between the tip and the root of the spine. Consequently, as the barb and main stem of the spine continue to capture and merge droplets, the merged large droplet commences its journey to the root under the combined influence of the Laplace pressure difference and the surface energy gradient. It is also noteworthy that this collection mechanism involves the rapid absorption of droplets upon contact with the hairlike structures of the spine root as most current dew collection systems incorporate an open collection vessel, which inevitably leads to some water evaporation loss.

Inspired by the cactus's dew collection mechanism, focusing on droplet capture and removal (transportation), the fabrication of tapered collection spines with dual gradients of structure and wettability can enhance the self-propulsion of captured droplets and expedite the regeneration of collection sites under the dual gradients of surface energy and structure. We achieved this by a precipitation post modification method, where a gradient change in wetting properties was obtained from the hydrophobic apex to the hydrophilic root of the modified superhydrophobic copper needle by gradient deposition of hydrophilic micronanoparticles on the substrate. When a droplet encounters a surface with a gradient of structure and wettability, it can undergo antigravity transportation despite being in a negative angle state, propelled by the driving force of the gradient of structure and surface energy. Thanks to the efficient transportation of droplets, this dew collection device can rapidly regenerate the dew collection surface and accelerate the dew collection cycle, thereby demonstrating outstanding dew collection performance [61]. Liu and coworkers, by mimicking the cactus's tapered spine structure, fabricated a flexible magnetic responsive spine array to improve the capture and collection ability of fog droplets in low-fog flow rate environments [62]. Under the action of an external magnetic field, the likelihood of collision between the oscillating spine and the fog droplet increases, and due to the periodic oscillation of the tapered spine, the effective dew collection area of the spine is expanded. Therefore, the magnetically controlled flexible tapered spine array exhibits excellent dew collection capability in low fog flow environments, while in the same scenario, due to the lower fog speed, fog is difficult to capture without an external magnetic field, and fog collection efficiency can be negligible. The design of the flexible tapered spine shows that researchers have taken account of the impact of fog flow rate during the dew collection process on the efficiency of dew collection, and they are also paying attention to the double aspects of the fog capture process and the transport link.

Inspired by the strikingly diverse water droplet capture, condensation and transport systems exhibited by the Namib Desert beetle, spider silk, and cactus, we engineered multifunctional patterned surfaces, heterostructured fibrous threads, and conical needle prongs for efficient water collection. The synergistic interplay between structural gradients and surface energy gradients enables rapid droplet removal; however, beyond droplet removal, capturing, condensing, and mitigating evaporation before reaching the critical droplet detachment (transportation) volume are crucial considerations in achieving high-efficiency water collection. Integrating a single or dual specific water collection mechanism is insufficient to achieve an integrated, multifunctional water droplet capture, condensation, transportation, and collection process. Therefore, designing novel multifunctional superwetting devices or systems and establishing theoretical models to analyze the influence mechanisms of surface topography and wettability, and geometry on water droplet collection are areas that require our attention and effort. Through relentless exploration in the field of water droplet collection, several bioinspired, multifunctional water droplet collection systems have been reported, demonstrating an efficient cycle of droplet capture, condensation, and transport during the collection process. Lai and coworkers summarized the evolution of water droplet collection devices from single biomimicry to multiple biomimicry, suggesting that more organisms with efficient collection mechanisms may be discovered in the future [63]. At present, multiple biomimetic devices can be classified as double biomimicry, triple biomimicry, and even quadruple biomimicry. How to skillfully integrate the mechanisms of water droplet collection from different animals and plants and consider the three fundamental processes of water droplet collection fully embodies the cutting-edge development of the field of water droplet collection. In this section, we will briefly introduce the design concepts of several typical multiple biomimetic water droplet collection systems, aiming to provide readers with a basic understanding of the integrated design and advanced development of water droplet collection systems.

The upper and lower surfaces of lotus exhibit complete wettability, with the upper surface exhibiting superhydrophobicity and the lower surface exhibiting superhydrophilicity. Inspired by the asymmetric wetting properties of lotus leaves, a series of Janus materials with asymmetric wetting properties were designed and fabricated. For Janus materials, the asymmetric wetting properties on both sides generate a wetting driving force on droplets, causing them to transport unidirectionally from the hydrophobic side to the hydrophilic side but not in reverse. During the water droplet collection process, Janus materials can accelerate the unidirectional penetration and removal of captured droplets, improving the problem of slow regeneration of capture sites due to gravity-induced droplet shedding and large droplet evaporation losses. As a result, many teams have strategically incorporated Janus materials into a water droplet collection system along with other biomimetic devices to achieve higher water droplet collection efficiency than surfaces with homogeneous wetting properties. For example, by combining Janus materials with the hydrophilic–hydrophobic wetting characteristics of desert beetles, we successfully prepared hydrophilic–hydrophobic/superhydrophilic copper foam through hydrothermal growth and local modification [64]. This Janus copper foam balances the three basic water droplet collection processes. The hydrophilic–hydrophobic surface facilitates the capture and condensation growth of water droplets, while the asymmetric wetting property further promotes the transport and removal of water droplets. Therefore, compared to homogenous wetting copper foam mainly relying on gravity-induced droplet removal, Janus foam copper exhibits superior water droplet collection performance. Similarly, integrating the tapered structure of cacti and Janus properties is also a design approach for a water droplet collection device [65].

The utilization of fluid-percolation surfaces in achieving low contact angle hysteresis and rapid droplet motion, benefiting from their ability to regenerate nucleation sites and enhance condensing heat transfer, has become a promising strategy for designing efficient fog collection systems. For example, we have successfully engineered an interdisciplinary four-dimensional bioinspired fog collection system by blending features from the bumpy structure and wettability distribution of desert beetles, conical structures of cactus spines, seed propulsion behavior of aquatic plants, and slippery surfaces of pitcher plants [66]. This multifunctional fog collection system is composed of superhydrophilic copper needles and regular hydrophilic bumps, with a zinc sheet substrate surrounding a hydrophobic slip surface. The top of the bump features regular circular holes, into which the superhydrophilic copper needle is inserted vertically and anchored to a sponge base. Numerous droplets captured by the slip surface are swiftly transported to the top of the bump due to the oil meniscus effect generated by the bump, significantly enhancing droplet transportation and regeneration of nucleation sites. Similar to the movement of seeds toward aquatic vegetation, the oil meniscus around the hydrophilic bump on the hydrophobic slip surface exerts capillary driving force on the droplets, causing them to move toward the bump in all directions. The structural gradient and superhydrophilicity of the tapered copper needle can help address the issue of hindering droplet removal due to the formation of a water film on the hydrophilic buffer, thereby minimizing the time required for water collection circulation. Therefore, this multifunctional bioinspired fog collection system not only facilitates droplet capture, unidirectional pumping, but also accelerates droplet removal and storage, demonstrating the comprehensive and coordinated nature of the three fundamental processes involved in fog collection.

It is worth noting that during the design and development of fog collection systems, considerable research attention has been paid to a crucial factor influencing fog collection – the velocity of the fog flow reaching the fog capture surface. When discussing the bioinspired fog collection concept using nanoscale cloth desert beetles, we observed how the geometry of the surface bumps could influence the flow distribution of the fog fluid, further impacting the supply of fog during the collection process. Thus, for two-dimensional fog collection surfaces, hydrophilic bumps and hydrophobic surrounding patterns are generally recognized as having enhanced fog collection capabilities. However, our findings reveal that compared to the impact of surface wettability, the diameter of one-dimensional copper wires has a more significant impact on fog collection efficiency [67]. In other words, it is imperative to recognize that when one-dimensional and two-dimensional materials are placed within the fog flow, they induce distinct levels of interference in the fog flow. While air flows over a two-dimensional plane, substantial entrainment occurs; however, for one-dimensional linear materials, this deviation is significantly reduced, resulting in higher energy fog droplets being captured by the surface. Therefore, the fog collection capacity of superhydrophobic copper wires, superhydrophilic copper wires, pristine copper wires, and copper wires with alternating hydrophilic and hydrophobic regions exhibits minimal differences. From these observations, it becomes evident that when designing fog collection devices, it is essential to consciously consider factors such as fog flow interference, for example, when designing tapered arrays on a two-dimensional plane, considering how the arrangement (square or hexagonal) of the array affects fog flow interference [68].

The above delineation succinctly prescribes biomimetic approaches in fog collection. The detailed design and fabrication of specialized fog collectors or systems will be thoroughly examined and introduced in subsequent chapters of this enlightening book. Subsequently, we will elaborate the supporting theories of fog collection from three interrelated dimensions: the surface nucleation mechanism of water molecules, the growth theory of surface droplets, and the fundamental principles of surface transport behavior. Currently, the design of fog collection systems is transitioning from a singular biomimicry approach to a multifaceted one. In the future, it is expected that more ingenious fog collection strategies derived from flora and fauna will be seamlessly integrated into these devices. On the basis of the currently recognized three fundamental processes of fog collection, there is an urgent need to establish more comprehensive theoretical models to analyze the influence of surface geometry and wettability on fog collection, thereby fostering a feedback loop between theory and practice for potential practical implications in the design and application of fog collection systems.

1.4 Biomimetic Fog Collection Performance Evaluation

While a variety of techniques exist for harvesting atmospheric water, it is challenging to compare them under uniform standards. The challenges are multifaceted. First, fog and dew collection are impacted by climatic conditions as well as location factors, with fog collection necessitating higher humidity environments than those suitable for adsorption-based atmospheric water collection. This implies that through research on diverse water collection methodologies, we must select appropriate collection methods in varying regions and environments to achieve site-specific benefits. Consequently, an evaluation of the performance of atmospheric water collection without regard to applicable environmental parameters is meaningless. Second, fog water collection ensues when minute droplets in the atmosphere collide with the surface of a fog collector and are subsequently captured and collected. There is no phase change involved in this process, nor does it require input of energy; it is a passive method of collection. For fog water collection, the momentum provided by airflow, the size and density of fog droplets, etc., are critical environmental factors influencing its efficiency. On the other hand, both dew collection and adsorption-based atmospheric water collection involve phase changes. These two types of atmospheric water collection can be classified according to whether external energy is required or not into cooling-based atmospheric water collection and adsorption-based atmospheric water collection, respectively. Therefore, for active collection requiring energy input, assessing its collection efficacy necessitates consideration of both the collection rate and energy consumption. Henceforth, we will succinctly analyze which environmental factors are pivotal determinants of the performance of various water collection technologies using the evaluation metrics currently employed, with the aim of providing a benchmark for test environments in evaluating atmospheric water collection, enabling a more informed comparison of the performance of similar water collection technologies.

In Wang et al.'s summary of progress in atmospheric water collection, they mention three metrics typically utilized to evaluate atmospheric water collectors – specific water production (SWP) per day per unit collector area, specific energy consumption (SEC) per unit mass water production, and feed air recovery ratio (RR). SWP is generally used to assess the water production capacity of passive atmospheric water collectors (which do not require energy input), while SWP and RR are more frequently applied to evaluate the energy efficiency and water vapor condensation effect of active atmospheric water collectors. For direct cooling water collection, SEC and RR are defined as follows:

$$\mathrm{SEC} = \frac{Q_{\mathrm{cond}}}{m_{\mathrm{H_2O}}} \approx C_p \left(\frac{\varepsilon_T}{\varepsilon_d}\right) \left(\frac{T_{\mathrm{i}} - T_{\mathrm{cond}}}{d_{\mathrm{i}} - d_{\mathrm{cond}}}\right) + h_{\mathrm{fg}}, \tag{1.6}$$

$$\mathrm{RR} = \varepsilon_d \left(1 - \frac{d_{\mathrm{cond}}}{d_{\mathrm{i}}}\right). \tag{1.7}$$

In the formulation, T_{i} signifies the temperature of the inlet airstream in the condenser and d_{i} indicates its humidity ratio. T_{cond} denotes the condensing temperature. This indicator measures the water production rate per unit mass of dry air (kilograms per kilogram). The parameters ε_{T} and ε_{d} signify the heat transfer and mass exchange efficiencies of the condenser, respectively. The total cooling load Q_{cond} of moist air encompasses both sensible heat load associated with the temperature change of wet air and latent heat load related to the condensing enthalpy (h_{fg}). Evidently, a lower sensible heat load may result in a smaller SEC, implying that inlet air necessitates a higher RH. SEC and RR definitions suggest that an optimal inlet air condition exists with low T_{cond}, low T_{i}, and high d_{i}. Similarly, Broday et al.'s moisture harvesting index (MHI) , which has gained widespread acceptance as a metric for quantifying the energy required for dew collection [69], although it does not quantify the actual system's energy consumption, considers the ratio between the minimum energy requirement for condensation and the specific energy and latent heat of each unit mass of water at a given inlet air condition [10]. MHI is defined as

$$\text{MHI} = h_{\text{fg}}/q. \tag{1.8}$$

In this context, $q = (h_o - h_i)/(r_o - r_i)$, where h_o and h_i mean the enthalpy in air at the outlet and inlet states, respectively, while r_o and r_i represent the humidity ratio (kg water per kg air) of these respective conditions [70]. The maximum value for MHI is 1, corresponding to pure saturated steam. Conventionally, the exhaust state of the damper system is set at a temperature of 4° C, with a humidity ratio of 5 g water per kg air. When MHI falls below zero, it means that either the inlet temperature or humidity is below the assumed outlet condition. A higher MHI generally facilitates dew formation, indicating reduced latent cooling demand and increased water production from incoming air [10]. MHI < 0.3 is considered highly impractical for dew collection conditions, even when using advanced dew systems under standard operating conditions [70]. Bagheri has tested commercial dew systems under various environmental conditions, yielding results consistent with predictions; the SEC of the dew system is highly dependent on ambient humidity and temperature [69]. SEC can be linked to MHI through a simple relationship: SEC × MHI = H_{fg}[12]. Based on the above analysis, the key environmental factors influencing the performance of atmospheric water collection based on cooling are humidity and temperature. Currently, the water yield ratio used to express dew collection is typically expressed as kg per day (active cooler) or kg m^{-2} day (passive radiative cooler). For atmospheric water collection based on adsorption, equilibrium vapor absorption in materials (water absorption per kilogram dry adsorbent, kg kg^{-1}) identified from adsorption isotherms is often used as an indicator of water collection water yield.

In fact, fog collection currently uses a single SWP as its evaluation metric. Commercial humidifiers are frequently used in laboratory simulations of foggy environments to evaluate the performance of biomimetic fog collectors. Generally, parameters such as temperature, RH, fog flow rate (cm s^{-1}), fog flux (mg s^{-1}), sample tilt angle, and distance between the sample and humidifier (x) can be controlled. How these parameters affect fog collection performance and how the reasonableness of parameter settings is determined can be measured by examining the collection environments of organisms and plants capable of collecting fog in nature. After Parker et al. discovered the fog collection capacity of desert beetles, they simulated the collection environment of desert beetles (desert wind speed of 5 m s^{-1}, tilt angle of 45°), utilizing Euler's first law. At this V wind speed, droplets roll down,

$$V = \sqrt{\left(\frac{4\rho_{\text{water}}}{3\rho_{\text{air}}} Rg \sin\theta\right)}. \tag{1.9}$$

Here, R is the droplet radius, g is the gravitational constant (9.8 m s^{-2}), θ is the tilt angle, and ρ is the medium's density ($\rho_{\text{air}} = 1$ kg m^{-3}, $\rho_{\text{water}} = 1000$ kg m^{-3}).

In this experimental setting, a spherical droplet must exceed a diameter of 5 mm to roll down. The desert beetle observed in actual conditions has droplets rolling off its back with diameters between 4 and 5 mm, validating the simulation results. Consequently, it is evident that wind speed equivalent to fog flow rate influences the size of droplets relying on gravity for descent, thereby affecting the efficiency of fog collection. Moreover, a specific fog flow rate also affects the kinetic energy imparted by airflow to droplets reaching the surface of the fog collector. Only sufficient kinetic energy can allow droplets to collide with the collector surface and be captured. As the airflow transports the droplets closer to the collector surface, some of the kinetic energy will be converted into viscous loss energy W due to friction resistance. According to Chandra et al.'s work, the viscous loss can be estimated as [71]

$$W \sim \varphi V t_c. \tag{1.10}$$

Here, V is the volume of fog droplets, and φ is a loss function, given by the following equation:

$$\varphi \sim \mu (U/r)^2. \tag{1.11}$$

Substituting into our equation, one observes that μ denotes the fluid's dynamic viscosity and U exemplifies the droplet's speed, while r signifies the radius of the droplet. Now, let us designate t_c, or time for complete coalescence, as the time required for a droplet to experience its utmost deformation postimpact with a rigid boundary at velocity U. We opt for the peak value,

$$t_c \sim r/U. \tag{1.12}$$

In addition, the kinetic energy of fog droplets is

$$\Delta E_k = 0.5\, mU^2. \tag{1.13}$$

In this context, m pertains to the mass of the fog droplet. Pertaining to the capture process, it is necessary for the system to surmount an energy barrier known as E in order to secure the attachment of fog droplets to the surface of a fog collector. From the energy point of view during the capture period, it becomes imperative that the fog droplet can be successfully captured by the material surface if the following relationship is confirmed: $E_k > E + W$. Basically, the fog droplet requires additional energy beyond the viscous loss during collision to be effectively captured on the surface of the fog collector.

Specifically, the fog flow degrades when fog droplets deviate from the material surface, thereby significantly reducing their kinetic energy; at such times, the impact of the fog flow rate on collection effectiveness also comes into play. Boundary layer theory can be used to describe the velocity gradient formed before fog droplets reach the catcher's surface during dynamic fog capture [72]. When fog droplets encounter the object's surface, there exists a velocity gradient along the vertical direction of the surface, which is referred to as the boundary layer [63]. The speed of the fog droplet will be influenced by the boundary layer upon impact on the object. The thickness of the boundary layer shows a negative correlation with the flow rate of fog. The speed U

of the fog droplet in the fog stream can be considered equivalent to the air current speed (70 cm s^{-1}). The Reynolds number (*Re*) is expressed as

$$Re = \frac{\rho U d}{\mu'}. \tag{1.14}$$

In this context, ρ and μ' represent the density and viscosity of the fog stream, respectively. The standard length, d, of a flat, smooth plate can be defined as unity. The calculated Reynolds number (roughly equivalent to 5×10^4) is lower than the standard value (5×10^5), indicating that the fog stream may be considered laminar. For laminar flows, the plate boundary layer thickness (δ) can be determined using the following equation:

$$\delta = 5xRe^{-0.5}. \tag{1.15}$$

In such experiments, x represents the distance between the fog collector and the fog flow (normally 0 to 20 mm). As shown in Eq. (1.15), the calculated δ is below 45 μm, indicating a significant deceleration of droplets as they pass through the fog collection zone. The droplet's reduced velocity results in fewer collisions with the substrate per unit time, making it challenging for kinetic energy to be completely exhausted. On surfaces with protuberances, droplets exhibit higher velocities on surfaces with an uneven, thin boundary layer structure, which escalates the frequency of droplet–substrate collisions, rapidly depleting the kinetic energy of droplets, thereby achieving superior fog collection efficiency.

In addition, the RH and fog flow rate directly affect the density of fog droplets in the airflow, which subsequently affects the effectiveness of fog collection. An increase in the sample distance from the humidifier and an expansion of the spray area of the humidifier result in smaller droplet diameters reaching the surface of the fog collector and a decrease in droplet density. From this analysis, it becomes apparent that temperature has minimal impact on the performance of biomimetic fog collectors tested in laboratories due to the use of commercial humidifiers typically employed in high-humidity conditions during testing. Temperature is merely noted as an objective environmental condition, primarily at room temperature.

Experimental conditions and environmental parameters reported in the current literature vary significantly, making it difficult to compare the overall performance of these fog collectors [63]. SWP (mg cm^{-2} h^{-1}) is commonly used in literature as a measure of single fog collection. The fog collection efficiency of the fog collector is defined as

$$\eta_H = M_H / M_F \tag{1.16}$$

Thereafter, M_H represents the amount of fog harvested by the device and M_F represents the overall amount of fog permeating through the collector. Attention is drawn to the discrepancy in environmental conditions under which bioinspired fog collection tests have been conducted across various literature sources, making it challenging to directly compare their collection efficiencies. To address this issue, we include here two key parameters that influence the assessment of bioinspired fog collection performance (i.e. SWP and fog collection efficiency). Taking into account environmental factors that can significantly impact these metrics,

$$\text{SWP} = f(U) * f(\text{RH}) * f(x). \tag{1.17}$$

Specifically, $f(U)$ represents the function describing the velocity of fog droplets, f(RH) represents the humidity condition (which equates to the effect on the flow rate of fog), and f(L) refers to the distance condition between the surface of the fog collector and the humidifier.

References

1 Kummu, M., Guillaume, J.H.A., de Moel, H. et al. (2016). *Scientific Reports* 6: 38495.
2 Liu, J., Yang, H., Gosling, S.N. et al. (2017). *Earth's Future* 5: 545.
3 Alsdorf, D.E., Rodríguez, E., and Lettenmaier, D.P. (2007). *Reviews of Geophysics* 45.
4 Boretti, A. and Rosa, L. (2019). *npj Clean Water* 2: 15.
5 He, C., Liu, Z., Wu, J. et al. (2021). *Nature Communications* 12: 4667.
6 Liu, Z., Ying, J., He, C. et al. (2024). *Landscape Ecology* 39: 10.
7 Thimmappa, R., Gautam, M., Bhat, Z.M. et al. (2021). *Cell Reports Physical Science* 2: 100627.
8 Wang, H. (2018). *Nature Nanotechnology* 13: 273.
9 Liu, Z., Wu, B., Zhu, B. et al. (2019). *Advanced Functional Materials* 29: 1905485.
10 LaPotin, A., Kim, H., Rao, S.R., and Wang, E.N. (2019). *Accounts of Chemical Research* 52: 1588.
11 Ejeian, M. and Wang, R.Z. (2021). *Joule* 5: 1678.
12 Tu, Y., Wang, R., Zhang, Y., and Wang, J. (2018). *Joule* 2: 1452.
13 Guo, J., Huang, W., Guo, Z., and Liu, W. (2022). *Nano Letters* 22: 3104.
14 Wang, W., Pan, Q., Xing, Z. et al. (2022). *Water Research* 211: 118029.
15 Rao, A.K., Fix, A.J., Yang, Y.C., and Warsinger, D.M. (2022). *Energy & Environmental Science* 15: 4025.
16 Rajaram, M., Heng, X., Oza, M., and Luo, C. (2016). *Colloids and Surfaces A: Physicochemical and Engineering Aspects* 508: 218.
17 Shi, W., Anderson, M.J., Tulkoff, J.B. et al. (2018). *ACS Applied Materials & Interfaces* 10: 11979.
18 Wang, X., Guo, Z., and Liu, W. (2023). *Advanced Materials Interfaces* 10: 2202123.
19 Lei, J. and Guo, Z. (2020). *Nanoscale* 12: 6921.
20 Rivera, J.d.D. (2011). *Atmospheric Research* 102: 335.
21 Wenzel, R.N. (1949). *The Journal of Physical and Colloid Chefogry* 53: 1466.
22 Zhu, H., Guo, Z., and Liu, W. (2016). *Chemical Communications* 52: 3863.
23 Si, Y. and Dong, Z. (2020). *Langmuir* 36: 667.
24 Zhang, H., Wang, F., and Guo, Z. (2024). *Advances in Colloid and Interface Science* 325: 103097.
25 Schemenauer, R.S., Fuenzalida, H., and Cereceda, P. (1988). *Bulletin of the American Meteorological Society* 69: 138.
26 Schemenauer, R.S. and Cereceda, P. (1991). *Ambio* 20: 303.
27 Parker, A.R. and Lawrence, C.R. (2001). *Nature* 414: 33.

28 Seely, M.K. and Hamilton, W.J. (1976). *Science* 193: 484.
29 Gao, Y., Wang, J., Xia, W. et al. (2018). *ACS Sustainable Chefogry & Engineering* 6: 7216.
30 Xu, C., Feng, R., Song, F. et al. (2018). *ACS Sustainable Chefogry & Engineering* 6: 14679.
31 Knapczyk-Korczak, J., Ura, D.P., Gajek, M. et al. (2020). *ACS Applied Materials & Interfaces* 12: 1665.
32 Sarkar, D., Mahapatra, A., Som, A. et al. (2018). *Advanced Materials Interfaces* 5: 1800667.
33 Bai, H., Wang, L., Ju, J. et al. (2014). *Advanced Materials* 26: 5025.
34 Chen, W. and Guo, Z. (2019). *Nanoscale* 11: 15448.
35 Rykaczewski, K., Jordan, J.S., Linder, R. et al. (2016). *Langmuir* 32: 9335.
36 Cao, M., Ju, J., Li, K. et al. (2014). *Advanced Functional Materials* 24: 3235.
37 Li, X., Yang, Y., Liu, L. et al. (2020). *Advanced Materials Interfaces* 7: 1901752.
38 Chen, D., Li, J., Zhao, J. et al. (2018). *Journal of Colloid and Interface Science* 530: 274.
39 Chen, H., Ran, T., Gan, Y. et al. (2018). *Nature Materials* 17: 935.
40 Dai, X., Sun, N., Nielsen, S.O. et al. *Science Advances* 4: eaaq0919.
41 Pan, Z., Pitt, W.G., Zhang, Y. et al. (2016). *Nature Plants* 2: 16076.
42 Nishimura, R., Hyodo, K., Mayama, H. et al. (2019). *Communications Chefogry* 2: 90.
43 Yang, H., Zhu, H., Hendrix, M.M.R.M. et al. (2013). *Advanced Materials* 25: 1150.
44 Ni, F., Xiao, P., Qiu, N. et al. (2020). *Nano Energy* 68: 104311.
45 Xing, C., Huang, D., Chen, S. et al. (2019). *Advanced Science* 6: 1900531.
46 Li, R., Shi, Y., Shi, L. et al. (2018). *Environmental Science & Technology* 52: 5398.
47 Ni, F., Xiao, P., Zhang, C. et al. (2019). *ACS Applied Materials & Interfaces* 11: 15498.
48 Wang, M., Sun, T., Wan, D. et al. (2021). *Nano Energy* 80: 105569.
49 Li, R., Shi, Y., Alsaedi, M. et al. (2018). *Environmental Science & Technology* 52: 11367.
50 Xu, J., Li, T., Chao, J. et al. (2020). *Angewandte Chemie International Edition* 59: 5202.
51 Gou, X. and Guo, Z. (2020). *Langmuir* 36: 64.
52 Yang, X., Liu, X., Lu, Y. et al. (2016). *The Journal of Physical Chefogry C* 120: 7233.
53 Feng, J., Zhong, L., and Guo, Z. (2020). *Chemical Engineering Journal* 388: 124283.
54 Wang, B., Zhang, Y., Liang, W. et al. (2014). *Journal of Materials Chefogry A* 2: 7845.
55 Zhang, L., Wu, J., Hedhili, M.N. et al. (2015). *Journal of Materials Chefogry A* 3: 2844.
56 Ping, Z., Sun, Q., Yi, J. et al. (2021). *ACS Applied Materials & Interfaces* 13: 49556.
57 Zhao, J., Deng, Y., Dai, M. et al. (2022). *Journal of Water Process Engineering* 46.
58 Liu, Y.F., Yang, N., Li, X. et al. (2020). *Small* 16.

59 Tian, X.L., Chen, Y., Zheng, Y.M. et al. (2011). *Advanced Materials* 23: 5486.

60 Hou, Y.P., Chen, Y., Xue, Y. et al. (2012). *Soft Matter* 8: 11236.

61 Tang, X., Huang, J.X., Guo, Z.G., and Liu, W.M. (2021). *Journal of Colloid and Interface Science* 604: 526.

62 Peng, Y., He, Y.X., Yang, S. et al. (2015). *Advanced Functional Materials* 25: 5967.

63 Yu, Z.H., Zhu, T.X., Zhang, J.C. et al. (2022). *Advanced Functional Materials* 32.

64 Zhou, H., Jing, X.S., and Guo, Z.G. (2020). *Journal of Colloid and Interface Science* 561: 730.

65 Zhou, H., Zhang, M.X., Li, C. et al. (2018). *Small* 14.

66 Zhou, H., Jing, X.S., Li, S.P., and Guo, Z.G. (2021). *Chemical Engineering Journal* 417.

67 Zhong, L.S., Zhang, R.C., Li, J. et al. (2018). *Langmuir* 34: 15259.

68 Ju, J., Yao, X., Yang, S. et al. (2014). *Advanced Functional Materials* 24: 6933.

69 Bagheri, F. (2021). *Water Resources and Industry* 25.

70 Gido, B., Friedler, E., and Broday, D.M. (2016). *Atmospheric Research* 182: 156.

71 Bussmann, M., Mostaghimi, J., and Chandra, S. (1999). *Physics of Fluids* 11: 1406.

72 Zhong, L.S., Zhu, H., Wu, Y., and Guo, Z.G. (2018). *Journal of Colloid and Interface Science* 525: 234.

2

Natural Prototype of the Bionic Fog Collection

In nearly 4.5 billion years of evolution and natural selection, nature itself has developed into a highly mature system that can use minimal resources to achieve maximum functions [1]. Almost all natural creatures have unique attributes, showing the coordination and unity between the structure and function to help them survive in the harsh environment [2]. Inspired by these natural organisms, biomimetic materials with special structures and functions have correspondingly appeared in numerous studies [3].

Human production and life cannot be separated from water. Without water, every life does not exist [4]. Many parts of the earth are severely short of freshwater resources, and recycling water from nature in an efficient and clean way is important to overcome the problem of freshwater shortage. In moist air, many microdroplets exist as fog. Inspired by the biology that can collect fog-collecting water from the air, the functional interface materials with the ability of fog-collecting water have attracted wide attention from the researchers. In extreme environments, such as deserts and the Gobi, where water resources are severely limited, certain organisms have developed highly specialized adaptations to meet their hydration needs. Fog and dew water are the two main forms of atmospheric precipitation. As a potential freshwater resource, it has already shown the great potential to solve the freshwater crisis in the arid areas [5]. In coastal desert areas, such as the Namib Desert, annual rainfall is low, but fog is a common phenomenon. Local residents use the designed meshes to effectively capture and collect fog to meet the daily needs of fresh water [6]. This simple and economical approach is important for alleviating the freshwater shortage in these arid areas.

With the rise of bionics in the twenty-first century, the study of traditional fog collection was further developed. In nature, many animals and plants have evolved special structures and functions to collect water from air to meet their own needs for water. The beetles from the Namib Desert, for example, can collect fog in the wind through alternating hydrophilic bulges and hydrophobic valleys on their backs [7]. Spider capture filaments enable continuous transport and efficient collection of water droplets through their natural arrangement of periodic spindle knots [8]; cacti

Bioinspired Materials for Fog Harvesting: Principles, Design, Preparation, and Applications, First Edition. Zhiguang Guo and Weimin Liu.

originating from the Chihuahua Desert can continuously capture and directionally transport fog through their conical spines and trichome structures [9]. The catchment behavior and mechanisms of these organisms have attracted widespread interest. Many results show that the surface roughness gradient, pyramidal curvature, and surface interface wetting changes play a decisive role in the formation and transport of condensed water droplets [10]. Inspired by these organisms, the researchers have developed a series of biomimetic fog collection materials, including man-made fiber materials inspired by spider silk [11] and cactus thorn-inspired cone structure materials, etc. [12]. In addition, the study of interfacial materials also comes from nature. Due to the survival law of natural selection, organisms are constantly evolving to adapt to the changeable living environment. As shown in Figure 2.1, with the passage of time and the progress of science and technology, many organisms with superhydrophobic and superhydrophilic wetting characteristics and their surface microstructures have been revealed successively [8, 14]. The researchers are inspired by natural creatures, took the epidermal structure with unique wetting organisms as a template, and combined with the theoretical thought of bionics, and prepared a variety of engineering materials. In this book, desert beetles, spider silk, pitcher plants, and cactus are introduced and analyzed in detail.

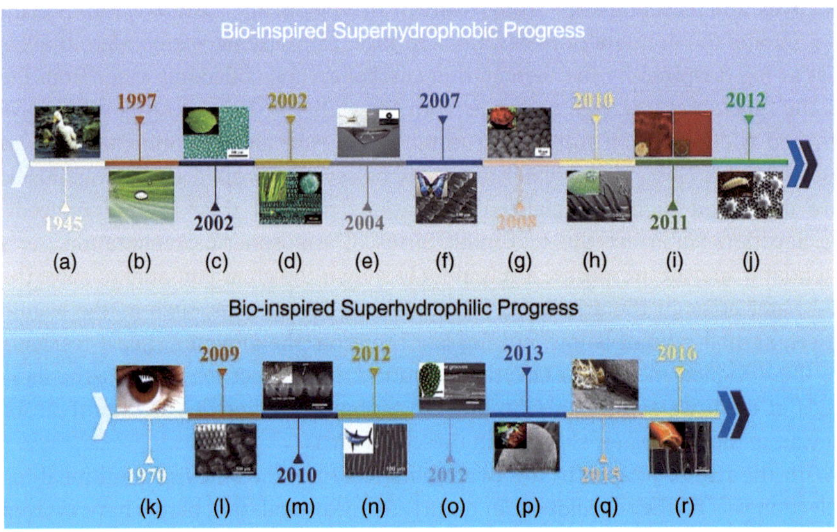

Figure 2.1 A wide variety of biological surfaces with special wetting properties and their microstructures have been successively discovered. (a) Duck feather. (b) Lotus effect. (c) Micro-nano cooperative lotus effect. (d) Rice leaf. (e) Legs of water striders. (f) Butterfly wings. (g) Rose effect. (h) *Salvinia molesta*. (i) Bacterial biofilm. (j) Collembola. (k) Precorneal tear film. (l) Fish scale. (m) Spider silk. (n) Shark skin. (o) Cactus. (p) Tree fog. (q) Lizard skin. (r) *Nepenthis alata*. Source: [13] / American Chemical Society / CC BY-NC 4.0.

2.1 Fog Collection of Bionic Prototype Plants

2.1.1 Cactus

The ability of cactus to survive in arid deserts, in addition to long-term natural selection to reduce water loss, is also related to its multi-structured and multi-functional fog collection system. After reporting the water collecting behavior of spider silk, the fog collection behavior of desert cactus thorns was studied [9]. On the stem of the cactus, the clustered needles and trichome structures grow evenly on its surface (Figure 2.2). The barbed structure at the tip helps to capture the tiny water droplets in the air. The presence of the groove structure in the middle of the sashimi induced a surface energy gradient and higher surface energy for finer grooves. More importantly, the barbed structure is conical and the sashimi diameter increases from the tip to the end. Similar to spider silk, the structure also produces a wettability driving force and a Laplace additional pressure driving force. The wettability driving force refers to driving the droplet movement on the special wetting surface. The wettability driving force is essentially generated by the chemical wettability gradient, which is embodied in the difference of contact angles. When the contact angles of different regions of the surface differ, the contact lines will be subjected to abnormal direction wettability force. If this force is large enough to overcome the friction resistance at the interface, the droplet moves directionally. The Laplacian additional pressure driving force was discovered by the Laplacian in 1805, which comes from the surface tension of the liquid. After generating the Laplacian pressure gradient, the droplets spontaneously move to the area of low curvature on the conical structure (such as the spine of the cactus).

The groove structure of the sashimi surface allows the droplet to have a contact angle hysteresis on the surface (caused by a certain degree of rough or uneven chemical composition on the real solid surface), making the droplet easier to move in the direction of the prick. The droplets are driven by the driving force from the beginning of the tip to the end, and the transported water can also be stored by the tiny villi at the end of the needle structure. The thorn structure and fog collection system of cactus can inspire the fog collection system with synergistic effect and further improve the efficiency of fog collection. As shown in Figure 2.2, fog in the air can accumulate on the barbs of the tip and reach the middle of the body. Directed barbs form asymmetric structures resembling butterfly wings, allowing unidirectional droplet transport. In addition, the unique conical structure and microgrooves in the middle of the sashimi can generate Laplacian pressure gradients and wettability gradients, respectively, under the synergy of the cactus surface can realize the capture and directional collection of water droplets. At the same time, the roughness of the tip of the conical structure is greater than the root, making the contact angle and surface energy of the tip and the root different, and the water droplets are easier to move to the root. At the same time, the gradient groove also has a certain surface energy gradient, which can produce the surface

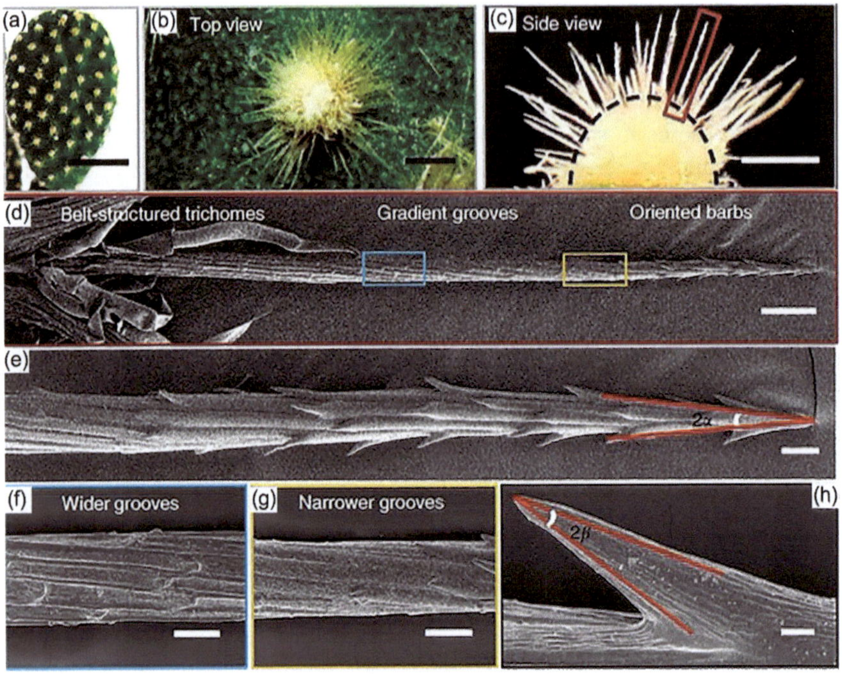

Figure 2.2 (a–c) Optical image of cactus plants with tapered spines. (d) SEM images of a single thorn are divided into three regions: tip with a directional barb, middle part with a gradient groove, and banded trichomes. (e) Tip with a directional barb, (f) middle part with a gradient groove, and (g) bottom with structural trichomes. (h) Magnified images of a single barb. Scale bar, 5 cm in (a), 500 mm in (b–c), 100 mm in (d), 20 mm in (e–g), and 2 mm in (h). Source: [9] / Springer Nature / CC BY-NC-SA 3.0.

wettability gradient force. In addition, the radius of the cactus tip is smaller than the radius of the root, so the Laplace pressure of the tip is larger than the root, and the Laplace pressure difference will drive the water droplets to move from the tip to the root. The surface energy gradient and the Laplace pressure difference of the cone thorn tip and root play a crucial role in the fog collection process of the cactus. The multifunctional structure of the cactus will integrate the capture, collection, transportation, and storage in the fog, forming an efficient directional fog and fog collection system (Figure 2.3).

Inspired by the excellent water absorption ability of the cactus stem [15], Lee et al. [16] reported a three-dimensional cylindrical double-structure water management system and was designed for fog collection in arid areas. The device is composed of a superhydrophobic copper mesh and superhydrophilic penetrating polymer network hydrogel, showing super fog collection performance with a fog collection efficiency of 209 mg^{-1} cm^{-2} h. The device not only has the water storage capacity of anti-evaporation functions but also has the ability to release water on demand. An artificial fog collection material was prepared by dispersing a large number of small zinc oxide lines over a large zinc oxide wire [12c]. The length of the main line and branch lines are controlled at 1 and 0.1 mm, respectively. At the same time, the diameter of these small ZnO lines gradually decreases from

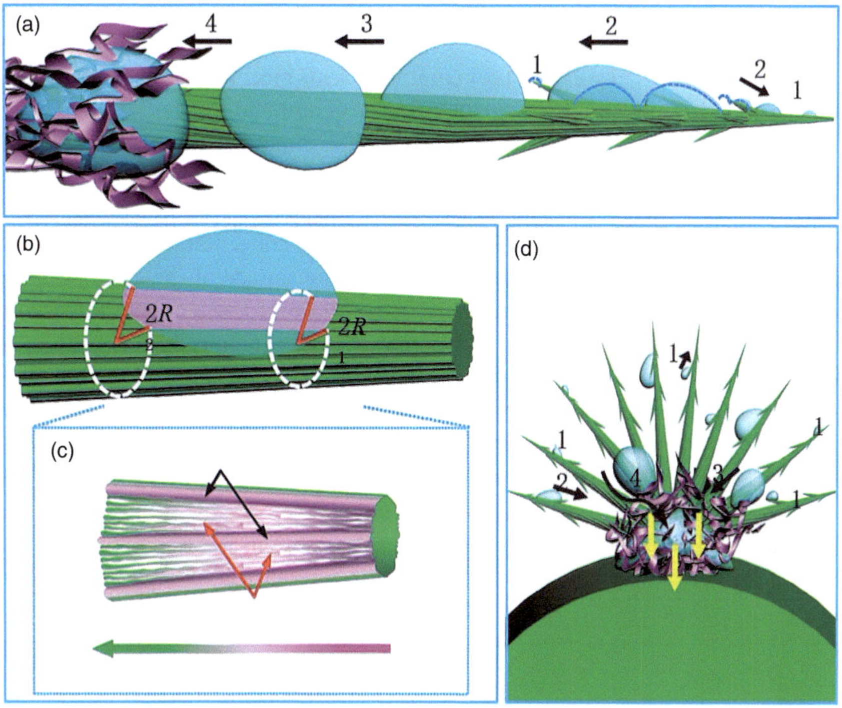

Figure 2.3 Summary of the fog collection mechanism of cactus thorns. (a) Schematic diagram of the unidirectional transport of water droplets on a single spur; (b) Laplace pressure gradient and surface free energy gradient form the driving force for the directional transport of water droplets; (c) enlarged zoom composed of multiple spines and trichomes; (d) the cactus obtains water from the air under the synergistic action of the spine surface. Source: Reproduced from [9] / with permission of Springer Nature.

the bottom to the top, forming a conical structure that can spontaneously move toward the bottom when the water droplets are deposited at the tip of ZnO. In an environment of 100% relative humidity, they performed two sets of experiments to study the nature of the artificial material and the fog collection of the cactus thorn, which showed that the fog collection capacity was several times higher than the cactus thorn.

A four-level layered wedge track inspired by the cactus leaf veins was designed to enhance the directional fog harvesting effect. Lin et al. [17] used alkali-assisted surface oxidation at room temperature to grow copper hydroxide nano-wires on a flexible substrate, and the nano-wires on the surface facilitate droplet absorption and provide a rich capillary channel for droplet transport. Then, the stratified wedge track has a Laplacian pressure difference that allows the water to flow spontaneously to a certain accumulation area. The fog collection performance of the bionic fog collector is not only related to the wettability and structure of the sample surface but also depends on the wind speed in the fog-containing areas [18]. Fog collectors function effectively only in areas with natural wind or under controlled artificial fog flows at specified wind speeds. However, in some fog areas, the fog with low flow capacity will stop in the air and on the ground, which will greatly reduce the collection efficiency of the fog collector [19]. Similarly, inspired by the structure of the cactus thorn, Peng et al. [12e] combined with mechanical perforation and template replication technology, cactus spine structure and magnetic response of flexible conical arrays together, preparation of polydimethylsiloxane cone arrays, specifically as shown in Figure 2.4, by adding quantitative cobalt magnetic particles in the hole, and then under the action of the magnetic field, the conical mold swing, fog collection. Magnetic induction-induced fog collection performed under windless conditions demonstrated the significant advantage of the cone array in collecting fog under windless conditions. The surface of the bionic cactus thorn structure with a wetting gradient was prepared electrochemically [12a]. Conical copper wires with a wettability gradient were obtained by electrochemical corrosion and chemical modification, allowing water to be spontaneously and efficiently transported on its surface. In order to build a conical structure on the copper wire and achieve electrochemical gradient corrosion, they coated a thin, flat gold NP on the surface, then placed the treated copper wire vertically in the reaction solution with the modifier, and slowly raised the reaction vessel at the bottom. In this process, the electrochemical corrosion and the low surface energy component

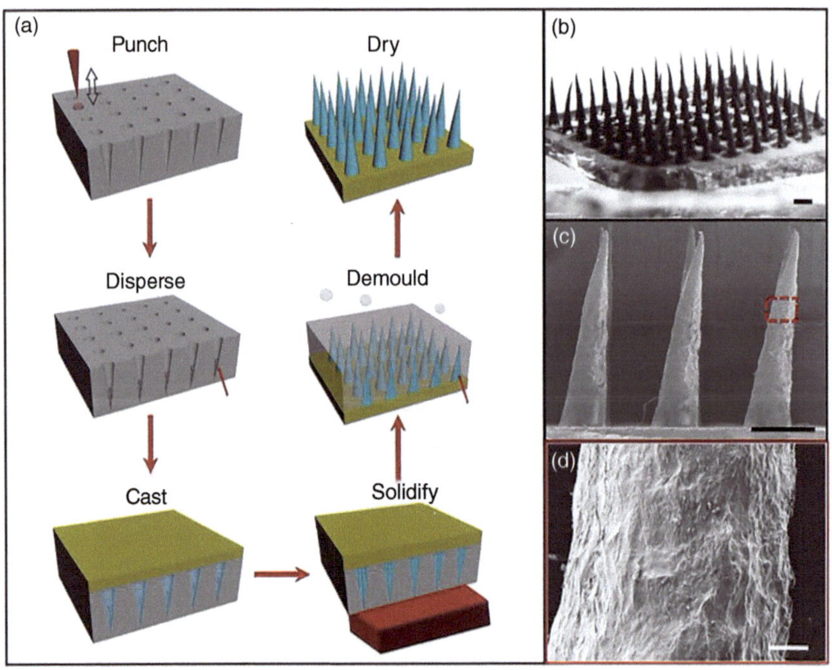

Figure 2.4 Schematic of the production of the cactus-inspired, ordered magnetic cone array and the corresponding optical and SEM images. (a) Preparation process of the magnetic cone array; (b) Optical photo and (c) SEM images of the orderly and neatly aligned magnetic cone array on the non-magnetic substrate layer; (d) SEM enlarged view for the middle of the magnetic cone. Source: Reproduced with permission from [12e] / John Wiley & Sons.

modification occur simultaneously, forming a comprehensive gradient force effect. The researchers observed a rapid transfer of water droplets on the structural surface and finally collected at the tip of the conical copper wire. Nowadays, chemical or electrochemical corrosion methods and template molding methods are one of the most common ways to prepare cactus thorn structure. However, because these structures have to be conical, these structures face the problem of easy folding and breaking. Therefore, ensuring the stable mechanical and chemical performance of the bionic cactus thorn structure is a vital factor to ensure its practical engineering.

2.1.2 Nepenthes

In nature, the Nepenthes genus have unique leaves and nutrient-absorbing trap cages, a surface structure with macroscopic and microscopic radial ridges able to lure and capture arthropods such as insects and further decompose and digest prey [20]. In air, the inner surface of Nepenthes leaf is moistened by water to form a water film. The droplets show strong slip characteristics on their surface. At a small tilt angle, the insect feet can quickly slide down from the surface, due to the presence of liquid film on the structured surface [21]. First, the condensation of rain or fog will moisten the surface of the pitcher plant opening, and then the nectar secreted at the opening can form a fluid film (Figure 2.5a). These nectary pores are located between

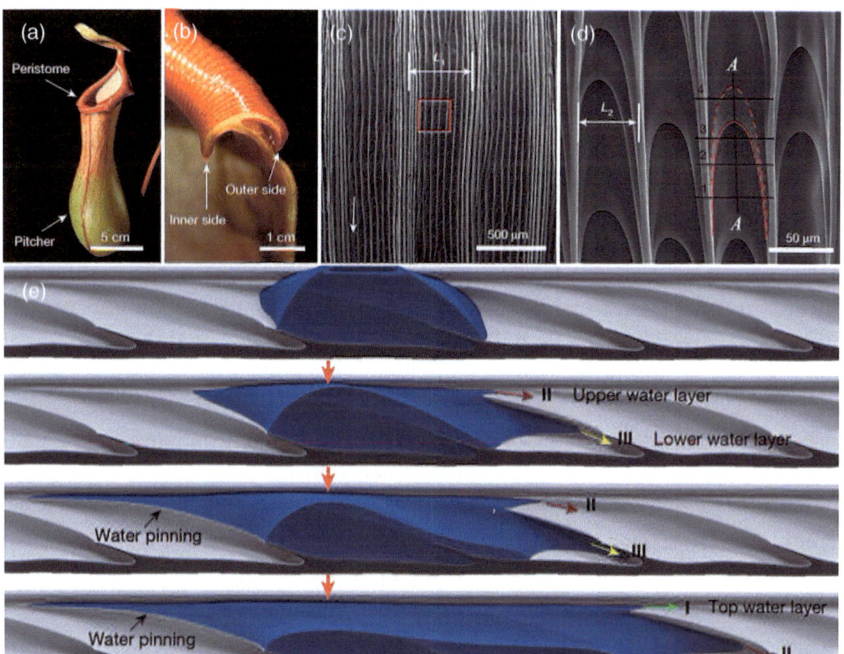

Figure 2.5 (a) Optical photo of the pitcher cage; (b) cross-section image of the edge of the cage; (c) the microgroove at the surface of cage; (d) the stacked ridge structure on the surface of cage; (e) schematic diagram of the droplet transport at the edge. Source: Reproduced with permission from [22] / Springer Nature.

the toothed processes at the inner edge of the opening (Figure 2.5b), which facilitates the spread of nectar on its surface. At the same time, it was observed that the structure of the opening was composed of regular straight epidermal cells forming a radial ridge structure (Figure 2.5c,d), and this directional structure facilitates nectar transport from the internal nectary to the edge of the surface. When the edge of the wet tank is "activated" by the water, the insect will "slide" into the trap, the effectiveness of the activation depends on the surface wrinkle of the edge and the layered microtopography, and the passive water activation of the trap must meet the following conditions: First, the water film must be continuous along the channel to the tank to prevent the sliding insect to flee; second, the water film must be stable to prevent dehydration and direct contact between the attachment pad and the surrounding, leading to the insect "water slip." Together, these two mechanisms form an efficient trapping mechanism that allows pitchers to capture some of the most skilled climbers in nature [23]. Further research indicates that due to the micro-nano scale of the epidermal layered structure at the edge of the pitcher cage mouth edge, the transfer of water in the direction of capillary ascent is enhanced, allowing the liquid to continue to positively fill the layered structure while being fixed in place by any reversed-moving leading edge of the water, preventing backflow and reverse flow. At the edge of the pitcher cage, smooth epidermal cells stack vertically to form a radial ridge, presenting a hierarchical structure along the direction of the trap cage. The structure allows the transport of droplets from the inside of the epidermis at the mouth edge of the pitcher plant, while preventing the transport of water in the opposite direction (Figure 2.5e). After a careful study of the feeding cages of the carnivorous plant pitcher plants, Jiang and Coworkers [22] found that water is continuously transported in the epidermis at the edge of the pitcher cage. Therefore, the structure of the opening of pitcher plants and its wetting surface characteristics provide potential ideas for preparing engineering materials for fog collection.

Zheng et al. [24] have designed hydrophobic wedge with an oil layer to achieve droplet transport. Spontaneous directional transport occurring on the surfaces of these structural gradients can be divided into two types: bulk motion and partial motion that leaves moist traces. Based on the structural features at the opening of pitcher plants, surfaces with structural gradients are widely used in various fields, such as oil–water separation, fog collection, and microfluidics [25]. The surrounding surface of the bionic pitcher plant is prepared, and the main bionic morphology is three-dimensional wedge angles and sharp edge in the microgroove to achieve high-density particle arrangement, which enables the ultrafast generation of microdroplets. In conjunction with the functional elements of the Nanoparticles (NPs), the investigator placed a row of semielliptically inclined microcavities in the parallel microchannel. Using the capillary effect, parallel microgrooves uniformly distribute particles into the microchannel, and the semielliptical oblique microcavity accurately captures individual particles or sequentially generates an array of microdroplets with a specific tilt direction. The generation of ultrafast microdroplets was achieved by simply adjusting the forward/backward contact angles on the biomimetic NPs (Figure 2.6). This bionic pitcher plant design also provides a potential design idea for engineering materials for fog collection. At present, the

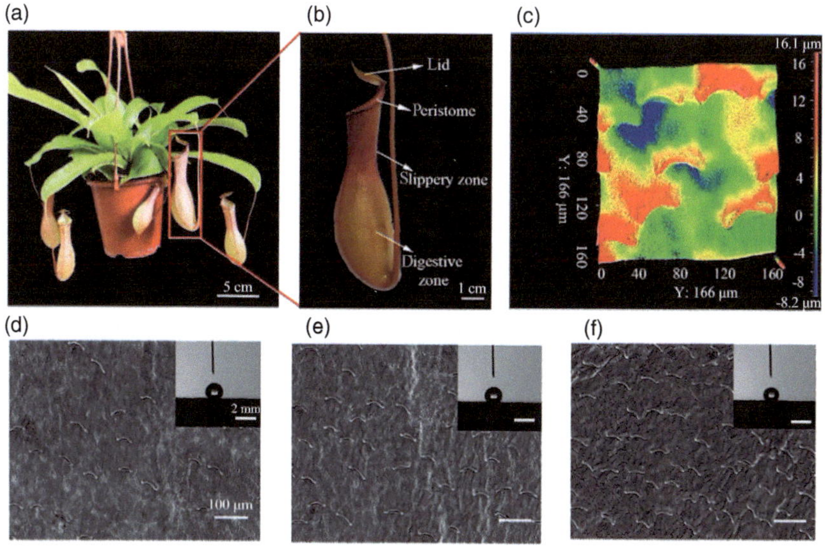

Figure 2.6 Relative images of the slippery zone of Nepenthes. (a) The mature Nepenthes alata. (b) The four areas of the pitcher division. (c) The crescent structures in the inner wall of the slippery zone. (d) SEM image of the inner surface in the slippery zone near peristome. The contact angle was 143.5°. (e) SEM image of the inner surface in the slippery zone near the middle zone. The contact angle was 135.7°. (f) SEM image of the inner surface in the slippery zone near the digestive zone. The contact angle was 128.9°. Source: Reproduced with permission from [26] / American Chemical Society.

bionic pitcher plant has been widely used in the operation of droplets, particles, and cells. However, most existing technologies require complex processes, fussy operations, or external settings, which deserve further improvement.

2.1.3 Iron Orchid

Iron orchid (Tillandsia), also known as the atmospheric pineapple, is a kind of perennial epiphytic herb, which has no stems, and its leaves are arranged in a rosette formation. There are many kinds of iron orchid, such as common lavender pineapple, silver leaf pineapple, and purple pineapple, mainly distributed in Ecuador, America tropical, and subtropical areas. Because of the lack of functional roots, ferric orchids rely on large multicellular trichomes on the surface of the leaf and stem epidermis to absorb water and nutrients to meet their water and nutrient needs, and the trichome structure is exposed to hyperarid conditions in the desert to prevent evaporation [27]. In the case of iron land in the Chilean Atacama Desert, their ability to capture water from air is almost entirely dependent on the trichomes covering the leaves (Figure 2.7a–c), which are composed of dead cells with abnormally thick outer cell walls on the dome cells (Figure 2.7d–f). The dome cells connect the outer shield and the mesophyll of the leaf through the short stalk of two living cells. Furthermore, the thick cuticle of the leaves forms a tubular structure around the dome cell and the hair stalk, thereby directing water from the surface to

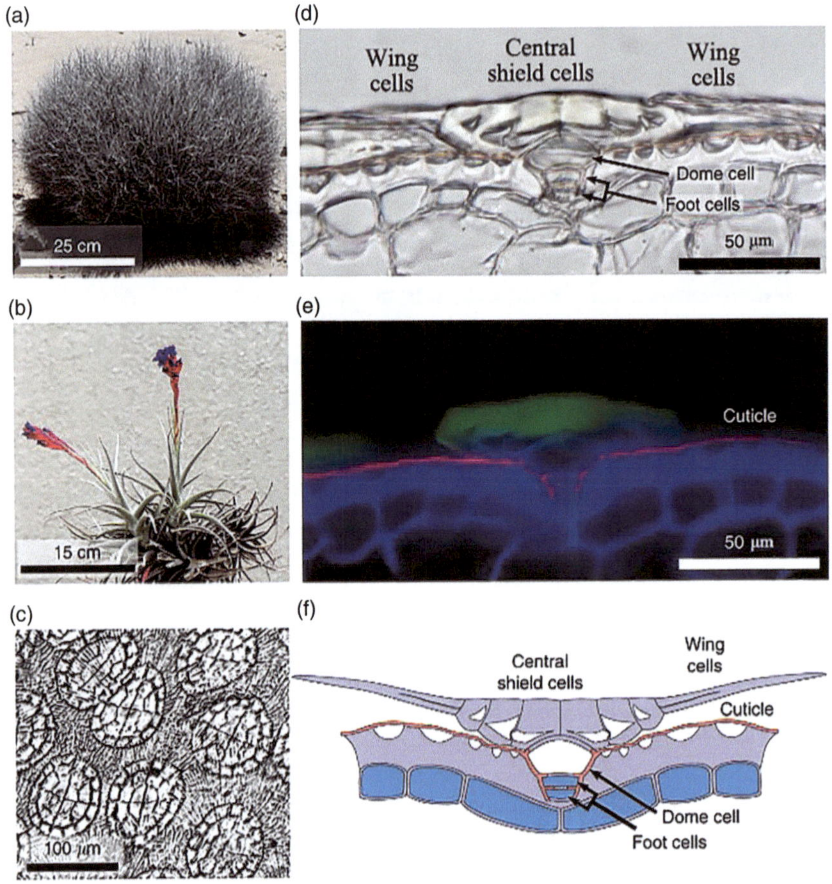

Figure 2.7 [(a) and (b)] Iron orchid in the desert; (c) surface view of dense trichomes; (d) transmission optical micrographs of trichsheets and (e) composite fluorescence images; (f) structure of trichomes. Blue indicates the living cells, and white indicates the lumen of dead cells. Source: Reproduced with permission from [28] / John Wiley & Sons.

the podocyte, which is taken up inside the cell [29]. The researchers observed that when the environment was dry, the central shielding cell sank like an impermeable plug into the space around the dome cell, thus preventing the water from spreading out. When the environment is wet, the trichomes expand, pushing them above the epidermis, opening a channel for the inward flow of liquid water [30]. At the cutting edge of the iron orchid branch, the researchers tracked the distribution of the water droplets on the outer surface. The experimental results confirmed that the diffusion of added water on the outer surface of the "atmospheric" epiphyte existed. In contrast, in plants with morphologically and physiologically similar epidermal trichomes, water applied to the cut end of the stem flows through the xylem within the interior rather than along the epidermis. Thus, the movement of water on the surface of iron orchid is mainly through capillary interactions between trichomes, which form dense pilose structures on the surface, resulting in water not being transported internally through the xylem. Therefore, the trichomes of the iron orchid are crucial in water transport. Water uptake through the trichomes is the only way that plants can adopt [31]. "Atmospheric" epiphytic lianas that occupy arid microhabitats also benefit from hydraulic conductivity through the low shoot xylem, thereby reducing water loss. The mechanism in which the trichomes of "atmospheric" epiphytic vines can switch between high and low water conductance can be used to inspire and design intelligent engineering materials for fog collection. For example, Dumais and Coworkers [32] mimicked the catchment structure of the iron orchid, and using artificial films, they proposed an asymmetric water transport strategy that can be used to design and prepare grids that capture fog. The rapid water transport phenomenon in the root of iron mosses revealed a unique multiscale structure, revealing the mechanism of its fog collection and transport [29]. These water transport mechanisms derived from iron orchid species are expected to inspire the design of fog collection devices.

2.1.4 Other Bionic Plant Objects

2.1.4.1 Arthmoss

Plants living in desert areas derive multiple survival strategies to obtain water, such as minimizing leaf area to reduce water loss [32], leaf structure used for obtaining water [33], plant geometry and surface chemical energy gradient to transport water droplets to favorable locations in the plant [8, 9], and the strong root system architecture that maximizes groundwater absorption [34]. Different from the survival strategies of these plants in desert areas, this desert moss (erythragus) has a unique "top-down" multi-scale integrated fog collection system, which does not rely on the root to absorb groundwater but uses the special sharp structure on the leaves to absorb water from the air [29]. As shown in Figure 2.8a, there are white wheat awns on the top of the leaf, and there are multiscale grooves and protrusion structures ranging from nanometer, micron to centimeter. The surface geometric features at different scales correspond to different mechanisms of fog trapping and transport. The white spike is mainly composed of densely covered tooth sting clusters and microscale and nanoscale grooves. Under the action of capillary tubes, the water in the air can be quickly captured by the superhydrophilic tooth sting, and the uneven distributed clusters of tooth sting on the surface of the spike is used as the nucleation site, which is the first step to capture the water source. The small droplets then condense and grow into larger droplets that are stored in the grooves of the awn tip. Finally, when the droplet is large enough, it is transported from the tip of the awn to the base at a higher speed. Once the transportation of the droplet is completed, the next stage of droplet capture, growth, and transportation on the plant surface begins. The fog collection and transport system of "small structure and large function" provides new ideas and insights for designing and preparation of bionic fog collection devices in arid areas.

The multifunctional hierarchical water catchment mechanism of desert moss is shown in Figure 2.9 [29]. When the atmosphere reaches its dew point, the

Figure 2.8 Plants of the fog collection phenomenon in nature. (a) Arthus. Source: Reproduced with permission from [30] / Springer Nature. (b) Cotula fallax. Source: [35] / American Chemical Society. (c) Bermuda grass. Source: Reproduced with permission from [36] / Elsevier. (d) Larch. Source: Reproduced with permission from [37] / Elsevier. (e) Wheat grass. Source: Reproduced with permission from [40] / Elsevier / CC BY 4.0.

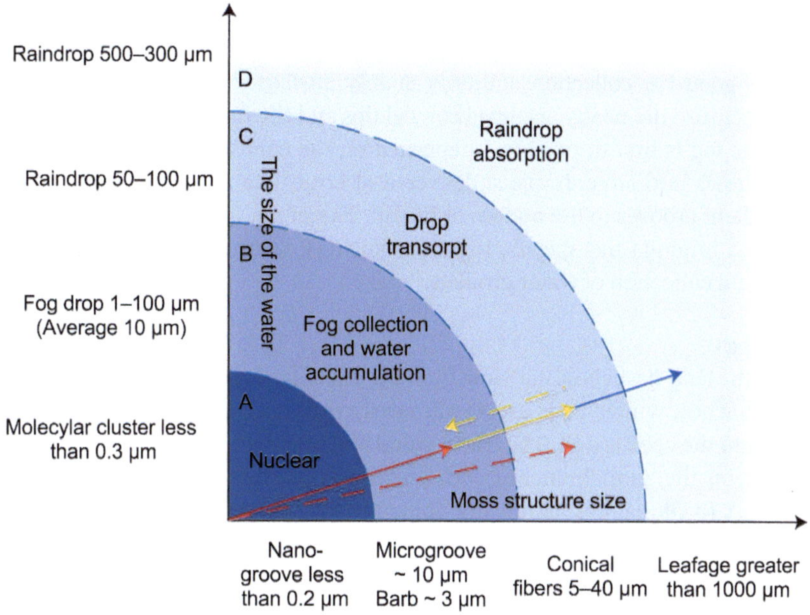

Figure 2.9 Multifunctional hierarchical water catchment mechanism of moss awn. Source: Reproduced from [29] / with permission of Springer Nature.

clusters of water molecules (region A) in the gas phase condense on the nanotrough of moss, forming a thin water film to fill the nanotrough or microtrough. When water is present in the form of droplets (area B), the droplets are collected on the microgrooves and barbs and aggregated into larger droplets. The uneven distribution of barbs on desert moss forms water areas and prepares larger water droplets (yellow arrows) for water transport. Rapid water transport (area C) occurs when the droplet becomes large enough and moves from the barb cluster to the bottom of the awn. When the droplet size exceeds the critical value, the droplet can be at 10–20 mms^{-1}. The speed is transmitted from the tooth cluster to the leaf (from the tip to the leaf within 50–200 ms). Once the droplet reaches a large enough volume, it moves from the cluster to the leaf, and then the moist cluster will again be ready to continue collecting, returning to the stage of fog collection and water accumulation (yellow dotted arrows).

2.1.4.2 Cotula Fallax

The shrub originates from South Africa and is composed of dense leaves and fine villi covering it. This multi-level structure gives it efficient fog collection performance (Figure 2.8b). Under high humidity, small fog beads can gather and grow at the top of the leaf until the leaf cannot support and fall off into the main body of the plant, and merge with other falling droplets. The larger droplets formed after merging are preserved in the plant structure under the common support of the leaves and villi until they grow further and fall into the lower soil without balance.

2.1.4.3 Bermuda Grass

Bermuda grass is a ubiquitous organism (Figure 2.8c). It is a tropical grass that has a certain drought tolerance and a certain ability to collect fog in the air [38]. The plant's good fog collection ability is mainly attributed to two characteristic surface structures: the neatly arranged conical tips and the interconnected gradient grooves. The fog is first trapped and deposited on the conical tip spines and then gradually grows into large droplets, delivered at critical values through interconnected gradient grooves to the bottom of the tip. Therefore, the surface structure of hundred and mududa has specific functions such as capture, growth, directional transport, and collection of water droplets.

2.1.4.4 Larch

Larch is with long branches and slow-growing short branches from axillary buds on long branches; winter buds are small, nearly spherical, bud scales are closely arranged, and the apex is dull. The leaves spiral and scattered on the long branches, and cluster on the short branches with narrow strips, flat, and thin four prism (Figure 2.8d). In China, it is mainly distributed in the Greater Hinggan Mountains, Changbai Mountains, northwest of Liaoning and so on. The leaves and the fine hairs covering the leaves constitute a unique three-dimensional hierarchy that relies on which to collect fog to adapt to arid environments [39]. The unevenly arranged fine hairs on the leaves can form a cushion structure supporting the droplets, first wrapping the droplets in a three-dimensional way, and then fixing the droplets through a fine nanoscale groove structure, while also preventing the droplets from falling off .

2.1.4.5 Dry Wheat Grass

Wheat grass is a short-lived grass. It is an important part of the desert and desert steppe vegetation, which is common in the spring. In the north of Xinjiang, it is distributed in the desert and desert grassland on the north slope of Tianshan Mountain and the alluvial fan plain. The leaves of the plants rely on their multi-scale layered surface structure to exhibit asymmetric-anisotropic fog collection behavior under arid climate conditions (Figure 2.8e) [40]. The plant surface structure includes macroscale grooves, microscale cones (tilted in the direction of water flow) and superimposed nanotextured structures. Biomimetic surfaces with similar structures were prepared by soft lithography combined with nanocoating deposition or functional nanoimprinting, and experiments demonstrated excellent fog collection [41].

2.2 Fog Collection of Bionic Prototype Animals

2.2.1 Desert Beetles

One of the typical biomimetic objects of the fog collection is the beetles in the Namib Desert. In 2001, Parker and Lawrence [7] observed the Namib Desert beetle in Nature, they found that the beetles had a hydrophilic and hydrophobic

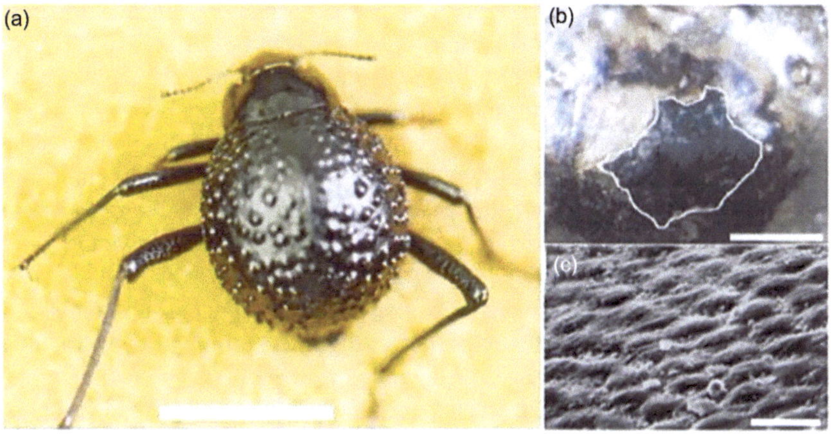

Figure 2.10 (a) hydrophilic protrusions and hydrophobic groove structures distributed on the back of adult female desert beetles; (b) the top is the non-waxy coating hydrophilic area, the side of the protrusion and the hydrophobic areas covered with wax layer; (c) SEM images of the rough surface of the groove area; Scale bar is (a) 10 mm, (b) 0.2 mm, and (c) 10 μm, respectively. Source: Reproduced with permission from [7] / Springer Nature.

pattern surface structure and demonstrated that it had efficient fog collection in fog-containing weather conditions. It is because of this particular wetting surface that beetles are also able to capture water in extremely arid deserts. Researchers found that the surface of the wing sheath was randomly distributed with processes of about 0.5 mm of diameter, and the spacing between the processes was about 0.5–1.5 mm (Figure 2.10). The top surface of the protrusion is smooth, forming a hydrophilic area of the non-waxy coating. And the surface of the groove area between the side of the protrusion is covered with a uniform hydrophobic wax layer, thus forming a special surface chemical wetting. When the droplet size gathered in the hydrophilic area reaches a critical value, the droplet overcomes the surface capillary force by rolling along the tilted dorsal surface to the mouth region of the beetle. In addition, the wind speed also has an influence on the droplet movement trajectory. Using the control form of Euler's first law, one can predict the wind speed when the droplet will roll along the surface:

$$V = \sqrt{\left[\left(\frac{\rho_{\text{water}}}{\rho_{\text{air}}}\right)\left(\frac{4}{3}\right) Rg\sin(\theta)\right]}, \tag{2.1}$$

where V is the wind speed, R is the droplet radius, and g is the gravity constant ($9.8\,\text{m}\,\text{s}^{-2}$), θ is the surface inclination angle, and ρ is the dielectric density ($\rho_{\text{air}} = 1\,\text{kg}\,\text{m}^{-3}$, $\rho_{\text{water}} = 1000\,\text{kg}\,\text{m}^{-3}$). The results indicate that when the wind speed is at $5\,\text{m}\,\text{s}^{-1}$ Under the conditions (Namibia), the diameter of droplet should exceed 5 mm to roll on the surface with an inclination angle of 45°, which coincides with the fog collection behavior of the beetle in nature, and the droplet diameter on the surface is about 4–5 mm to carry out the continuous fog collection process. The discovery of special wetting on the back of desert beetles provides ideas for the preparation of fog collector with excellent performance, which is of great significance to solve the problem of water shortage in arid areas.

Since then, many studies on the surface preparation method of simulating desert beetle and its fog collection performance have been reported. The key to functional surface of the desert beetle is the structure of hydrophilic and hydrophobic phase. For such a special surface, at least two different wetting regions (hydrophilic and hydrophobic) are constructed on surface. Today, there are four common preparation methods: affinity, water mixing, direct printing, mask-based lithography, and response to wettability transition.

First, the water mixing method is to combine the hydrophobic substances and the hydrophilic substances directly mixed together, forming a hydrophilic and hydrophobic alternate surface. Wang and coworkers [42] reported a simple, easy to operate, and inexpensive method to prepare hydrophilic and superhydrophobic materials. The surface of the beetle model was prepared by mixing method from Guo group [43]. Iron and cobalt NPs with different concentration ratios are evenly coated on fabric, and then the surface is modified with thiol solution to obtain the biomimetic hydrophilic and hydrophobic alternating fog collection surface. Because thiols can modify cobalt particles rather than iron particles, cobalt particles act as hydrophobic substances on the resultant surface, and iron particles act as hydrophilic substances, so that the mixed composite surface resembles the dorsal structure surface of beetle. Hybrid method is a simple and fast method to prepare hydrophilic and hydrophobic surface of beetle, which can greatly save production cost and be widely used in practical production.

Second, the inkjet printing method can accurately deposit the desired spray solution and form a designed pattern. For example, the prepared dopamine is sprayed on the superhydrophobic surface [44]. Through in situ polymerization of dopamine, the sprayed area becomes superhydrophilic, and the small superhydrophilic area is successfully prepared on the superhydrophobic substrate, so the surface of the superhydrophobic-superhydrophilic bionic material is formed. This method not only realizes the pattern design and high stability on superhydrophobic surfaces on a large scale but also realizes the design of complex patterns on superhydrophobic surfaces in the future.

Third, photography with a mask can also design the desired shape and size in advance, and design different shapes and sizes by changing the shape of the mask and the distance to the light [45]. Photolithography is a more complex method to prepare biomimetic beetle model surfaces, which includes multiple steps including mask preparation, pattern transfer, and adjustment of pattern wettability.

Fourth, the response to wettability transition method, the biomimetic desert beetle surface is constructed with two different wettabilities zone. In general, the substances or regions of hydrophilic and hydrophobic properties are constructed on the same surface simultaneously. Among the super-wetting materials, there is an intelligent responsive material that can realize the wettability transition. These intelligent materials can undergo a wettability transformation under the action of external stimuli, such as light, pH, temperature, electric field, magnetic field, humidity, etc. [46, 47]. However, the preparation of biomimetic surfaces of

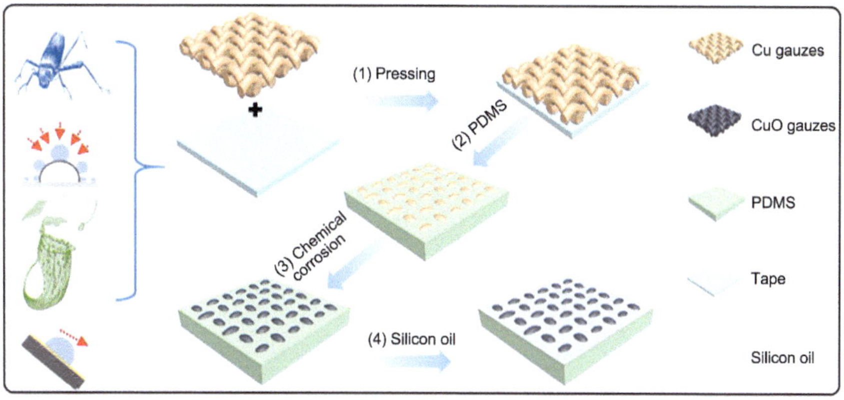

Figure 2.11 Study on the application of bionic composite surface for fog collection. Source: Reproduced from [47] / with permission of John Wiley & Sons.

desert beetles still starts from a more precise micro/nano-size patterns as well as a more reasonable hydrophobic/hydrophilic ratio (Figure 2.11).

In the past decade, this mode of spontaneous unidirectional transport of droplets from the hydrophobic side to the hydrophilic side under the action of external energy input, has been widely used in fog collection engineering due to its potential value. The surface of desert beetle biomimetic composite material has higher fog collection efficiency; therefore, we can use the advantages of biomimetic composite material design and auxiliary manufacturing technology to design low-cost, efficient, sustainable and energy-saving biomimetic fog collection materials.

2.2.2 Spider Silk

Spider silk is a filament composed of hydrophilic flagellar proteins with humidity sensing. Previously, the scientist mainly focuses on its excellent biomechanical properties [48]. But it wasn't until 2010 that its ability to collect water in humid air first caught the attention of researchers such as Zheng [8]. The surface of natural spider silk has a unique wet reconstruction structure such as spindle joints and junctions (Figure 2.12). Under the joint influence of the surface energy gradient of the spindle joints and joints and the Laplace pressure, the spider silk surface can realize the constant condensation and directional collection of water droplets. In the dry state, the capture filament of the sieve spider consists of fluffy and linked structures periodically distributed on two spindle fibers, and its main component is a high humidity-sensitive hydrophilic protein. In humidity, these fluffy structures shrink into opaque protrusions that eventually transform into periodic spindle structures with thicker diameter and link structures with thinner diameter (Figure 2.12).

The diameters of the spindle and link structures are approximately $21.0 \pm 2.7\,\mu m$ and $5.9 \pm 1.2\,\mu m$, respectively, and the spacing between them is approximately $89.3 \pm 13.5\,\mu m$ (Figure 2.12). Among these, the spindle structure consists of highly random nanofibers and the link structure consists of nanofibers relatively parallel to the filament fiber axis. Although the chemical components of the two surfaces

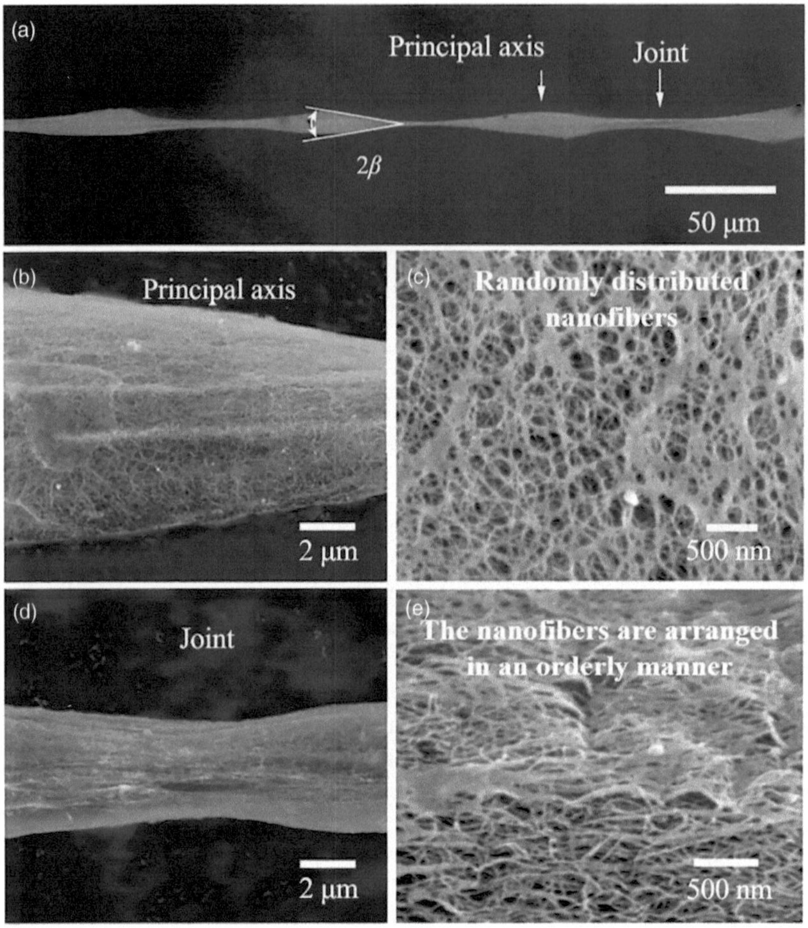

Figure 2.12 SEM images of the spider silk structure. (a) The periodic spindle joint of elongated joint with the vertex angle (2β) of approximately 19°. The low (b) and high (c) plots show that the spindle segments are randomly interwoven by nanofibers; (d–e) low (d) and high (e) joint maps, consisting of nanofibers arranged relatively parallel to the filament axis. Source: Reproduced with permission from [8] / Springer Nature.

differ little, the spindle structure exhibits a greater roughness. Furthermore, the diameter of each spindle segment of the spider silk varied in a gradient. According to the Wenzel equation [49],

$$\cos\theta_w = r\cos\theta. \tag{2.2}$$

And r is the surface roughness, θ and θ_w are apparent and intrinsic contact angles, respectively. For surface-hydrophilic spider silk, the region of the spindle structure is rougher and therefore more hydrophilic than the linked region [50]. The force generated by the surface energy gradient caused by the difference in surface roughness can be expressed as [51]

$$F = \int_{L_j}^{L_K} \gamma(\cos\theta_A - \cos\theta_R)\,dl. \tag{2.3}$$

γ is the surface tension of the water, and θ_A and θ_R are the forward and backward angles of water droplets on the spider silk, respectively, and dl is from the link structure (L_j) to the spindle structure (L_k). This force would drive the droplet moving from the linked region to the more hydrophilic spindle structure region, enabling directional fluid transport. The second possible force driving the droplet-directed transport originates from the geometrical feature of the spindle structure. The spindle segments can be viewed as two conical objects that are relatively bent and joined together, and this cone with a curvature gradient causes a Laplacian pressure difference that drives the directional transport of water droplets. The Laplacian pressure difference on the spindle structure can be expressed as [8, 9, 52]

$$\Delta P = \int_{r_1}^{r_2} \frac{2\gamma}{(r + R_0)^2} \sin \beta \, dz. \tag{2.4}$$

And r is the local radius of the taper, $R_0 = (3\,V/4\pi)^{1/3}$ is the droplet radius, V is the droplet volume, γ is the surface tension of water, β is 1/2 of the top angle of the spindle node, and dz is the integral variable of the spindle node diameter. Laplace pressure in the link region with large curvature (local radius r_1) greater than the spindle structure region with less curvature (the local radius is r_2). Thus, the Laplacian pressure difference generated within the droplet drives the droplet from the linked region toward the spindle structure region.

The combined pattern of spindle and link structures conferred unique surface energy gradient and geometric structure gradient to spider silk. Both gradients can provide a driving force for droplet-directed transport. The surface energy gradient provides the surface wetting driving force, allowing the droplet to move toward a more hydrophilic region, and the geometric gradient can generate a Laplace additional pressure to drive the droplet toward a larger diameter. The hydrophilicity of spider silk and the nanoscale fiber structure on the surface further enhance the ability of this droplet for directional transport (Figure 2.13). This specific fog collection mechanism provides a theoretical basis and experimental basis for achieving efficient fog collection and plays an important role in guiding the preparation of one-dimensional biomimetic superwetting materials.

Currently, some representative research results of bioinspired fibers in the past 10 years (Figure 2.14) [54], there are four widely reported methods of preparing bionic spider silk materials, namely the dip coating method [11a, d], the electrodynamic method [48b, 49a], microfluidic method [11f], fluid spraying method [49c, 55]. In a study, the researchers probed four polymers: polyethyl acetate, polymethacrylate (PMMA), polystyrene (PS), and polyvinylidene difluoride (PVDF), coated on artificial spindle joint, prepared, artificial filament, the filament from the nature and structure are similar to spider silk, have good ability to capture water [11c]. To further investigate the fog collection capacity of the periodic spindle segments, some bionic fibers with polyvinylidene difluoride spindle segments were synthesized . Compared with ordinary fibers, the fog collection on the bionic fibers is better, and the spindle joints on its surface can serve as the deposition and collection of water molecules. Tian et al. [54] combined electrospinning technique and electrospraying. Hydrophilic PEG imprint on hydrophobic polystyrene

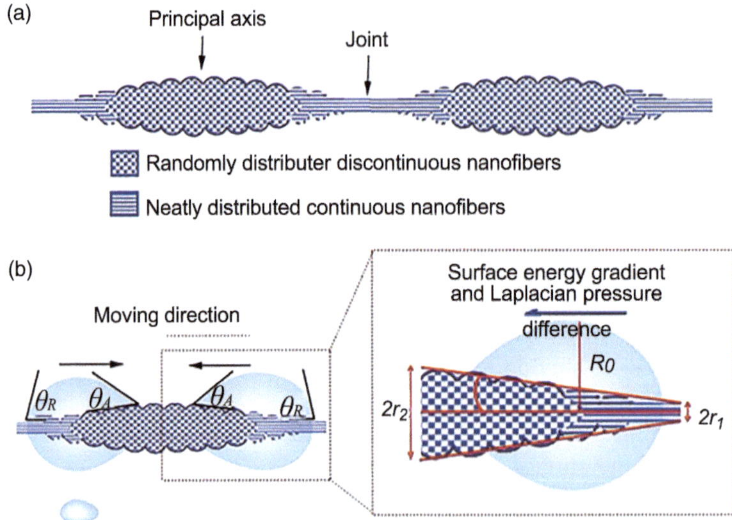

Figure 2.13 Mechanism of directional collection of water droplets on spider filaments after wetting and remodeling. (a) Schematic diagram of spindle joints (spindle) and link structure (joints); (b) the surface energy gradient and the Laplace pressure difference model of the water droplets. Source: Reproduced from [8] / with permission of Springer Nature.

fiber lines formed spindle segment structures similar to those on spider silk. Meanwhile, similar bead-like fiber materials were also prepared by the electrospinning technique, mainly combining polymethylmethacrylate (PMMA) and titanium tetrachloride ($TiCl_4$) [56].

In a recent study report, Chu et al. [11g] prepared spider filamentous microfibers using a simple and fast microfluidic technique. The preparation method is through a series of chemical cross-linking, the solution evaporation and drying, the microfluidic produced liquid ejected the corresponding template to make the magnetic ferric oxide grow on the calcium alginate microfibers, and finally formed the spider filamentary spindle strands and thin threads. The size of the spindle segments and fine lines varies with the rate size of the ejected fluid. Due to the presence of triiron oxide on the spindle segments, the magnetic field can regulate the structure and size of the spindle segments, thus obtaining biomimetic structures with better fog collection functions. Of course, fluid spraying technology can also be used to obtain cheap and tough bionic fiber on a large scale. As a common method, fluid spraying technology has been used for a long time. So far, it has been used in a number of fields, such as corrosion resistant metal wire, high mechanical performance fiber, and the candle manufacturing industry. Using this approach, Jiang and Coworkers [11a] prepared bionic fibers on a large scale, with periodic spindle segments on surface. They attached a capillary with an inner diameter of 400 μm tightly to a nylon fiber and soaked in a container containing a polymer solution of polymethyl methacrylate (PMMA) and dimethylamide (DMF), connecting an engine at the other end of the nylon fiber. Once the engine is started, the fiber is pulled out, and the polymer is also

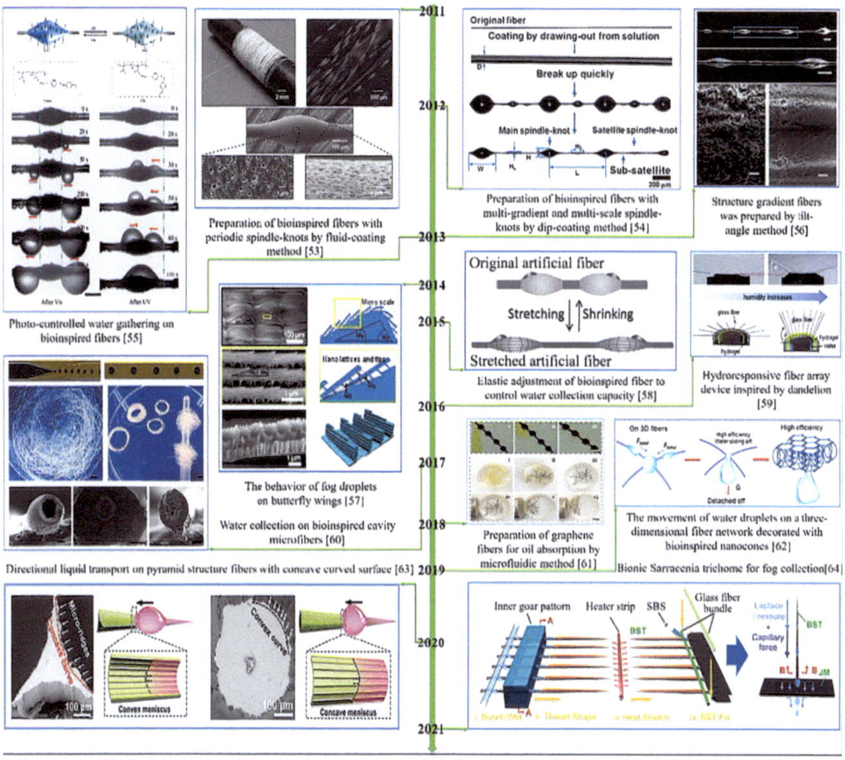

Figure 2.14 Some representative research of bioinspired fibers in the past 10 years. Source: Reproduced with permission from [51, 53] / Elsevier.

coated on surface. Finally, it was placed in air and formed by the Reynolds instability, similar to the globular polymer spindle segments with beads on spider filaments. The ratio of PMMA and DMF and the rate of filament line pulling influence the spindle segment formation, thus determining the final fog collection efficiency of this final biomimetic structure.

2.2.3 Desert Lizards

In nature, organisms living in arid environments have developed a variety of fog collection strategies, such as lizards that have evolved a variety of epidermal structures, whose folds, grooves and cracks can serve as fog collection systems and water transport channels, allowing lizards to obtain water from nature (Figure 2.15a) [10a]. During the night, when the outer skin decreases with the temperature of the external environment, water deposits on the hydrophilic skin surface and delivers to the mouth through the interconnected capillary system formed in their wrinkled epidermis. This process is a catchment process that does not require additional external energy to drive [59]. Desert lizards, including the Australian thorn devil and Texas horn lizards, consist of three layers from the scale epidermis: β -keratin, mesos, and α -keratin. The outer layer of β -keratin is composed of rigid

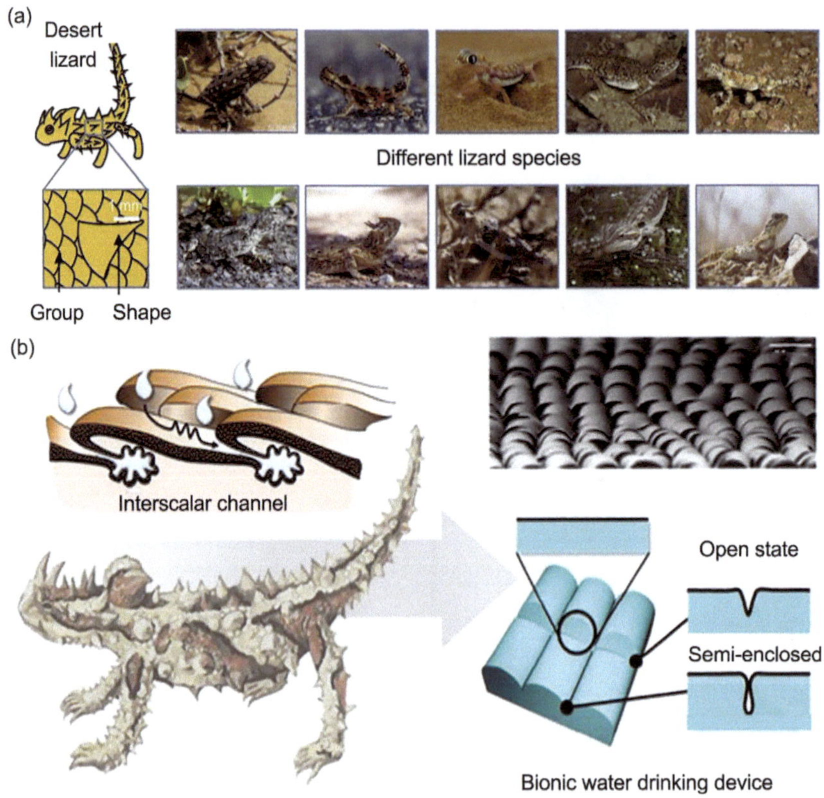

Figure 2.15 (a) Schematic diagram of surface scales of desert lizards and water-eating desert lizard species. Source: Reproduced with permission from [62, 63] / Elsevier. (b) The design of the fog collection structure was inspired by the epidermis of desert lizards. Source: Reproduced with permission from [62] / American Chemical Society.

proteins, whereas the inner layer of α -keratin is composed of flexible proteins [10b]. From the layered epidermis of lizards, a folded structure with scales is formed on the whole layered epidermis. The interscalar channels on these epidermises and the micro-and nano-sized structures are connected with the half-tube capillary system, realizing the capillary drive control of water flow in synergy (Figure 2.15b). Fog collection in desert lizards includes two elements: specific behaviors combined with particular body postures, and specific epidermal forms that allow the delivery and collection of water to the oral cavity [60]. The typical body postures of most hygroscopic desert lizards include similar hindlimb extension and bowing, in which gravity is used to support skin microstructure and channel function, guaranteeing membrane impermeability that protecting the lizards from dryness [59]. Meanwhile, water is transported in the passage between scales rather than simply spreading on the skin surface, thus most scale surfaces are not covered by water during transport. Transport of water in the pipeline avoids wetting most of the surface, thus water loss a smaller volume [61]. Two kinds of structures have been found, allowing lizards to store water while drinking water. First, in the layered channel structure, the cavity can rapidly absorb water into its large-capacity channel system, while the capillary structure extends the transport distance. Second, a combination of an asymmetric waterway geometry and interconnection network structure, the presence of a single channel between the micro/nanosized structures allows oriented liquid delivery locally, while specific interconnections facilitate water delivery to the oral cavity [62].

Existing biomimetic designs pay little attention to lizards. Most of them based on bionic studies of pitcher plants and cactus, but the authors believe that the study of lizards' epidermal structure is promising. The semi-open, semi-tubular channel at the scale junction of the desert lizard is similar to the circumferential surface of the carnivorous plant *Arabidopsis thaliana*, and this naturally occurring, well-defined strategy of water transport and wetting control is instructive for fog collection, and dew collection. Inspired by the epidermis of lizards, a mixed structure of fluid networks was designed by controlling the formation of fissure and fold structures [58]. After photography, the channel structure is formed by stress and partial oxidation, and finally a semi-closed channel is formed by the mark groove on the surface. Because cracks and folds are interconnected, water can be transported spontaneously away from the surface cracks to the internal fold channel through the capillary action. Moreover, given the tunable shape of the cracks and folds, the fluid network can be regulated by applying mechanical strain. This modulation on a fluid network can be used to control the water flow rate. These studies suggest that the mechanism of fluid transport control in lizards can act as a mechanism for fog collection and water transport.

There are also framed processes on the epidermis of desert lizards, and this structure can promote the transport direction to rise with the capillaries and prevent reflux. This asymmetric liquid diffusion principle realizes the unidirectional liquid transport of the bionic microgroove structure [64]. In addition, this thorn, similar to the cactus, can serve as a condensation focus (Figure 2.16). Microdecoration

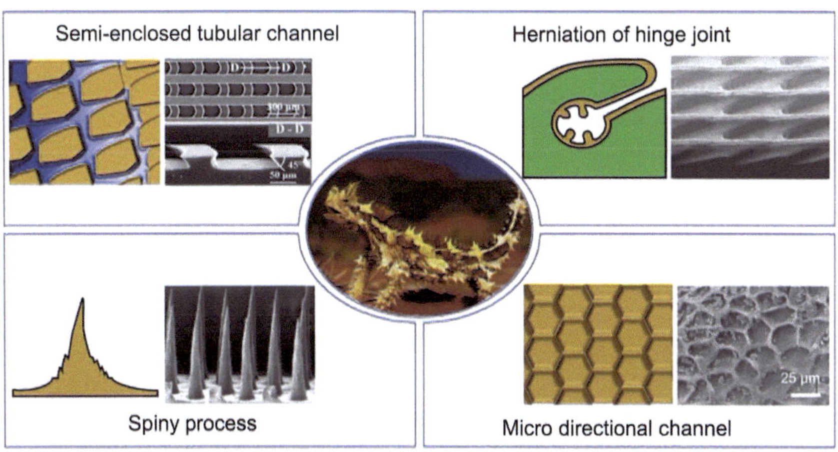

Figure 2.16 Water management properties and preparation methods of bionic surfaces inspired by desert lizards. Source: Reproduced with permission from [57, 63] / Elsevier.

of the skin surface can increase the surface hydrophilicity. While understanding the principle of lizard fog collection also provides a design strategy for preparing the engineering structure and system of fog collection.

2.2.4 Other Animal Inspiration

2.2.4.1 Musick

As a potential resource with many uses, the study of mussel adhesive properties plays an important role in the development of fog-water collecting surfaces. Mussel adhesion is produced by foot filaments contained in its biological structure, each composed of foot filament threads and foot filament patches. The inside of the foot filament thread is the core of the foot filament, and the outside is the cuticle, and the end of the foot filament thread expands to form foamy foot filament patches, which are also wrapped by the foot filament epidermis. The accessory gland (accessory gland), collagen gland (collagen gland) and phenol gland (phenol gland) are responsible for the molecular synthesis and storage of the epidermis, core and plaque, respectively. The secreted protein (Mfp) of these glands is the main reason for the adhesion of foot filaments [65]. We have shown that Mfp has strong adhesion and cohesion, so Mfp can adhere to the surface of various organic or inorganic substances without tearing, and the chemical bonds that Mfp attaches to different types of substrates include hydrogen bonds, metal-catechol coordination bonds, oxidative cross-linking and electrostatic interactions [64].

In the fog collection project, mussels have a hydrophilic biological adhesive substance dopamine (DA) solution, dopamine, which can adhere to the Cassie superhydrophobic surface. Then, a hydrophilic polydopamine pattern was built on a superhydrophobic surface. Therefore, combined with the self-polymerization of DA, special wettability can form. For example (Figure 2.17), by prewetting with dichloromethane (DCM), hydrophilic and bioadhesive dopamine solutions from mussels can drop onto a Cassie superhydrophobic copper surface with 0.1 μL of ultra-low volume, so that the low surface tension DCM will "mask" the high surface tension DA. As DCM volatilize, DA attaches to the Cassie superhydrophobic surface and then self-polymerizes into hydrophilic polydopamine regions, thus inspired by mussels that can successfully develop hydrophilic/superhydrophobic patterned surfaces with efficient water absorption. This preparation method is widely used in a large number of superhydrophobic materials in Cassie state, such as iron plate, zinc plate, copper wire, cotton fabric, copper mesh and nickel foam, which is expected to achieve large-scale production of simple and cheap bionic fog catching surface in industry [66].

2.2.4.2 The Caterpillars and the Earthworms

There are caterpillars of different shapes and sizes in nature, which is caused by various factors such as feeding on different plant species and sites. But the vast majority of caterpillar body is slender tubular, soft and segmented, composed of head, chest, abdomen three parts, body surface more markings, lines and body hair. The various parts of the body often have bristles, trichoma, hair pinch, hair process and spines.

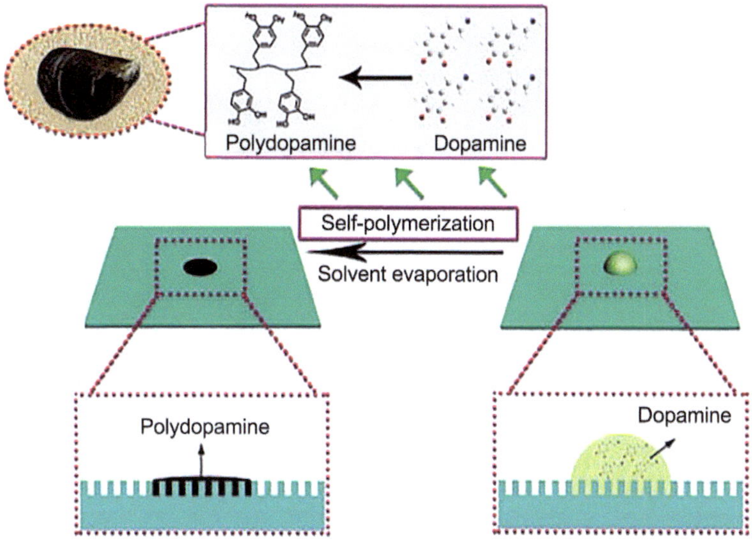

Figure 2.17 Surface design of mussel-inspired, highly efficient fog collection materials. Source: Reproduced with permission from [65, 66] / The Royal Society of Chemistry.

They are called trichoma; the hairs are long and dense in tuft or tuft. Some species have thorns, thorns on the branches called branches. These body hair structures can provide inspiration for the design of fog-collecting surfaces.

Earthworm, the whole is cylindrical, slender, similar body nodes, node and node between the internode grooves. The body wall of the whole body of the earthworm is composed of cuticle membrane, epithelium, ring muscle layer, longitudinal muscle layer and body cavity epithelium. The outermost layer is the single layer of columnar epithelial cells, and the secretions of these cells form the cutin membrane. This membrane is extremely thin and consists of collagen fibrous and non-fibrous layers with small holes in it. Columnner epithelial cells are mixed between glandular cells, divided into mucous cells and protein cells, can secrete mucus to make the body surface moist. Earthworms encounter severe stimulation, mucus cells secrete a large number of wrapped around the body into a mucus membrane, has a protective effect. Based on their complex self-lubrication mechanism of epidermal glands, mucus is constantly secreted under external mechanical stimulation. The rough skin composed of macroscopic ring space and microlines can stabilize the secreted mucus and form a thick smooth layer, showing a lubrication characteristic that can pass through sticky soil without producing stains [67]. This lubrication characteristic can not only be applied in wear resistance, resistance reduction and energy saving, but also quickly transfer and remove droplets in the process of fog collection, and then improve the efficiency of fog collection.

Inspired by the surface of caterpillars and earthworms, Professor Guo research group [68] designed a synergistic porous surface of superhydrophilic-superhydrophobic-smooth injection lubricant with gradient wetting, and caterpillar-inspired fiber trichomes were replaced with rough Fe_3O_4 NPs are covered to allow

Figure 2.18 A fibrous trichome with special wetting properties prepared from caterpillars and earthworms for efficient fog collection. Source: Reproduced with permission from [66, 68] / The Royal Society of Chemistry.

tiny fog droplets to be captured from moist air by superhydrophilic tips. The middle region of the superslip region constructed inspired by the earthworm epidermis was then rapidly removed and delivered by rapid droplets obtained by immersion in the lubricant (Figure 2.18). In addition to excellent fog collection performance, porous surfaces inspired by caterpillars and earthworms have continuous antifouling capability in mud, which expands its potential applications such as self-driven water harvesting, self-floating oil cleaning, reduced friction and wear resistance, and efficient antifouling.

2.2.4.3 Cicada Wing

Cicadies are a family in the suborder Hemptera, with more than 2000 species, widely distributed in temperate to subtropical areas and dwelling in deserts, grasslands and forests. Most cicada adults are 2–7 cm long and have two pairs of hymenoptera with basically the same shape, mostly transparent or translucent, with distinct wing veins. Cicwings are specialized flight organs and highly variable in size, with an adult wingspan of 4.5–17.5 cm. Most cicada wing consists of a series of longitudinal veins, crossing veins, and areas surrounded by these wing veins, and the main components are proteins, chitin, and wax [69]. The nanostructure of cicada wing surface is similar to that of a single small eye of the insect compound eye. The morphology of each nanomastoid structure is basically the same and has similar spacing in all directions (200–1000 nm) in all directions. The mastoid structure on the surface of cicada wing has a large aspect ratio, with a slight arc at the top, and a height between 225 and 250 nm. In addition, a similar structure is found on the wing veins of cicada wings [70]. The morphology of the nanostructures is roughly cylindrical and conical, and the slight differences in the morphology of the wetting and optical properties of the wings [71]. These nanostructures also provide inspiration for designing special wetting surfaces.

For example, water droplets, pollen or dust can be easily removed from the wing veins, and the phenomenon of rapid droplet removal can be applied in fog collection engineering. For the first time, Xie et al. [70] revealed the gradient wetting state of the cicada wing veins. That is, from the wingfoot to the wingtip, the transition from the Wenzel state (hydrophilic) to the Cassie-Baxter state (superhydrophobic) is due to the different radius, height and gap of the NPs on its surface (Figure 2.19). On the cicada wing, the superhydrophobic zone has less adhesion to the droplets than the hydrophilic zone. Therefore, the accumulated droplets can easily roll from the superhydrophobic region to the hydrophilic area and can maintain the light weight on the wing with slight jitter. This nanopillar and gradient wetting state of the natural wing surface plays a crucial role in fog collection and droplet transport processes.

2.2.4.4 Water Bird Beak

Waterbirds is a general term for living and active birds on the water surface. The common waterbirds include red-crowned cranes, herons, mandarin ducks, egrets and thrushes, etc. Waterbirds mainly feed on small crustaceans and other invertebrates. The feeding mechanisms of these waterbirds were found to transport prey with the transport of millimeter-sized droplets caused by surface tension, and repeatedly opening and closing its mouth like tweezers. The bird moves the droplet from the beak tip into the mouth in a way similar to a ratchet rotation. When the beak is in an open state, a conical liquid channel is provided, generating an asymmetric Laplacian pressure to drive the directional movement of the droplets. When the beak is in the closed state, the capture droplets suffer from directional diffusion through the capillary transport. The repeated opening and closing of the "beak" successfully send the water droplets into the waterbird's mouth [69].

Inspired by this observation, Ma et al. [72] applied a similar principle in a "switch" of a hydrogel (Figure 2.20a–d). Combined with the hydrogel characteristics with different expansion rates in the wet/dry state, the two copper wires can reversibly switch from the cone structure to parallel structure according to the humidity change (Figure 2.20g). Under dry conditions, the Laplacian pressure difference generated by the conical structure can drive the directional movement of the

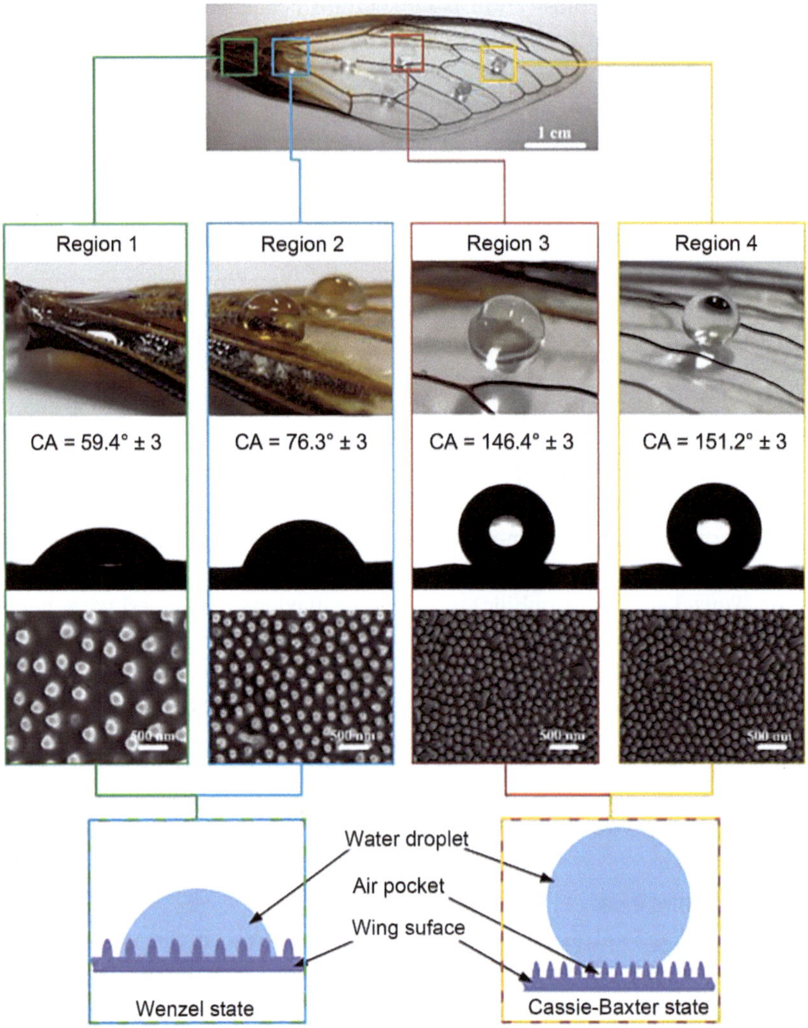

Figure 2.19 Changes in the wetting state of the different wing veins sections. Source: Reproduced with permission from [70, 72] / Elsevier.

microdroplets; and change the wetting by controlling the chemical properties and roughness of the solid surface to facilitate the effective sliding of the droplets (Figure 2.20e,f). In the wet state, the switch is off and the droplet is nailed due to the parallel arrangement of two microfibers to the surface. In the study, the on–on motion of the switch can also be reversed by changing the direction of the hydrogel drive system, with the switch on during hydration and the switch off during drying. The intelligent switch can control the directional delivery of liquid droplets through changes in humidity. Therefore, the switch in this design can control the directional delivery of droplets through the change of humidity, thus realizing the functions such as fog collection and liquid manipulation.

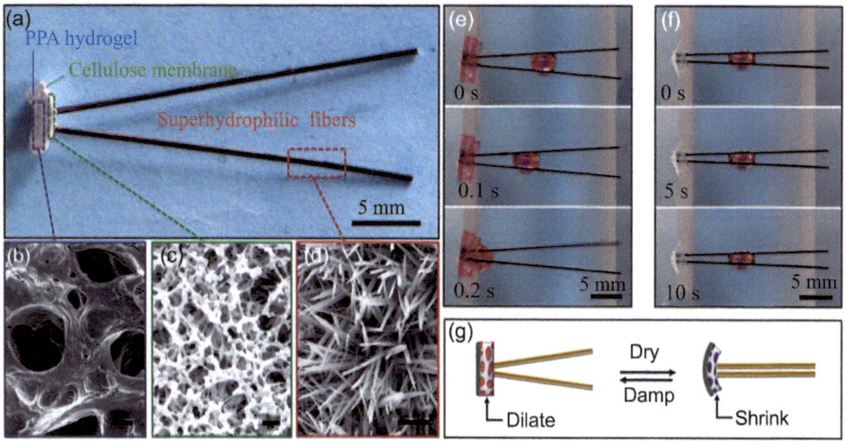

Figure 2.20 Bionic directional liquid delivery switch imitating water beak (a–d) the optical diagram of the open light and the SEM image of each parts; (e) the top view of the switch in the "open" state and the rapid directional delivery of water droplets when the hydrogel expands; (f) the top view of the microdroplet in the nail when the hydrogel shrinks (scale: 5 mm); (g) the responsidiagram of the switch: the switch is in the open and closed state when expanding. Source: Reproduced with permission from [72, 73] / The Royal Society of Chemistry.

In brief, many biological species have developed elegant schemes for water-harvesting [74]. Combining their mechanisms and design strategies is crucial to fabricate interfacial materials with efficient water collection. This chapter extracts the water-collecting principles and characteristics of Namib Desert beetles, cactus and spider silk, and so on. Various bionic water collecting strategies are reviewed under the guidance of these typical creatures and will be served for the coming content.

References

1 (a) Meyers, M.A., Chen, P.-Y., Lin, A.Y.-M., and Seki, Y. (2008). *Progress in Materials Science* 53: 1. (b) Woodyer, R., Chen, W., and Zhao, H. (2004). *Journal of Chemical Education* 81: 126. (c) Peisajovich, S.G. (2012). *ACS Synthetic Biology* 1: 199. (d) Gou, N., Yuan, S., Lan, J. et al. (2014). *Environmental Science & Technology* 48: 8855.

2 (a) D'Alessio, L. and Rigol, M. (2015). *Nature Communications* 6: 8336. (b) Wise, E.L. and Rayment, I. (2004). *Accounts of Chemical Research* 37: 149.

3 (a) Walther, A., Bjurhager, I., Malho, J.-M. et al. (2010). *Nano Letters* 10: 2742. (b) Bixler, G.D. and Bhushan, B. (2012). *Soft Matter* 8: 11271. (c) Sakhavand, N. and Shahsavari, R. (2015). *Nature Communications* 6: 6523. (d) Razzak, M. and De Brabander, J.K. (2011). *Nature Chemical Biology* 7: 865. (e) Marguet, M., Bonduelle, C., and Lecommandoux, S. (2013). *Chemical Society Reviews* 42: 512.

(f) Liu, Y., Lu, Y., and Liu, Z. (2012). *Chemical Science* 3: 1467. (g) Hou, X., Guo, W., and Jiang, L. (2011). *Chemical Society Reviews* 40: 2385. (h) Balmert, S.C. and Little, S.R. (2012). *Advanced Materials* 24: 3757. (i) Wen, L., Tian, Y., and Jiang, L. (2015). *Angewandte Chemie International Edition* 54: 3387. (j) Shao, Y. and Fu, J. (2014). *Advanced Materials* 26: 1494.

4 (a) Stadler, A.M., Embs, J.P., Digel, I. et al. (2008). *Journal of the American Chemical Society* 130: 16852. Ridoutt, B.G. and Pfister, S. (2010). ACS Publications, 44(16): 6019-21; (c) Gawor, A. and Wania, F. (2013). *Environmental Science: Processes & Impacts* 15: 1671. (d) Gilbertson, L.M., Zimmerman, J.B., Plata, D.L. et al. (2015). *Chemical Society Reviews* 44: 5758. (e) Mansur, A.R. and Oh, D.-H. (2015). *Food Microbiology* 51: 154. (f) Carroll, H.A., Davis, M.G., and Papadaki, A. (2015). *Nutrition Research* 35: 865.

5 (a) Klemm, O., Schemenauer, R.S., Lummerich, A. et al. (2012). *Ambio* 41: 221. (b) Shanyengana, E.S., Henschel, J.R., Seely, M.K., and Sanderson, R.D. (2002). *Atmospheric Research* 64: 251. (c) Kaseke, K.F. and Wang, L. (2018). *GeoHealth* 2: 327.

6 (a) Cereceda, P., Larrain, H., Osses, P. et al. (2008). *Atmospheric Research* 87: 312. (b) Hiatt, C., Fernandez, D., and Potter, C. (2012). *Atmospheric and Climate Sciences* 2 (4): 7. (c) Estrela, M.J., Valiente, J.A., Corell, D., and Millán, M.M. (2008). *Atmospheric Research* 87: 324.

7 Parker, A.R. and Lawrence, C.R. (2001). *Nature* 414: 33.

8 Zheng, Y., Bai, H., Huang, Z. et al. (2010). *Nature* 463: 640.

9 Ju, J., Bai, H., Zheng, Y. et al. (2012). *Nature Communications* 3: 1247.

10 (a) Malik, F.T., Clement, R.M., Gethin, D.T. et al. (2014). *Bioinspiration & Biomimetics* 9: 031002. (b) Park, K.-C., Kim, P., Grinthal, A. et al. (2016). *Nature* 531: 78.

11 (a) Bai, H., Ju, J., Sun, R. et al. (2011). *Advanced Materials* 23: 3708. (b) Bai, H., Sun, R., Ju, J. et al. (2011). *Small* 7: 3429. (c) Bai, H., Tian, X., Zheng, Y. et al. (2010). *Advanced Materials* 22: 5521. (d) Hou, Y., Chen, Y., Xue, Y. et al. (2012). *Langmuir* 28: 4737. (e) Dong, H., Wang, N., Wang, L. et al. (2012). *ChemPhysChem* 13: 1153. (f) Chen, Y. and Zheng, Y. (2014). *Nanoscale* 6: 7703. (g) He, X.-H., Wang, W., Liu, Y.-M. et al. (2015). *ACS Applied Materials & Interfaces* 7: 17471. (h) Tian, Y., Zhu, P., Tang, X. et al. (2017). *Nature Communications* 8: 1080.

12 (a) Ju, J., Xiao, K., Yao, X. et al. (2013). *Advanced Materials* 25: 5937. (b) Ju, J., Yao, X., Yang, S. et al. (2014). *Advanced Functional Materials* 24: 6933. (c) Heng, X., Xiang, M., Lu, Z., and Luo, C. (2014). *ACS Applied Materials & Interfaces* 6: 8032. (d) Cao, M., Ju, J., Li, K. et al. (2014). *Advanced Functional Materials* 24: 3235. (e) Peng, Y., He, Y., Yang, S. et al. (2015). *Advanced Functional Materials* 25: 5967. (f) Bai, F., Wu, J., Gong, G., and Guo, L. (2015). *Advanced Science* 2: 1500047. (g) Zhou, H., Zhang, M., Li, C. et al. (2018). *Small* 14: 1801335.

13 Si, Y., Dong, Z., and Jiang, L. (2018). *ACS Central Science* 4: 1102.

14 (a) Barthlott, W., Schimmel, T., Wiersch, S. et al. (2010). *Advanced Materials* 22: 2325. (b) Feng, L., Li, S., Li, Y. et al. (1857). *Advanced Materials* 2002: 14. (c) Feng, L., Zhang, Y., Xi, J. et al. (2008). *Langmuir* 24: 4114. (d) Epstein, A.K.,

Pokroy, B., Seminara, A., and Aizenberg, J. (2011). *Proceedings of the National Academy of Sciences* 108: 995. (e) Zheng, Y., Gao, X., and Jiang, L. (2007). *Soft Matter* 3: 178. (f) Nickerl, J., Helbig, R., Schulz, H.-J. et al. (2013). *Zoomorphology* 132: 183.

15 (a) Tan, X., Shi, T., Tang, Z. et al. (2016). *Journal of Bionic Engineering* 13: 364. (b) Yi, S., Wang, J., Chen, Z. et al. (2019). *Advanced Materials Technologies* 4: 1900727.

16 Lee, S.J., Ha, N., and Kim, H. (2019). *ACS Sustainable Chefogry & Engineering* 7: 10561.

17 Lin, J., Tan, X., Shi, T. et al. (2018). *ACS Applied Materials & Interfaces* 10: 44815.

18 Schemenauer, R.S. and Cereceda, P. (1992). *Journal of Applied Meteorology and Climatology* 31: 275.

19 Kylstra, J.A. (1968). *Scientific American* 219: 66.

20 (a) Bohn, H.F. and Federle, W. (2004). *Proceedings of the National Academy of Sciences* 101: 14138. (b) Gorb, E.V., Purtov, J., and Gorb, S.N. (2014). *Scientific Reports* 4: 5154. (c) Gorb, E.V., Baum, M.J., and Gorb, S.N. (2013). *Scientific Reports* 3: 3078.

21 Wei, Q., Schlaich, C., Prévost, S. et al. (2014). *Advanced Materials* 26: 7358.

22 Chen, H., Zhang, P., Zhang, L. et al. (2016). *Nature* 532: 85.

23 Labonte, D., Robinson, A., Bauer, U., and Federle, W. (2021). *Acta Biomaterialia* 119: 225.

24 Zheng, Y., Cheng, J., Zhou, C. et al. (2017). *Langmuir* 33: 4172.

25 (a) Li, C., Wu, L., Yu, C. et al. (2017). *Angewandte Chemie International Edition* 56: 13623. (b) Morrissette, J.M., Mahapatra, P.S., Ghosh, A. et al. (2017). *Scientific Reports* 7: 1800. (c) Ody, T., Panth, M., Sommers, A.D., and Eid, K.F. (2016). *Langmuir* 32: 6967.

26 Peng, Z., Chen, Y., and Wu, T. (2020). *ACS Applied Materials & Interfaces* 12: 47299.

27 Jing, X., Si, W., Sun, J. et al. (2022). *Advanced Materials Interfaces* 9: 2101231.

28 Herppich, W.B., Martin, C.E., Totzke, C. et al. (2019). *Plant, Cell & Environment* 42: 1645.

29 Rundel, P., Palma, B., Dillon, M. et al. (1997). *Revista Chilena de Historia Natural* 70: 341.

30 Pan, Z., Pitt, W.G., Zhang, Y. et al. (2016). *Nature Plants* 2: 16076.

31 Carl, M. (1904). *Physiologische Bromeliaceen-Studien.*

32 Raux, P.S., Gravelle, S., and Dumais, J. (2020). *Nature Communications* 11: 396.

33 Beadle, N.C.W. (1966). *Ecology* 47: 992.

34 (a) Roth-Nebelsick, A., Ebner, M., Miranda, T. et al. (1965). *Journal of the Royal Society Interface* 2012: 9. (b) Gürsoy, M., Harris, M.T., Downing, J.O. et al. (2017). *Colloids and Surfaces A: Physicochemical and Engineering Aspects* 529: 195.

35 Andrews, H.G., Eccles, E.A., Schofield, W.C.E., and Badyal, J.P.S. (2011). *Langmuir* 27: 3798.

36 Sharma, V., Sharma, M., Kumar, S., and Krishnan, V. (2016). *Flora* 224: 59.

37 Zhang, X., Yu, P., Wang, D., and Xu, Z. (2022). *Agricultural and Forest Meteorology* 325: 109134.
38 Wang, Y., Zhang, L., Wu, J. et al. (2015). *Journal of Materials Chefogry A* 3: 18963.
39 (a) Gibbens, R.P. and Lenz, J.M. (2001). *Journal of Arid Environments* 49: 221. (b) Markle, M.S. (1917). *Botanical Gazette* 64: 177.
40 Gürsoy, M., Harris, M.T., Carletto, A. et al. (2017). *Colloids and Surfaces A: Physicochemical and Engineering Aspects* 529: 959.
41 Wang, B., Zhang, Y., Liang, W. et al. (2014). *Journal of Materials Chefogry A* 2: 7845.
42 Zhang, L., Wu, J., Hedhili, M.N. et al. (2015). *Journal of Materials Chefogry A* 3: 2844.
43 de Gans, B.-J., Duineveld, P.C., and Schubert, U.S. (2004). *Advanced Materials* 16: 203.
44 (a) Ye, Q., Zhou, F., and Liu, W. (2011). *Chemical Society Reviews* 40: 4244. (b) Takei, G., Nonogi, M., Hibara, A. et al. (2007). *Lab on a Chip* 7: 596. (c) Laure, W., Woisel, P., and Lyskawa, J. (2014). *Chefogry of Materials* 26: 3771. (d) Shen, G., Liang, B., Wang, X. et al. (2011). *ACS Nano* 5: 6148. (e) Yang, H., Zhu, H., Hendrix, M.M.R.M. et al. (2013). *Advanced Materials* 25: 1150.
45 Lyu, P., Zhang, X., Peng, M. et al. (2021). *Advanced Materials Interfaces* 8.
46 (a) Vollrath, F. and Edmonds, D.T. (1989). *Nature* 340: 305. (b) Vollrath, F. (2000). *Reviews in Molecular Biotechnology* 74: 67. (c) Liu, Y., Shao, Z., and Vollrath, F. (2005). *Nature Materials* 4: 901.
47 Wenzel, R.N. (1936). *Industrial & Engineering Chefogry* 28: 988.
48 (a) Becker, N., Oroudjev, E., Mutz, S. et al. (2003). *Nature Materials* 2: 278. (b) Peters, H.M. (1984). *Zoomorphology* 104: 96. (c) Quéré, D. (2008). *Annual Review of Materials Research* 38: 71.
49 (a) Edmonds, D.T. and Vollrath, F. (1992). *Proceedings of the Royal Society of London. Series B: Biological Sciences* 248: 145. (b) Daniel, S., Chaudhury, M.K., and Chen, J.C. (2001). *Science* 291: 633.
50 Lorenceau, E. and Quéré, D. (2004). *Journal of Fluid Mechanics* 510: 29.
51 Chen, S., Yang, F., and Guo, Z. (2022). *Advances in Colloid and Interface Science* 309: 103017.
52 Oda, M., Suga, S., Yoshii, H., and Furuta, T. (2008). *Asia-Pacific Journal of Chemical Engineering* 3: 63.
53 Srivastava, P. and Yadav, U. (2023). *Nanobiotechnology for Bioremediation*, 273–296.
54 Tian, X., Bai, H., Zheng, Y., and Jiang, L. (2011). *Advanced Functional Materials* 21: 1398.
55 Zhao, L., Song, C., Zhang, M., and Zheng, Y. (2014). *Chemical Communications* 50: 10651.
56 Sherbrooke, W.C., Scardino, A.J., de Nys, R., and Schwarzkopf, L. (2007). *Zoomorphology* 126: 89.
57 He, G., Zhang, C., and Dong, Z. (2023). *iScience* 26: 105819.
58 Demartini, D.G., Errico, J.M., Sjoestroem, S. et al. (2017). *Interface* 14: 20170151.

59 Yenmisx, M., Ayaz, S.D., and Vesely, M. (2015). *Zoomorphology* 135: 137.

60 Comanns, P., Buchberger, G., Buchsbaum, A. et al. (2015). *Journal of the Royal Society Interface* 12.

61 He, G., Zhang, C., and Dong, Z. (2023). *iScience* 26.

62 Cha, J., Shin, H., and Kim, P. (2016). *ACS Applied Materials & Interfaces* 8: 28418.

63 Li, C., Yu, C., Zhou, S. et al. (2020). *Proceedings of the National Academy of Sciences of the United States of America* 117: 23436.

64 (a) Yang, J., Stuart, M.A.C., and Kamperman, M.J.C.R. (2014). 43, 8271; (b) Lee, H., Dellatore, S.M., Miller, W.M., and Messersmith, P.B.J. (2007). 318, 426.

65 Zhu, H., Duan, R., Wang, X. et al. (2018). *Nanoscale* 10: 13045.

66 Li, D., Wang, Z., Wu, D. et al. (2019). *Nanoscale* 11: 11774.

67 Tobin, M.J., Puskar, L., Hasan, J. et al. (2013). *Journal of Synchrotron Radiation* 20: 482.

68 Kesel, A.B., Philippi, U., and Nachtigall, W.J.C. (1998). *Medicine* 28: 423.

69 Watson, G.S. and Watson, J.A. (2004). *Applied Surface Science* 235: 139.

70 Xie, H., Huang, H.-X., and Mi, H.-Y. (2018). *Materials Letters* 221: 123.

71 Prakash, M., Quéré, D., and Bush, J.W.M. (2008). *Science* 320: 931.

72 Ma, X., Cao, M., Teng, C. et al. (2015). *Journal of Materials Chefogry A* 3: 15540.

73 Zhu, H., Guo, Z., and Liu, W. (2016). *Chemical Communications.* https://doi.org/10.1039/C5CC09867J.

74 (a) Bentley, P.J. and Blumer, W.F.C. (1962). *Nature* 194: 699. (b) Comanns, P., Effertz, C., Hischen, F. et al. (2011). *Beilstein Journal of Nanotechnology* 2: 204.

3

Basic Theory

Water is ubiquitous in the troposphere and can exist in gas, liquid, and solid states. The amount of water vapor contained in the air is usually described by relative humidity (RH), usually in %, which is defined as the ratio of its partial pressure to saturated vapor pressure at the same temperature (T) multiplied by 100 [1],

$$RH(\%) = p/P_s \times 100\%, \tag{3.1}$$

where p is the partial pressure of water vapor and P_s is the saturated vapor pressure of water. Under subsaturated conditions with RH below 100%, some water vapor will adhere to the solid surface and nucleate. The balance between nucleating and condensing water attached to the surface of the solid phase and water vapor in the gas phase depends on the RH, temperature, and type of dust particles. When the RH is higher than 100%, that is, under supersaturated conditions, the ability of water vapor particles to be activated to condense to form droplets depends on the nucleation diameter and condensation activity, which is determined by its chemical composition and mixing state. The above two processes are called moisture collection and fog collection [2]. Heterogeneous nucleation reactions involve the interactions between gas and solid phases. For instance, during cloud formation, the activation of mineral dust particles into cloud droplets can initiate multiphase reactions within these water-laden droplets. Such heterogeneous reactions – both direct and indirect – can significantly alter the concentrations of trace gases, thereby influencing tropospheric chemistry. Furthermore, these reactions may lead to modifications in the surface and overall chemical composition of mineral dust particles, which in turn affects their interaction with water and the efficiency of water collection processes [1–3]. This chapter primarily focuses on the fundamental principles of water collection, exploring two key methods: fog harvesting and moisture absorption. By examining these approaches, we aim to provide a solid theoretical foundation for advancing our understanding of water collection techniques.

Bioinspired Materials for Fog Harvesting: Principles, Design, Preparation, and Applications,
First Edition. Zhiguang Guo and Weimin Liu.

3.1 The Theory of Fog Harvesting

3.1.1 Fog Capture

In order to understand the process of droplet capture, it is essential to first examine the behavior of droplets at the surface, specifically their state upon contact. Furthermore, the phenomenon of droplet nucleation at the surface can be clarified through established thermodynamic principles.

3.1.1.1 Young's Equation and Wenzel–Cassie Theory

Wetting is characterized by the three-phase interface involving a gas, a liquid, and a solid. Consider a droplet of liquid placed on an ideal solid surface. The angle formed by the intersection of the solid–liquid interface and the gas–liquid interface is referred to as the contact angle (CA) (θ) (Figure 3.1a). The interfacial tensions involved can be described using Young's equation, which is instrumental in studying solid–liquid interactions. And the three-phase interfacial tension generally obeys the following Young equation [5]:

$$\gamma_{SG} = \gamma_{SL} + \gamma_{LG}\cos\theta. \tag{3.2}$$

In this context, the interfacial tensions are represented by γ_{SG} for solid–gas, γ_{SL} for solid–liquid, and γ_{LG} for gas–liquid. The wettability of a surface is determined by the CA of the liquid on that solid surface. When the angle (θ) is less than 90°, the liquid is able to wet a portion of the solid surface. Conversely, if the angle falls between 90° and 150°, the surface is classified as hydrophobic. If the angle exceeds 150°, the surface is categorized as superhydrophobic (SHB).

Surface roughness plays a critical role in the wettability of materials. It is defined as the ratio of the true surface area to the apparent surface area of a rough surface (Figure 3.1b,c) [6]. As a droplet advances over a surface, the apparent surface area of the liquid interface increases by dS, while the true area of the solid–liquid interface is augmented by rdS. In an equilibrium state, the true area of contact between the solid and liquid decreases by rdS, while the apparent area of the gas–liquid interface grows by dS cos θ_w,

$$\sigma_{SL} r\mathrm{d}S - \sigma_{LG}\mathrm{d}S\cos\theta_w + \sigma_{SG}\cos\theta = 0, \tag{3.3}$$

$$\cos\theta_w = \frac{r(\sigma_{SG} - \sigma_{SL})}{\sigma_{LG}} = r\cos\theta_y. \tag{3.4}$$

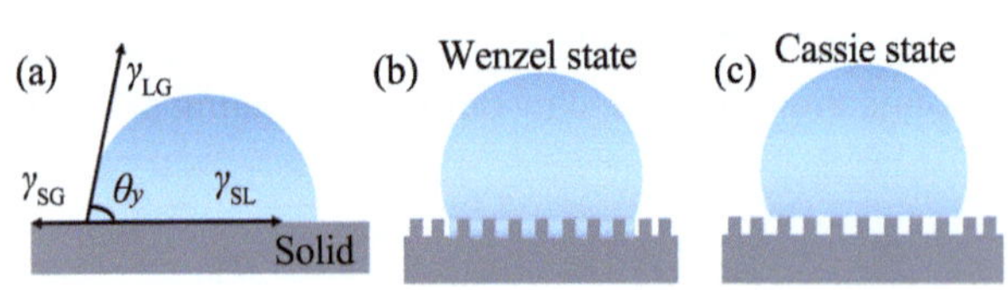

Figure 3.1 (a) The contact angle of a droplet at equilibrium on an ideal solid surface. (b and c) The possible state of a droplet on a rough surface. Source: [4] / The Royal Society of Chemistry.

The surface morphology of real solids exhibits significant complexity. In 1944, acknowledging the critical role of surface chemistry, Cassie and Baxter introduced a more intricate model by integrating additional coefficients into the governing equations [7]. This model accounts for the complex interplay between surface morphology and chemical properties, highlighting the impact of various chemical factors on wettability [7].

Furthermore, Cassie and Baxter proposed a combined model that incorporates multiple materials with distinct surface characteristics. They defined "n" as the number of different types of materials that can be randomly distributed across the surface. Each material possesses its own specific surface tensions, denoted as $\gamma_{i,\mathrm{SL}}$ and $\gamma_{i,\mathrm{SA}}$, correlating to the solid–liquid and solid–air interfaces, respectively. The area fraction of each material is represented by f_i, which must satisfy the following equation:

$$f_1 + f_2 + \ldots + f_n = 1, \gamma_{\mathrm{SA}} = \sum_i^n f_i(\gamma_{i,\mathrm{SA}}), \gamma_{\mathrm{SL}} = \sum_i^n f_i(\gamma_{i,\mathrm{SL}}),$$

$$\cos\theta_{\mathrm{CB}} = \sum_i^n f_i(\gamma_{i,\mathrm{SA}} - \gamma_{i,\mathrm{SL}})/\gamma_{\mathrm{LA}} = \sum_i^n f_i \cos\theta_i, \tag{3.5}$$

analyzing a simplified case where the composite surface is composed of two different materials. After applying the Cassie and Baxter equation, the apparent CA of the material surfaces can be extrapolated to the following:

$$\cos\theta_{\mathrm{CB}} = f_1 \cos\theta_1 + f_2 \cos\theta_2. \tag{3.6}$$

3.1.1.2 Thermodynamic Perspective – Nucleation of Fog Droplets

One hundred years ago, Gibbs et al. introduced the classical nucleation theory, which is grounded in the principles of thermodynamics. This theory posits that the phase change nucleation process is contingent upon the formation of a critical nucleus. Specifically, a nucleus must exceed a certain critical size for a spontaneous phase transition to occur within the system. In the context of fog harvesting, the nucleation phenomenon involves the condensation of water vapor in a gaseous state into droplets within the liquid phase. This process occurs on the surface of a solid medium, termed the "third phase" [6–8]. The conditions necessary for the condensation and nucleation of water molecules can be described as follows: When the gas–liquid mixture reaches a supersaturated state, water molecules spontaneously form droplets on the surface of the third-phase solid. From a classical thermodynamic perspective, the condensation process is understood as a reduction in the system's free energy after phase nucleation, ultimately driving the system toward an equilibrium or critical state. In the framework of statistical mechanics, the thermodynamic behavior of gaseous water molecules at the interface with the solid phase fluctuates continually. When these fluctuations comply with the second law of thermodynamics, the thermodynamic energy barrier is surmounted, resulting in the spontaneous condensation of water molecules and the subsequent transition of the system to a more stable thermodynamic state [9].

The foundational thermodynamics of condensation nucleation can be summarized by the Gibbs free energy criterion for phase change nucleation, expressed mathematically as $\mathrm{d}G < 0$, where G represents the Gibbs free energy. This criterion indicates that for nucleation to be favorable and for a phase transition to proceed, the change in Gibbs free energy must be negative,

$$\mathrm{d}U = T\mathrm{d}S - p\mathrm{d}V + \mu \mathrm{d}n_{\mathrm{A}}, \tag{3.7}$$

where U represents the internal energy, S represents the entropy, V refers to the volume, T refers to the temperature, p refers to the pressure, μ refers to the chemical potential, and n_{A} refers to the number of particles of component A in the system. In addition, the Gibbs free energy of the system can be further expressed as

$$\mathrm{d}G = V\mathrm{d}p - S\mathrm{d}T + \mu \mathrm{d}n_{\mathrm{A}}. \tag{3.8}$$

In the process of condensation and nucleation of water molecules, entropy is a crucial factor influencing the free energy associated with the phase transition. From a thermodynamic perspective, an increase in entropy tends to facilitate the nucleation of the condensed phase. This can be understood in terms of statistical mechanics, where phase transformation is associated with a transition toward configurations that encompass a greater number of accessible microstates within the system. This discussion provides a macroscopic view based on thermodynamic principles. Next, we will examine the chemical potential of water molecules in both the gaseous and condensed phases, focusing on the changes that occur during the nucleation process. For the purposes of this analysis, we will make two key assumptions: First, the gaseous water molecules can be approximated as ideal gases prior to condensation, and second, the ambient temperature and pressure of the system remain constant throughout the condensation process. Moore's Gibbs free energy criterion is

$$\Delta G_{\mathrm{m}} = G_{\mathrm{m}}^{\mathrm{l}} - G_{\mathrm{m}}^{\mathrm{g}} \le 0 \tag{3.9}$$

By introducing parameters such as chemical formulas, the Kelvin formula can be derived as follows [6, 10]:

$$p = p^{*} e^{V_{\mathrm{m}}^{\mathrm{l}} 2\gamma / rRT}, \tag{3.10}$$

where p represents the total pressure of the system, p^{*} represents the normal pressure, V_{m} represents the molar volume under constant temperature conditions, γ represents the surface tension of the curved surface, and r refers to the radius of the core.

In the process of gas–liquid condensation, the nucleation phase leads to the formation of small spherical droplets, which establish a curved gas–liquid interface. This curvature induces additional Laplace pressure, subsequently influencing the vapor pressure within the gas phase. Under certain temperature conditions, steam exhibits thermodynamic instability relative to liquid water, resulting in the prediction that vapor will condense and nucleate into liquid droplets. Initially, the system can be characterized by a multitude of water molecules aggregating to form minuscule droplets. Due to their small size and high vapor pressure, these nascent droplets do not contribute to an increase in water concentration; rather, the surrounding water molecules tend to evaporate. This phenomenon results in a stabilization of atmospheric water vapor as the propensity for condensation and nucleation is countered by an enhanced tendency for evaporation. This condition is referred to as the supersaturated state of the gas phase. While the supersaturated state is thermodynamically unstable compared to bulk liquid water, it remains stable in relation to the formation of small droplets. Consequently, the condensation of water vapor into minuscule droplets is thermodynamically unfavorable before the establishment of a significant bulk liquid phase. Therefore, in the design of fog collection surfaces, it is imperative to create areas with high surface energy to effectively capture and retain fog droplets.

3.1.2 Fog Droplet Coalesce

The process of droplet formation following the critical nucleation of vapor-phase water molecules on a surface is influenced by a variety of complex factors related to continued enrichment and growth. Initially, the growth of small droplets is often impeded by enhanced evaporation effects, which arise due to the size constraints of these droplets. To counteract this evaporation tendency and promote droplet growth, two primary mechanisms can be identified. The first is the spontaneous nucleation, during which a substantial number of water molecules can rapidly aggregate and condense to form larger droplets. The formation of these larger droplets mitigates the impact of enhanced evaporation, allowing them to continue growing despite the initial constraints. The second is the condensation on particulate surfaces. Alternatively, the presence of tiny particles or foreign matter within the system serves as nucleation sites for water molecules. These particles facilitate the attachment and aggregation of water molecules, creating additional surfaces for condensation to occur, thereby promoting droplet growth. In essence, nucleation provides critical surfaces that enable water molecules to attach, subsequently enhancing condensation rates. According to the classical nucleation theory, the pre-exponential factor δ of the molecular dynamics of the nucleation process can be expressed as [11]

$$\delta = \mathcal{N}_{nc}\mathcal{Z}_{nc}\mathcal{Q}_{kin}, \tag{3.11}$$

where $\mathcal{N}_{nc}$ is the number of possible nucleation sites within unit volume; $\mathcal{Z}_{nc}$ refers to the Zeldovich factor [12], which means that some critical clusters or critical nuclei will not continue to grow and nucleate due to evaporation trends or dissociation of other factors; and $\mathcal{Q}_{kin}$ refers to the kinetic correction factor of the nucleation process [13], which represents the frequency at which clusters of water molecules in the gas phase reach the vicinity of the critical nucleus and rearrange themselves in the form of condensation. In short, $\mathcal{Q}_{kin}$ is related to the mobility of gas-phase water molecule clusters moving to the surface and the mobility on the surface, rather than the self-diffusion coefficient of particles $\mathcal{D}$, a kinetic factor that can be accurately expressed by a parameter [14]. Water molecule clusters do not always participate stably in the growth process of droplets on the surface. To study the migration of water molecule clusters to the critical core surface of the solid surface for enrichment and growth without returning to the gas–liquid mixed phase, it is necessary to consider the behavior of water molecule clusters on the surface. The nucleation rate η_{nc} is given as

$$\eta_{nc} = \rho_{nc}^{c} A_{nc}^{c} \omega, \tag{3.12}$$

where ρ_{nc}^{c} is the critical core density, A_{nc}^{c} refers to the area where the critical core is hit, and ω refers to the probability of water molecule clusters hitting the critical core. The critical nuclear density ρ_{nc}^{c} is expressed according to the critical free energy,

$$\rho_{nc}^{c} = n_s \exp(-\Delta G_c / k_B T), \tag{3.13}$$

where n_s is the density of the total core point. The critical nuclear density is fundamentally influenced by temperature and supersaturation levels. However, the growth of droplet enrichment is inherently unstable. When clusters of water molecules collide with the critical nucleus, there is a specific residence time during which these clusters remain at the surface before departing and returning to the gas–liquid mixed phase. The residence time τ of water molecule clusters on the surface is given as

$$\tau = \frac{1}{\vartheta} \exp(E_{des} / k_B T), \tag{3.14}$$

where ϑ refers to the vibration frequency of water molecule clusters adsorbed on the critical core surface and E_{des} refers to the desorption energy of water molecule clusters detaching from the surface. According to the theory of gas molecule motion, the number of gas molecules hitting the solid surface per unit area and unit time is the collision flux Z_w,

$$Z_w = p / \sqrt{2\pi m k_B T} = p / \sqrt{2\pi M k_B T / N_A}, \tag{3.15}$$

where p is the pressure, N_A is Avogadro's constant, and m is the mass of the gas molecules. According to the residence time and the number of collisions of gas phase particles, the number n_{ad} of water molecule clusters adsorbed on the critical core surface can be calculated as

$$n_{ad} = \tau p N_A / \sqrt{2\pi M R T}, \tag{3.16}$$

where R is the gas constant, $R = k_B N_A$, and M is the molar mass of the gas molecules. Considering the diffusion frequency ϑ_D of water molecule clusters on the surface,

$$\vartheta_D = \vartheta \exp(-E_D / k_B T), \tag{3.17}$$

where E_D refers to the diffusion activation energy of water molecule clusters on the solid surface. Therefore, according to the diffusion frequency ϑ_D of the surface and the number n_{ad} of water molecule clusters adsorbed on the critical core surface, the probability ω of water molecule clusters hitting the critical core can be deduced as

$$\omega = n_{ad} \times \vartheta_D = \tau p N_A \vartheta \exp(-E_D / k_B T) / \sqrt{2\pi M R T}. \tag{3.18}$$

Considering that the critical core shape of the droplet is a regular hemisphere, the area of the impacted core A_{nc}^{c} can be obtained as

$$A_{nc}^{c} = 2\pi r_c a_0 \sin\theta, \tag{3.19}$$

where a_0 is the particle size of water molecule clusters hitting the surface of the critical core. Therefore, the further nucleation rate η_{nc} can be expressed as

$$\eta_{nc} = 2\pi r_c a_0 \sin\theta n_s p N_A \vartheta \exp\left(\frac{E_{des} - E_D - \Delta G_c}{k_B T}\right) / \sqrt{2\pi mRT}. \tag{3.20}$$

Since χ is the nucleation kinetics correction factor. The mobility and nucleation rate of water molecule clusters in the system determine the kinetic equation of droplet growth,

$$\varphi \propto \chi J \eta_{nc}. \tag{3.21}$$

3.1.3 Droplet Transport

The movement behavior of droplets can generally be categorized into two distinct types. The first type encompasses the movement of droplets along the surfaces of materials, a phenomenon observable in nature, such as the movement of droplets on butterfly wings and the fog collection capabilities of spider silk. The second type involves directional transport behavior occurring in the cross-sectional direction, which is particularly significant for fog collection applications. To facilitate directional droplet transport, a driving force is essential. Recent research has identified three primary categories of driving forces: Laplace pressure, wettability gradients, and external energy input. In the context of advancing technologies for water fog collection, droplet manipulation, drug delivery systems, and microfluidics, numerous researchers have designed experiments aimed at enhancing the performance of directional droplet transport. By investigating the various driving forces that influence droplet behavior, we can significantly improve our understanding of how to accelerate droplet transport and optimize fog collection performance. This exploration not only informs our current work but also lays the foundation for innovative approaches to harnessing these forces in practical applications.

3.1.3.1 Laplace Driving Force

Laplace's seminal work in 1805 focused on understanding the pressure dynamics within a droplet, which results from the surface tension of the liquid. This phenomenon of internal and external pressure differentials is commonly observed in various natural systems. For instance, two interconnected bubbles exhibit a tendency to exchange size, where one bubble becomes smaller, while the other expands. Similarly, capillary wetting is a ubiquitous occurrence across different environments. These phenomena can be viewed as manifestations of Laplace's principles concerning surface tension. To derive the Laplace pressure equation, we can employ a simplified model. Consider, for example, an oil droplet (o) suspended in water (w) with an initial radius denoted as R. If we assume that this droplet is to expand outward by an infinitesimal amount of dR, we can calculate the energy required for this system transformation as

$$\delta W = -p_o \mathrm{d}V_o - p_w \mathrm{d}V_w + \gamma_{ow} \mathrm{d}A, \quad (3.22)$$

where $\mathrm{d}V_o = 4\pi R^2 \mathrm{d}R = -\mathrm{d}V_w$ is the increased volume of the system and $\mathrm{d}A = 8\pi R \mathrm{d}R$ is the newly increased oil-water interface area. External work is needed to overcome the volume work and provide the energy to create a new interface. p_o and p_w are the corresponding pressures of oil and water, and γ_{ow} is the interfacial tension between oil and water. When the system reaches equilibrium again, $\delta W = 0$, and then

$$\Delta P = p_o - p_w = 2\gamma_{ow}/R. \quad (3.23)$$

Inspired by the unique adaptations of flora in extreme environments, cactus species have evolved specialized structures to effectively harvest moisture in arid and semiarid regions. Notably, cactus spines are characterized by their directional barbed form, gradient grooves, and band-like trichomes, all of which contribute to their fog collection efficiency. The spines exhibit a conical structure, where the diameter gradually increases from the apex to the base. This unique morphology includes a secondary arrangement of gradient grooves on the surface of the cone (Figure 3.2). When water droplets collect on the conical surface, they encounter varying Laplace pressures on opposing sides due to this geometric configuration. The resulting surface energy gradient, created by the gradient grooves, generates a driving force that directs water droplets toward the base of the spine.

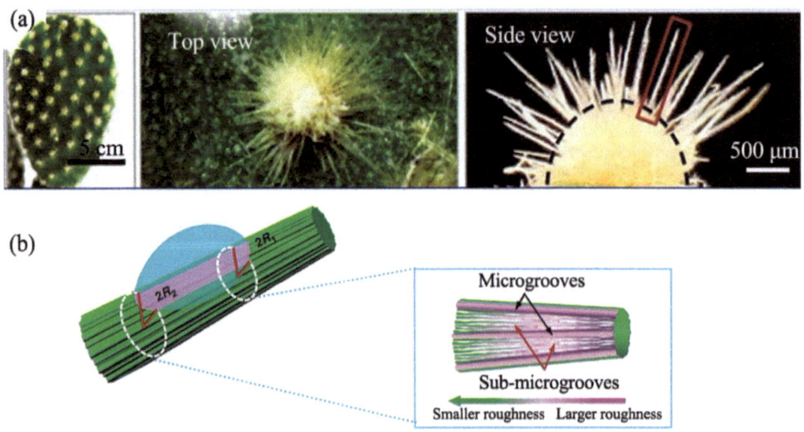

Figure 3.2 (a) Digital images of the hierarchical structure of the cactus in different scales [15]. (b) Cactus spines have a conical structure with grooves with gradients in the conical spines. The conical structure causes the diameter of the spines to increase from the tip to the bottom, which causes a difference in the CA of the droplets in the spines, which causes a difference in the curvature of the front and back of the droplets, the difference in curvature generating a Laplace surcharge pressure directed toward the bottom, and the gradient in surface energy generated by the grooves of the gradient generating a driving force, directed toward the bottom, and these forces drive the droplets toward the bottom of the spines. Source: [15] / Springer Nature / CC BY-NC-SA 3.0.

Specifically, the pressure differential induced by the Laplace effect compels droplets on the conical surface to migrate toward the side with a larger radius. Thus, cactus spines can be conceptualized as conical structures with symmetrically arranged grooves, promoting the accumulation and transport of moisture. This tapered geometry leads to a significant Laplace pressure difference (ΔP) between the tip and the base, facilitating efficient water harvesting in challenging environments [15–17],

$$\Delta P = -\int_{R_1}^{R_2} \frac{2\gamma}{(R+R_0)^2} \sin \alpha \mathrm{d}r, \tag{3.24}$$

where R represents the radius of both ends of the cone (the local radius of the spine on both the front and rear sides of the water droplet) and γ is the surface tension of water. R_0 represents the radius of the water droplet, α is the half vertex angle of the cone, and $\mathrm{d}r$ is the incremental radius of the cone. The Laplace pressure experienced by the droplet near the tip of the spine (small radius R_1) is greater than the Laplace pressure near the tail of the spine (large radius R_2), as shown in Figure 3.2b. The difference in force on the two ends of the water droplet creates a driving force that causes the water droplet to move along the spine of the cactus spine from the tip to the tail. In other words, the Laplace driving force formed by the difference in curvature on both sides of the droplet on the one-dimensional conical structure can induce the directional transport behavior of the droplet.

3.1.3.2 Wetting Gradient Driving Force

With the advancement of multifunctional wettability surfaces, the development of surface chemical gradients to facilitate droplet transport has emerged as a significant research focus [18–20]. While the static CA is a widely used metric for assessing the wetting performance of solid surfaces, a comprehensive evaluation of droplet transport on these surfaces necessitates consideration of dynamic factors. One critical parameter in this evaluation is the rolling angle, which is defined as the inclination angle at which a droplet of a specific volume or mass begins to roll down an inclined plane [18, 21]. A smaller rolling angle indicates greater hydrophobicity of the solid surface. Upon contact with a solid surface, droplets exhibit contact angle hysteresis (CAH), which arises from the differing CAs at the leading (front) and trailing (rear) edges of the droplet. Several factors contribute to this disparity, including the tilt angle of the surface and the chemical composition of the surface materials. Initially, as the solid surface begins to incline, the droplet remains stationary until a critical tilt is reached, at which point it begins to roll. The CAs at the two ends of the droplet during this process are referred to as the advancing CA (θ_A) and the receding CA (θ_R). The difference in cosine values between these two angles defines the CAH of the surface, which is an important indicator of the surface's wettability characteristics,

$$\cos \theta_A - \cos \theta_R = \mathrm{CAH}. \tag{3.25}$$

The resistance due to contact hysteresis (F_{CAH}) is

$$F_{CAH} = \pi R_0 \gamma (\cos \theta_A - \cos \theta_R), \tag{3.26}$$

where R_0 represents the contact radius of the three-phase contact line. When the surface is tilted at a certain angle, and driven by gravity, the droplets begin to roll down. If the CAH is larger, the droplets will be less likely to roll off the solid surface. On the contrary, if the CAH is smaller, it is easier for the droplet to roll off the solid surface.

On another note, Eq. (3.23) indicates that an inhomogeneous chemical composition of a surface can lead to asymmetric effects on the contact line of surface droplets. The Wenzel equation elucidates how surface structure influences the apparent CA of water droplets, thereby highlighting the impact of surface chemical properties. The roughness coefficient (r) is defined as the ratio of the actual surface area of a rough surface to its geometric projected area. This coefficient reflects the surface energy characteristics of the material; specifically, a larger (r) value corresponds to increased surface energy, which can alter the wettability of the surface. Consequently, this alteration affects both the CA of the droplet and the forces acting on the contact line. A gradient in surface roughness can induce a corresponding wettability gradient. As a result, droplets tend to spontaneously move from hydrophobic regions toward hydrophilic regions, driven by the gradient in surface wettability. This phenomenon, driven by variations in surface roughness or chemical properties, is collectively referred to as the chemical gradient driving force. The disparity in the roughness coefficient (r) between different regions of the surface leads to variations in surface free energy, thereby generating a net driving force (F) that influences the movement of water droplets. For instance, the differences in roughness between the spindle nodes and the connecting points on spider silk result in distinct surface energy gradients, which facilitate the directional transport of droplets from the connecting points toward the spindle nodes [17]. The expression for this driving force can be formally described as follows:

$$F = \int_{L_\mathrm{j}}^{L_\mathrm{k}} \gamma(\cos\theta_\mathrm{A} - \cos\theta_\mathrm{R})\mathrm{d}l, \tag{3.27}$$

where γ represents the surface tension of the droplet, θ_A and θ_R represent the advancing and receding angles of the droplet movement, respectively, and $\mathrm{d}l$ represents the directional transport distance of the droplet. In summary, the changes in surface energy induced by chemical gradients play a crucial role in the rapid transport of droplets from hydrophobic to hydrophilic surfaces.

In nature, the upper surface of the lotus leaf exhibits SHB properties, while the water-facing side is hydrophilic and displays characteristics of underwater superoleophobicity [22]. This remarkable contrast in wettability on opposing surfaces of the lotus leaf is referred to as the "Janus phenomenon" [23]. Janus materials, characterized by their asymmetric structures, properties, or compositions on both sides, can manifest in two- and three-dimensional porous forms. Inspired by the lotus leaf, a variety of materials with asymmetric wettability are being developed. Janus membranes, in particular, demonstrate wettability gradients that facilitate complex unidirectional permeation behavior, which differs significantly from the behavior of droplets migrating across a homogeneous solid surface. Specifically, a droplet can transition from a hydrophobic surface to a hydrophilic surface, whereas the reverse is effectively barred as the liquid cannot penetrate in the opposite direction. The unique unidirectional transport behavior observed in Janus materials arises from the opposing wettability exhibited by each side, using the wettability gradient in the membrane's cross section as the primary driving force.

To illustrate this concept, we can consider a Janus membrane composed of micropillars, as described by Yang et al. [23]. They provide a theoretical framework for understanding the anisotropic breakthrough pressure associated with such structures. Initially, the droplets cannot penetrate the micro-pillar array unless the extrusion pressure – the pressure differential acting on the liquid-gas interface – exceeds a critical threshold. This threshold is dependent on several factors, including the geometric microstructure of the surface, the properties of the membrane, and the wetting characteristics and surface tension of the liquid involved. The extrusion pressure may be derived from various sources, such as the mechanical pump pressure applied to the liquid, the Laplace pressure induced by the liquid, or other forces like gravity. For instance, as depicted in Figure 3.3a,b, the local liquid–gas interface is squeezed between two cylindrical surfaces. When a specific pressure is applied, the three-phase contact line (solid–liquid–gas) moves along the cylinder's surface until it reaches an equilibrium position, at which point the extrusion pressure balances the pressure differential across the liquid–gas interface. Assume that the fixed position of the three-phase contact line has a fixed angle φ (the angle along the tangent plane of the cylinder and the horizontal plane), its value is $\pi \sim 0$, and the extrusion angle is $\alpha = \theta(\varphi) - \varphi$, where, $\theta(\varphi)$ is the local CA. The curvature radius of the solid–liquid interface after extrusion can be expressed as

$$r = d/2\sin\alpha = \frac{d_0 + 2r(1 - \sin\varphi)}{2\sin[\theta(\varphi) - \varphi]}. \tag{3.28}$$

From this, the tolerated pressure difference ΔP can be obtained as

$$\Delta P = \frac{\gamma}{r_1} = \frac{2\gamma\sin[\theta(\varphi) - \varphi]}{d_0 + 2r(1 - \sin\varphi)}, \tag{3.29}$$

where γ is the surface tension of the liquid. The parameter defining the cylinder spacing ratio is $k = d_0/2r$; then the above equation can be rewritten as

$$\Delta P = \frac{\gamma}{r} \times \frac{\sin[\theta(\varphi) - \varphi]}{k + 1 - \sin\varphi}. \tag{3.30}$$

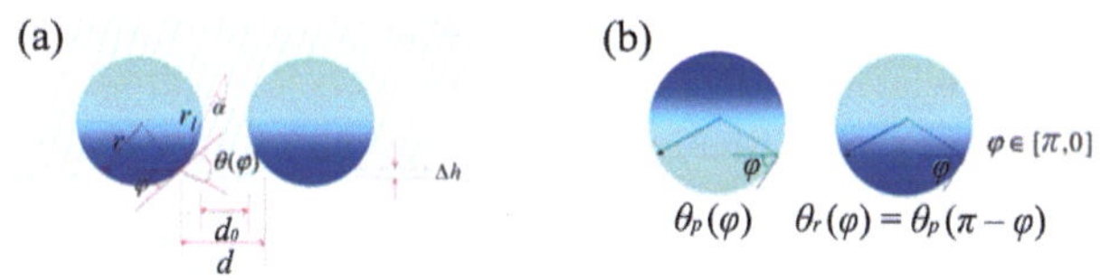

Figure 3.3 (a,b) The solid–liquid–gas interface between two cylinders. The three-phase contact line spacing on the neighboring cylinders is $d = d_0 + 2r(1 - \sin\varphi)$, where d_0 and r are the distance and radius between the two cylinders. Source: [24] / The Royal Society of Chemistry.

Here, if the extrusion pressure is provided by gravity, the equation applies when the extrusion height $\Delta h(< d/2)$ and the characteristic dimensions of the cylinder (i.e. r and d_0) are much smaller than the capillary length. If the anisotropy ratio of the critical breakthrough pressure of the droplet becomes larger, the droplet can penetrate in the cross-sectional direction, but not vice versa. In practical applications, the fabric can be regarded as spaced cylinders with rough substructures on the surface, and the influence of the substructure is manifested as local CA. Therefore, the column membrane model can be used to describe the fabric structure and analyze the liquid penetration phenomenon.

3.1.3.3 Energy Input Driving Force

The role of structural and chemical driving forces in droplet transport remains relatively limited. To enhance the efficiency of droplet movement, researchers are exploring the integration of external energy sources, such as light, electricity, and magnetism. This approach aims to facilitate droplet responses through external field stimulation. However, it is important to recognize that each method of field stimulation has its own specific advantages and disadvantages; thus, selecting the most appropriate technique should be determined based on the particular context and requirements of the application.

Traditionally, light irradiation on a solid surface has been perceived as providing negligible driving force for droplet transport. Nevertheless, the energy absorbed at the interface of a wettable surface due to light can be converted into chemical energy or other forms, enabling directional transport of droplets [25–27]. For instance, Li et al. [28] have proposed a strategy that effectively leverages light to manipulate the motion of binary droplets on high-energy surfaces, as illustrated in Figure 3.4a. When a focused light beam is directed onto a substrate, the substrate serves as a localized heating source through photothermal conversion. This phenomenon allows the binary droplets to be manipulated along the high-energy surface via mechanisms such as free contact line pinning, thereby demonstrating light-driven droplet transport. The underlying principle is that the localized photothermal effect induces a surface tension gradient at the droplet interface, which drives its movement. The remarkable versatility of light facilitates dynamic control over the speed, direction, and trajectory of binary droplet transport, particularly when assisted by two-dimensional channels on high-energy surfaces. This capability enables complex droplet transport, controlled merging of droplets, and even antigravity motion. Moreover, the platform shows promise for the directional transport of biological samples, including DNA molecules and cells, as well as in sequential microreactor applications. In a related study, Zhao and coworkers [29] introduced a programmable graphite-impregnated porous graphene film (Figure 3.4b). Thanks to the superior photothermal properties of this graphene material, the paraffin within the film undergoes a state transition between liquid and solid when exposed to near-infrared (NIR) light. This transition results in the formation of droplet transport channels, thus facilitating directional transport. The film can dynamically and reversibly transition between smooth and rough surfaces, allowing for remote adjustments of wettability and other properties.

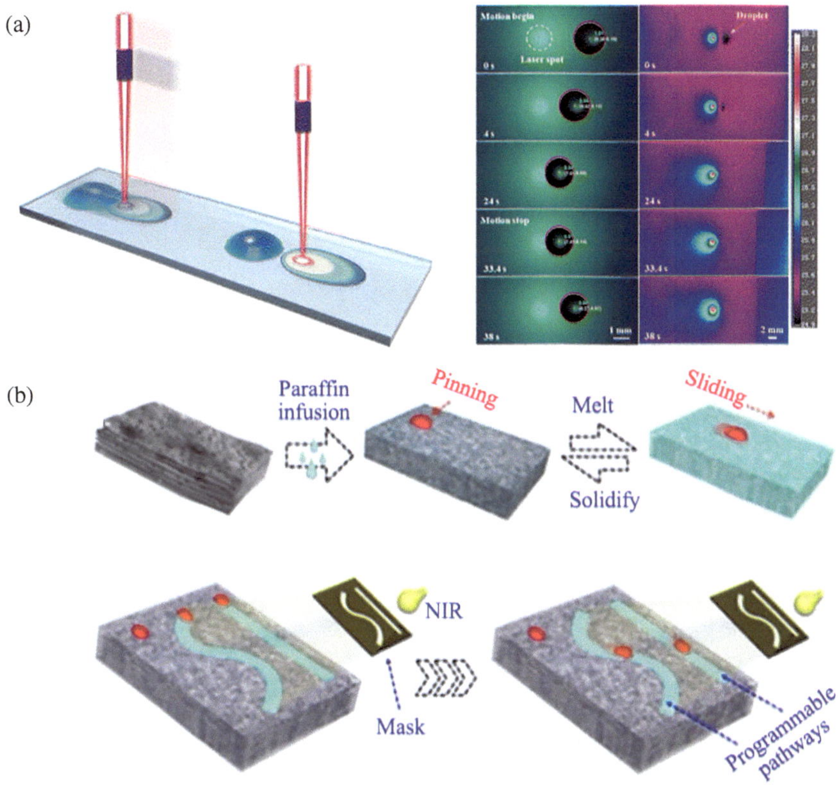

Figure 3.4 (a) Schematic illustration of light-activated binary droplet transport on a high-energy surface and the thermal images of top-view snapshots. Source: [28] / with permission of The Royal Society of Chemistry. (b) Schematic diagram of porous graphene membranes impregnated with paraffin wax to form droplet transport pathways induced by NIR light. Source: [29] / American Association for the Advancement of Science – AAAS / CC BY-NC 4.0.

In summary, droplet transport can be effectively achieved through external field stimuli, particularly light response mechanisms. The key advantage of employing light is its ability to transcend spatial limitations while enabling remote manipulation. However, achieving precision in manipulation remains a challenge that requires further exploration.

Zhang et al. [30] developed an innovative asymmetric amphiphilic surface that employs self-driven triboelectric adsorption to harvest clean water from the atmosphere, as illustrated in Figure 3.5a. The study focuses on the electrostatic forces between liquid and solid materials, which facilitate electrostatic adsorption through the spontaneous interfacial triboelectric charge generated between droplets and the collector surface. Their approach demonstrates remarkable efficiency by converting the mechanical energy produced by falling droplets into electrical energy. Furthermore, the electrostatic adsorption process is significantly enhanced through the application of an external charge. In another study, Ben et al. [31] constructed SHB magnetic microcoria array surfaces that feature switchable

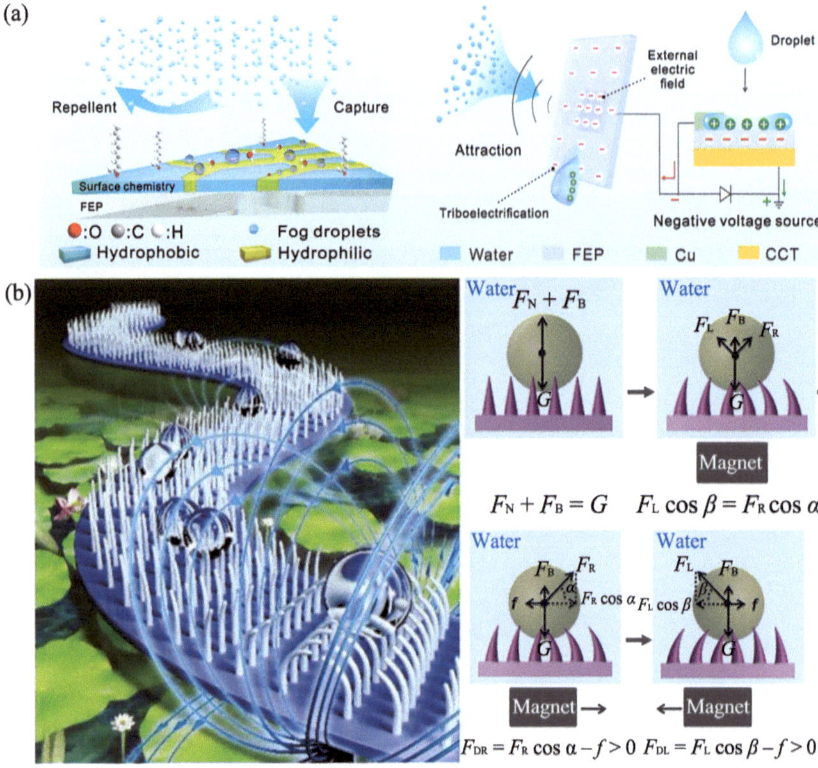

Figure 3.5 (a) Schematic diagram of the biomimetic hydrophobic–hydrophilic coating. The fog was repelled in the hydrophobic area and attracted in the hydrophilic area, and electrostatic-assisted enhancement of water adsorption, where the electrostatic sources including triboelectrification between the droplet and FEP as well as an additional charge from the D-TENG. Source: [30] / Springer Nature / CC BY 4.0. (b) Analysis of the mechanism of the oil droplet movement process on the underwater superoleophobic magnetic microcilia array (MMA) surface. Source: [31] / John Wiley & Sons / CC BY-NC 4.0.

structures activated by an external magnetic field, as depicted in Figure 3.5b. These SHB surfaces enable continuous and directional manipulation of water droplets in the air, while the underwater superoleophobic magnetic microcilia array surfaces control oil droplets in aquatic environments under the influence of an external magnetic field. This research highlights the potential of nondestructive droplet transport mechanisms for various applications, including liquid transport, droplet reactions, and microtubule transport. Such advancements pave the way for practical applications of droplet manipulation in smart microstructured surfaces.

3.2 The Theory of Atmospheric Water Harvesting

Molecules and atoms can adhere to surfaces via two distinct mechanisms: physisorption and chemisorption. In physisorption, the interaction between the adsorbate and substrate is primarily governed by van der Waals forces, which are characterized by their long-range yet relatively weak nature. When a particle is physisorbed, the energy released during this process is comparable to the enthalpy of condensation. For the adsorbed molecule to remain bound to the surface, this released energy must be adequately dissipated. If the energy remains within the system, the adsorbed molecule is likely to dissociate from the surface. Fortunately, the energy involved in physisorption is sufficiently low to be dissipated into the vibrations of the substrate lattice, ultimately leading to its transformation into thermal motion. In contrast, chemisorption involves the formation of strong chemical bonds between the adsorbate and the surface of the solid. In this process, adsorbates tend to occupy sites that provide optimal coordination with the substrate. The enthalpy of chemisorption is significantly higher than that of physisorption, which means that the range of interactions is typically short range. This notable difference in the magnitudes of enthalpy highlights the fundamental distinctions between these two adsorption processes and their respective interaction strengths with the surface.

Upon the adsorption of gaseous water molecules onto a solid surface, a dynamic equilibrium is established. This equilibrium is characterized by the delicate balance between free water molecules in the surrounding environment and those adsorbed onto the surface. The extent of surface coverage is intrinsically linked to the prevailing pressure and temperature conditions of the gaseous phase. Simultaneously, the degree of surface coverage significantly influences the enthalpy of adsorption – a thermodynamic parameter that governs the likelihood of an order–disorder transition within the ensemble of adsorbed molecules. This complex interplay between the surface coverage and enthalpy of adsorption manifests as an ongoing dynamic process, where adsorption and desorption events occur in continuous equilibrium. The intricate dynamics of this phenomenon are closely tied to the principles of thermodynamics and surface science. When gaseous water molecules interact with the solid surface, they undergo a phase transition, resulting in a delicate balance of forces that dictate their behavior and spatial distribution. This equilibrium state is highly sensitive to fluctuations in environmental conditions, such as variations in pressure or temperature, which can disrupt the balance and trigger rearrangements or desorption events within the adsorbed molecular ensemble. The degree of surface coverage is usually expressed as a fraction (χ),

$$\chi = N_{\text{occupied}}/N_{\text{available}}, \tag{3.31}$$

where N_{occupied} is the number of adsorption sites of the adsorbed substrate occupied by water molecules and N_{occupied} is the number of adsorption sites of the adsorbed substrate available for occupation by water molecules. Alternatively, it is expressed as adsorption volume,

$$\chi = V/V_{\infty}, \tag{3.32}$$

where V is the volume of adsorbed adsorbate and V_{∞} is the volume of adsorbate corresponding to a complete monolayer coverage. These two volumes are the free gas-phase water molecules measured at the same temperature and pressure conditions, not the volume occupied by the adsorbed water molecules when attached to the solid surface. The expression for the variation of surface coverage (χ) with pressure at selected temperatures is called the adsorption isotherm. With the accumulation of experimental data and the history of adsorption, there are several main adsorption models which will be elaborated in detail.

3.2.1 Langmuir Model

The Langmuir model is predicated on the assumption that adsorption occurs on a uniform surface, whereby each molecule possesses constant enthalpies and a fixed sorption activation energy [32]. In practice, the validity of Langmuir isotherms is evaluated by measuring the relationship between surface coverage and pressure. Moreover, the Langmuir adsorption isotherm is built upon several foundational assumptions. First, it posits that the adsorption process cannot exceed monolayer coverage, making it solely applicable to scenarios involving monolayer adsorption. Second, it assumes that all adsorption sites on the surface are energetically equivalent. Third, it maintains that each molecule can occupy only a single vacancy on the surface. Finally, it asserts that the probability of adsorption at any site is independent of the occupancy of neighboring sites, implying that no interactions occur between adsorbed molecules [33, 34].

Based on these assumptions, the adsorption rate is directly proportional to both the collision rate at the surface and the partial pressure (p) of the gas molecules being adsorbed. Additionally, the number of available vacancies significantly influences the adsorption rate as adsorbed molecules can only occupy these unoccupied sites. The unoccupied number is $S(1-\chi)$, where S is the total occupied number and χ is the coverage fraction. Therefore, the rate of change of surface coverage ($\mathrm{d}\chi/\mathrm{d}t$) due to adsorption can be expressed as

$$\frac{\mathrm{d}\chi}{\mathrm{d}t} = k_{\mathrm{ad}} p S(1-\chi). \tag{3.33}$$

The rate of change in surface coverage ($\mathrm{d}\chi/\mathrm{d}t$) due to desorption is proportional to the number of adsorbed molecules and can be presented as follows:

$$\frac{d\chi}{dt} = -k_{\mathrm{de}} S\chi, \tag{3.34}$$

At equilibrium, the sum of the two rates must be zero, indicating no net change in χ. Therefore, the Langmuir adsorption isotherm equation can be derived from Eqs. (3.33) and (3.34),

$$\chi = \frac{kp}{1+kp}, \tag{3.35}$$

where $k = k_{\mathrm{ad}}/k_{\mathrm{de}}$ and k is a temperature dependent quantity. Using the Gibbs–Helmholtz equation to relate k to the standard Gibbs energy of adsorption ($\Delta_{\mathrm{ad}}G^{\circ}$), then, we can derive the standard enthalpy of adsorption ($-\Delta_{\mathrm{ad}}H^{\circ}$),

$$\frac{\mathrm{d}(-R\ln(kp^{\circ}))}{\mathrm{d}T} = \frac{\mathrm{d}(\Delta_{\mathrm{ad}}G^{\circ}/T)}{\mathrm{d}T} = \frac{-\Delta_{\mathrm{ad}}H^{\circ}}{T^2}. \tag{3.36}$$

Furthermore, there is the following relationship:

$$\frac{dln(kp^{\circ})}{dT} = \frac{-\Delta_{ad}H^{\circ}}{RT^2}. \tag{3.37}$$

The isothermal standard enthalpy of adsorption ($\Delta_{ad}H^{\circ}$) is limited by the surface coverage of the adsorbed molecules; therefore, we can obtain following equation:

$$\left(\frac{\partial \ln(kp^{\circ})}{\partial(1/T)}\right)_{\chi} = \frac{-\Delta_{ad}H^{\circ}}{R}. \tag{3.38}$$

3.2.2 Brunauer-Emmett-Teller Adsorption Isotherm Model

If a monolayer of a Langmuir adsorption isotherm were to act as a substrate for subsequent adsorption, the volume of gas adsorbed at high pressures would not plateau at the level indicative of complete monolayer coverage. Instead, it would continue to increase indefinitely. To address this phenomenon, Stephen Brunauer, Paul Emmett, and Edward Teller formulated the Brunauer-Emmett-Teller (BET) isotherm, which is the most widely utilized model for multilayer adsorption. The BET isotherm equation can be expressed as

$$\frac{V}{V_{mono}} = \frac{cz}{1 - z[1 - (1 - c)z]}, \tag{3.39}$$

where $z = p/p^*$ (p^* is the vapor pressure of the pure liquid substrate) and V is the volume of the adsorbed gas. The constant c is the characteristic parameter of the system, and $c = k_0/k_1$, where $k_0 = k_{ad,0}/k_{de,0}$ is the ratio of the rate constants of adsorption and desorption from the matrix and $k_1 = k_{ad,1}/k_{de,1}$ is the ratio of the rate constants of subsequent layers. The characteristic constant c of the BET adsorption isotherm can be expressed in terms of enthalpies as

$$c = \frac{k_0}{k_1} = \frac{e^{\Delta_{des}G^{\circ}/RT}}{e^{\Delta_{vap}G^{\circ}/RT}} = \frac{e^{\Delta_{des}H^{\circ}/RT}e^{-\Delta_{des}S^{\circ}/R}}{e^{\Delta_{des}H^{\circ}/RT}e^{-\Delta_{vap}S^{\circ}/R}}. \tag{3.40}$$

The entropies of desorption and vaporization can be assumed to be equal as they correspond to similar processes involving the escape of the condensed adsorbate into the vapor phase. Thus, c can be expressed as

$$c = e^{(\Delta_{des}H^{\circ} - \Delta_{vap}H^{\circ})/RT}, \tag{3.41}$$

where $\Delta_{des}G^{\circ}$ represents the standard Gibbs free energy of desorption, $\Delta_{des}G^{\circ} = RT\ln(k_0p^{\circ})$ or $k_0p^{\circ} = e^{\Delta_{des}G^{\circ}/RT}$ represents the standard Gibbs free energy of desorption/vaporization, and $\Delta_{vap}G^{\circ} = RT\ln(k_1p^{\circ})$ or $k_1p^{\circ} = e^{\Delta_{vap}G^{\circ}/RT}$. When the enthalpy of desorption ($\Delta_{des}H^{\circ}$) for the first adsorbed monolayer is significantly larger than the enthalpy of vaporization ($\Delta_{vap}H^{\circ}$) for the liquid adsorbate, the characteristic constant c is also large. When c is much greater than 1, the BET adsorption isotherm model can be simplified to

$$\frac{V}{V_{mono}} = \frac{z}{1 - z}. \tag{3.42}$$

The BET adsorption isotherm model agrees well with experimental observations over a limited range of pressures, but it underestimates the extent of adsorption at low pressures and overestimates adsorption at high pressures.

3.2.3 Freundlich Adsorption Isotherm Model

The Freundlich isotherm model posits that adsorption takes place on heterogeneous surfaces, where interactions among adsorbed molecules can occur. Unlike simpler models that assume the formation of a monolayer, the Freundlich model accommodates the possibility of multilayer adsorption. Additionally, it suggests that the adsorption energy is not constant; rather, it decreases exponentially as the adsorption process progresses. Consequently, the Freundlich adsorption isotherm model provides a more nuanced understanding of adsorption behaviors in heterogeneous systems. The Freundlich adsorption isotherm model is

$$\chi = c_1 p^{1/c_2}. \tag{3.43}$$

Equation (3.14) corresponds to a logarithmic change. The isotherms attempt to account for the interactions between adsorbed molecules on the surface. The different adsorption isotherm models above are in broad agreement with experiments over a limited pressure range, but they remain largely empirical.

3.2.4 Dubinin–Radushkevich Adsorption Isotherm Model

The Dubinin–Radushkevich (D–R) adsorption isotherm model provides a comprehensive framework for understanding the adsorption mechanism by characterizing it through a Gaussian energy distribution on heterogeneous surfaces. This model enables the categorization of the adsorption process into two distinct types: physical adsorption, characterized by an energy of adsorption E_{ad} less than 8 kJ mol^{-1}, and chemical adsorption, which is identified by an E_{ad} value ranging from 8 to 168 kJ mol^{-1}. Here, E_{ad} signifies the energy required to displace a molecule from its position on the adsorbent's surface to a state of infinite separation. The D–R adsorption isotherm model plays a pivotal role in elucidating the interactions between the adsorbate and adsorbent, particularly in systems where surface heterogeneity is prevalent. By incorporating these energy distributions, the model facilitates a deeper understanding of the underlying processes involved in adsorption phenomena, thus enhancing its applicability in various fields, including environmental engineering and materials science. The Dubinin–Radushkevich adsorption isothermal model is [34]

$$\chi = \chi_t \exp(-k_{ad}\varepsilon^2), \tag{3.44}$$

where χ_t is the theoretical isothermal saturation and adsorption capacity, k_{ad} is the Dubinin–Radushkevich isotherm constant, and ε is the Polanyi potential energy, which can be calculated by the following equation:

$$\varepsilon = RT \ln\left(1 + \frac{1}{c_{eq}}\right), \tag{3.45}$$

where c_{eq} is the adsorption concentration at equilibrium conditions; furthermore, the energy of per molecule of the adsorbate E_{ad} is

$$E_{ad} = (\sqrt{2k_{ad}})^{-1}. \tag{3.46}$$

3.2.5 Temkin Adsorption Isotherm Model

The Langmuir adsorption isotherm model posits that all adsorption sites are both equivalent and independent, indicating that the enthalpy of adsorption remains constant regardless of surface coverage (χ). However, experimental observations frequently demonstrate that the enthalpy of adsorption tends to decrease as χ increases. This trend suggests that the most energetically favorable adsorption sites are preferentially occupied first. To address this variability in adsorption energetics, the Temkin adsorption isotherm model introduces an alternative expression that accommodates the dependence of adsorption energy on surface coverage, which is expressed as follows:

$$\chi = c_1 \ln(c_2 p). \tag{3.47}$$

In Eq. (3.18), c_1 and c_2 are constants that correspond to the assumption that the enthalpy of adsorption varies linearly with pressure. The Temkin adsorption isotherm model is based on the assumption that the decrease in the heat of adsorption of all the molecules in the surface layer is linear, and that the adsorption is characterized by a homogeneous distribution of the binding energy, which is derived assuming that the decrease in the heat of adsorption is linear [34].

3.2.6 MOF- and COF-Based Water Adsorption-Release Mechanism

Metal–organic frameworks (MOFs) and covalent organic frameworks (COFs) are typically obtained in the form of polycrystalline powders. This characteristic limits the application of sophisticated techniques designed for the characterization of single-crystalline MOFs, particularly in the study of in situ water sorption processes. Experimental observations, logical deductions, and computational simulations collectively contribute to our empirical understanding of water sorption mechanisms in MOFs and COFs. The proposed mechanism encompasses three key stages: (i) the adsorption of water molecules at polar sites, (ii) the formation of water clusters or adsorption layers, and (iii) pore filling. While these processes are generally reversible, irreversible capillary condensation may occur if the pore diameter exceeds a critical threshold (D_c), which can be derived from the equation provided below,

$$D_c = \frac{4\sigma T_c}{T_c - T}, \tag{3.48}$$

where σ is the van der Waals diameter of the adsorbate, and T_c and T refer to the critical temperature of the adsorbate and the sorption temperature, respectively. At 298 K, D_c of water is determined to be 2.076 nm. Capillary condensation would inevitably cause hysteresis in the water isotherm and increase the energetic cost. In recent years, microporous MOFs and COFs are recommended for water harvesting processes. This subsection presents a description of several typical MOF adsorption and desorption mechanisms.

The (M)MOF-74 series, which includes isostructural compounds such as M = Mg^{2+}, Zn^{2+}, Ni^{2+}, Co^{2+}, is notable for its relatively high water stability [35]. Upon activation through vacuum heating, the removal of terminal ligands results in a substantial number of open metal sites. These sites exhibit varying binding strengths to water molecules, which is systematically reflected in the three distinct stages of dehydration observed during heating (Figure 3.6a). Research indicates that the high thermal energy required for the complete regeneration of (M)MOF-74 materials is attributable to the strong interactions of water with the open metal sites of the secondary building units (SBUs). Consequently, the elevated regeneration temperatures – up to 195 °C for (Zn)MOF-74 – pose challenges for applications reliant on efficient adsorption–desorption cycles. The robust binding of water to these open metal sites leads to significant Henry's constant values and results in type I isotherms (Figure 3.6e).

In microporous, hydrophilic adsorbents, the physisorption of water initiates through nucleation at primary adsorption sites, subsequently followed by the growth of water clusters rather than the formation of single or multilayered adsorptions [37]. Typically, the primary adsorption sites are located near polar, hydrophilic centers within the structure, such as SBUs. The adsorbed water molecules can function as additional sites, facilitating the formation of water clusters – a phenomenon evidenced by the observed stepwise isotherms. The reversible formation of these water clusters in MOF-801 has been investigated using X-ray single crystal and neutron powder diffraction techniques. At low relative pressures, water adsorption occurs within two tetrahedral pores, with the highest occupancy observed at sites proximate to the SBU-OH groups (sites I and II). Additionally, a third site (III) was identified within one tetrahedral pore, where a water molecule is hydrogen-bonded to three site II molecules, leading to the formation of a cubic water cluster. The distinct adsorption sites within the tetrahedral and octahedral pores of MOF-801 are illustrated in Figures 3.6 b–d.

Conversely, the pore filling mechanism in hydrophilic microporous MOFs such as MOF-801, which relies on reversible and cluster-mediated adsorption, contrasts sharply with capillary condensation – a predominantly irreversible and hysteretic process. The resulting "S"-shaped isotherm typically exhibits hysteresis between the adsorption and desorption branches (Figure 3.6f,g). For MOFs possessing pore sizes of less than 20 Å, hysteresis is not solely attributable to irreversible capillary condensation; rather, it often results from framework deformation caused by host–guest interactions, MOF degradation, or the formation of water superclusters [40, 41].

3.2.7 Hydrogel-Based Water Adsorption-Release Mechanism

When a hydrogel is immersed in water, the ionic groups within the material undergo complete or partial ionization, resulting in the generation of free dissociated ions

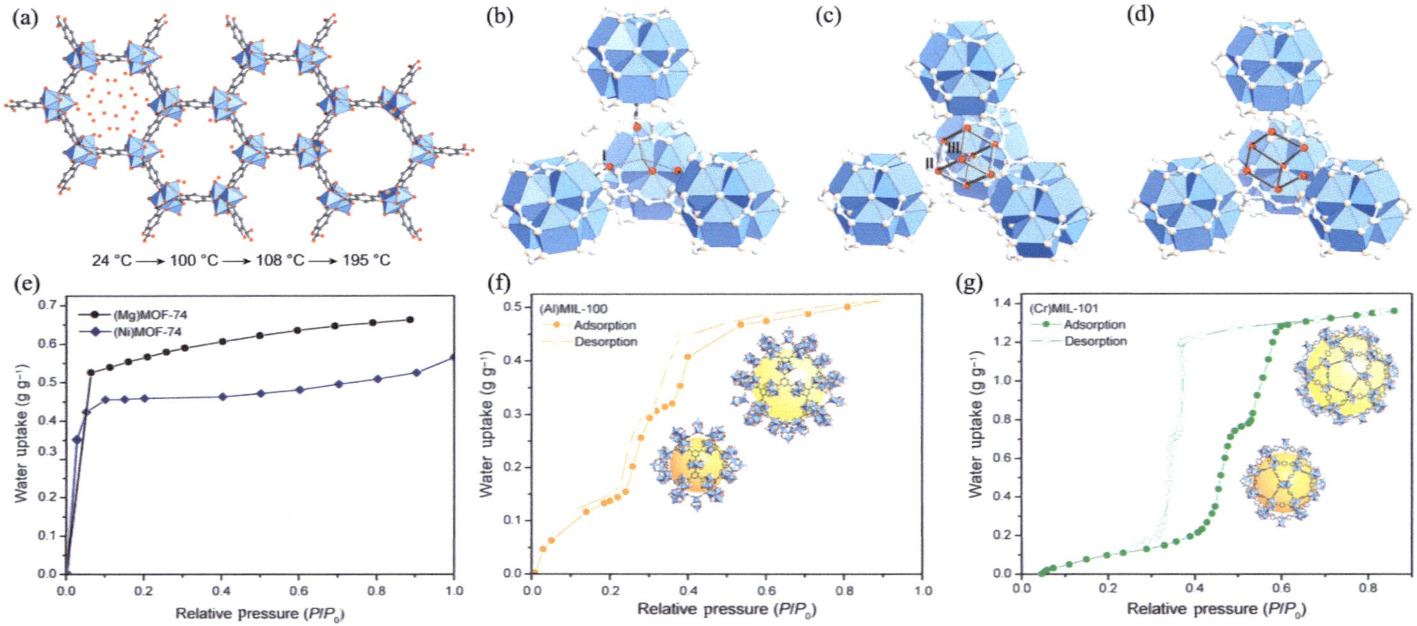

Figure 3.6 (a) Loosely bound water in the center of the pores is removed under mild conditions (24–100 °C), strongly adsorbed water is removed at 108 °C, and water chemisorbed on the SBUs is removed at high temperature (195 °C). Source: [36] / John Wiley & Sons. (b–d) Initial water adsorption upon exposure to humidity occurs through hydrogen-bond formation to the SBUs (site I and II) and among adsorbed water molecules (site III) as determined by single-crystal diffraction. (d) Crystals exposed to increased levels of humidity (100% RH) show similar arrangements as well as incomplete cubic clusters. Source: [37] / American Chemical Society. (e) The resulting water adsorption isotherm is of Type I, and the difference in capacity between (Ni) and (Mg)MOF-74 is due to the difference in formula weight. Source: [38] / American Chemical Society. (f and g) Water-sorption isotherms of (Al)MIL-100 and (Cr)MIL-101. Source: [39] / John Wiley & Sons.

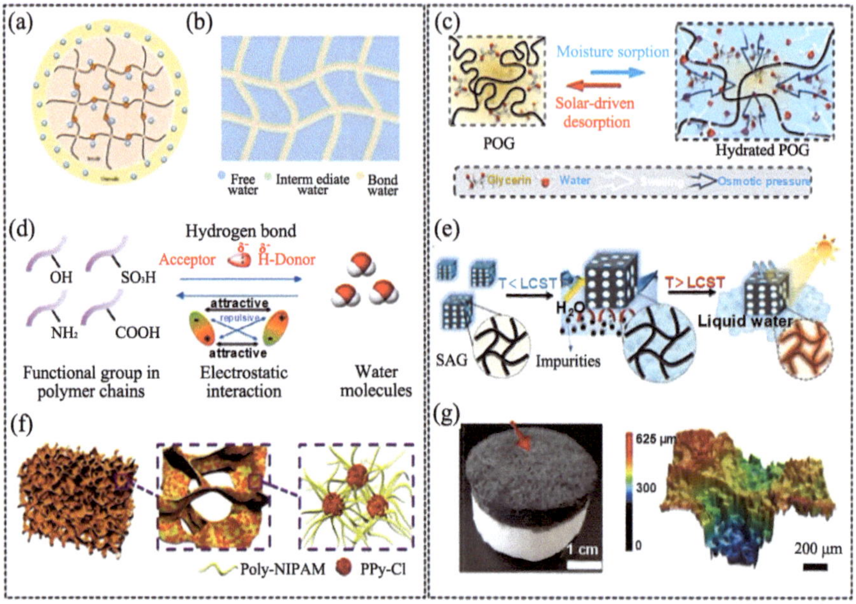

Figure 3.7 The mechanism of water absorption and desorption: (a) The network structure of hydrogel. Source: [42] / with permission of Elsevier. (b) The state of water inside hydrogel. (c) Noncovalent interaction between water molecules and functional groups in the hydrogel network. Source: [43] / American Chemical Society. (d) Internal gaps, micron channels, and molecular meshes in the hydrogel network. Source: [44] / John Wiley & Sons. (e) Schematic diagram of the water desorption under natural sunlight. Source: [42] / with permission of Elsevier. (f) Clean water is harvested from the lake and released clean water by solar energy. Source: [45] / John Wiley & Sons. (g) Digital images and morphology of the steam generator. Source: [46] / John Wiley & Sons.

alongside counter ions that remain fixed within the gel network. This leads to a significant electrostatic attraction between the dissociated ions and the oppositely charged polymer, which effectively balances the diffusive forces of the ions with the attraction exerted by the counter-charged network, as illustrated in the theoretical model presented in Figure 3.7a [47]. The ionization and swelling behavior of the gel is influenced by the ionic strength of the surrounding electrolyte solutions. Specifically, as ionic strength increases, the water absorption capacity of the gel diminishes due to a reduced osmotic pressure difference between the internal (gel) and external (solution) phases [48]. From a thermodynamic perspective, water will continue to penetrate the gel until the chemical potential difference between the gel and the water reaches equilibrium – this tendency is amplified in the presence of polar ionic groups [49].

On the other hand, the swelling of hydrogels can be limited under certain conditions. In linear polymers, for example, crystalline regions serve as physical cross-links that restrict water expansion. However, most hydrogels are characterized by covalent cross-linking, which further influences their swelling properties. Flory's elastic gel theory provides a foundational mathematical framework for understanding the swelling capacity of these gels [50],

$$Q^{5/3} = \frac{\left(\frac{i}{2}V_u S^{\frac{1}{2}}\right) + \left(\frac{1}{2}-X_1\right)/V_1}{V_e/V_0}, \tag{3.49}$$

where Q is the water absorption rate, V_e/V_0 is the cross-linking density, $\left(\frac{1}{2}-X_1\right)/V_1$ is the gel affinity for water, V_u is the volume of the structural unit, I/V_u is the density of the fixed charge, and S is the external solvent ionic strength of the liquid electrolyte.

The formula illustrates the intricate relationship between the water absorption capacity of the hydrogel and several pivotal factors, namely the degree of cross-linking, water affinity, external ionic strength, and charge density. The first term of the equation accounts for the osmotic pressure, while the second term reflects the affinity of the hydrogel for water molecules. The overall water absorption capacity is represented by the sum of these two terms, with the ratio V_e/V_0 indicating the cross-linking density. A decrease in the amount of cross-linking agent will result in a lower V_e, which, in turn, enhances the hydrogel's water absorption capacity at a consistent grafting rate. Additionally, it is important to note that when the hydrogel interacts with saltwater, the ionic strength (I/V) is significantly higher than that associated with freshwater. This elevated ionic strength leads to a pronounced reduction in the hydrogel's water absorption capacity [51, 52].

Hydrogels represent a class of polymers characterized by their exceptional hydrophilicity and significant moisture absorption capacity. These materials are primarily composed of cross-linked polymers containing electrolyte and hydrophilic monomers. Structurally, hydrogels form three-dimensional networks with low levels of cross-linking, featuring an abundance of hydrophilic functional groups such as carboxyl, hydroxyl, and acyl groups. These hydrophilic groups are synthesized through chemical or physical cross-linking of intertwined polymer chains. Consequently, upon water absorption, hydrogels function as ion-conducting networks consisting of polymer electrolytes and entrapped water molecules. The state of water within hydrogels after absorption is illustrated in Figure 3.7b [43], which categorizes water into bound water, intermediate water, and free water. Within the ion-conducting network, mobile ion pairs, derived from the polymer electrolytes, play a crucial role. In its dry state, the long polymer chains are entangled and interconnected, forming a solid network that lacks ionized species. However, upon hydration, the interaction between the hydrophilic groups and water molecules leads to the expansion of the polymer network, creating a concentration gradient of ions throughout the structure [53].

Assuming the polymer network contains a defined quantity of hydrophilic ions (Figure 3.7c), the resulting osmotic pressure difference between the internal and external environments drives water molecules into the hydrogel network. It is important to note that when the absorbed water contains salts, both the osmotic pressure and the capacity for water absorption may be diminished. The hydrophilic functional groups within the polymer network are thus essential for facilitating the expansion of the network and enhancing water infiltration. On one hand, the presence of hydrophilic ion pairs is vital for initiating the overall water absorption process in hydrogels. On the other hand, the unique internal voids, micron-scale channels, and hydrated ions (e.g. PPy—Cl) within the network significantly increase both the water absorption capacity and the rate of absorption (Figure 3.7d) [44].

In recent years, significant advancements have been made in enhancing the water release rate of hydrogels through structural and compositional modifications, notably the incorporation of solar absorbers. Among various materials, carbon-based substances have gained prominence as effective light-absorbing components due to their inherent porous structure, large specific surface area, and exceptional capability for light-to-heat conversion. For instance, Wang's research team developed a hydrogel that demonstrated impressive atmospheric water harvesting capabilities by incorporating carbon nanotubes into a sodium alginate matrix. This innovative hydrogel achieved a remarkable water evaporation efficiency, collecting 5.6 g of water per gram of material under one sun illumination (Figure 3.7e) [42]. Moreover, researchers have explored the unique properties of specific materials to further increase the water release rate. Temperature-sensitive polymers, such as poly(*N*-isopropylacrylamide) (NIPAAm), are noteworthy examples, as illustrated in Figure 3.7f. These thermoresponsive hydrogels undergo rapid shrinkage and expulsion of water once the temperature surpasses their lower critical solution temperature [45].

In addition to integrating solar absorbers, scholars have sought to optimize the water release rate of hydrogels by minimizing heat losses associated with reflection, radiation, and conduction. For example, some researchers have effectively reduced reflective losses by enhancing the surface roughness of light-absorbing materials [54]. Expanding on this concept, Li et al. designed a hydrogel featuring a rough surface paired with a core-shell structure (Figure 3.7g). The rough surface significantly improved light absorption by diminishing sunlight reflection, while the core-shell architecture facilitated a continuous water supply. Simultaneously, this design mitigated conductive heat losses, thus ensuring sustained water provision.

In summary, we have analyzed the factors related to the interconversion of the three states at a superficial level by starting from the knowledge of physical chemistry in terms of the three forms in which water exists. Subsequently, based on the above foundation, the focus of atmospheric water collection is placed on both fog collection and hygroscopic collection for discussion. In fog collection, from the theoretical analysis, it seems that the main factors affecting the efficiency of fog collection are the efficiency of droplet capture, the rate of droplet growth, and the rate of droplet shedding. These three are complementary to each other. Therefore, the subsequent preparation of related bionic fog collection engineering materials needs to consider the coordinated role of these three. Through the theoretical analysis, it is also possible to analyze the water harvesting function of various plants and animals from the macroscopic phenomenon to the microscopic level of understanding, for example, the hydrophilic and hydrophobic hybridized surface on the back of a beetle, the spindle knot structure on a spider's silk, and the conical spines on a cactus. The theoretical foundation adds up to a more profound understanding of the structure of water collection by plants and animals, and lays the foundation for the subsequent design of bionic fog collection engineering materials. However, fog collection is limited by geographic and climatic factors. The development of more widely used materials is imminent. Consequently, the theory related to the latest technology (atmospheric water collection) in recent years is analyzed. A new direction is constructed for bionic atmospheric collection engineering materials. Among many materials, MOFs, hygroscopic salt-hydrogel composite adsorbents, and COF atmospheric water harvesting efficiency performances are more remarkable. The theory of hygroscopicity desorption of related materials was carried out. Relevant theoretical manuals are provided for the subsequent construction of bionic atmospheric water harvesting engineering materials. It also provides a strong theoretical support for the subsequent enhancement of atmospheric water collection efficiency improvement.

References

1 Mingjin, T., Cziczo, D., and Grassian, V. (2016). *Chemical Reviews* 116: 4205–4259.
2 Tu, Y., Wang, R., Zhang, Y., and Wang, J. (2018). *Joule* 2: 1452–1475.
3 Thanh, N.T., Maclean, N., and Mahiddine, S.J. (2014). *Chemical Reviews* 114: 7610–7630.
4 Zhan, D. and Guo, Z. (2023). *Materials Horizons* 10: 4827–4856.
5 Atkinson, D. (1998). *Scientific Discourse in Sociohistorical Context: The Philosophical Transactions of the Royal Society of London, 1675-1975*. Routledge.
6 Wenzel, R.N. (1936). *Industrial Engineering Chemistry* 28: 988–994.
7 Whyman, G., Bormashenko, E., and Stein, T. (2008). *Chemical Physics Letters* 450: 355–359.
8 Volmer, M. and Weber, α. (1926). *Zeitschrift für Physikalische Chemie* 119: 277–301.
9 Matsumoto, M., Saito, S., and Ohmine, I. (2002). *Nature* 416: 409–413.
10 Yang, Q., Sun, P., Fumagalli, L. et al. (2020). *Nature* 588: 250–253.
11 V. I. Kalikmanov, *Nucleation Theory*, Dordrecht: Springer Netherlands 2012, pp. 17-41.
12 Vehkamäki, H., Määttänen, A., Lauri, A. et al. (2007). *Atmospheric Chemistry and Physics* 7: 309–313.
13 Auer, S. and Frenkel, D. (2001). *Nature* 409: 1020–1023.
14 Sosso, G.C., Chen, J., Cox, S.J. et al. (2016). *Chemical Reviews* 116: 7078–7116.
15 Ju, J., Bai, H., Zheng, Y. et al. (2012). *Nature Communications* 3: 1247.
16 Tang, X. and Guo, Z. (2020). *New Journal of Chemistry* 44: 20495–20519.
17 Zheng, Y., Bai, H., Huang, Z. et al. (2010). *Nature* 463: 640–643.
18 Chaudhury, M.K. and Whitesides, G.M. (1992). *Science* 256: 1539–1541.
19 Park, K.-C., Kim, P., Grinthal, A. et al. (2016). *Nature* 531: 78–82.
20 Zhang, M., Wang, L., Hou, Y. et al. (2015). *Advanced Materials* 27: 5057–5062.
21 Singh, R.A., Yoon, E.S., Kim, H. et al. (2007). *Surface Engineering* 23: 161–164.
22 Cheng, Q., Li, M., Zheng, Y. et al. (2011). *Soft Matter* 7: 5948–5951.

23 Yang, H.C., Hou, J., Chen, V., and Xu, Z.K. (2016). *Angewandte Chemie Internation Edition* 55: 13398–13407.
24 Tian, X., Li, J., and Wang, X. (2012). *Soft Matter* 8: 2633–2637.
25 Ichimura, K., Oh, S.K., and Nakagawa, M. (2000). *Science* 288: 1624–1626.
26 Kwon, G., Panchanathan, D., Mahmoudi, S.R. et al. (2017). *Nature Communications* 8: 14968.
27 Xiao, Y., Zarghami, S., Wagner, K. et al. (2018). *Advanced Materials* 30: 1801821.
28 Li, W., Li, D., Zhu, X. et al. (2023). *Lab on a Chip* 23: 4287–4301.
29 Wang, J., Gao, W., Zhang, H. et al. (2018). *Science Advanced* 4: eaat7392.
30 Zhang, S., Chi, M., Mo, J. et al. (2022). *Nature Communications* 13: 4168.
31 Ben, S., Zhou, T., Ma, H. et al. (2019). *Advanced Science* 6: 1900834.
32 Senthil Kumar, P., Ramalingam, S., Senthamarai, C. et al. (2010). *Desalination* 261: 52–60.
33 Foo, K.Y. and Hameed, B.H. (2010). *Chemical Engineering Journal* 156: 2–10.
34 Wibowo, E., Rokhmat, M., Sutisna et al. (2017). *Desalination* 409: 146–156.
35 DeCoste, J.B., Peterson, G.W., Jasuja, H. et al. (2013). *Journal of Materials Chemistry A* 1: 5642–5650.
36 Dietzel, P.D.C., Johnsen, R.E., Blom, R., and Fjellvåg, H. (2008). *Chemistry – A European Journal* 14: 2389–2397.
37 Furukawa, H., Gándara, F., Zhang, Y.-B. et al. (2014). *Journal of the American Chemical Society* 136: 4369–4381.
38 Schoenecker, P.M., Carson, C.G., Jasuja, H. et al. (2012). *Industrial & Engineering Chemistry Research* 51: 6513–6519.
39 Kalmutzki, M.J., Diercks, C.S., and Yaghi, O.M. (2018). *Advanced Functional Materials* 30: 1704304.
40 Coudert, F.-X., Boutin, A., Fuchs, A.H., and Neimark, A.V. (2013). *Journal of Physical Chemistry Letters* 4: 3198–3205.
41 AbdulHalim, R.G., Bhatt, P.M., Belmabkhout, Y. et al. (2017). *Journal of American Chemical Society* 139: 10715–10722.
42 Zhang, Z., Fu, H., Li, Z. et al. (2022). *Chemical Engineering Journal* 439: 135756.
43 Zhou, X., Guo, Y., Zhao, F., and Yu, G. (2019). *Accounts of Chemical Research* 52: 3244–3253.
44 Zhao, F., Zhou, X., Liu, Y. et al. (2019). *Advanced Materials* 31: 1806446.
45 Ni, F., Qiu, N., Xiao, P. et al. (2020). *Angewandte Chemie International Edition* 59: 19237–19246.
46 Li, N., Qiao, L., He, J. et al. (2021). *Advanced Functional Materials* 31: 2008681.
47 Borges, F.T.P., Papavasiliou, G., and Teymour, F. (2020). *Biomacromolecules* 21: 5104–5118.
48 Bussamra, B.C., Sietaram, D., Verheijen, P. et al. (2021). *Separation and Purification Technology* 255: 117636.
49 Li, Z., Liu, Z., Ng, T.Y., and Sharma, P. (2020). *Extreme Mechanics Letters* 35: 100617.
50 Wall, F.T. (1954). *Science* 119: 555–556.

51 Turan, E., Demirel, G., Zengin, M. et al. (2006). *Journal of Applied Polymer Science* 102: 1624–1630.

52 Li, J., Hu, Y., Vlassak, J.J., and Suo, Z. (2012). *Soft Matter* 8: 8121–8128.

53 Sekine, Y. and Ikeda-Fukazawa, T. (2009). *The Journal of Chemical Physics* 130: 034501.

54 Chen, Y.W., Zhan, H., and Wang, J.N. (2021). *Nanoscale* 13: 11878–11886.

4

Factors of Fog Harvesting

4.1 Mesh Characteristic

In Chapter 3, we examined the physicochemical knowledge pertaining to the three forms of the existence of water and undertook a cursory analysis of the factors associated with the interconversion of the three states. Subsequently, the discussion turns to the topic of atmospheric water collection, with a focus on both fog collection and hygroscopic collection. With regard to fog collection, the theoretical analysis reveals that the primary factors influencing the efficiency of this process are droplet capture efficiency, droplet growth rate, and droplet shedding rate. In light of this, this chapter examines the impact of these factors on the potential for enhancing fog collection efficiency. It presents a range of approaches employing different bionic organisms that could be employed to achieve this goal.

4.1.1 Biological Fog Trap

Although the collection of fog with biological characteristics has sparked the interest of contemporary researchers, utilizing grid systems for fog collection is not a recent development. Historical evidence indicates that various ancient civilizations have been harvesting fog for centuries [1]. For instance, in the sixteenth century, indigenous people on the Canary Island of El Hierro utilized local laurel Garoé trees (*Ocotea foetens*) for fog collection [2]. The broad, waxy leaves of these "Fountain Trees" are naturally adept at capturing moving fog, allowing droplets to fall to the ground. The island's inhabitants later built cisterns beneath these trees to store the collected water for agricultural and drinking purposes. This method is still in use today, and the trees are proudly represented on the island's flag. In Peru, researchers Dr. Anne Lummerich and Dr. Kai Tiedemann replicated similar tree structures using she-oaks. Additionally, tree-based fog harvesting systems have been established in the Sultanate of Oman with olive trees. The fine, structured leaves of these trees, along with other plants like the South African Cotula Fallax, effectively capture and channel fog droplets to their base and roots (Figure 4.1).

Bioinspired Materials for Fog Harvesting: Principles, Design, Preparation, and Applications,
First Edition. Zhiguang Guo and Weimin Liu.

Figure 4.1 Images of Cotula fallax in South African (a) in dry condition, (b) after collecting fog droplets, and (c) later in the day after collecting fog droplets (scale bar: 20 mm). (d) A diagram illustrating the fog collection process throughout the day. (e) She-oak tree harvesters designed and created in Peru by researchers Anne Lummerich and Kai Tiedemann. The leaves are slender and linear, guiding droplets down to the base of the tree. Source: [1] / John Wiley & Sons / CC BY-NC-ND 4.0.

There is a widely accepted belief that the current science of fog harvesting materials originated from mesh netting experiments conducted at the Catholic University of the North in Chile in 1956. These nets were hung from basic structural frames aligned with the direction of the prevailing fog, sometimes reinforced with cables to endure strong winds. During fog events, airborne droplets passing through the nets would be captured on the mesh fibers, similar to how droplets are collected by trees and other plant species. As more droplets were caught, they would increase in size until gravity caused them to fall into a collection gutter for immediate use, storage, or transport to nearby communities. In the late 1980s, further experiments were carried out by Otto Klemm, Robert Schemenauer (the founder of the NGO Fog Quest), and their team using Raschel Mesh, a mass-produced knitted textile that is still in use today [3]. Since then, fog harvesting installations have been deployed around the world, in coastal Africa, the Middle East, and South and North America, notably through non-profit organizations like Fog Quest. The number of such deployments has steadily increased. At present, there are three types of representative modern fog collectors, including large mesh fog collectors, "harp" fog collectors, and bionic fog collectors.

4.1.2 Mesh Materials

The passive fog collection grid can enhance its water collection efficiency by incorporating additional biomimetic designs [4], and the grid's structure significantly influences droplet capture and movement. For this reason, conventional fog collectors frequently use a grid structure [5]. Over the past 50 years, fog harvesting has been carried out using mesh collectors in various countries [5a, 6], including Chile, Croatia, Ecuador, Eritrea, Morocco, Namibia, Nepal, Peru, and Spain, among others.

Most effective methods for collecting fog involve using a grid made of fibers or wires to capture it. This grid is positioned vertically to face the airflow, allowing the wind to carry fog droplets toward it. When the droplets hit the grid, they adhere to it, combine, and increase in size until they become heavy enough to fall into a container due to gravity. The first significant developments in creating an artificial fog collector took place in South Africa in 1969. The fog collector is made up of two sizable plastic grids measuring $28 \times 3.6\,m^2$, and it collects $31\,m^3$ of water per month, i.e. $\approx 11\,L\,m^{-2}\,d^{-1}$ [7]. Due to the collision of droplets while the fog flow passing through the grids, they cluster together to create larger droplets. This causes water to flow from the grids through the drains into the collection channel [8]. A large-scale fog collection experiment was conducted in a coastal village of 330 people in Chile in 1987 [3, 9]. In Chile, the typical diameter of droplets at high altitudes ranged from 8 to 12 μm, with droplet concentrations usually between 100 and 400 droplets per square. The droplets were gathered using a straightforward impaction method. A large polypropylene mesh was positioned in the droplets' path, causing some droplets to bypass the mesh while others hit the surface and were collected. Numerous studies have been carried out to enhance the design of the fog collector and the materials used for fog collection to boost efficiency [10]. The Raschel mesh is the most commonly utilized fog-collecting mesh globally, achieving an efficiency of approximately 5% [5a, 11]. This indicates that, depending on various weather conditions, approximately 3–10 L of water can be gathered for each square meter of mesh each day. The primary raw material for making Raschel mesh is polyethylene [12], which features a mesh-like structure. It includes a water film area at the top and a supporting framework linked to a water tank underneath. When fog flow contacts the grid surface, fog droplets will collide with the surface and deposit onto it. Over time, the droplets grow and aggregate, and when they reach a certain size, they fall off. Raschel grids are affordable, readily available, and suitable for use in different weather conditions. Nonetheless, there is still room for enhancing their efficiency [13]. The effectiveness of fog collection is typically measured by the fog collection rate, which indicates the volume of water gathered per unit time and area of the collector. At present, a major challenge in fog collection is the low collection rate, particularly in situations with minimal wind speed or fog density. With the current grids [8, 10], the fog collection rate is approximately $1\text{–}10\,L\,m^{-2}\,d^{-1}$.

4.1.3 Mesh Porosity

The mesh filters utilized for fog collection are generally made from textile materials, including woven mesh or felt mesh [14]. Similar to the mesh filters utilized in rainwater harvesting, cross-woven mesh is also employed for fog collection, but it encounters challenges such as loss of liquid droplets due to flow resistance, blockage of mesh openings, and re-entrainment [6d]. The fundamental concept behind using grids for fog collection is that when a solid structure is introduced into a fog flow, smaller fog droplets tend to be redirected by the air current, while only the larger, stickier droplets deviate from the airflow and collide with the surface. However, this theory has not been fully validated by observational data as the experimental results do not provide adequate proof of effective fog capture by the grids. In fact, on superhydrophobic surfaces, fog droplets quickly adhere to the highly wetting areas, leading to ineffective fog interception by the grids. Additionally, these droplets coalesce to form larger droplets, which are subsequently removed.

Grid-shaped structures have some drawbacks, including the potential for clogging, where water droplets gather in the spaces between fibers, creating larger clusters that obstruct airflow. Additionally, re-entrainment can occur when water droplets that have settled detach and re-enter the airflow. The swift movement of droplets on fibers with circular cross-sections can further encourage re-entrainment. Given these issues, the numerical simulation analysis of clogging and re-entrainment may face significant challenges and uncertainties.

Before reaching a critical size, water droplets will gradually grow due to air resistance. If this resistance becomes greater than the surface tension, the droplets will return to the air. In contrast, untreated densely woven grids have higher surface energy, making it harder for water droplets to detach after they settle. When the volume of the water droplet is less than the critical volume, the grid exerts a force that keeps the droplets in place, which is stronger than the force of gravity. As a result, these grids cannot accumulate more water until the droplets become large enough to overcome the grid's retention or are blown away. Researchers have been exploring various solutions to improve fog collection efficiency, given the geometric limitations of the crossed grid pattern [6a]. Potential solutions include increasing the grid's porosity, treating the surface to lower the contact angle, and redesigning the grid to facilitate quicker and more efficient droplet removal.

4.1.4 Geometric Design of Mesh

To address the challenges posed by Raschel grids and 3D grids, researchers are investigating altered geometric designs for fog collection grids to prevent clogging and enhance the efficiency of fog water collection. The porosity of a grid refers to the ratio of the fiber surface area to the total external surface area of the grid. This is inversely related to the air permeability ratio or the opening area ratio, which measures the proportion of the grid's opening area to its total area. In certain contexts, this is also known as pore density. Porosity is typically expressed as a percentage, ranging from 0% to 100%, where 0% indicates no grid and 100% signifies a completely opaque grid that does not allow air flow. The shadow coefficient (s) of a grid can be calculated using Eq. (4.1),

$$s = 1 - \frac{A_f}{A_t}, \tag{4.1}$$

where A_f and A_t represent the opening area and total area of the grid, respectively. The thickness of the fibers, the size of the openings, the external dimensions, and the geometric shape of the mesh will all impact the porosity coefficient. Elshennawy et al. [16a] calculated the shading coefficients of the rectangular, trapezoidal, triangular, and hexagonal model grids, respectively (Figure 4.2).

Drawing inspiration from the effectiveness of California's coastal redwoods in capturing atmospheric water droplets, a new design called "pianistic" or "parallel" fog collectors has been developed [6h]. As illustrated in Figure 4.3, these "pianistic" fog collectors utilize finely meshed metal wires as "keys" to efficiently and consistently capture and drain fog. Vertical guide pipes aligned with the water outlet help minimize the stickiness of water droplets, even in micron-sized pipes [15].

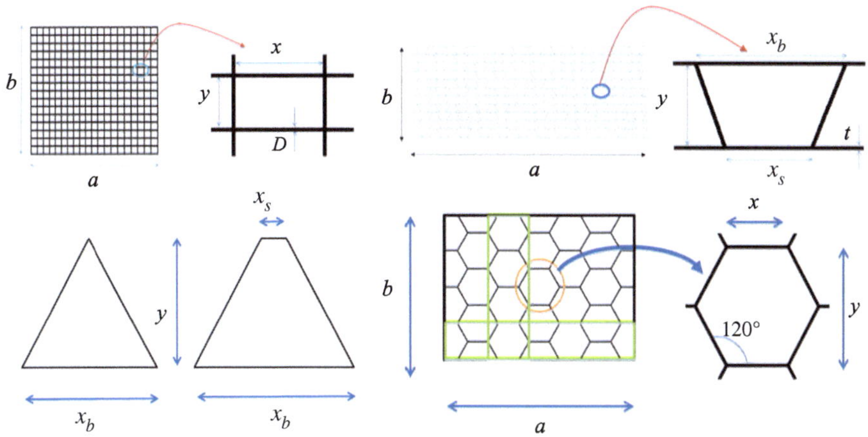

Figure 4.2 Diagrams of different geometric grid structures. Source: [10] / with permission of Elsevier.

Figure 4.3 The "harp" fog collector used for fog collection. Source: [6h] / American Chemical Society.

When compared to traditional grid fog collectors, "pianistic" fog collectors achieve a fog collection rate that is three times greater and significantly surpasses the 50% improvement seen with superhydrophobic grid fog collectors. Additionally, in smaller-scale applications, "pianistic" fog collectors are less prone to clogging due to the reduced adhesion of water droplets along the guide pipes, as opposed to the intersecting lines found in grids. This phenomenon is akin to how water droplets can easily slide along the surface of a superhydrophobic concavity [16] as both scenarios present the liquid–air interface in a longitudinal manner, avoiding interference from oblique intersecting lines.

Advanced textile manufacturing technologies, such as computer-aided knitting, embroidery, and fiber pressing, enable the creation of grids with significant potential.

Additionally, techniques like additive manufacturing and fiber functionalization provide remarkable opportunities to precisely control the placement of wire or thermoplastic materials, allowing researchers to explore innovative mesh designs like never before. For instance, building on the tangling issues associated with the fog harp's macrostructural designs, these technologies facilitate the design and precise fabrication of specialized "harp meshes" that leverage the best features of harps while eliminating the expensive need for specialized frames to tension the vertical wires. Current research, including some ongoing work by the authors, has focused on creating 3D-printed sheets of plastic mesh inspired by the geometries of thin white pine leaves, which could potentially achieve this goal (Figure 4.4). Currently, 3D printing with single nozzles is economically inefficient due to its slow

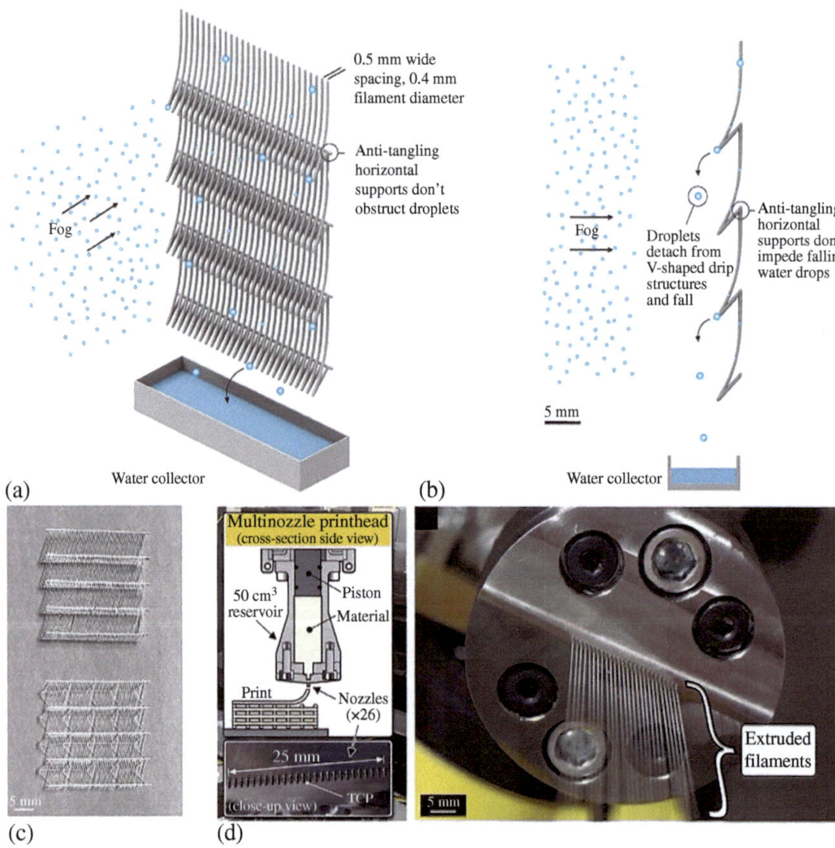

Figure 4.4 (a) A depiction of a 3D-printed harp mesh design inspired by sequoia leaves, featuring a "drip feature" to avoid horizontal supports that could cause fiber tangling. (b) A side view of the proposed sequoia leaf-inspired harp mesh concept. (c) Prototypes of a biomimetic fog harvesting mesh design inspired by sequoia leaves, created using a single nozzle 3D printer. (d) A 25-nozzle 3D printing machine capable of producing harp meshes more rapidly than traditional single nozzle machines, making additive manufacturing a more feasible option for mesh production. Source: Reproduced with permission from [17a] / Elsevier.

printing speed. However, emerging additive manufacturing technologies, such as multi-nozzle fused filament fabrication 3D printing, which utilizes 20 or more parallel nozzles, have the potential to significantly reduce the time required to effectively print sheets of optimized harp meshes [1, 17]. Various fabrication technologies that can print polymer–metal materials have been suggested. Additionally, related research has concentrated on creating spiderweb-like knots on fog harvesting mesh fibers, utilizing a bioinspired nanofibril-humped method or adjusting the deposition flow rate of 3D printers to increase the volume of material produced [18]. These characteristics could be integrated into the suggested multi-nozzle 3D printing methods. Additionally, 4D manufacturing, which involves programmable, shape-shifting, and responsive 3D-printed mesh designs, could provide enhanced control over the development of intricate and highly efficient geometric structures [19]. The meshes created with these new advanced technologies may one day rival the speed and scale of conventional textile weaving and extrusion machines that manufacture Raschel and other fog meshes. In line with the earlier suggestions by researchers Anne Lummerich and Kai Tiedemann that fog harvesting is close to becoming economically viable, the integration of advanced manufacturing methods and biological knowledge holds great promise for achieving this goal.

In addition to the similar devices in California, there are also similar devices in Israel that have been developed to collect fog water. These devices are designed to mimic the natural process of fog collection and have been found to be effective in capturing and collecting fog water. However, there are still opportunities to further optimize these devices by focusing on the design of the grid itself, such as (i) centered on droplet behavior, geometric designs of the mesh pattern can be improved to optimize the collection of fog water; (ii) surface functionalization of the material can improve the material's ability to collect fog water; (iii) the properties of the material can be optimized to improve the efficiency of fog collection. The main aim of this article is to summarize the development of biomimetic fog collectors to date, highlighting opportunities for improvement and defining the potential for further refinement of these devices. Another goal is to improve the durability of zip lines and alternative products, which are prone to stretching and tearing in strong winds [20]. In remote areas where fog collection is most effective and has the greatest impact on water-scarce communities, determining methods to reduce maintenance requirements will also meet the needs of these communities [3].

4.2 Collector Design

The process of collecting water from fog involves three main components: capturing, aggregating, and transporting [21]. The efficiency of fog collection is measured as the ratio of the water reaching the collector tank to the liquid water flux given to the collector grid. The efficiency of the fog capture process is significantly influenced by the structure and wettability of the capture surface. Hydrophilic materials can quickly gather fog water, but their strong adhesion to water droplets can lead to the rapid formation of larger droplets or water films, which reduces the number of sites available for trapping fog water. On the other hand, hydrophobic surfaces can facilitate the growth of small, trapped droplets into larger ones, which can then be easily removed by gravity, leaving the trapping site vacant. However, these hydrophobic surfaces are slower to capture surrounding fog droplets, and some microdroplets may bounce off when they strike the surface vertically, hindering efficient fog water collection. The choice of what kind of wetting material to carry out structural bionics is not only a choice question for scientific researchers but also a required question. It is not difficult to imagine that the removal of droplets facilitates the capture of new droplets, thus forming a continuous and efficient collection cycle, which requires the ability of the surface to transport the droplets. Hydrophilic surfaces are good for droplet trapping, but bad for droplet removal, while hydrophobic surfaces are better for droplet removal.

Over millions of years, plants and animals have evolved, showcasing nature's intelligence through their adaptations. Researchers studying the surface structures and chemical makeup of organisms that gather fog water have discovered that these species have developed a sophisticated system for capturing, transporting, and removing water droplets. The thoughtful design of bionic structures aids in advancing water collection technologies. Biological surfaces, such as spider silk [22], desert beetle [23], cactus [24], Nepenthes [25], and Syntrichia caninervis [26], use a variety of surface chemical and morphological characteristics to control interactions with water and to collect water from the atmosphere even in arid environments [27].

Desert beetles, for instance, utilize hydrophilic ridges to effectively capture fog and hydrophobic structures to remove droplets through their back features. Similarly, spider silk and cacti possess comprehensive systems for capturing, transporting, and collecting fog. In humid conditions, the periodic spindles and joint structures of spider silk not only trap droplets efficiently but also utilize the curvature and surface energy gradients to facilitate the directional movement of droplets. Cacti have developed a multilevel, functional structure in their spines to enhance the efficient collection of fog, allowing captured droplets to merge and be directed toward the trichomes at the base of the spines. These three organisms exemplify the most notable examples of fog-collecting adaptations in nature and are currently significant subjects of study for scientists interested in biomimicry. In addition to the desert beetles, spiders, and cacti mentioned above, Nepenthes also has the ability of continuous and directional transport of fog water, forming a complete fog water collection system [25]. In humid environments, pitcher plants rely on their highly hygroscopic nectar and microstructure to form a smooth water film on the surface, reducing the friction between insects and their surface, and thus being used to catch insects [28]. Inspired by pitcher plants, many slip perfusion surfaces have been prepared. Due to the characteristics of low surface energy and high nucleation density, droplets can condense quickly when the sliding surface is applied to fog water collection [29]. In addition, the injected lubricating oil layer makes the surface a continuous, flat surface with uniform atomic distribution, which is not only conducive to condensation water droplets but also has low contact angle hysteresis and high slippability of condensed water droplets on it, which also ensures the rapid removal of droplets and the formation of new nucleation sites [30]. Following the natural method, droplets on the surface of Arana tend to move in the direction of surface energy reduction. After being inspired, the prepared imitation Arana leaf surface can achieve one-way droplet transport under the condition of inverse gravity [31]. There are many other natural fog-trapping plants, including the dragon fruit plant, the Australian spinified devil lizard, the dragon fruit plant, etc. [32], all of which share a heterogeneous surface with different chemistry and morphology, although the final wettability is different [33]. The shape, size, material, and surface properties of these bionic fog collectors all affect their ability to capture and retain fog.

At present, the research on the mechanism of bionic fog collection mainly focuses on surface structure and surface chemistry. The surface structure and surface chemical properties can greatly affect the behavior of droplets on the surface of the material, so they can affect the process of fog water trapping and water transport on the surface of the material. The surface structure has a different influence on the wettability and fog-collecting process. In general, structural factors can be divided into microstructure and macrostructure. In terms of microstructure, the micro-nanostructure can influence the adhesion characteristics of droplets, with the Wenzel and Cassie states serving as the theoretical foundation for how point structures affect wettability. On a macroscopic level, a bionic one-dimensional fog water collection system inspired by cactus spines and spider silk demonstrates a high efficiency in fog collection. Zheng and coworkers [34] developed a conical copper wire featuring a periodic roughness gradient that can gather fog water and facilitate remote one-way droplet transport through the surface energy released during droplet coalescence, the Laplace pressure resulting from the geometric curve gradient, and the surface energy differences due to periodic roughness. Tian et al. [35] created a large-scale biostimulating fusion-node-microfiber that excels in fog collection, thanks to its spindle knot design that enhances the water driving force for directional transport. Drawing inspiration from the ultra-wettability of surfaces, a unique pattern of alternating hydrophobic and hydrophilic chemical regions found on the back of desert beetles led to the creation of various bionic two-dimensional surfaces. The hydrophilic surface can promote the capture rate of fog water, and the hydrophobic surface is conducive to the discharge of water [36]. In addition to this special hydrophilic and hydrophobic interphase chemical mode, the design of structural gradient gives the material surface anisotropic wettability driving force [37] and Laplace pressure [38], which can also improve the material's ability to collect fog. The composite structure and traditional three-dimensional structure influene the fog collecting process.

4.2.1 Micro-Nano Structure on the Surface

There are many methods to prepare wettable micro-nanostructures on the substrates, such as etching, electrostatic spraying, deposition, and sol–gel. When the micro-nanostructure of the surface is greater than a certain critical value, more air will be trapped between the droplet and the substrate surface, thus reducing the contact area between the substrate and the liquid. However, under the action of

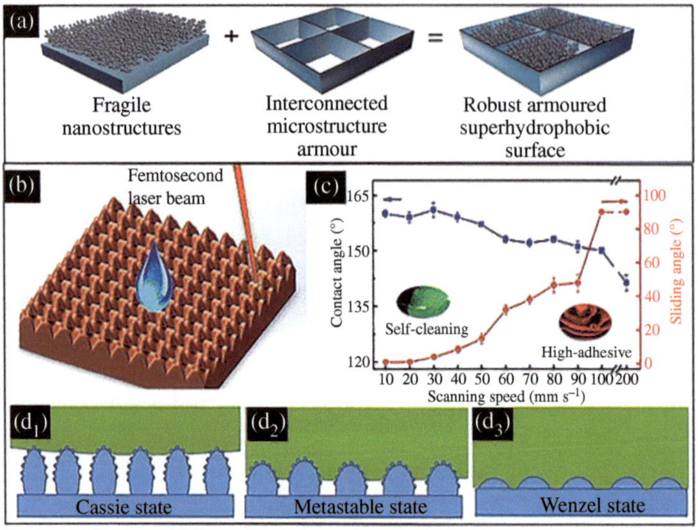

Figure 4.5 (a) Schematic diagram of placing a water-repellent nanostructure in a microstructure "armor" to enhance the mechanical stability of a superhydrophobic surface. Source: [39] / Springer Nature. (b) Schematic diagram of a surface microstructure changed by a femtosecond laser. (c) Average height of the micro-protrusions and average depth of the micro-pits as a function of the scanning speed of the femtosecond laser beam. The insets show the schematics of the microstructures on the surfaces. (d1), (d2), and (d3) Schematic illustration of a water droplet in the Cassie state, the metastable state, and the Wenzel state. Source: [40] / American Chemical Society.

mechanical loads, the surface micro-nanostructures are subjected to very high local pressure, which makes them very fragile and easy to wear. Zhao and coworkers [39] developed wettable functional surfaces by creating structures at two different scales. The nanostructure design imparts hydrophobic characteristics, while the microstructure design ensures durability. The microstructure acts like "armor," safeguarding the underlying nanostructures, as shown in Figure 4.5a. Thus, this approach not only utilizes microstructures to protect nanostructures but also allows for the control of wettability through the combination of microstructures and nanostructures. Future research could enhance this design and explore its application in fog collection, as well as address the durability of nanostructures.

Femtosecond laser ablation technology mainly changes the wettability of the specimen by changing the microstructure of the specimen surface. For example, Long et al. [40] affected the adhesion of the copper surface by changing the speed of femtosecond laser scanning, as shown in Figure 4.5b,c. Different surface morphology leads to different wettability, which leads to different water adhesion. In the Cassie state, water droplets are trapped and suspended in the gas layer of the structure, so the adhesion of the surface is extremely low, resulting in a slightly inclined surface and water droplets rolling out easily. In contrast, in the Wenzel state, the water droplets fully penetrated the groove, and the surface has a high adhesion that can hold the water droplets to the surface without moving. Surfaces composed of deep microstructures have greater roughness, a pattern that results in a layer of air between the surface and the droplet. In this case, the air portion of the surface can be considered to be completely moisture free. Thus, the result is a relatively low water adhesion. As the scanning speed increases, the surface of the microstructure formed by the femtosecond laser ablation becomes shallower, at which point droplets can partially enter the groove. Similarly, as the scanning speed continues to increase, droplets can eventually enter the groove completely, and this combination effect in turn increases the water adhesion properties, as shown in Figure 4.5d1–d3. In applications, differences in water adhesion can be exploited to transfer water without loss and fog collection.

In summary, the micro-nanostructure of a surface greatly influences the wettability of droplets. Currently, the typical techniques for creating these microstructures and nanostructures are chemical etching and physical etching. However, chemical etching tends to be more susceptible to wear and failure. On the other hand, physical etching can produce consistent micro-nanostructures with balanced forces, resulting in better wear resistance.

4.2.2 One-Dimensional Fog Collection Structure

One-dimensional (1D) designs are linear and made up of an infinite number of points, typically utilizing one-dimensional materials. Drawing inspiration from how spider silk and cactus spines gather fog droplets in nature, the researchers developed one-dimensional structures that possess fog-collecting capabilities by examining the mechanisms behind fog collection [41]. It is common for most materials to mimic spider silk, which can effectively collect fog due to its hydrophilic surface (HS), i.e. spindle knots and joints. Many research groups have successfully prepared filament-like materials through a series of methods such as electrospinning [42], fluid coating [43], microfluidic spinning [44], and dip coating, and studied the factors affecting their efficiency. Here, a single mimicry can be based on a specific organism. However, many studies have also combined other organisms to improve the efficiency of materials inspired by spider silk.

The Group's earlier research has validated these enabling factors. We examined the impact of one-dimensional linear micron structure copper wires versus two-dimensional surfaces on fog water collection. The findings revealed that fog on copper wires provided water at a rate 100 times greater than that on a two-dimensional surface. Additionally, by modifying the relationship between the radius of the copper wire and the disturbance caused by fog flow, we discovered that the fog capture efficiency on the copper wire surface can reach as high as 90%. The one-dimensional structure used for fog collection has two main advantages: First, the one-dimensional surface is placed in a foggy environment, and the relative contact area with fog droplets is large. As a result, the probability of droplet capture increases. Second, the one-dimensional design enables the directional movement of droplets, which allows for the rapid transfer of collected droplets to the storage area. This also initiates a new cycle of collection.

On this basis, a lot of research on fog collection performance of one-dimensional structures have been carried out in recent years. For example, Song et al. [45] prepared a series of biomimetic spider silk structures combined with polydimethylsiloxane (PDMS) and graphene oxide and then altered the wettability by changing the surface roughness by laser etching, as shown in Figure 4.6. By controlling the drawing rate, different sizes of spindles can be formed. Among them, the multi-scale structure produces a wettability gradient force, and the conical spindle structure produces a Laplace pressure difference, which work together to enhance the growth of water droplets. In addition, droplet convergence on the main axis causes droplet deformation, and the narrowing of the solid–liquid

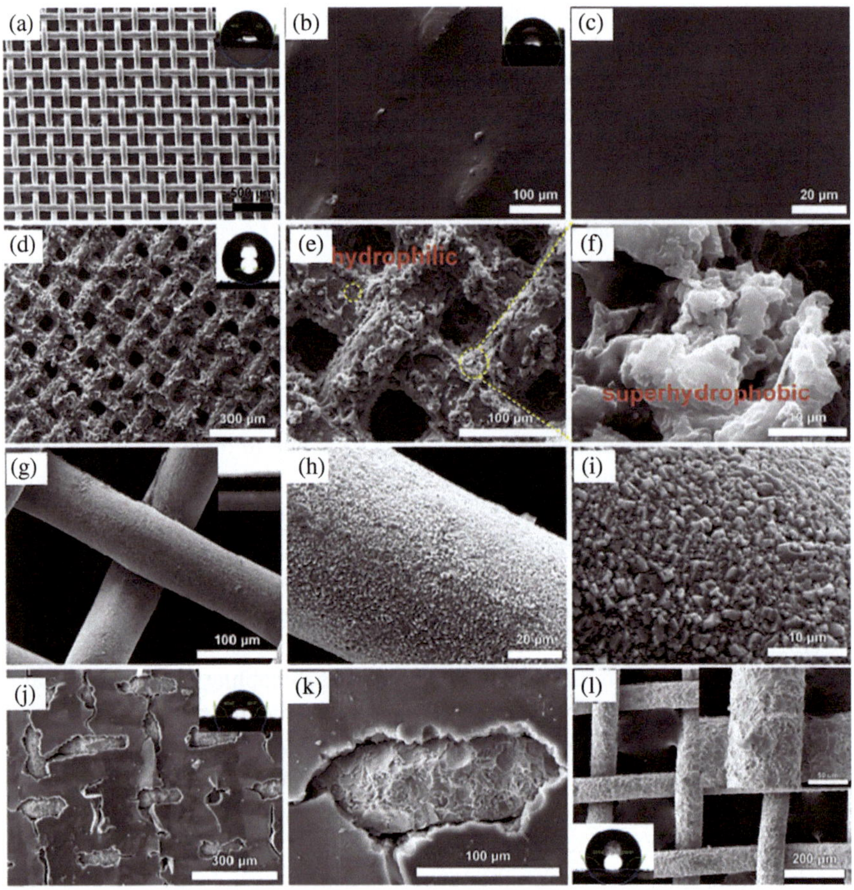

Figure 4.6 The micromorphology of the Cu mesh with different wetting coatings. (a) SEM images of the Cu mesh. (b–f) SEM images of the Cu mesh with a PDMS/G coating (b–c) before and (d–f) after the laser etching process and ultrasonic vibration. (g–i) SEM images of the Cu mesh with a superhydrophilic coating. (j–k) SEM images of the Cu/PDMS mesh after laser treatment and ultrasonic agitation. (l) SEM images of the Cu mesh after laser treatment and ultrasonic agitation. (The insets in a, b, d, f, j, and l are the corresponding water drop profiles on the copper mesh with different wetting coatings.) Source: [45] / The Royal Society of Chemistry.

interface releases the convergence energy, which enhances the directional droplet transport. By controlling different spindles, the maximum fog collection efficiency is 7.74 $g/(mn\ m)$.

Inspired by cactus spines, Jiang and coworkers [46] used PDMS cone arrays to conduct an efficient fog collection experiment. In the test, the authors fabricated the PDMS cones into hexagonal and quadrilateral arrays. On the surface of the hexagonally arranged PDMS cones, the staggered arrangement causes each cone to be completely enveloped by the airflow, while this does not happen on the surface of the quadrilateral arranged cones, as shown in Figure 4.7. Therefore, the larger and more complex the flow field around the cone, the larger the effective deposition area, and the more effective the deposition area.

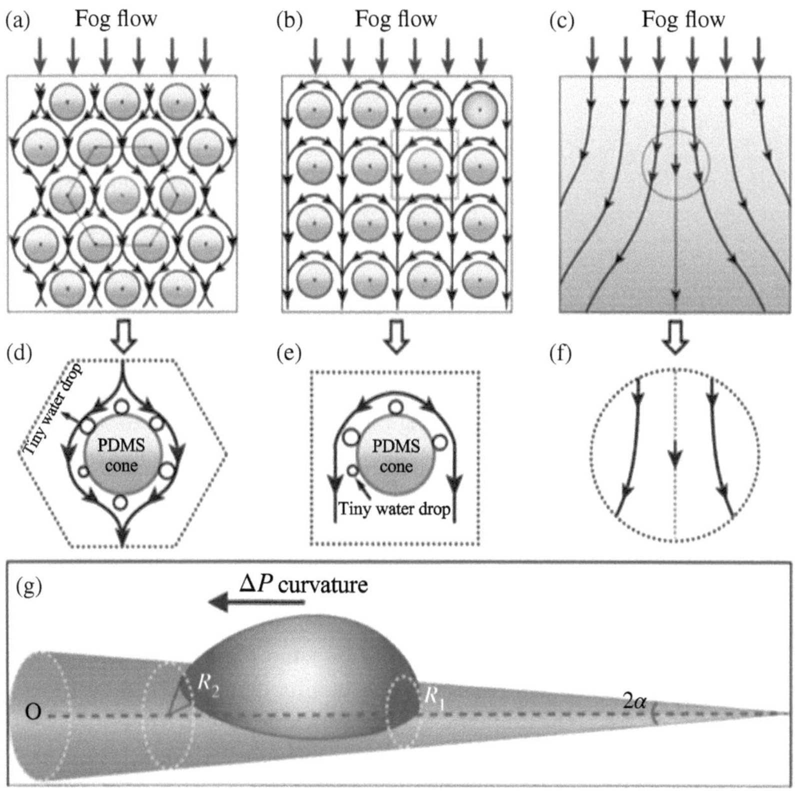

Figure 4.7 Mechanism of the different fog collection efficiency on different structured surfaces. (a–c) Different streamlines of fog flow around surfaces with hexagonally arranged cones, tetragonally arranged cones, and solid surface with microscale smoothness. (d–f) Magnified details of fog flow around each cone corresponding to (a–c). (d) Cones are entirely wrapped by the fog and tiny drops can deposit on both the windward and leeward part of the cone. (e) Front of the cone is wrapped leaving rear of it "vacuum of fog". Tiny water drops deposit on the windward of the cone but few on the leeward. (f) Fog flow bears larger resistance to pass through the solid surface and most of it flows around the side of the surface with a little portion sweeping across the top surface, resulting in fewer drops being deposited. (g) Water drops on the cone move directionally toward the region with larger curvature radius due to gradient of Laplace pressure $\Delta P_{curvature}$.

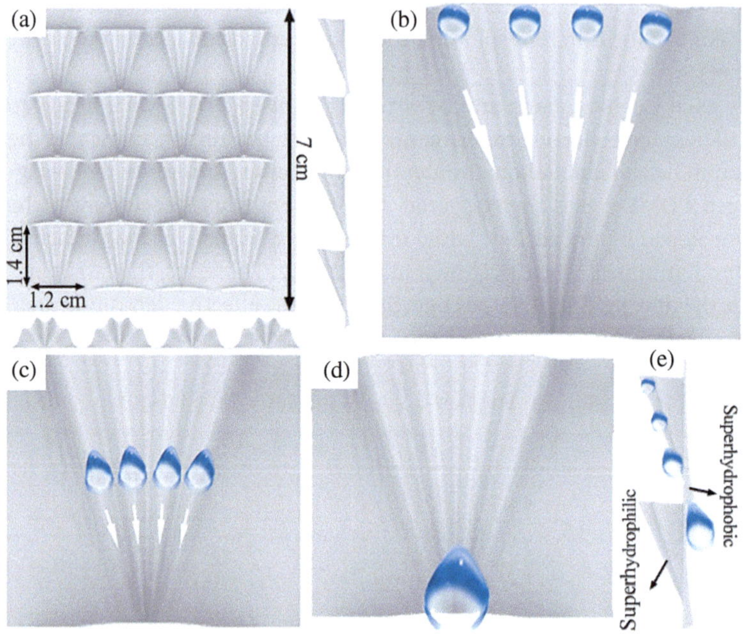

Figure 4.8 (a) Three views of the fog collector. (b–d) The nucleation, growth, and convergence of droplets in the fog collection process, respectively. (e) The movement of droplets during fog collection. Source: [47] / John Wiley & Sons / CC BY 4.0.

Wang et al. [47] proposed an effective fog collector combining wettability and spoke geometry, as shown in Figure 4.8. Wettability is achieved by anodizing zinc oxide (ZnO) nanowires on the front side of the system and applying superhydrophobic silica on the back side. The entire fog collection process is analyzed in detail. ZnO nanowires exhibit superhydrophilic properties, while silicon dioxide (SiO_2) is superhydrophobic. During atomization, droplets tend to form and grow on the superhydrophilic surface before rolling off the superhydrophobic surface. Water droplets traveling along a designated path on the fog collector can easily overcome the sticking effect at the end of the structure, detach from the surface, and be collected. The fog collection efficiency is 1.82 times greater than that of a flat zinc sheet. In conclusion, the low adhesion of the superhydrophobic surface to the droplet allows it to be released from the surface at a small rolling angle, rather than sticking to the superhydrophilic surface and evaporating. Additionally, the quantity and arrangement of bumps influence the droplet growth rate, with superhydrophobic protrusions demonstrating superior fog collection efficiency.

In addition to the preparation of raised structures on the sample surface, laser perforation on the sample surface can also improve the collection efficiency of fog. For example, Song et al. [48] punched small holes in the sample surface, which was 30% more efficient than flat fog collection surfaces. A small hole with diameter $d = 50 \pm m$ and periodic $p = 200 \pm m$ was drilled on an Al surface by the laser drilling technique. These holes are small in size and appear to have no effect on the

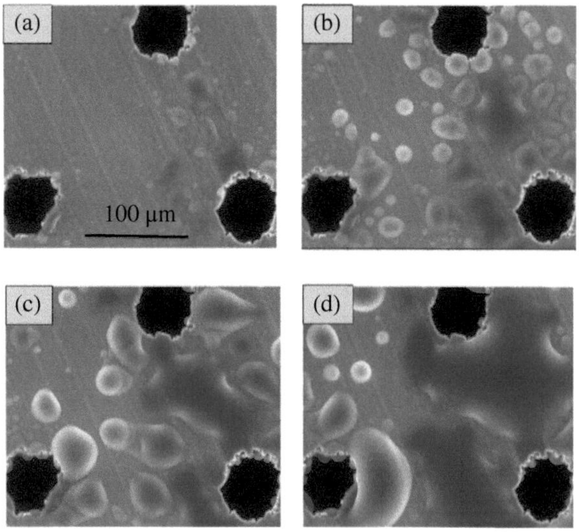

Figure 4.9 ESEM images of droplet condensation on perforated Al plates with a periodic array of holes with a diameter of $d = 50\,\mu m$ and a periodicity of $p = 200\,\mu m$. (a) Nucleation of droplets, (b) condensation of droplets, (c) formation of large condensates, and (d) coalescence of large condensates. Source: [48] / AIP Publishing.

surface but provide conditions for heterogeneous nucleation, as shown in Figure 4.9. Around the hole, water droplets preferentially nucleate at a lower energy due to man-made microscopic defects. In a humid environment, the nucleated droplets gradually grow and merge with neighboring droplets. The droplets are collected by gravity. At the same time, when the large droplet falls, the small droplet in its path (the droplet that cannot fall freely under gravity factors) coalesces, improving the collection efficiency. By adjusting the arrangement of micro hole arrays, the fog collection efficiency can reach 22.5 *L m*.

In addition to the preparation of small holes with heterogeneous nuclei, Ren et al. [49] also used laser to prepare slightly larger small holes for droplet penetration. Driven by the roughness gradient of the inner surface of the micropore and the Laplacian pressure, the water droplets on the upper surface transfer rapidly to the lower surface. The tiny droplets deposited on the top surface polymerize on the surface to form droplets larger than the micropore diameter. Under the action of surface tension, the water droplets approach the micro hole and rapidly transfer to the bottom surface within 0.363 s. Under the combined action of the wettability difference caused by the rough gradient and Laplacian pressure caused by the cone shape, water droplets can quickly pass through the micro hole, as shown in Figure 4.10. Compared with the hydrophilic membrane, the fog collecting performance of the conical microporous structure is improved by 209%, and the fog collection efficiency is 4.08 g cm^{-2}. In short, laser one-step preparation of pore structures is not only simple but also low production cost and can be widely used. In addition, by selecting a wear-resistant substrate, its durability can also be guaranteed. For example, in practice, it is widely used in copper and aluminum substrates.

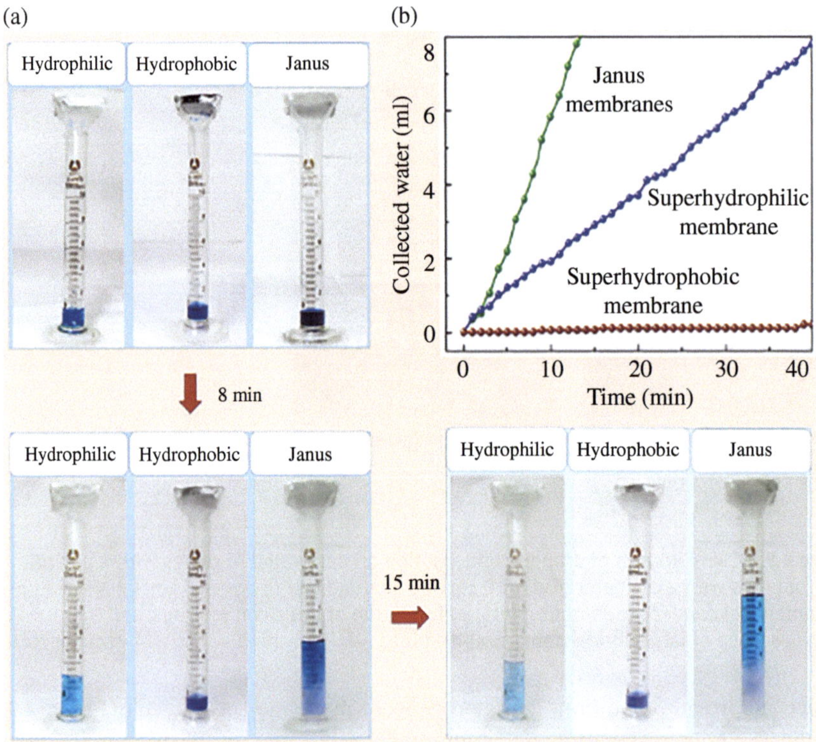

Figure 4.10 (a) The fog collection system at 0, 8, and 15 min. (b) The fog collection ability of Janus membrane comparison with two kinds of wetting membranes. Source: [49] / with permission of Royal Society of Chemistry.

Second, the cavity structure is applied to the fog set application. On the one hand, the cavity structure strengthens the deposition of the droplet; on the other hand, the chemical gradient and structural gradient inside the cavity enhance the directional transport performance of the droplet, which is conducive to the formation of new cyclic deposition on the top surface of the droplet.

4.2.3 Two-Dimensional Fog Collection Structure

In nature, the *Stenocara gracilipes* of the Tenebrionidae family, also known as the fog Stand beetle, is a beetle native to the Namib Desert in southern Africa that absorbs atmospheric moisture through the hydrophilic protrusion of its shell. It rises to the surface and is then transported to the mouth through a drain tank to supply water. Inspired by the shell structure of this desert beetle species, a large number of scholars have prepared various biomimetic structural surfaces that combine superhydrophilic/hydrophilic and superhydrophobic/hydrophobic regions. These called hybrid wetting surfaces (HWSs) have broad application prospects in water/fog collection [50], desalination [51] and heat exchange [52] due to their high condensing capacity and high drop stripping efficiency. Thus, the

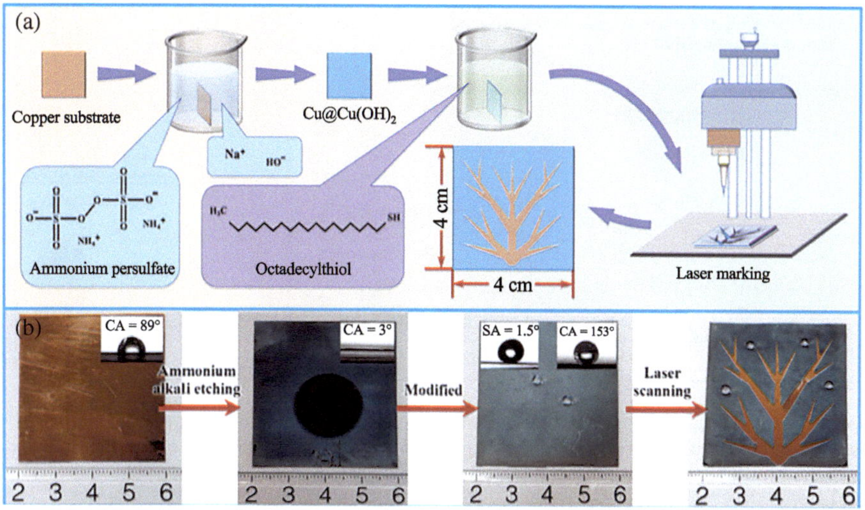

Figure 4.11 (a) Schematic diagram of the preparation process of a patterned surface. (b) Optical photographs of the original copper, SHL copper, SHB copper, and patterned copper. Source: [57] / The Royal Society of Chemistry.

HWS helps to collect water from the atmosphere. At present, the HWS is usually performed using reactive ion etchin [53], laser processing [6e, 54], photocatalysis [36a, 55] and inkjet printing [56].

Wang et al. [57] used 3D printing technology and laser scanning technology to prepare a hybrid surface with a venelike pattern on a copper substrate (Figure 4.11). The copper sheet was reacted in a mixed solution of 2.5 mol L^{-1} NaOH and 0.1 mol L^{-1} $(NH4)_2S_2O_8$. After removing the residue, a layer of $Cu(OH)_2$ nanowerns with a thickness of about 30 mm was grown on the surface, showing SHL state (WCA = 31°). The SHB surface with WCA = 153° and SA = 1.5° was obtained after the modification with nonamine alcohol. Under the effect of Laplace pressure gradient and wettability gradient, the superhydrophilic (SHL) vein pattern on the superhydrophobic (SHB) surface facilitates the directional transport of water droplets. The scheme combines wettability and surface pattern, and the water collection efficiency can reach 4258.59 mg cm^{-2} h^{-1}.

Zhang et al. [56] directly prepared superhydrophilic micropatterns on superhydrophobic surfaces by inkjet printing method (Figure 4.12) and compared the water collection efficiency of different sizes of patterns. The micropattern superhydrophobic surface integrates the hydrophilicity required for droplet nucleation and the superhydrophobicity required for droplet removal into one surface, which has better water collection performance than uniform superhydrophilic and superhydrophobic surfaces. They also investigated the effect of pattern size and spacing on water collection efficiency. Micropattern superhydrophobic surfaces with pattern sizes similar to the backs of desert beetles (i.e. pattern sizes of about 500 μm and spacing distances of about 1000 μm) have the highest water collection efficiency. The water collection efficiency of polydopamine patterned superhydrophobic substrates

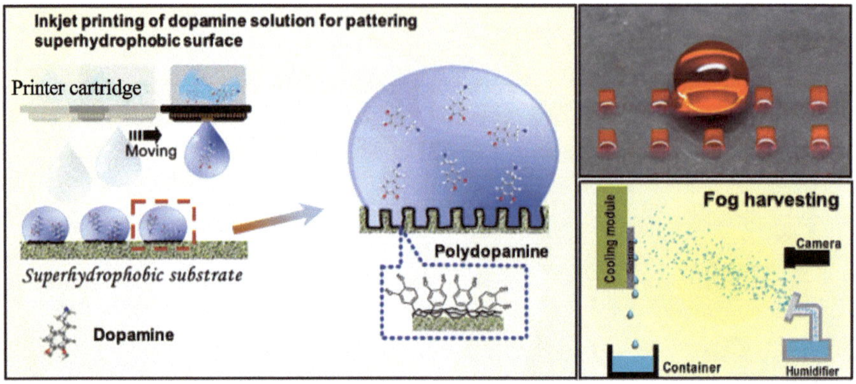

Figure 4.12 Direct micropatterning of superhydrophilicity on superhydrophobic surfaces is achieved by inkjet printing a mussel-inspired ink of dopamine solution on the superhydrophobic surface, followed by the formation of polydopamine via the in-situ polymerization. The micropatterned superhydrophobic surfaces exhibit enhanced fog harvesting efficiency. Source: [56] / with permission of Royal Society of Chemistry.

with a pattern size of about 200 μm and a pattern spacing of about 1000 μm is only slightly higher than that of uniform superhydrophobic surfaces. This may be due to the relatively small proportion of the superhydrophilic region (~5%) and the pattern size of about 500 μm and the pattern spacing of about 1000 μm (25%) compared to the polydopamine patterning superhydrophobic substrate with a pattern size of about 200 μm and pattern spacing of about 400 μm (16%).

Mahapatra et al. [54d] used carbon dioxide laser-processing techniques to create patterns on superhydrophobic aluminum surfaces, which promoted the collision of droplets on these surfaces. They strategically positioned a wettability model, featuring a superhydrophilic orbit, as shown in Figure 4.13a. This configuration encourages capillary-driven condensate discharge and improves heat transfer by minimizing the size of separated droplets. The wettability domain (hydrophilic and superhydrophilic) distributed on the metal surface is used to promote the droplet condensation of the former and the rapid condensate removal of the latter, thereby enhancing the overall condensation heat transfer of the entire pattern-based surface. Staggered conical superhydrophile wedges provide long wettability transition lines and capillary-driven condensate drainage to facilitate rapid condensate removal from metallic substrates. In sample 1, a fractional area design of 29.9% resulted in a maximum HTC enhancement of 34.4% and a maximum water collection rate enhancement of 30.5% compared to a mirror-polished surface. In sample 2, when area accounted for 35.9% of the plate, the maximum enhancement value was 18% and the maximum enhancement value of water collection rate was 15.4% (Figure 4.13b,c). The research offers quantitative insights into the effects of droplet and film condensation zones on overall heat transfer for this substrate. The optimal design is achieved by adjusting the ratio of the superhydrophilic track to the total surface area of the condenser and modifying the spatial arrangement of the track to optimize rapid drainage.

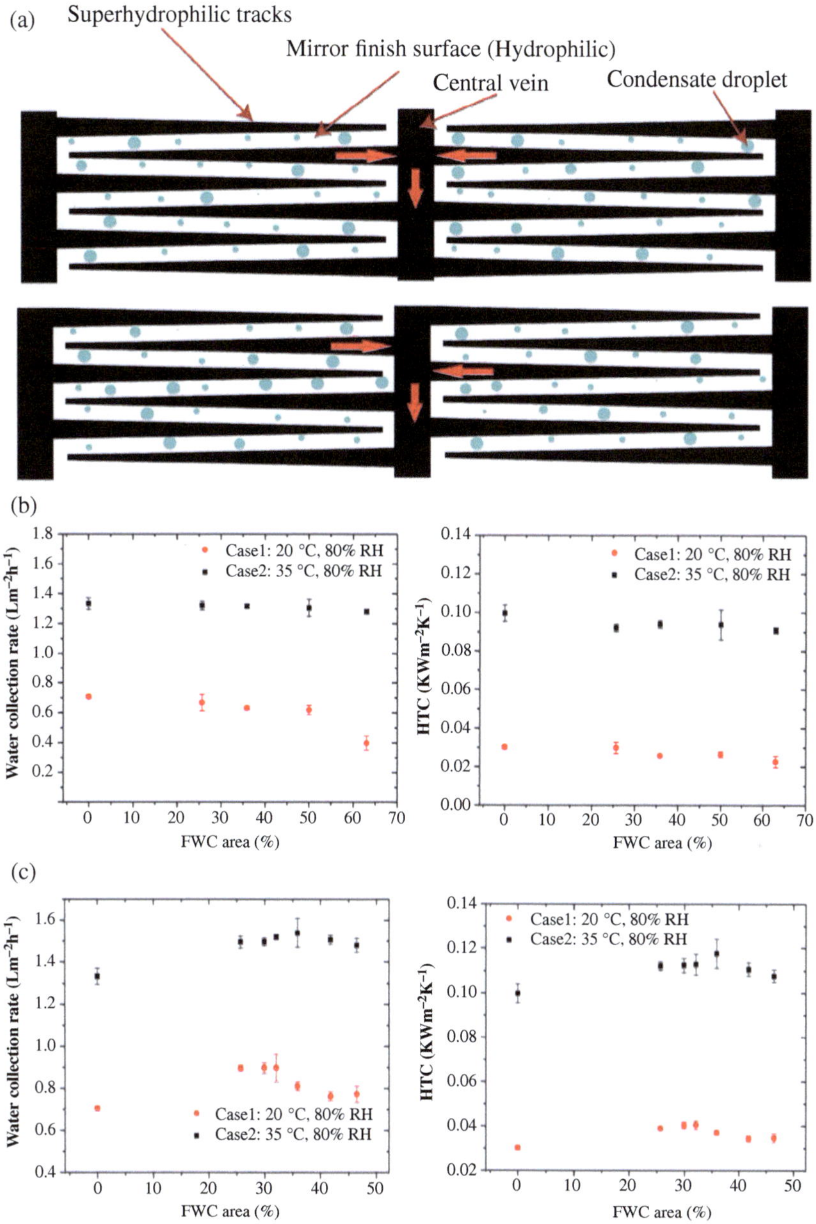

Figure 4.13 (a) Two designs of periodically patterned samples for condensation experiments. (b) water condensate collection rate and HTC with FWC area fraction of the first model. (c) water condensate collection rate and HTC with FWC area fraction of the second model.

4.2.4 Three-Dimensional Fog Collection Structure

When the sample is placed in the fog flow, the fog flow will produce a larger angle deviation. As a result, some droplets fail to reach the surface for two-dimensional fog structure. This led to the development of three-dimensional fog collectors. Current research categorizes 3D fog collectors into two main types [32b]: one type features a composite structure that combines a conical needle design with a functional surface, while the other type is designed to capture surrounding fog droplets. Many researchers have created composite structure fog collectors by integrating one-dimensional structures with two-dimensional surfaces. For example, Zheng and coworkers [34] designed a micro/nanostructured conical spine and Janus membrane Integrated system (MNCS + JM), as shown in Figure 4.14a. The MNCS is responsible for capturing, condensing, and transporting water, while the JM facilitates the rapid movement of water from its hydrophobic side to its hydrophilic side, ensuring quick circulation. Findings indicate that, in addition to the super-hydrophobic properties of the JM, there is a synergistic effect between the surface energy released by droplets and the Laplace pressure during coalescence, which significantly enhances fog collection rates. However, as droplets approach the JM, the Laplacian pressure driving force must overcome the hydrophobic force (HF) to reach the JM. Additionally, an energy barrier must be surpassed for droplets to transition from the hydrophobic side of the JM to the hydrophilic side.

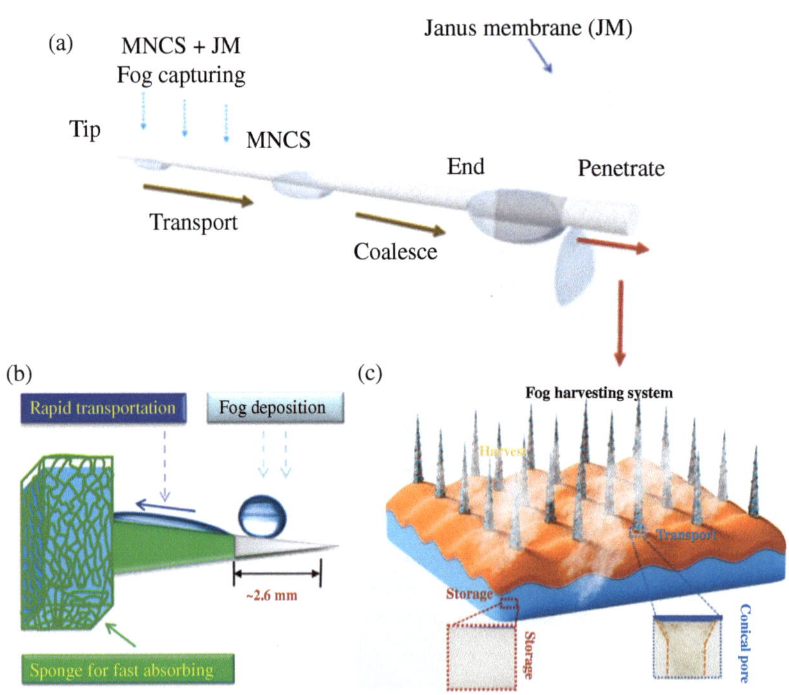

Figure 4.14 (a) An overview of the whole fog harvesting process of the MNCS + JM, including capture, transport, coalesce, penetrate, and collect. (b) The concept of the overall fog capture system. (c) An illustration of the integrated system for self-driven fog collection. Source: [58] / Elsevier.

Fu et al. [58] proposed a new combined fog and dew collector inspired by desert beetles. The collection part of the system has an alternating convex gradient structure, which gives it different chemical wettability. Janus sponges are used as storage components and mechanically assembled together to form a fog/dew collection system (Figure 4.14b). This integrated method enhances the efficiency of collecting water by modifying the behavior of liquids during the fog collection process. This includes the formation of droplets during capture, rapid movement through conical channels, and the separation and storage of droplets using Janus sponges. The combined effect of the cone trap and sponge separation significantly boosts fog collection efficiency. Additionally, under typical subcooling conditions, the system is capable of effectively gathering dew. In extremely cold environments, the system demonstrated impressive dew collection rates of $1.18\,g\,cm^{-2}\,h^{-1}$, a performance that many fog collection systems cannot match. Overall, the collection system, which encompasses the steps of capture, transport, separation, and storage, enables the portable collection of freshwater in remote coastal or inland areas. Zhao and coworkers [59] prepared a super hydrophilic and hydrophobic integrated conical stainless-steel needle (SHCSN) by a simple method, which is used to capture, condense and transport droplets, as shown in Figure 4.14c. The droplets are swiftly absorbed by a super-hydrophilic sponge substrate. The surface of the integrated needle works in conjunction with the rapid deposition of fog in the hydrophobic area, causing the water droplets to move quickly downward due to Laplacian pressure, where they condense into smaller droplets along their path. When a droplet reaches the super-hydrophilic asymmetric surface, it achieves maximum transport speed and is directed to the bottom of the stainless-steel needle. Upon contacting the super-hydrophilic sponge matrix, the droplet is quickly absorbed and stored within it. However, while the sponge matrix has a high droplet absorption rate, its efficiency decreases once saturated. Additionally, external forces are required to expel water droplets from the sponge base, which increases energy consumption.

In conclusion, the integration of one-dimensional conical needles with two-dimensional surfaces to create a composite three-dimensional structure offers distinct benefits for fog collection. The conical needle design facilitates rapid droplet deposition, while the primary Laplacian pressure, along with other adjustable driving forces, enhances the orientation and transport of droplets to the two-dimensional surface more efficiently. Additionally, the two-dimensional surfaces must adhere to two essential design criteria: they should quickly absorb droplets from the conical needles and efficiently release the collected droplets into a storage container.

The concave-convex structure has great influence on fog collection, which cannot be ignored. A 2D plane is cut into a 3D structure to collect water [59]. This 3D paper-cut structure can control the surface fog flow and form a stable reverse rotating vortex, which greatly improves the fog collection efficiency. First, the 3D structure has a larger surface area in direct contact with the fog stream than the 2D plane. Secondly, the fog flow is easy to flow quickly in the 2D plane, and the 3D structure can effectively reduce the fast flow of the fog flow. In recent years, researchers have been working on the preparation of bionic 3D fog collector, which provides the possibility to solve the dilemma of water shortage in arid areas. Inspired by the Bird's Nest fern, Li et al. [60] prepared a novel 3D wave-driven structure that

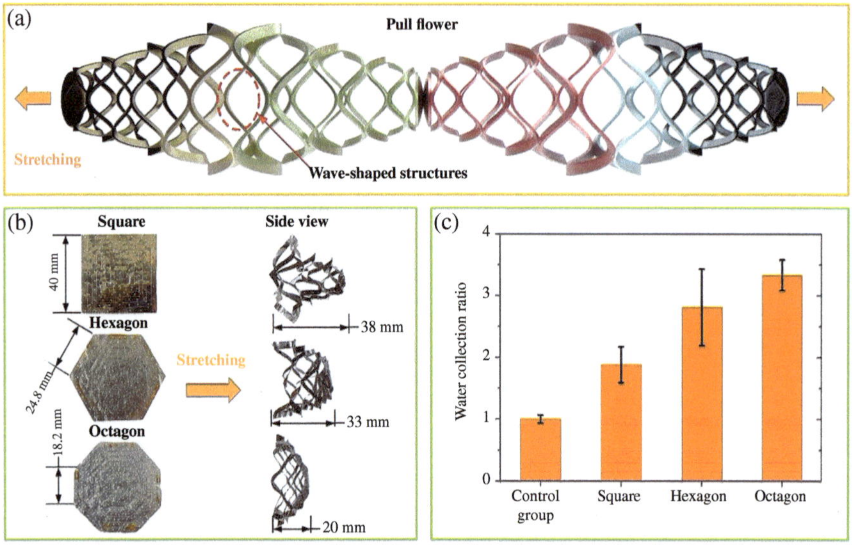

Figure 4.15 (a) The expansion of the wave-shaped bionic structures. (a) The pull flower. (b) The fabrication of the wave-shaped surfaces on polygonal structures with the same area. Source: [60] / Elsevier. (c) The water collection ratio of the polygonal structures.

can be fabricated by simple cutting and stretching (Figure 4.15a). They change the number of gaps around the center to create a series of polygonal structures with the same area, including squares, hexagons, and octagons, as shown in Figure 4.15b. The experimental test shows that the fog-harvesting efficiency of polygonal structures under the maximum stretch length increase with the number of sides (Figure 4.15c). The efficiency of the octagonal structure can reach 3.33 times that of the control group. And the reduction of the scale will increase the geometric gradient of the structure, thereby enhancing the fog-harvesting efficiency. It can be predicted that the efficiency will reach a limit when the polygon is approximated to a circle.

In the future, we will gain additional insights into nature and carry out extensive studies and optimizations related to fog water collection, including its supply, growth, droplet transport mechanisms, and collection methods. Bionics will serve as a key resource for enhancing the efficiency of fog water harvesting and addressing the issue of freshwater scarcity. It will also be a crucial approach to fully resolving the water needs in arid and desert regions.

4.3 Wettability of the Collector

A way to reduce clogging in fine mesh grids is to use surface designs treated with superhydrophobic or lubricating substances. This approach can enhance the movement of droplets in the Cassie state and greatly decrease contact angle hysteresis. Zhang et al. [61] deposited a layer of tannic acid on various substrates as a primer, self-assembled vinylsilane into interdigitated organosilicon nanowires, and then underwent thiol-ene click reaction with 1H,1H,2H,2H-perfluorodecanethiol. This coating demonstrated exceptional static superhydrophobic properties, excellent puncture resistance, and the ability to consistently generate stable water droplets at Wenzel sites. As a result, over a period of 192 h of continuous fog collection, the coating achieved a fog collection rate (FCR) of $2.13\,g\,cm^{-2}\,h^{-1}$, which is three times higher than that of the uncoated substrate and surpasses most previous research findings. Additionally, both experimental and modeling analyses of the system suggest that the main factors for attaining a high FCR with superhydrophobic coatings are the creation of condensate droplets with a critical radius of about 1 mm and a Wenzel site ratio between 0.3 and 0.4.

Superhydrophobic surfaces have reduced surface energy, resulting in weak adhesion forces with droplets. This allows droplets to quickly detach from these surfaces, which has been shown to enhance the transport and removal of droplets during fog collection [62]. While large droplets can be quickly removed from superhydrophobic surfaces, small droplets have difficulty merging into larger ones on these surfaces. This is because they tend to bounce off the superhydrophobic material, making it challenging for these surfaces to effectively collect nearby fog droplets [63]. Researchers have developed a range of mixed-wetting surfaces that incorporate the benefits of both hydrophilic and hydrophobic surfaces, including hydrophilic-superhydrophobic and superhydrophilic-superhydrophobic combinations [64]. On surfaces with mixed wettability, droplets can quickly be absorbed by the superhydrophilic parts and merge into larger droplets, which can then easily slide off the superhydrophobic sections of the surface [65].

It is widely recognized that Janus membranes improve defogging efficiency by promoting the removal of water from the hydrophobic side, as capillary action naturally moves it to the hydrophilic side. This enhancement in drainage due to capillary action also lowers the rate at which water re-evaporates, helping to reduce the amount of water that is lost after being collected [36a, 66]. Thanks to the capillary action of Janus membranes, different combinations of these membranes with unique wetting characteristics (like superhydrophobic/hydrophilic or hydrophobic/superhydrophilic) have been used to enhance the efficiency of fog collection [49, 67]. So far, many experiments have been carried out to create enhanced Janus structures aimed at boosting fog harvesting efficiency.

Hydrophobic coatings have solved the problem of clogging but have introduced a new issue. When large droplets collide while sliding, they may not merge, which disrupts mass accumulation and momentum, leading to bouncing. Additionally, maintaining the long-term stability of hydrophobic coatings is a major challenge [68, 69], particularly in the naturally humid environments of fog collectors. Surfaces treated with lubricants also struggle with durability, as oils can obscure detached water droplets, gradually wearing down the lubrication layer and potentially contaminating the collected water [70, 71]. Janus systems have been effectively utilized to direct droplets away from grids and prevent clogging, gaining significant interest due to the development of new micro/nanofabrication technologies [72]. Superhydrophobic micro/nanostructures can be created on meshes and fibers through various methods, including nanomanufacturing, photolithography, wet chemistry, and nanofiber electrospinning [73]. Similar techniques are also used to develop micro/nanostructures that enhance droplet collection [46, 74]. Current laboratory methods used in for creating specialized wetting surfaces often depend on fluorides or long-chain alkanes, which raise toxicity concerns. Recently, PDMS has gained popularity as a hydrophobic material, offering a substitute for fluorine modifications and the ability to engage in photocatalytic reactions with metal oxide particles like TiO_2, SiO_2, ZnO, Fe_2O_3, Ag_2O. Various specialized wetting materials can be prepared by grafting PDMS onto metal oxide surfaces through controlled photocatalytic reactions [75].

4.4 Collector Durability

Informal observations suggest that Raschel nets are susceptible to stretching and tearing in high winds [20]. In remote areas where fog collection is most effective and crucial for water-scarce communities, finding ways to minimize frequent maintenance is essential to support those who depend on these fog collection systems [54b]. However, since fog collection takes place outdoors, strong winds and sand can cause wear and tear on the equipment, potentially leading to permanent damage. Thus, enhancing the durability and stability of fog water collectors is a priority. Many bio-inspired fog collectors feature micro-nanostructured surfaces, and if these structures are compromised, the efficiency of fog collection could significantly decline or even fail entirely.

Fog collection systems are designed to gather water from fog, which can be a crucial resource in dry and semi-dry areas. However, the effectiveness of these systems can be affected by several factors, including the deterioration of collectors over time. This wear can result from physical damage, dirt accumulation, or other environmental influences, all of which can hinder the collection process. A significant challenge in fog collection is that the delicate structures are susceptible to damage and scaling when used outdoors for extended periods. This is especially relevant for structures that depend on nano/microscale features to improve droplet-substrate interactions. Physical wear can diminish the effectiveness of these features, leading to a reduction in overall collection efficiency. Another approach to enhance fog collection efficiency is through the use of electrostatic fog collectors, which utilize repelling electrodes to modify the electric field strength and improve collection rates. Furthermore, altering the surface properties of grids and the design of fog collectors can also boost their collection capabilities, including methods to mitigate the effects of wear. Regular cleaning and maintenance of collectors can enhance their performance, as the buildup of dust and dirt can negatively impact efficiency.

One of the primary issues affecting the sustainability of fog collection projects is the maintenance of large fog collectors (LFCs). LFCs frequently suffer damage from severe wind events, sunlight (UV radiation), and various environmental influences that impact their structure, grid, and other parts. In fact, numerous LFCs either fail or become nonfunctional when exposed to extreme wind pressures, which is the primary load affecting them. LFCs can fail for various reasons: some collapse due to

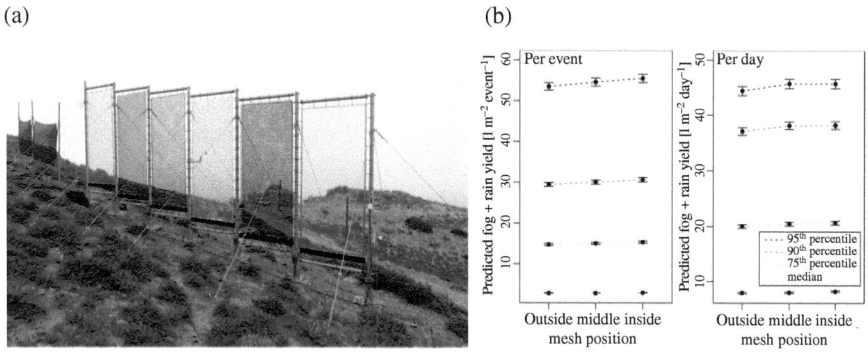

Figure 4.16 (a) Photograph of the pilot facility representing the novel fog collector design. Source: [76] / Springer Nature / CC BY 4.0. (b) Dependence of fog + rain yields on mesh positions. Shown are predicted values for yields for various levels of intensity (mean yield). For each position, the four points denote the expected yield under average conditions (median), under above average fog intensity (75th percentile), and under two extreme fog conditions denoted by the 90th and 95th percentiles.

structural support issues, while others fail because of foundation problems. Additionally, many LFCs become non-functional due to mesh tearing, which prevents them from collecting water. An innovative fog collector (Figure 4.16a) features a sturdy frame made from galvanized steel pipes, with vertical pipes measuring 114.3 mm in diameter and horizontal pipes at 88.9 mm, both having a wall thickness of 4 mm. It is secured to concrete foundations and pegs using stainless steel ropes. The vertical fog collection nets, which are 4.5 m tall and 2.0 m wide, are mounted on support grids and attached with flexible rubber expanders. This design allows for some flexibility to withstand high winds and gusts. The expanders serve as predetermined breaking points, making them easy to replace if overloaded, while keeping the main structure intact. Additionally, plastic gutters are incorporated into the lower mesh mounting system, and multiple frames can be connected to reduce material costs. Here, a team of researchers tested various types of woven plastic and metal meshes for fog harvesting performance (Figure 4.16b) [76], eventually culminating in the development of Aqualonis "3D Mesh" which is both thicker than Raschel and purportedly more durable in high winds, measured in their own internal tests. Supported by the German Government (WasserStiftung) these tests claim 36% greater SY over Raschel mesh [1].

4.5 Environmental Factors

The efficiency of fog collection is affected by various factors, including the geometric shape, size, and wettability of the collector, as well as the porosity of mesh collectors. Additionally, the characteristics of the fog droplets (such as size, size distribution, and content) and the airflow conditions (including air velocity, density, and viscosity) also play a significant role [6h, 10].

4.5.1 Temperature

Variations in temperature can influence the processes of condensation and evaporation of fog. Lower temperature can improve the effectiveness of fog collection, as they facilitate tiny droplets, resulting in fog that is easier to collect. Conversely, at higher temperature, fog may evaporate more rapidly, leading to a decrease in the volume of water that can be collected. The development of fog is closely linked to the relative humidity in the atmosphere; when the temperature drops and relative humidity reaches 100%, fog is more likely to form. Thus, changes in temperature directly affect how often and how intensely fog occurs.

Variations in temperature can influence the chemical makeup of fog. For instance, lower temperature might alter the solubility of certain substances in fog, which in turn affects its chemical characteristics. In certain areas, individuals take advantage of the seasonal patterns of fog and employ specific methods to gather fog, which aids in addressing water shortage problem. The success of these methods is directly influenced by environmental temperature.

Lower ambient temperature can exacerbate air pollution, leading to a decline in air quality, which in turn may affect the formation of fog and the chemical composition of fog. Most studies both domestically and internationally have found that urbanization leads to the heat island effect (temperature increase) and dry island effect (decrease in relative humidity) [77]. However, some areas with unique topography, such as Chongqing, exhibit a heat island accompanied by a wet island. The complex topography of Chongqing, located at the confluence of two rivers, can enhance the convergence and upward movement of water vapor between the urban area and the Yangtze River, creating a unique "heat island + wet island" phenomenon that favors the vertical development of fog. However, the high urban temperatures also hinder the formation of fog. As human activity increases, vegetation coverage in some parts of the world decreases, and changes in underlying surface properties and reduced plant transpiration may hinder the formation of fog [78].

4.5.2 Wind Speed and Relative Humidity

The relative humidity in the air affects the condensation and capture of fog. Under high humidity conditions, the production of fog may increase, thereby enhancing its collection potential. Fog collection requires conditions of high humidity (approaching 100% RH), which limits the practical application of fog collection technologies. In mountainous, arid, and coastal areas, humid air masses are prone to forming conditions conducive to dense fog. However, in areas facing water scarcity, it is extremely challenging to create such conditions, thus lacking the basic requirements for fog collection. These areas, which are precisely the focal points of water collection efforts, are also the ones facing the most difficulty in forming the necessary conditions for fog collection.

In terms of bio-inspired fog collection, although bio-inspired structures provide us with many inspirations for designing fog collection devices or systems, current bio-inspired fog collection processes or major scientific research are still confined to testing in laboratory high-humidity environments (exceeding 60% RH). Before applying this fog collection technology to real-life situations, many issues need to be addressed.

Humidifiers are commonly used in laboratories to simulate fog environments, so the simulated humidity, fog flow, distance between sample and humidifier directly impact experimental data. The collection rate is always a necessary parameter for evaluating fog collection performance. In addition to the performance of the fog collector, factors such as fog flow and humidity also affect the efficiency of the fog collector. When fog droplets are sprayed onto a functional surface, they adhere to the surface under the influence of adhesion forces, then absorb other droplets from the fog flow, condensing and growing. Moreover, appropriate fog flow helps overcome surface adhesion forces and promotes droplet transport. Furthermore, as the distance between sample and humidifier increases, the spraying area of the humidifier enlarges, and the droplet size decreases, resulting in a decrease in droplet density.

The efficiency of fog collection is influenced not just by the size of the droplets and the characteristics of the surface, but also by the speed of the wind carrying the fog. Wind speed impacts how often droplets come into contact with the collector and how long they stay there, with optimal wind speeds promoting droplet accumulation on the collector. If the wind speed is too high, it can impede the capture of droplets, while if it is too low, it can decrease the time droplets spend in contact with the collector, ultimately affecting the efficiency of the collection process.

In areas with dense fog, the wind speed that carries the fog is often quite low, which greatly hampers the efficiency of fog collection systems. To address the limitations of current fog harvesting methods, Peng et al. [77] developed a magnetic-responsive surface by combining structures resembling cactus spines with a flexible conical array. They utilized a template approach with PVC polymer sheets featuring conical holes, and then added cobalt microparticles (Co MPs), PDMS prepolymer, and a curing agent in sequence. To achieve vertical alignment of the Co MPs within the holes, the sample was positioned on top of a permanent magnet (Figure 4.17). The presence of the magnet causes the Co MPs to accumulate at the tops of the cones, creating a varied terrain on the cone ridges, resulting in a rougher surface at the top compared to the bottom.

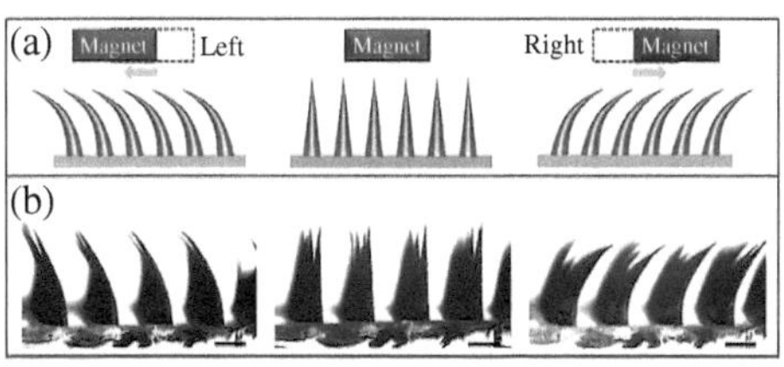

Figure 4.17 The magnetic responses of the cactus-inspired array under an external magnetic field. (a) Schematic of the magnetically induced conical array responses to the magnetic field. (b) The responses of the magnetically conical array under the magnetic field were recorded by a high-speed CCD camera. Scale bar: 1 mm.

4.5.3 Direction of Fog Flow

In addition to surface chemistry and morphological properties promoting fog collection, certain species of the nanostructure desert beetle also exhibit specific fog-basking behavior. For example, beetles of the *Onymacris unguicularis* species orient their posterior upward toward the mist-laden wind atop sand dunes in the early morning, allowing fog to condense on their exoskeletons, with collected water rolling into their mouths [79]. This behavior suggests that when designing artificial fog collectors, both the material used and the orientation of the substrate in relation to the fog are crucial factors.

Breuer et al. treated paper with a superhydrophobic coating, composed of a combination of low molecular weight wax and hydrophobic cellulose-based polymer. An experimental setup was developed to measure the effect of fog tilt angle on fog collection for three surfaces. To observe the impact of droplet adhesion on the surface and droplet rolling efficiency, fog was processed in two different surface orientations: at tilt angles of 90° and 30° relative to the fog flow. It can be observed that tilt angles of 30° and 90° did not significantly affect water collection on glass or polyethylene (PE) substrates. However, for the superhydrophobic paper substrate, the amount of water collected at a 90° tilt angle was nearly twice that collected at a 30° tilt angle. As previously mentioned, this difference can be explained by the surface's excellent rolling behavior, enabling droplets to flow off the surface, where gravity must overcome capillary adhesion forces generated by interaction with the surface.

4.5.4 Liquid Water Content of Fog

Fog is made up of small liquid droplets that are suspended in the air, usually measuring between 1 and 30 μm in diameter, and is a more easily collectible form of atmospheric water than water vapor. Essentially, fog can be thought of as clouds that are touching the ground, and its classification depends on the physical processes that create it. When clouds move over a mountain, they create a blanket of fog, known as upslope fog, which typically contains more liquid water than fog that forms over land or sea [13, 80]. This fog at high altitudes is essential for supplying water to dry areas. Fog harvesting involves capturing tiny water droplets from low-lying clouds (fog), which are then gathered after the droplets merge, grow, and fall off. The liquid water content (LWC) of fog greatly influences the efficiency and success of collecting fog water. A higher water content theoretically enables the collection of more water resources. As the quality of fog improves, the LWC also rises, indicating that more water can be collected under better fog conditions. However, an increase in fog water content can lead to a wider range of droplet sizes, particularly larger droplets, which may impact the collection process [81]. There is a notable inverse relationship between horizontal visibility and relative humidity during the formation and persistence of dense fog; as visibility decreases, the water content of the fog increases. This suggests that in conditions of low visibility, or denser fog, the water content is higher, potentially improving collection efficiency. Moreover, an increase in the number of fog droplets larger than 10 μm is a key factor in the rise of water content, as these larger droplets tend to settle more quickly, facilitating collection. Nonetheless, the actual success of fog water collection is influenced by various factors such as the frequency of fog, its concentration, and wind strength, which can differ by location. Therefore, even with high fog water content, if fog occurs infrequently or wind conditions are not conducive to collection, the total amount of water gathered may still be limited.

4.5.5 Drop Size Distribution

As the fog density increases, the amount of fog droplets and water content rises significantly. When visibility drops below 200 m, both the average and maximum droplet sizes increase notably. This suggests that in thick fog conditions, there is a higher number of larger fog droplets, which may improve collection efficiency. Smaller droplets tend to remain suspended in the air longer and are harder to capture, while larger droplets are more easily collected due to gravity, leading to better capture rates on fog collection surfaces. Additionally, smaller droplets, having a greater surface area relative to their volume, are more likely to evaporate, which can diminish collection efficiency. The distribution of droplet sizes also influences how they merge on collection surfaces. Larger droplets can facilitate the merging of smaller ones, resulting in bigger droplets that are easier to collect. By understanding the characteristics of droplet size distribution, specific designs for fog collectors, such as hydrophobic fibers or hydrophilic surfaces, can be developed to enhance droplet capture and transport [82]. Wind speed influences how fog moves and how quickly droplets settle. When there is a broad range of droplet sizes, particularly with a significant number of larger droplets, collection efficiency can improve at higher wind speeds. This is because larger droplets are more easily carried by the wind and can be deposited onto collection surfaces more effectively.

4.5.6 Terrain, Elevation and Climate

The distribution and density of fog can be influenced by terrain and altitude, which in turn affects the efficiency of collection efforts. Certain locations, particularly mountainous areas, may experience more frequent or denser fog, making them better suited for collection. Early initiatives for collecting fog water were based on meteorological principles, and related studies have enhanced understanding of how fog water forms and distributes in arid regions. Climatic events like rainfall and storms can temporarily change fog distribution and limit collection opportunities. Field studies conducted in Hofuf, Oman, during the summer monsoon seasons of 1989 and 1990, as well as in Peru in 1990, showcased the potential for fog water collection in regions with specific climatic and geographical characteristics [13]. In a later study by Robert Schemenauer and Pilar Cereceda, the authors highlighted three key factors for effective fog harvesting: (i) fog should be common and last for a significant duration throughout the year; (ii) fogs at higher elevations, which contain a considerable amount of liquid water, are particularly important for water collection initiatives in dry regions; (iii) the presence of wind is necessary to enhance the efficiency of fog collection. Typically, fog harvesting takes place in coastal and mountainous areas to meet these requirements [6e].

Internationally, the longest accumulation of fog observation data is likely in the Po River Valley in northern Italy [83] and in the central valley regions of California, USA [84]. Extensive field studies on fog have taken place in these two regions over the last 30 years, revealing a marked decline in fog water content. There has been a clear reduction in the occurrence of fog and dense fog (with visibility under 200 m), likely attributed to a decrease in condensation nuclei resulting from rising temperatures and better air quality. The Namib Desert, situated on the Atlantic coast of southern Africa, is among the oldest and driest deserts globally. It has experienced hot and arid weather for many years, with little to no rainfall. Nevertheless, about every two weeks, a thick fog rolls in from the sea, carried by the wind. Unfortunately, due to the absence of mountains in the area, the fog quickly disperses, making it challenging to gather any water. To endure in these extreme conditions, one must be adept at harvesting fog.

In summary, the environment helps the creatures to evolve to survive by harvesting water from the fog. Inspired by nature, there were many ingenious structures and designs fabricated by researchers all around the world. In this chapter, we discussed the effects of the mesh characters, the design of the collector, and the surface wettability on the fog harvesting efficiency.

References

1 Kennedy, B.S. and Boreyko, J.B. (2023). *Advanced Functional Materials* 2306162.

2 (a) Gioda, H.Z. and Gonzáles, E. (1995). *Advances in Horticultural Science* 9: 112–118. (b) Cereceda, P. and Schemenauer, R.S. (1994). *Fog Collection's Role in Water Planning for Develoing Countries*, vol. 18. Oxford, UK: Blackwell Publishing Ltd.

3 Cereceda, P. and Schemenauer, R.S. (1991). *Ambio* 303–308.

4 Malik, F.T., Clement, R.M., Gethin, D.T. et al. (2014). *Bioinspiration & Biomimetics* 9: 031002.

5 (a) Park, K.C., Chhatre, S.S., Srinivasan, S. et al. (2013). *Langmuir* 29: 13269–13277; (b) Cereceda, P. and Schemenauer, R.S. (1994). *Journal of Applied Meteorology and Climatology* 33: 1212–1322.

6 (a) Ghosh, R. and Ganguly, R. (2018). *Applications Paradigms of Droplet and Spray Transport: Paradigms and Applications*, Springer Nature, 237–266. (b) Schemenauer, J.P. and Wetterdienst, D. (1988). *Annals from the German Meteorological Society* 25. (c) Abdul-Wahab, S.A., Al-Hinai, H., Al-Najar, K.A., and Al-Kalbani, M.S. (2007). *Journal of Water Supply: Research and Technology-AQUA* 56: 275–280. (d) Rivera, J.d.D. (2011). *Atmospheric Research* 102: 335–342. (e) Klemm, O., Schemenauer, R.S., Lummerich, A. et al. (2012). *Ambio* 41: 221–234. (f) Batisha, A.F. (2015). *Sustainability of Water Quality and Ecology* 6: 1–10. (g) Ghosh, R., Ray, T.K., and Ganguly, R. (2015). *Energy* 89: 1018–1028. (h) Shi, W., Anderson, M.J., Tulkoff, J.B. et al. (2018). *ACS Applied Materials & Interfaces* 10: 11979–11986.

7 Di Bitonto, M.G., Kutlu, A., and Zanelli, A. (2022). *Water SA* 28: 349–360.
8 Fessehaye, M., Abdul-Wahab, S.A., Savage, M.J. et al. (2014). *Renewable and Sustainable Energy Reviews* 29: 52–62.
9 Schemenauer, R.S., Fuenzalida, H., and Cereceda, P. (1988). *Bulletin of the American Meteorological Society* 69: 138–147.
10 Azeem, M., Noman, M.T., Wiener, J. et al. (2020). *Environmental Technology & Innovation* 20: 101169.
11 Kowalski, N.G., Shi, W., Kennedy, B.S., and Boreyko, J.B. (2021). *ACS Applied Materials & Interfaces* 13: 38826–38834.
12 Feld, S.I., Spencer, B.R., and Bolton, S.M. (2016). *Journal of Sustainable Water in the Built Environment* 2.
13 Fernandez, D.M., Torregrosa, A., Weiss-Penzias, P.S. et al. (2018). *Aerosol and Air Quality Research* 18: 270–283.
14 (a) Algarni, S. (2018). *Water and Environment Journal* 32: 301–309. (b) Noman, M.T., Militky, J., Wiener, J. et al. (2018). *Ultrasonics* 83: 203–213.
15 Tu, Y., Wang, R., Zhang, Y., and Wang, J. (2018). *Joule* 2: 1452–1475.
16 (a) Elshennawy, A.A., Abdelaal, M.Y., Hamed, A.M., and Awad, M.M. (2023). *Water Resources Management* 37: 6107–6126. (b) Kusumaatmaja, R.J.V.H., Bastiaansen, C.W.M., and Yeomans, J.M. (2008). *Langmuir* 24: 7299–7308.
17 (a) B. D. Chauvette J F, Hia I, L, *Additive Manufacturing* 2021, 47, 102294. (b) Uzel, S.G.M., Weeks, R.D., Eriksson, M. et al. (2022). *Advanced Materials Technologies* 7: 2101710. (c) Mhatre, P.S. (2019). *Process Planning for Concurrent Multi-nozzle 3D Printing*. Rochester Institute of Technology. (d) Cleeman, J., Bogut, A., Mangrolia, B. et al. (2022). *Additive Manufacturing* 56.
18 (a) Liu, Y., Yang, N., Gao, C. et al. (2020). *ACS Applied Materials & Interfaces* 12: 28876–28884. (b) Zhong, L., Zhu, L., Li, J. et al. (2021). *Molecular Systems Design & Engineering* 6: 986–996.
19 Taylor, J.M., Luan, H., Lewis, J.A. et al. (2022). *Advanced Materials* 34: e2108391.
20 (a) Rivera, J. d., D. and Lopez-Garcia, D. (2015). *Atmospheric Research* 151: 250–258. (b) Holmes, R., Rivera, J.D.D., and de la Jara, E. (2015). *Atmospheric Research* 151: 236–249.
21 (a) White, B., Sarkar, A., and Kietzig, A.-M. (2013). *Applied Surface Science* 284: 826–836. (b) Park, K.-C., Kim, P., Grinthal, A. et al. (2016). *Nature* 531: 78–82. (c) Zhong, L., Zhu, H., Wu, Y., and Guo, Z. (2018). *Journal of Colloid and Interface Science* 525: 234–242.
22 Zheng, Y., Bai, H., Huang, Z. et al. (2010). *Nature* 463: 640–643.
23 Parker, A.R. and Lawrence, C.R. (2001). *Nature* 414: 33–34.
24 Ju, J., Bai, H., Zheng, Y. et al. (2012). *Nature Communications* 3: 1247.
25 Chen, H., Zhang, P., Zhang, L. et al. (2016). *Nature* 532: 85–89.
26 Pan, Z., Pitt, W.G., Zhang, Y. et al. (2016). *Nature Plants* 2: 16076.
27 (a) Zhang, S., Huang, J., Chen, Z., and Lai, Y. (2017). *Small* 13: 1602992. (b) Henschel, J.R. and Seely, M.K. (2008). *Atmospheric Research* 87: 362–368. (c) He, G., Zhang, C., and Dong, Z. (2023). *iScience* 26: 105819.
28 Bohn, H.F. and Federle, W. (2004). *Proceedings of the National Academy of Sciences* 101: 14138–14143.

29 (a) Adera, S., Alvarenga, J., Shneidman, A.V. et al. (2020). *ACS Nano* 14: 8024–8035. (b) Jing, X. and Guo, Z. (2019). *ACS Applied Materials & Interfaces* 11: 35949–35958.
30 Anand, S., Paxson, A.T., Dhiman, R. et al. (2012). *ACS Nano* 6: 10122–10129.
31 Feng, S., Zhu, P., Zheng, H. et al. (2021). *Science* 373: 1344–1348.
32 (a) Comanns, P., Esser, F.J., Kappel, P.H. et al. (2017). *Royal Society Open Science* 4: 170591. (b) Zhong, L., Zhang, R., Li, J. et al. (2018). *Langmuir* 34: 15259–15267. (c) Song, D. and Bhushan, B. (2019). *Philosophical Transactions of the Royal Society A: Mathematical, Physical and Engineering Sciences* 377: 20180335.
33 Brown, P.S. and Bhushan, B. (2016). *Philosophical Transactions: Mathematical, Physical and Engineering Sciences* 374.
34 Zhou, H., Zhang, M., Li, C. et al. (2018). *Small* 14: e1801335.
35 Tian, Y., Zhu, P., Tang, X. et al. (2017). *Nature Communications* 8: 1080.
36 (a) Bai, H., Wang, L., Ju, J. et al. (2014). *Advanced Materials* 26: 5025–5030. (b) Gao, Y., Wang, J., Xia, W. et al. (2018). *ACS Sustainable Chemistry & Engineering* 6: 7216–7220.
37 Bai, H., Ju, J., Zheng, Y., and Jiang, L. (2012). *Advanced Materials* 24: 2786–2791.
38 (a) Song, Y.Y., Liu, Y., Jiang, H.B. et al. (2018). *Nanoscale* 10: 3813–3822. (b) Luo, H., Lu, Y., Yin, S. et al. (2018). *Journal of Materials Chemistry A* 6: 5635–5643.
39 Wang, D., Sun, Q., Hokkanen, M.J. et al. (2020). *Nature* 582: 55–59.
40 Long, J., Fan, P., Gong, D. et al. (2015). *ACS Applied Materials & Interfaces* 7: 9858–9865.
41 (a) Ju, J., Zheng, Y., and Jiang, L. (2014). *Accounts of Chemical Research* 47: 2342–2352. (b) Jiang, Y., Savarirayan, S., Yao, Y., and Park, K.-C. (2019). *Applied Physics Letters* 114.
42 (a) Dong, H., Wang, N., Wang, L. et al. (2012). *ChemPhysChem* 13: 1153–1156. (b) Tian, X., Bai, H., Zheng, Y., and Jiang, L. (2011). *Advanced Functional Materials* 21: 1398–1402.
43 Bai, H., Sun, R., Ju, J. et al. (2011). *Small* 7: 3429–3433.
44 Hu, X., Tian, M., Pan, N. et al. (2019). *Carbon* 152: 106–113.
45 Song, Y.Y., Liu, Y., Jiang, H.B. et al. (2018). *Nanoscale* 10: 16127–16137.
46 Ju, J., Yao, X., Yang, S. et al. (2014). *Advanced Functional Materials* 24: 6933–6938.
47 Wang, X., Guo, Z., and Liu, W. (2023). *Advanced Materials Interfaces* 10.
48 Song, K., Kim, G., Oh, S., and Lim, H. (2018). *Applied Physics Letters* 112.
49 Ren, F., Li, G., Zhang, Z. et al. (2017). *Journal of Materials Chemistry A* 5: 18403–18408.
50 (a) Yang, X., Song, J., Liu, J. et al. (2017). *Scientific Reports* 7: 8816. (b) Yu, Z., Yun, F.F., Wang, Y. et al. (2017). *Small* 13. (c) Kostal, E., Stroj, S., Kasemann, S. et al. (2018). *Langmuir* 34: 2933–2941.
51 Hou, Y., Miao, Y., Chen, X. et al. (2015). *ACS Nano* .
52 (a) Wang, M., Liu, Q., Zhang, H. et al. (2017). *ACS Applied Materials & Interfaces* 9: 29248–29254. (b) Alwazzan, M., Egab, K., Peng, B. et al. (2017). *International Journal of Heat and Mass Transfer* 112: 991–1004.

53 (a) Mishchenko, L., Khan, M., Aizenberg, J., and Hatton, B.D. (2013). *Advanced Functional Materials* 23: 4577–4584. (b) Lee, J., Hwang, S., Cho, D.-H. et al. (2017). *Applied Surface Science* 394: 543–553.

54 (a) Zhang, D., Chen, F., Yang, Q. et al. (2012). *ACS Applied Materials & Interfaces* 4: 4905–4912. (b) Muller, F.A., Kunz, C., and Graf, S. (2016). *Materials (Basel)* 9. (c) Bachus, K.J., Mats, L., Choi, H.W. et al. (2017). *ACS Applied Materials & Interfaces* 9: 7629–7636. (d) Mahapatra, P.S., Ghosh, A., Ganguly, R., and Megaridis, C.M. (2016). *International Journal of Heat and Mass Transfer* 92: 877–883.

55 (a) Kamegawa, T., Shimizu, Y., and Yamashita, H. (2012). *Advanced Materials* 24: 3697–3700. (b) Lai, Y., Huang, J., Cui, Z. et al. (2016). *Small* 12: 2203–2224.

56 Zhang, L., Wu, J., Hedhili, M.N. et al. (2015). *Journal of Materials Chemistry A* 3: 2844–2852.

57 Wang, Q., Yang, F., and Guo, Z. (2023). *Chemical Communications* 59: 6048–6051.

58 Fu, Y., Wu, L., Ai, S., and Guo, Z. (2023). *Nano Today* 52: 101979.

59 Chen, D., Li, J., Zhao, J. et al. (2018). *Journal of Colloid and Interface Science* 530: 274–281.

60 Li, S., Zhu, J., Liu, C. et al. (2023). *Chemical Engineering Journal* 464.

61 Zhang, J., Li, B., Zhou, Z., and Zhang, J. (2024). *Small* e2312112.

62 (a) Bai, H., Ju, J., Sun, R. et al. (2011). *Advanced Materials* 23: 3708–3711. (b) Tian, X., Chen, Y., Zheng, Y. et al. (2011). *Advanced Materials* 23: 5486–5491. (c) Yu, T.S., Park, J., Lim, H., and Breuer, K.S. (2012). *Langmuir* 28: 12771–12778. (d) Rajaram, M., Heng, X., Oza, M., and Luo, C. (2016). *Colloids and Surfaces A: Physicochemical and Engineering Aspects* 508: 218–229.

63 Li, J., Gao, C., Pei, W. et al. (2020). *ACS Applied Materials & Interfaces* 12: 48049–48056.

64 (a) Zhang, J., Zhang, Y., Yong, J. et al. (2022). *Optics & Laser Technology* 146. (b) Feng, J., Zhong, L., and Guo, Z. (2020). *Chemical Engineering Journal* 388. (c) Liu, J., Xiong, J., Huang, Q. et al. (2022). *Journal of Cleaner Production* 350. (d) Su, Y., Chen, L., Jiao, Y. et al. (2021). *ACS Applied Materials & Interfaces* 13: 26542–26550.

65 Zhu, H., Huang, Y., Lou, X., and Xia, F. (2019). *Materials Today Nano* 6.

66 Ding, Y., Tu, K., Burgert, I., and Keplinger, T. (2020). *Journal of Materials Chemistry A* 8: 22001–22008.

67 (a) Yin, K., Yang, S., Dong, X. et al. (2018). *ACS Applied Materials & Interfaces* 10: 31433–31440. (b) Li, J., Li, W., Han, X., and Wang, L. (2021). *Journal of Colloid and Interface Science* 581: 545–551.

68 Torresin, D., Tiwari, M.K., Del Col, D., and Poulikakos, D. (2013). *Langmuir* 29: 840–848.

69 Paxson, A.T., Yague, J.L., Gleason, K.K., and Varanasi, K.K. (2014). *Advanced Materials* 26: 418–423.

70 Lalia, B.S., Anand, S., Varanasi, K.K., and Hashaikeh, R. (2013). *Langmuir* 29: 13081–13088.

71 Smith, J.D., Dhiman, R., Anand, S. et al. (2013). *Soft Matter* 9: 1772–1780.

72 Lee, J.H., Lee, Y.J., Kim, H.Y. et al. (2022). *ACS Applied Materials & Interfaces* 14: 21713–21726.

73 Knapczyk-Korczak, J., Szewczyk, P.K., and Stachewicz, U. (2021). *RSC Advances* 11: 10866–10873.

74 Yi, S., Wang, J., Chen, Z. et al. (2019). *Advanced Materials Technologies* 4.

75 (a) Jing, X. and Guo, Z. (2019). *Nanoscale* 11: 8870–8881. (b) Luo, Z., Li, Y., Duan, C., and Wang, B. (2018). *RSC Advances* 8: 16251–16259.

76 Schunk, C., Trautwein, P., Hruschka, H. et al. (2018). *Aerosol and Air Quality Research* 18: 240–253.

77 Peng, Y., He, Y., Yang, S. et al. (2015). *Advanced Functional Materials* 25: 5967–5971.

78 Sachweh, M. and Koepke, P. (1995). *Geophysical Research Letters* 22: 1073–1076.

79 Nørgaard, T. and Dacke, M. (2010). *Frontiers in Zoology* 7: 1–8.

80 Schemenauer, R.S. and Joe, P.I. (1989). *Atmospheric Research* 24 (24): 53–69.

81 Tang, X. and Guo, Z. (2020). *New Journal of Chemistry* 44: 20495–20519.

82 Zhou, M., He, X., Wu, X. et al. (2020). *Applied Materials Today* 21.

83 Zimmermann, L. and Zimmermann, F. (2002). *Journal of Hydrology* 256: 166–175.

84 Herckes, P., Marcotte, A.R., Wang, Y., and Collett, J.L. (2015). *Atmospheric Research* 151: 20–30.

5

Micro-Structures of Fog Harvesting Materials

In natural environments, numerous organisms have evolved the remarkable ability to collect water from fog, even under extremely harsh conditions, which offers a source of inspiration for the creation of innovative functional fog harvesting materials, as explained in Chapter 4. [1]. In this chapter, the focus would be the micro-structures the nature inspired organisms. It is well-documented that numerous water-harvesting surfaces have been engineered, drawing inspiration from the intrinsic properties of plants and animals observed in natural environments, such as desert beetles [2], cacti [3], spider silk [4], and the like that live in water-scarce environments. Butterfly wings possess distinctive, asymmetrical racket-like nanostructures that guide the movement of droplets in specific directions. The dogwood uses its tapered ridges to condense water droplets. Desert grasses have two-meter-high rigid rods that transport water to their roots. Through the observation and study of these plants and animals, researchers have invented a plethora of water-collecting materials. A closer look reveals that they all have very similar but not identical microstructures on their surfaces, which all enable the surface to capture droplets when in contact with water fog, and to allow the droplets to grow and condense and slide down. Therefore, it can correspond to the three key phases of capture, condensation, and transport in fog water collection [2a]. Rational structural design and microstructural fabrication techniques can significantly enhance the efficiency of water-harvesting surfaces.

The morphology of the microstructure of a water-fog collecting surface affects its ability to capture droplets, and a significant factor affecting the capacity of a water-collecting surface to capture droplets is the wettability of the material's surface. Over the course of a long period of natural selection and evolution, many plants and animals have acquired the ability to capture various droplets. For example, butterflies have evolved asymmetrical racket-like microstructures on their wings, and this asymmetry allows for the directional transport of droplets. Among natural materials, desert beetles, cacti, and spider silk stand out as particularly amazing, whose microscopic fog-trapping structures greatly enhance their water-harvesting capacity [2a]. Biomimetic water collection materials can be divided into two-dimensional or three-dimensional materials in terms

Bioinspired Materials for Fog Harvesting: Principles, Design, Preparation, and Applications, First Edition. Zhiguang Guo and Weimin Liu.

of structure. Although the methods of capturing droplets are different, they all have the ability to capture, collect and store tiny airborne droplets [5]. Although the unique biomimetic structure improves the efficiency of fog water collection, the single functionality of most materials is limited to a certain extent, so it is necessary to balance the collection efficiency and the diversity of use scenarios.

5.1 Microscopic Water-Collecting Structures in Desert Beetles

Many plants and animals utilize capillary action and gravity to direct cohesive droplets over specific surface structures. Desert beetles have evolved the ability to capture airborne droplets due to the large temperature difference between day and night in desert regions. Desert beetles exemplify a group of organisms that employ superhydrophobic (SHB) structures to capture water droplets, and Parker et al. [6] first found that desert beetles use protrusions spaced between hydrophilic phases on their dorsal shells to trap water (Figure 5.1a1). The body is covered with many small, rounded bumpy structures with varied wettability. These hydrophilic bumps quickly capture water spray and condense it into droplets (Figure 5.1a2 and a3). When the gravitational force of the condensing liquid droplets exceeds the capillary force, the droplets converge at the apex, undergo transfer, and roll along the surface of the beetle's elytron (dorsal carapace). Eventually, they enter the beetle's oral cavity, thereby accomplishing the acquisition of water resources. It is also possible to combine microscopic projections on the surface of the beetle with other structures, such as the spines of a cactus are needle-like and conical structures, with their diameters gradually increasing from the tip to the base. This structure simultaneously gives rise to a filamentary drive and a Laplace-induced additional pressure drive. This makes it easier for water droplets to move in the direction of the punctate spines. The dorsal surface of the Desert Shell Beetle consists of two types of surfaces, namely hydrophilic and hydrophobic ones, both of which are hump-slotted surfaces. The hydrophilic/hydrophobic region is highly efficient in fog collection under fog-wind conditions. To enhance the fog collection efficiency, hydrophilic protrusions and transport grooves were designed on the conical needles across the entire system, and these were demarcated by SHB regions (Figure 5.1b,c). Nanoscale hydrophobic silica was applied to the chemically-etched zinc sheet via the mask method. After undergoing chemical etching, the surface of the zinc flake exhibited rod-like microstructures (Figure 5.1d). This indicates that zinc oxide was successfully grown on the surface, forming a twisted rod-like structure, which was then encapsulated by hydrophobic spherical nanosilica (Figure 5.1d). The surface of the micro-nanostructures obtained at the unobtained mask site has a SHB wettability of 152°, while the mask bump and groove locations remain superhydrophilic structures close to 0°.

It has been experimentally determined that the water-collection efficiency of a hydrophobic surface is directly proportional to both the water contact angle (WCA) and the roll-off angle of the surface. There are numerous techniques for the

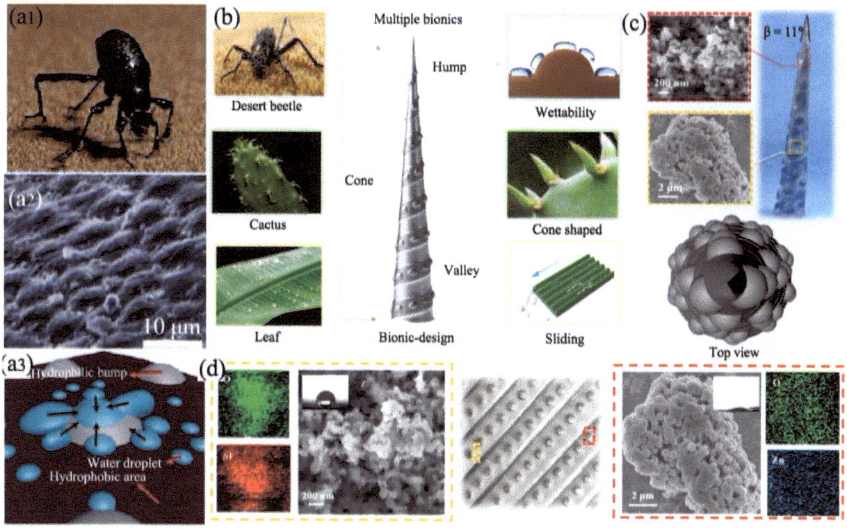

Figure 5.1 (a1) The Namib Desert beetle. (a2) SEM image on the dorsal shell of a desert beetle. (a3) The mechanism of directional water collection on the back of the beetle. Source: Reproduced with permission from [2] / Elsevier. (b) A schematic diagram depicting an efficient fog-capturing conical needle that draws inspiration from a variety of fog-collecting organisms. This conical needle is designed based on the unique fog-collecting mechanisms of different organisms, aiming to improve the efficiency of fog capture. (c) The surface morphology featuring hump-groove structures, the hybrid wettability, and the top view of the conical needle. (d) The surface topography of the collection part within the fog-collection system, along with the scanning electron microscopy (SEM) images and energy-dispersive spectroscopy (EDS) images of the hump/groove-hybrid wettable surfaces. Source: (b–d) Reproduced with permission from [7] / Elsevier.

fabrication of biomimetic desert beetle microstructures [12]. Guo et al. employed the 3D printing technology to design and fabricate a modular fog-collection system. This system is composed of a water-collection module and a tank module, which can be assembled in a Lego-like manner within reasonable limits. Combined with a nano-fabric beetle-inspired hybrid-patterned surface, the system shows remarkable fog collection capabilities [8] (Figure 5.2a). In the experiments, it was found that although higher superhydrophilic density can result in an enhanced propensity for water condensation, it is not conducive to the removal of water droplets. In order to achieve the greatest possible efficiency in the collection of water, it is necessary to optimize the superhydrophilic density. Zheng and coworkers [9] designed nanostructured surfaces with different pattern shapes (Figure 5.2b). The experimental results demonstrated that the collection efficiencies of both pure superhydrophilic surfaces and pure superhydrophobic surfaces were relatively low. In contrast, the star-patterned surface exhibited the highest collection efficiency. Further research indicated that the five-point star model exhibited higher efficiency compared to other (drawable) star-shaped designs. It was discovered that the design of the surface pattern exerted a significant

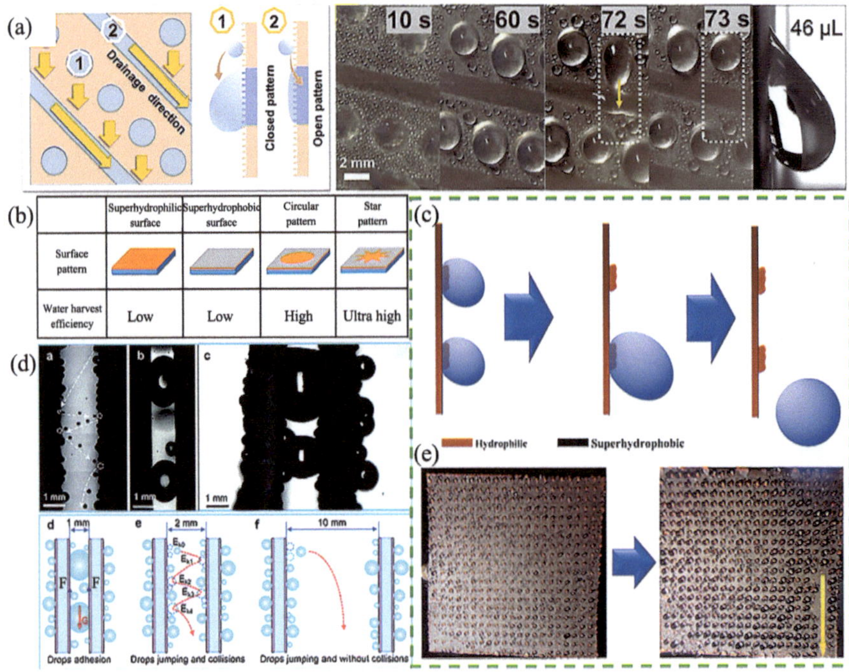

Figure 5.2 (a) Schematic diagram of the desert beetle sector and macro image of water droplets moving across the patterned surface during fog collection. Source: Reproduced with permission from [8] / The Royal Society of Chemistry. (b) Table showing the water harvesting efficiency of surfaces with different patterns. Source: [9] / John Wiley & Sons. (c) Schematic representation of the driving force on the surface of a released beetle due to the surface energy gradient and a schematic representation of droplet coalescence and detachment from the surface. Source: Reproduced from [10] / with permission of Elsevier. (d) Water collection is achieved through the collision of droplets between two parallel and opposite surfaces (collision surfaces). Source: [11] / Royal Society of Chemistry / CC BY-NC 3.0. (e) Schematic diagram of the droplet removal method. Source: Reproduced with permission from [10] / Elsevier.

influence on the collection efficiency. The hexagonal shape enables droplets to move directionally along the structure rapidly. Additionally, creating patterns within the micro/nanoscale range facilitates alterations in the specific surface area, which promotes surface adsorption and enhances the fog-trapping efficiency [13]. Pang et al. successfully prepared a series of bionic ultra-wet copper surfaces with excellent fog collection efficiency by simple etching, modification, laser scanning and electroplating methods inspired by desert mesopods [10] (Figure 5.2c–e). The patterned matrix and selective functionalization enhanced the water-collection performance. Moreover, the hexagonal surface shape promoted the rapid movement of droplets. Both of these factors made contributions to the improvement of water-collection efficiency. The results suggest that the construction of appropriate hydrophilic regions on SHB surfaces has the potential to enhance fog collection.

Besides, when creating bionic beetle structures, the effect of surfaces with protrusions on water harvesting performance should be addressed. Indeed, the periodic inhomogeneous protruding surface structure of beetles can improve water harvesting efficiency by reducing water evaporation during transportation [19]. Shahrokhian et al. conducted water harvesting efficiency experiments on samples with the same surface chemistry but different surface morphology. According to the results of the study (Figure 5.3a1), the volume of water collected increased nearly three times with the addition of millimeter-scale bumps on a spherical surface. Ahmad et al. conducted a statistical examination of droplet shedding from multiple SHB bumps and discovered that the shape, diameter, height, inclination, and wettability of these bumps exerted an influence on droplet shedding (Figure 5.3a2). The experimental results showed that the shedding volume of hemispherical protrusions was smaller than that of cylindrical protrusions. The volume of the shed droplets increased as the diameter of the protrusions increased, yet it decreased with the increase in the patch contact angle (CA) and the surface inclination. The critical

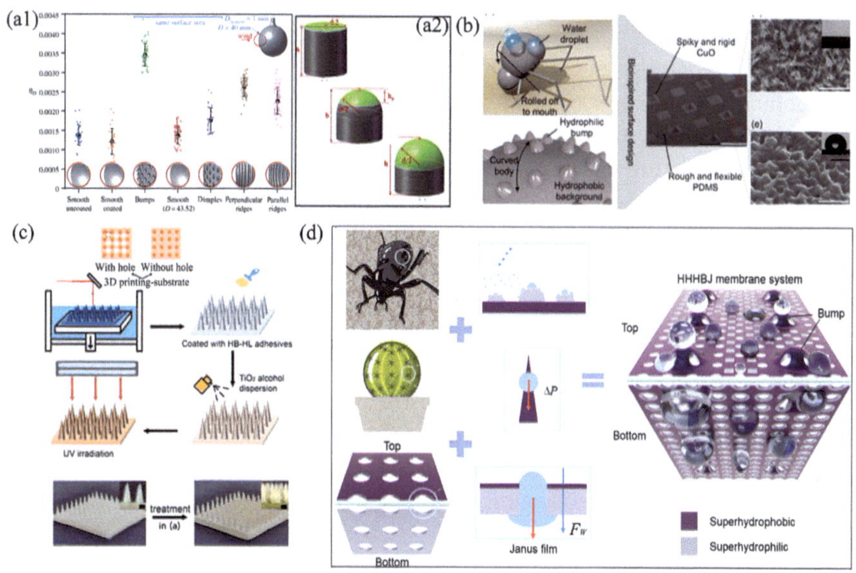

Figure 5.3 (a1) Deposition efficiency of the target surface with a diameter of 40 mm with different surface characteristics. All rights reserved. Source: [14] / The Royal Society. (a2) The hydrophilic and hydrophobic distributions of different shapes. The areas of the hydrophilic patches are presented in green, while the surrounding SHB areas are shown in grey. Source: Reproduced from [15] / with permission of Elsevier. (b) 3D Copper Oxide Patterns with Superhydrophilicity Formed on a Hydrophobic Rough Elastomer Background. Source: Reproduced from [16] / with permission of Elsevier; Reproduced with permission from [16] / MDPI / CC BY 4.0. (c) The schematic diagram of the fabrication process for cone arrays with SHB–SHL properties, along with photographs of the samples before and after fabrication. Source: [17] / The Royal Society of Chemistry / CC BY-NC 3.0. (d) Water collection membrane inspired by desert beetles and cactus thorns, combined with Janus-effect vegetation. Source: Reproduced with permission from [18] / American Chemical Society.

protrusion height follows a similar pattern. The larger the critical volume, the more water evaporates and the less favorable for collection, providing a new design idea to improve the water collection efficiency of the surface. The current research on biomimetic water harvesting is no longer constrained to a single biomimetic structure. Instead, it is evolving to encompass the rational combination of the structural elements of multiple organisms, resulting in the creation of multifunctional structures. For example, the hybrid hydrophilic and hydrophobic surface of a beetle is combined with the structure of a cactus. Peng et al. used UV-induced controlled diffusion combined with 3D printing to construct biomimetic water harvesting materials with different three-dimensional structural parameters, which have the hydrophobic and hydrophilic properties of a desert beetle and the tapered array shape of a cactus (Figure 5.3c), resulting in the improvement of 109% for the water harvesting efficiency. Su et al. designed and prepared a graded hydrophilic/hydrophobic/convex Janus membrane on an aluminum sheet using laser etching. When compared with a standard Janus membrane (Figure 5.3d), the water collection efficiency exhibited an increase of over 250%. Notably, when the fog stream impinged on the surface at an incidence angle of 45°, the collection efficiency soared by as much as 600%.

Freshwater shortage poses a serious threat to the long-term progress and development of human society in the twenty-first century Overall, the study of water-collecting structures of desert beetles has contributed to the improvement of the microstructure of water-fog-collecting materials. A significant number of researchers have designed the microstructures and macroscopic patterns of fog-collecting surfaces by employing diverse preparative materials and methods, and these efforts have made contributions to the development of fog-collection technologies. In addition, beetles have been effectively combined with cacti, spider silk [20], hogweed [21], lotus leaves [3] and other structures to create a range of multifunctional water fog collection surfaces. It can be noted that various organisms have evolved to possess distinctive microstructures and three-dimensional morphologies, which inspired the development of fog and water harvesting materials [20]. For instance, mimicking the structure of a cactus greatly contributes to the development of bionic fog collection, providing new perspectives for bionic surface design and preparation.

5.2 Microscopic Water-Collecting Structures in Cacti

Cacti are able to survive in desert environments due to their inherent drought resistance, [22], more precisely, the good water-harvesting capacity and low water loss (Figure 5.4a1). Studies of cacti have demonstrated that their hydrophilic conical spines, along with the oriented tiny barbs on the spine surfaces, create an outstanding gradient-wetting water-harvesting system relying on the synergistic effect of Laplace pressure differences and wettability differences (Figure 5.4a2). As a result of the Laplace pressure difference, water droplets flow down the conical spines [25]. The hair-like hydrophilic fibers, which grow at the base of the spines,

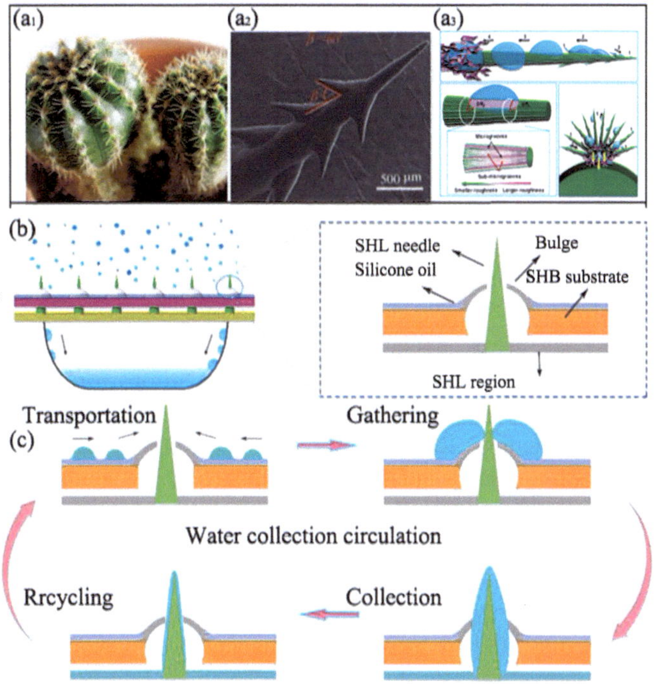

Figure 5.4 (a1) The physical characteristics of the cactus. (a2) SEM of cactus spines. (a3) The mechanism underlying water collection by the cactus. Source: Reproduced with permission from [23] / John Wiley & Sons. (b) Schematic diagram of the water collection mechanism and system on the bionic cactus fog collector. (c) Schematic diagram of the fog collection process of the fog collector system: water droplet capture-coalescence-conveyance-removal. Successful completion of the circulating water collection process. Source: (b,c) Reproduced from [24] / with permission of Elsevier.

absorb the collected water and transfer it to the stem of the cactus (Figure 5.4a3). The cactus serves as the primary source of inspiration and the basis for the development of the multifunctional fog collection system. The cactus-inspired collectors have greatly improved the efficiency due to the automatic and continuous manner for collecting, transporting and storing water. Therefore, it is important to study the bulbous structure of the cactus for the advancement water harvesting technologies, such as the structural influence of on the overall water harvesting [2a].

Gradient-wetting microspines resembling cactus conical spines can now be fabricated by many methods. Metal wire chemical etching, electrochemical etching, X-ray etching [26], and replica molding methods [27] have been successfully employed to fabricate conical spines and their arrays. However, most of these methods are inefficient, limiting the practical application of bionic cactus collectors. Due to mold and process limitations, replica molding methods are not suitable for tips with large length and aspect ratios. Therefore, the development of functional materials with efficient and stable production techniques is essential for the practical application of cactus structures with fog-collecting functions.

Zhou et al. proposed a distinctive design approach: a multi- biologically inspired water-harvesting slip-surface. This surface draws inspiration from Namib Desert beetles, cacti, aquatic plants, and Nepenthes. It synergistically integrates the functions of fog capture, condensation, and transportation, and demonstrates superior performance compared to other synthetic surfaces [24] (Figure 5.4b,c). For the first time, a conical microneedle tip array was constructed, and a cactus-inspired integrated fog collector was successfully fabricated. By means of the enhanced magnetic particle-assisted molding (MPAM) technology, polydimethylsiloxane (PDMS) was combined with magnetic particles (MPs) to create the conical microneedle tip arrays. Based on the experimental data, the optimal weight ratio of PDMS to MPs was determined to be 2 : 1. The researchers fabricated a large cactus-shaped spherical fog collector by integrating hydrophobic microtip arrays with a hydrophilic sponge matrix. This offers a novel perspective on the functionality of cactus-like structures. With the advancement of 3D printing technology, the preparation of cactus conical spines is becoming less challenging, and this technology offers the benefit of enabling the creation of complex structures and personalized items under controlled conditions. Furthermore, a variety of shapes can be produced with ease and efficiency. Fog set structures and optimization were accomplished, eliminating the limitations of the complex process. The experimental results indicated that, compared with the smaller cones, the larger cones exhibited lower water collection efficiency and lower water droplet velocity at the cone tip. The higher the height of the cone the better the water collection effect. When the cone is vertical, the water collection effect is similarly improved. The experimental results demonstrate that the hexagonal cluster shape contributes to the improvement of the fog flow direction around the 3D-printed pin-prick. Moreover, the printed pricks with a vertex angle of 10° and a hydrophobic coating exhibit the highest collection efficiency. In addition, Peng et al. utilized the UV-induced controlled diffusion method in combination with 3D printing to fabricate bionic materials with different structural parameters with the hydrophobicity of a desert beetle and a cactus-like tapered array, which improved the water collection efficiency by 109% [28] (Figure 5.5a). It can be noted that electrochemical etching, 3D printing and replica molding have become the most prevalent techniques for the fabrication of bionic cactus conical needles.

The droplets are transported unidirectionally by capillary force and Laplace pressure difference. Wang et al. combined the advantages of cactus conical spines with microbarbs and trichomes of the genus Sand Snake with hierarchical microchannels to rapidly fabricate spines with barbs and hierarchical channels using laser direct structuring (Figure 5.5b). Furthermore, during the water-collection process, the asymmetric growth of water droplets gives rise to a significant contact-angle hysteresis [31]. This contact-angle hysteresis then serves to drive the water flow in the direction of unidirectional transport [32]. However, the water collection performance of micro-barbs with different orientations exhibits notable differences. The results indicate that the water collection efficiency of conical spines with barbs is higher than that of those without barbs, and among them, the conical spines with rear-located barbs exhibit the highest water collection efficiency. Therefore, the

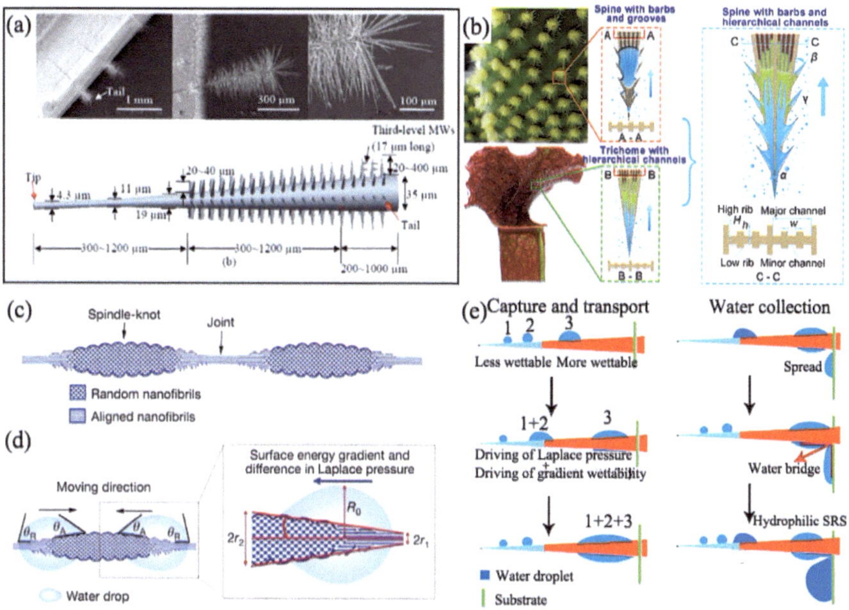

Figure 5.5 (a) A perspective view presenting three such structures, a close-up view of one among the three structures, and a magnified view of its tail section. Source: Reproduced with permission from [28] / American Chemical Society. (b) The Design of a Bio-inspired Spine Incorporating Backward-Oriented Microbarbs and Hierarchical Microchannels Aimed at Fog Harvesting. Source: Reproduced with permission from [29] / American Chemical Society. (c) Spindleknots consist of nanofibers that are highly randomly interwoven, whereas joints are composed of nanofibers arranged in a relatively aligned manner. (d) The anisotropy of the surface structure gives rise to a surface energy gradient. Meanwhile, the conical spindle-knot generates a Laplace pressure difference from the high-curvature region (joint) to the low-curvature region (spindle-knot). These two factors collaborate to propel water droplets from the joint to the spindle-knot. (e) Regarding the fog-collecting process of the SRS + SHB-HL CCN fog-collection system, it is important to note that during the collection stage, when water droplets come into contact with the hydrophilic smooth and rough surfaces, a water bridge is formed. This water-bridge formation is conducive to the rapid removal of water droplets. Source: (c–e) Reproduced from [30] / with permission of American Chemical Society.

design of rear barbs can be improved to enhance their water collection performance. Most researchers have focused their endeavors on the development of wettability-gradient surfaces, while giving scant consideration to the similarity of such nanostructures. It is postulated that the groove structure exerts a crucial influence in the water transportation process. To facilitate the fabrication of such grooved structures, electrospinning (ES) has emerged as a convenient option for the construction of nanostructures. According to Wenzel's formula, due to the fact that the surface of the spindle node is rougher than that of the joint, the surface energy at the spindle node is higher. Droplets move under the driving force stemming from the surface-energy gradient, migrating from the region with lower hydrophilicity to the region with higher hydrophilicity. Owing to the conical geometry of the spindle-junction, a Laplace pressure difference is generated at this location for the droplet. This pressure difference propels the droplet from the union to the spindle-junction, as depicted in Figure 5.5c,d. The experimental results show that the prepared needles with grooves have high droplet transportation capacity at different inclination angles. The findings indicated that the transport velocity of the water droplet could reach over 360 $\mu m\ s^{-1}$, presenting a similarity to the cactus-like conical spines that were elaborated upon previously. In the meantime, they fabricated bare silver needles and smooth PI fibers and carried out experiments within the same controlled environment. Water droplets were only capable of fusing at the tip of the silver needle and failed to be transported in the intended direction. The water droplets moved at a speed of only 159 $m\ s^{-1}$ on the smooth PI fiber. This suggests that the nano-trench structure can greatly improve the efficiency of transporting water. It can be attributed to the capillary forces that prompt the fibers within the spine to absorb the accumulated water [33]. A cactus-shaped model was constructed by using 180 artificial cactus spines and a spherical sponge. In the course of 15 minutes, this device was able to collect 1.3 ml of water. Furthermore, when the Janus film is selected as the substrate, although the asymmetric wettability on the two sides of the film generates a wettability gradient that promotes the unidirectional transfer of droplets, the process of the droplets penetrating the film still requires a certain amount of time. On the contrary, the hydrophilic smooth-rough surface, owing to its strong affinity for water, exhibits outstanding water-absorption properties and lubrication properties, which further expedite the water transport process (Figure 5.5e).

In a bio-inspired system, there exist two distinct purposes. Firstly, it has a water-absorbing function integrated with an anti-evaporation function. Secondly, it features a water-releasing function that can be utilized as required. All of the above formulations can be achieved in cactus-like conical spines with 3D structures, which, however, are relatively complex and cumbersome to prepare. The transformation from a 3D structure to a 2D structure can contribute to cost management and improve the efficiency of fog collection. Song et al. simplified the three-dimensional conical structure of the cactus into a two-dimensional triangle. Droplets can also form Laplace gradients [34] (Figure 5.6a). Experimental data showed that the hydrophilic triangle design bounded by a SHB band was efficient in self-transport. Further tests showed that collected water propagated slower with a larger angle of entrapment in the triangular pattern of H. Yue et al., but larger droplets were transferred to the storage layer [35] (Figure 5.6b). The collection rate increased as the pattern length decreased [38]. The two-dimensional triangles are conducive to the fabrication of an efficient fog-collection system. Moreover, the two-dimensional structure offers novel insights into the preparation and structural design of water-collecting materials. In the oriental art, "kirigami [33]" and "origami" have interesting geometric structures obtained by folding and cutting paper. Therefore, in order to efficiently collect fog, we created and presented a cactus kirigami [33]. Inspired by blue orchids and cacti, Wang et al. designed fog collectors with three different bifurcated configurations that are independent of the direction of the fog flow. The structures are schematically shown in Figure 5.6c [36]. The choice of paper-based

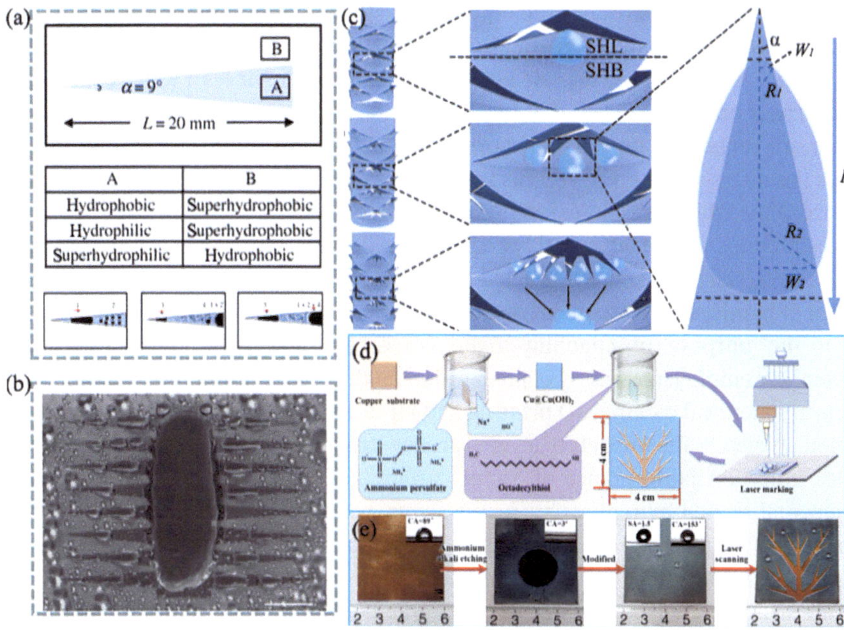

Figure 5.6 (a) A schematic illustration depicting a surface featuring triangular patterns, along with the nucleation and movement of droplets on a hydrophilic triangular pattern that is encircled by a SHB region. Source: Reproduced from [34] / with permission of The Royal Society of Chemistry. (b) The water-gathering area in the reservoir is encircled by a triangle pattern. Shown is a photograph of the water collected in a reservoir that is surrounded by triangle designs. Source: Reproduced with permission from [35] / The Royal Society. (c) A schematic illustration of the origami-based waterlogging sample. Source: Reproduced with permission from [36] / Elsevier. (d) A schematic representation of the preparation procedure for a patterned surface. (e) Optical photographs of raw copper sheets, superhydrophilic copper, SHB copper, and patterned copper. Source: (d,e) Reproduced with permission from [37] / The Royal Society of Chemistry.

substrate was based on due consideration of performance, cost and environmental friendliness, while the use of laser cutting technology allows for large-scale preparation of origami structures with high-density ridges. The porous structure inherent in cellulose endows the paper-based substrate with superhydrophilicity. However, once wetted by fog, this superhydrophilic property often leads to a degradation in the mechanical strength of the substrate [39]. Consequently, the surface of the fabricated origami was coated with hydrophobic wax. This treatment not only guaranteed the purity of the chemical composition but also imparted long-term hydrophobicity. Moreover, the rough surface resulting from laser ablation is smoothened. The entire process is not only convenient but also time-efficient. Approximately two or three spines can be fabricated within 1 s. In addition, Wang et al. prepared hydrophilic and hydrophobic hybridized surfaces with vein-like patterns on copper substrates using 3D printing and laser scanning techniques [37]. On the SHB surface, the superhydrophilic (SHL) vein-like pattern plays a crucial role in facilitating the directional transport of water droplets under the joint action of the Laplace pressure gradient and the wettability gradient. This proposed scheme integrates wettability and surface patterning. As depicted in (Figure 5.6d,e), it achieves a water collection efficiency of up to 4258.59 mg cm^{-2} h^{-1}. The microchannels arranged on the cactus spines are capable of rapidly transporting water via the sliding of the liquid film, while the micro-barbs on the cactus spines can effectively capture the fog. By making use of the fabricated conical spines, a two-dimensional spider-web-like structure water-trapping device and a three-dimensional cactus-structure water-trapping device were designed and constructed [40]. This provides a new perspective on the design of bionic water catchment devices.

For the purpose of enabling high-volume, low-cost mass production, the researchers employed a variety of preparation methods for this microstructure, such as electrochemical etching, 3D printing and replica molding. To enhance water collection efficiency, the impacts of layered groove structures and oriented barbs on the water-collection performance of cactus surfaces, along with the water-collection systems fabricated by replicating cactus stems, were also explored. Moreover, the transformation from 3D cactus spines to 2D structures streamlines the preparation procedure and expedites the advancement of water-harvesting technologies. In reality, the structure of cactus spines has been integrated with diverse other biological characteristics, each having its unique array of applications. The current state of research is multi-biomimicry and the development of multifunctional structures will contribute to the development of water harvesting technologies.

5.3 Inspired by the Spindle Knot Structure of Spider Silk

A multitude of plants and animals exhibit distinctive micro- and nanostructures at the micron and nanometer scales, which impart unique wettability characteristics. Desert beetles and cacti, for example, have unique water-harvesting skills. In the early morning, condensation occurs on spider silk. Spider silk is composed of hydrophilic flagellar proteins [41]. These proteins are highly sensitive to humidity and possess outstanding mechanical properties [42]. Moreover, silk exhibits an outstanding water-collection effect. The structure of spider silk was analyzed by Zhen et al., which revealed that the fluff consists of nanofibers and two spindle fibers (Figure 5.7a). These randomly distributed hydrophilic nanofibers augment the wettability of the spider silk, thereby facilitating the condensation of water droplets upon it [45]. When coming into contact with the dry spider silk, the water droplet experienced a structural transformation. Droplets condensed on the translucent fluff. As the droplets grew, the fluff contracted, and eventually, periodic spindle-shaped junctions were formed prior to the commencement of directional fog collection (Figure 5.7b). Through the analysis of the water-collection process in spider silk, as depicted in (Figure 5.7c), the fundamental principle of directional water collection was determined to be the periodic structure of the spindle-shaped elements and their joints.

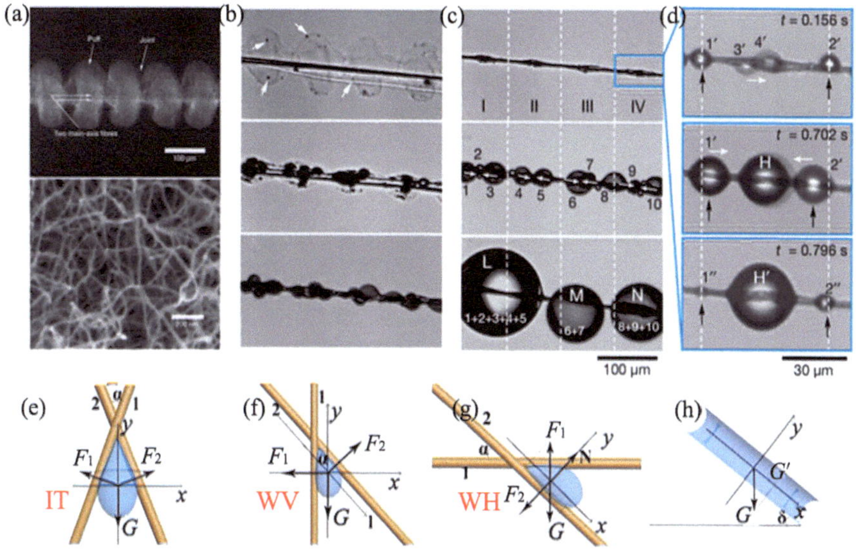

Figure 5.7 (a) Structures of dry capture silk. (b) In the fog, the dry spider silk takes shape, with spindle-knots being connected by joints. (c) The directional water collection on the wet-rebuilt multistructural spider silk. (d) More detailed directional drop movement on individual spindle knot. Source: (a–d) Reproduced with permission from [43] / Springer Nature. (e–g) Analysis of the force on the water drop when the copper wires cross at an angle of α. (g) Force on a water drop when two copper wires cross a horizontal line forming an isosceles triangle. (h) Force component of gravity along the angle of inclination when the tube is inclined in the shape of a drop of copper wire. Source: (e–h) Reproduced from [44] / with permission of Elsevier.

Studies of the water-collection principle of individual spindle junctions (Figure 5.7d) have found that water droplets condense mainly at the junction, or accumulate at the spindle, and that the next round of water condensation and transportation may begin at the junction when the first water droplet leaves the spider web junction. Nevertheless, this phenomenon is exclusive to the "wet reconstruction" of spider silk. Other silk and nylon fibers do not display such a property. As depicted in Figure 5.7e–h, drawing inspiration from the anisotropic wettability of butterfly wings, Liang et al. devised an integrated lattice micromechanical device featuring dual different super-wettability trajectories . This device is capable of mechanically adjusting the integrated wettability through the rational rotation of one layer of superhydrophilic and one layer of superhydrophobic trajectories, and can further govern the pinning and sliding of droplets [44]. The SEM image of the spindle on the spider silk well describes the principle of spider silk fog collection (Figure 5.8a). It can be observed that the spider silk is composed of alternating spindle-shaped junctions. These spindle-shaped junctions are made up of highly randomized nanofibers, resulting in a microscopic rough structure on its surface. Secondly, the nanofibers in the spliced part are relatively parallel to the spindle. This leads to the formation of a relatively smooth morphology and an anisotropic alignment morphology. Consequently, a surface energy gradient and a difference

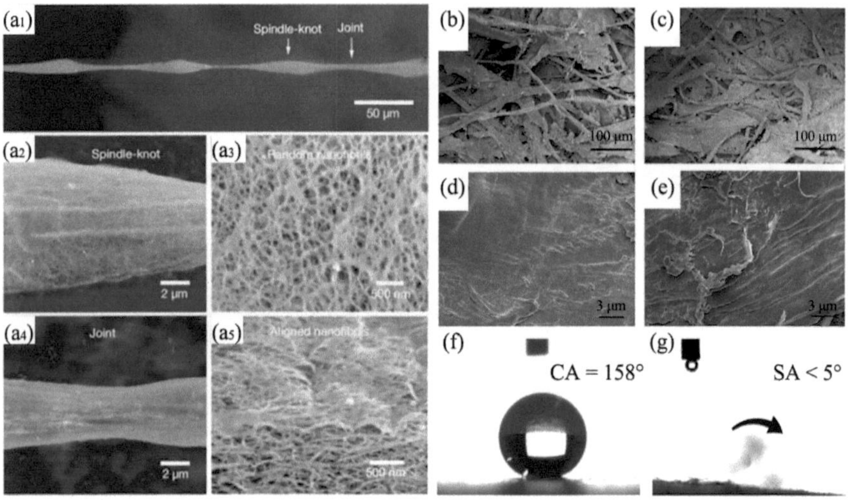

Figure 5.8 (a) Structure of wet-rebuilt spider silk. Source: Reproduced with permission from [43] / Springer Nature. (b–e) Bionic writable multifunctional recycled paper with uniform superhydrophobicity inside and outside SEM. (f,g) Contact and sliding angles of droplets on SHB bionic writable multifunctional recycled paper. Source: (b–g) Reproduced with permission from [46] / The Royal Society of Chemistry.

in the Laplace pressure gradient are produced, thereby creating conditions for the directional transport of droplets. Sub-millimeter-sized droplets can be propelled by either surface energy gradients or Laplace pressure differences. In contrast, micrometer-sized droplets cannot be driven in the same way. This is due to CA hysteresis, which becomes more significant as the droplet size shrinks and adhesion increases [47]. The wet-reconstructed spider silk can surmount the CA hysteresis by leveraging the Laplace pressure gradient and the surface energy gradient to drive micrometer-sized droplets. Simultaneously, this structure intensifies the hysteresis effect, enabling the droplets to be transported directionally.

Building upon this foundation, the research team led by Bai [48] delved into the underlying principles of water-collecting structures inspired by spider silk. Initially, they fabricated a range of artificial spider silks featuring spindle-shaped bodies. During this process, they meticulously designed the chemical compositions and nanostructures of these artificial silk materials. Upon research, it was discovered that through optimizing the curvature, deformation, and roughness gradient of the fiber surface, the control over the driving direction of droplets could be realized. As a result, droplet actuation has been studied with greater clarity, particularly the impact on surface nanostructures. This discovery holds great significance for the fabrication of smart materials that have the ability to drive micrometer-sized droplets. Ever since then, the research and development of spider-silk-based water-collection technology have been on the rise. The preparation of artificial spider silk provides a potential solution to ease the water shortage problem in arid areas. In evaluating the potential use of bionic materials, it is essential to consider the environmental impact and manufacturing costs, which may, however, limit the application of the materials. In previous work, microfibers of artificial spider silk were prepared by dip-coating method [48], which showed good water harvesting effect, and the fabrics are more adaptable to the environment, more versatile and economical to use compared to planar materials. Nevertheless, the mass production is restricted. In recent years, electrostatic spinning technology has emerged as an effective approach for fabricating microscopic materials [49]. During the preparation process, beaded fibers are usually regarded as an intermediate state between the products of electrostatic spinning and electrospray. Generally, they need to be avoided. The reason is that beaded fibers possess a periodic knot-like structure, which bears a strong resemblance to the structure of natural spider silk. Bead-like fibers are commonly considered as an intermediate state between the products of electrostatic spinning and electrospray. Usually, efforts are made to avoid them. This is because bead-like fibers have a periodic knot-like structure that closely resembles the structure of natural spider silk. In the past, although some researchers prepared beaded fibers through electrostatic spinning, they overlooked the advantages of this beaded structure. Dong et al. [50] used coaxial electrostatic spinning by using the instability effect to generate spins on the fibers, forming a novel artificial spider silk fiber. During the preparation process, it was found that the flow rates and concentrations of the two liquids directly affected the morphology of the product. Thus, the selection of optimal flow rates and concentrations enables the production of ideal regular fibers and the realization of directional water collection. Consequently, electrostatic spinning offers an efficient and low-cost method for the large-scale preparation of water-collecting fibers. The viscosity of the spinning solution not only exerts a great influence on the morphology of the fibers but also affects the size of the spindle nodes, which in turn impacts the efficiency of fog collection. Inspired by nature and traditional papermaking knowledge, Si et al. creatively prepared a multifunctional SHB paper with uniform inside and outside by utilizing waste paper for secondary use, which is not only highly water repellent to various liquids, but also has excellent self-cleaning, stain-resistant and oil-absorbent capabilities. Notably, the SHB recycled paper retains excellent writability, flexibility, foldability, and cutability. These properties endow it with the capacity to satisfy the requirements of daily use. The unique uniform superhydrophobicity inside and out allows this paper to withstand any degree of wear and tear (Figure 5.8b–e). Upon the contact between liquid water and the surface, the liquid water instantaneously diffuses and wets the paper. Nevertheless, as depicted in Figure 5.8g, the superhydrophobic silica-grafted paper derived from superhydrophobic recycled paper (srp) exhibits remarkable superhydrophobicity. Its CA reaches as high as 158°. A 5-μL water droplet can maintain a near-spherical shape while standing on the paper surface and remain in that state for an extended duration. In addition, when the SRP is placed on the tabletop at an angle, even if it is less than 5°, a drop of water (about 12 mL) from a micro-syringe can be rolled down easily and quickly, as shown in Figure 5.8f.

Meanwhile, water collection experiments on the prepared spindle junction fibers revealed that the larger spindle junctions have higher water collection efficiency than the smaller ones, which is of great significance for the practical application of water collection. Prior to the utilization of bionic spider silk in a practical production context, the issue of large-scale production must be addressed. By optimizing the viscosity of the solution, the surface tension of the solution, and the pull-out speed, it becomes feasible to guarantee the acquisition of regular and flawless spindle-knot fibers. The dip-coating method necessitates that the fiber be immersed in water and subsequently withdrawn. This would be limited by discontinuities and would not be possible for mass production [48]. Therefore, large-scale preparation is carried out using fluid coating method and electrodynamic technique. The fluid coating method has a high propensity to generate diverse biomimetic fibers, such as glass fibers and copper wires. In contrast, fiber-bead heterostructures are fabricated through the electrodynamic technique. Lee et al. [51] developed microfluidic technique to fabricate fibers with complex morphology and tunable composition, which can be useful for water harvesting, drug delivery, etc. The fabricated biomimetic functional fibers featuring unique wettability can be applied to regulate the directional transport of droplets and boost the efficiency of water collection. Moreover, they can enhance solid–liquid adhesion. Fibers with periodic spindle-shaped junctions exhibit stronger adhesion to droplets compared to smooth fibers. This phenomenon is attributed to the three-phase contact surface induced by the spindle-shaped structure on the fiber surface [52].

Stranded-fiber heterostructures have garnered significant interest in the manufacturing domain as the core components of advanced micro/nano devices. In addition to directed water harvesting and catalysts [57], a number of methods have been developed for the preparation of micro- and nano-heterostructures, such as microfluidic and aerodynamic [57] or electro-mechanical spin devices, as well as wet reconstruction inspired by spider silk. Zhao et al. fabricated biomimetic hetero-structured string-bead fibers (BHBFs) by employing hydrolyzed nanoparticles of polymethylmethacrylate (PMMA) and titanium tetrachloride (TiCl4). And they were prepared using a wet assembly system, including PMMA electrostatic spinning, nanoparticle atomization and multistage aqueous agglomeration. As a result, the relevant parameters could be controlled to efficiently prepare BHBFs with excellent heterogeneous structures Figure 5.9a). In the course of the experiments, the wet assembly of BHBFs was associated with the humidity-driven aggregation of particles within the droplets. Once PMMA fibers containing NPCTi are positioned in a humid environment, water condenses on the hydrophilic NPCTi, giving rise to small droplets. Subsequently, these small droplets grow and coalesce with one another to form larger droplets. At this stage, the nanoparticles within the droplets are arranged periodically along the PMMA fibers. Subsequently, the water evaporates, leading to the formation of beads (Figure 5.9b). Then $TiCl_4$ was hydrolyzed and NPCTi was secondarily trapped and secondarily assembled to form dendritic structures on the beads. In the experimental procedures, droplets gradually aggregated on the fibers, mainly due to the recapture and interception of nanoparticles caused by surface roughness. The surface energy and curvature of the bead-shaped and string-shaped structures differed significantly. This led NPCTi to directly aggregate around the beads. After the multi-stage wet-assembly process, larger beads were formed (Figure 5.9c). Furthermore, Yue and colleagues trapped

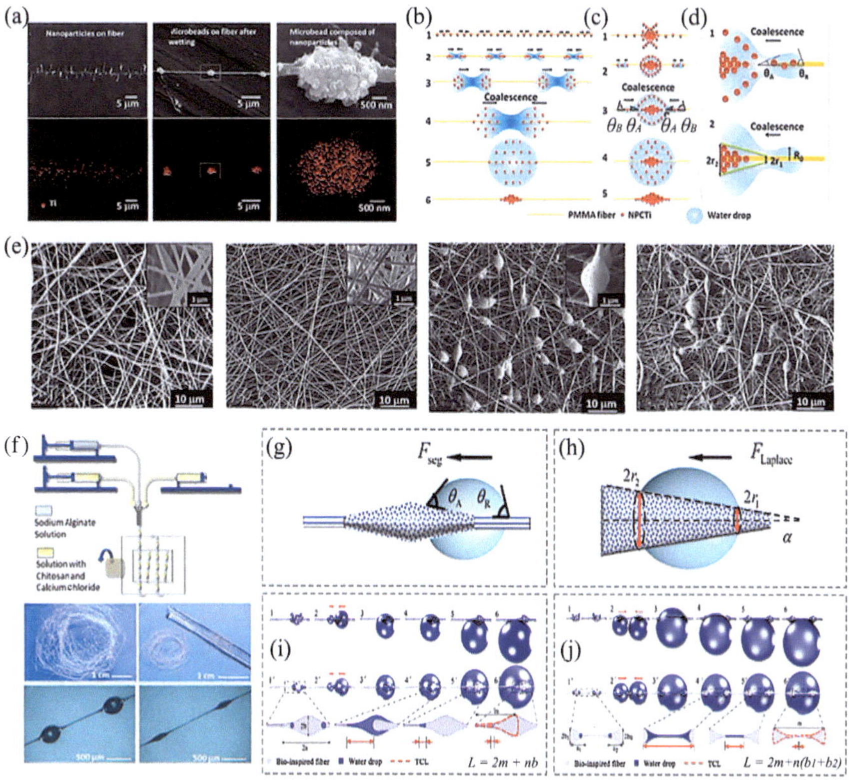

Figure 5.9 (a) Scanning Electron Microscopy (SEM) images and Energy-Dispersive Spectroscopy (EDS) mapping analysis of the PMMA-Ti fiber both prior to and subsequent to wetting. (b) Formation of the "bead-on-string" structure via a single-stage process. (c) The control of the "bead" size in a multi-stage process. (d) The mechanism on formation of larger beads. Source: (a–d) Reproduced with permission from [53] / The Royal Society of Chemistry. (e) Scanning electron microscopy (SEM) micrographs that are representative of pure polyvinylidene fluoride (PVDF) fibers, pure poly(*N*-isopropylacrylamide) (PNIPAM) fibers, and PNIPAM-PVDF bead-on-string fibers, as well as the SEM micrographs of PNIPAM-PVDF bead-on-string fibers subsequent to the water condensation test, are presented. Source: Reproduced with permission from [54] / John Wiley & Sons. (f) The fabrication of bioinspired microfibers through the microfluidic approach. Presented here are the optical images of a large quantity of collected dehydrated microfibers. Additionally, microscopic images of the microfiber prior to and after dehydration are provided. Source: Reproduced with permission from [55] / John Wiley & Sons. (g) The mechanism of directional droplet transport on a spider web. (h) Surface energy gradient. Source: (g,h) Reproduced with permission. [56] Copyright 2021, American Chemical Society. (i) TCL of a droplet suspended over an axial node. (j) TCL of a droplet suspended between two spindle nodes. Source: (g–j) Reproduced from [56] / with permission of American Chemical Society.

nanoparticles during the process (Figure 5.9d1). The NPCTi concentration was higher on the "beads" than on the "strings", endowing the "beads" with greater hydrophilicity. This generated a surface-energy gradient within the fibers, propelling the droplets to migrate from the strings to the beads. Secondly, the beads are akin to two conical components (Figure 5.9d2) that curve in opposing directions and are joined together. This arrangement generates a Laplace pressure gradient that drives droplets to shift from the "string-like" part to the "bead" region. The combined effect of the Laplace pressure gradient and the surface gradient formed by the chemical composition plays a crucial role in promoting the growth of droplets. This not only explains the principle of water evaporation-induced self-assembly but also makes it possible to precisely fabricate BHBFs at the micro- and nano-scales. As a consequence, it offers support for the design and advancement of intelligent hetero-structured materials.

In addition to other functions, beaded fibers can enhance the efficiency of water fog collection. Thakur et al. [54] utilized the electrostatic spinning method to manufacture beaded poly(*n*-isopropylacrylamide) (PNIPAM)/poly(vinylidene fluoride) (PVDF) layered fibers. The structural characteristics of these fibers bear resemblance to those of natural spider silk. Fog collection experiments were also conducted with the original PNIPAM and PVDF fibers to investigate the effect of the wettability of the beaded fibers on the water collection efficiency. PVDF fibers are characterized by stable mechanical properties, remarkable hydrophobicity, and excellent chemical resistance [57]. Given that PNIPAM is hydrophilic, it can combine with PVDF fibers to establish hydrophilic-hydrophobic regions. The surface energy gradient that comes into being as a result facilitates the directional movement of droplets. By contrast, the surfaces of pure PNIPAM fibers and PVDF fibers are sleek. Core-shell electrostatic spinning was employed to fabricate fiber-on-bead fibers, during which the flow rates of the two solutions were optimized, and a consistent spider spindle structure was achieved (Figure 5.9e). It was demonstrated by the water collection experiments that the layered hydrophilic and hydrophobic fiber membranes could capture more water at room and elevated temperatures. The reason why the water collection performance of PNIPAM-PVDF fibers has been enhanced is the wetting gradient formed due to the variation in surface composition along the fiber length. The prepared beaded fibers collected water better than the individual pristine fibers and most of the studies, suggesting that the samples have superior fog collection capability. Huang et al. integrated the characteristics of spider silk and desert beetles and utilized electrostatic spinning to produce beaded nanofibers with both hydrophobic and hydrophilic natures. Two-component nanofibers made up of polyacrylonitrile (PAN) and expanded graphite (EG) were developed. PAN and EG functioned as the column and header components respectively. The differential wettability of these two materials demonstrated the successful fabrication of the desert-beetle-like structure. Scanning electron microscopy (SEM) and WCA measurements confirmed the existence of a beaded structure. In water-harvesting experiments, the water-harvesting efficiency of the prepared beaded fibers was 91% greater than that of the raw materials, and their enhanced mechanical properties met the basic needs for water-harvesting.

As shown in Figure 5.9f, the droplets enlarge and approach the spindle junction. Subsequently, they start to merge and ultimately form larger droplets. The process of water collection commences with the wet reconstruction of the spider silk [58]. Finally, the droplets are suspended on the mandrel node. In this process, the spindle node can serve as a condensation point and droplet collection point. Water undergoes condensation at the junction, giving rise to a trajectory that extends from the junction to the spindle node. The condensation area and the main collection unit function in unique manners. They work in conjunction with the joint and the spindle junction to enable the directional collection of water. The subsequent process gets underway when the droplets combine and are concentrated specifically in the spindle junction and the joint. By regulating the flow rate and component content of the fluid, the fibers can reach their optimal water collection efficiency. Furthermore, an electrostatic attraction occurs between the sodium alginate solution and chitosan, two polyelectrolytes. This electrostatic force results in the establishment of a stable connection between the spindle junction and the main fibers. Even in acidic and alkaline environments, the material maintains robust mechanical properties and excellent corrosion resistance. Notably, one of the greatest advantages is the ability to control the morphology of the spindle junction by altering the concentration of the calcium chloride complex in the fluid. This cost-effective and adaptable preparation method has promoted the development of HRSF preparation technology. As depicted in Figure 5.9g, surfaces with high roughness possess distinct surface energies compared to those with low roughness. Thus, augmenting the surface roughness gives rise to a surface energy gradient. The surface free energy of the spindle junction exceeds that of the bond. An advantage lies in the fact that nanofibers expand the surface area of the hump-shaped fiber. When BNF fibers are in a wet condition, the existence of anisotropic capillary channels among the nanofibers facilitates the internal transport of droplets, significantly enhancing the collection efficiency. Owing to the wetting effect resulting from the capillary force in the nanofiber-hump channels and nanofiber-joint channels, along with the internal transfer effect of the Laplace pressure difference, BNF effectively addresses the issue of rapid small-droplet transport. The internal transport capacity of microdroplets demonstrated by BNF further propels the development of water collection technology.

Considering the specific properties of spider silk, wet reconstructed spider silk exhibits characteristic shape memory effects triggered in humid environments [59]. In addition, spider silk has a high strength-to-mass ratio with a diameter of only a few micrometers. Venkatesan et al. [60] dedicated themselves to the research and development of artificial spider silk with the properties of spider captured silk (SCS) . Before their work, Tian et al. [52] were the first to report the preparation of PMMA-coated nylon fibers featuring the SCS structure. Subsequently, numerous researchers modified the polymer coating. They adjusted the nanostructures, surface energy, and the size of spindle junctions of the fibers to regulate their water-harvesting efficiency [61]. Moreover, most of these materials are hydrophobic, indicating that nylon fibers have poor water-harvesting properties. Nylon has a similar chemical structure to silk in terms of its main chain, but its properties are quite different due to the diversity of its composition. Degummed silk is characterized by a high modulus, excellent hygroscopicity, and remarkable thermal stability. On the contrary, nylon has low moisture absorption capacity, weak thermal stability, is non-biodegradable, and poses harm to the environment. Besides, silk lacks the periodic spindle junctions similar to those of SCS, and no directional water collection phenomenon has been detected. eMaSp2 demonstrates a high level of hydrophilicity and sensitivity to water molecules. All-silk protein fiber has formed structural and chemical resemblances to natural SCS. This novel fiber is 252 times lighter compared to polymer-coated nylon fibers. It enhances the fiber's droplet coagulation capacity and water collection efficiency. Moreover, the material is environmentally friendly and causes no pollution to the environment.

Advances in bionic spider silk have facilitated advancements in the field of water harvesting technology. Having carried out an inspection of the structure of spindle junctions and spindle fibers within spider silk, biomimetic spider silks were created by impregnation, microfluidic and electrostatic spinning methods. Through the optimization of the manufacturing process and the design of the spindle junction and spindle fiber structures, specifically, by adjusting the curvature, chemical composition, and roughness gradient on the fiber surfaces, numerous biomimetic spider silk-like structures were produced. These structures demonstrated excellent water-harvesting performance (as shown in Figure 5.9i and Figure 5.9j). In addition, it is important to highlight the distinctive memory effect of spider silk, as it plays a significant role in advancing our understanding and study of this material. To sum up, the research mentioned above offers fresh insights into integrating bionic spider silk with other bionic structures. This integration can yield multifunctional structures, thus promoting the progress of water harvesting technologies.

5.4 Fog Collection Material Inspired by Lotus Leaves

It has been well-known for ages that some natural surfaces possess unique wettability features. The lotus leaf is a highly renowned instance, with remarkable hydrophobicity. Zhou Dunyi, a Song-Dynasty poet, described the lotus leaf as remaining pure despite growing in silt [62].

This is due to the very complex micro and nanostructures on the surface of the lotus leaf. Many tiny papillae can be clearly seen on the surface under a super-resolution microscope. The average dimension of these papillae is around 10 μm, and the average distance between them is about 12 μm. Each papilla is equipped with numerous minuscule protrusions with a diameter of around 100 μm. The surface of the leaf is overlaid with a layer of biomass wax with low surface energy. The unique microscale and nanoscale structures, together with the low-surface-energy substance covering them, enable the upper surface of the lotus leaf to display excellent superhydrophobicity [63]. By emulating the micro-nanostructure on the upper surface of the lotus leaf and applying low-surface-energy modifiers, numerous materials with outstanding hydrophobic properties have been developed. [64] As depicted in Figure 5.10a,b, Li et al. designed

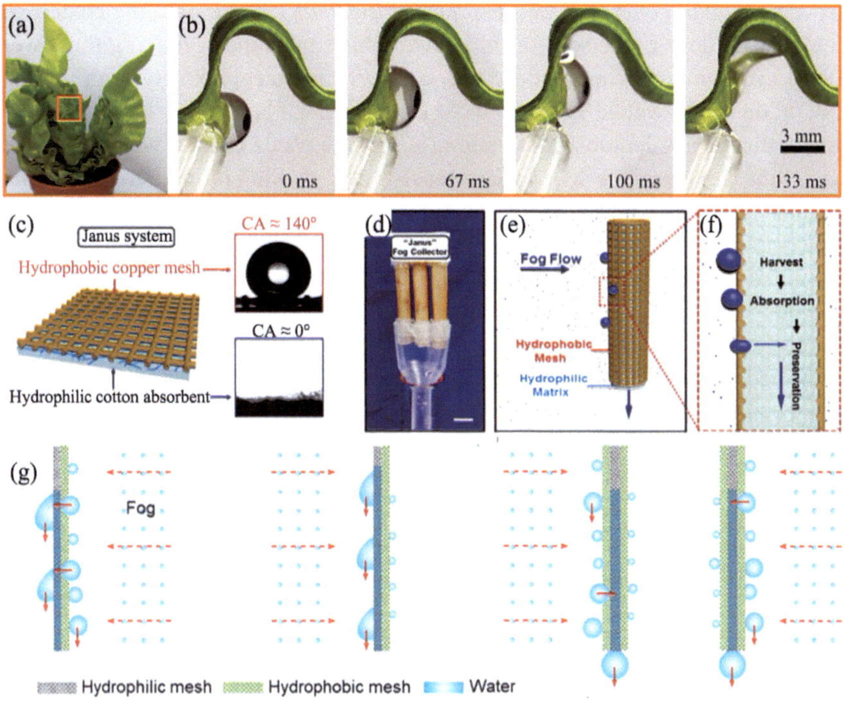

Figure 5.10 (a,b) The design of the kirigami structure for effective fog collection was inspired by the wave-shaped leaf of the Bird's Nest Fern. The top-view image presenting the spontaneous movement of a droplet on the wave-shaped leaf of the Bird's Nest Fern is provided. Source: Reproduced with permission from [65] / Elsevier. (c) Illustration of a cooperative hydrophobic/hydrophilic Janus water-collection system. A hydrophobic net, which has a CA of 140°, is closely combined with a hydrophilic cotton absorbent. (d) The continuous Janus fog collector is composed of a modified copper mesh cylinder in combination with hydrophilic absorbent cotton. (e,f) The continuous fog-collection process of the Janus collector encompasses three key steps: efficient collection, directional absorption, and rapid storage. Source: (c–f) Reproduced with permission from [66] / John Wiley & Sons. (g) Diagrams are presented to illustrate the structures of the Janus fog collector and the layer-structured fog collector. Source: Reproduced from [67] / with permission of Elsevier.

a 3D kirigami structure inspired by the innovative corrugation-driven structure in bird's nest ferns. It can be fabricated via simple shearing and stretching techniques. The water collection ratio of this structure increases as the external load grows, presenting a unique load-responsive feature. The 3D array configurations of the kirigami structures contribute to higher flow rates and enhanced droplet condensation efficiency. Merely by altering the geometric shape, the water collection ratio can reach up to 3.45 times that of the pure kirigami structure. When a wetting gradient is introduced, this ratio can even increase up to 4.58 times. Just like the upper surface of the lotus leaf, the lower surface also has such micro-and nanostructures made up of many flat, faintly raised papillae [42, 68], However, low-surface-energy wax crystals do not cover the lower surface of the lotus leaf. Besides, the epidermal glands on the lower side secrete some hydrophilic substances, making the lower surface hydrophilic. (When underwater, it shows superoleophobicity, with the CA of oil on the water-wetted lower surface of the lotus leaf being over 150°). A super-hydrophilic surface demonstrates a strong affinity for water, causing water to adhere firmly to it. Conversely, a SHB surface repels water, making it prone to float at the air-water interface. Consequently, the distinct wettability of the upper and lower surfaces of lotus leaves enables them to attach stably to the air-water interface and keep afloat even in inclement weather. Inspired by the lotus leaf, numerous Janus materials have been engineered through a superhydrophobic/superhydrophilic dual-cooperation strategy. Similar to the lotus leaf's structure, Janus materials are constituted by a SHB upper surface and a superhydrophilic lower surface, which allows them to maintain stability at the air/water interface. Janus materials not only possess excellent anti-overturning capabilities like lotus leaves but also have an edge over conventional materials in mass transfer. This advantage is primarily ascribed to their surface wettability, layer thickness, and asymmetric pore structure. By modulating the distribution of surface wettability, layer thickness, and pore size within the thickness dimension of Janus materials, a synergistic effect can be generated. This effect can result in diverse mass-transfer behaviors, meeting the requirements of specific application scenarios. Thus, Janus materials hold great promise in the field of fog collection. [69] For instance, by modifying the thickness and pore size on both sides of the Janus material with asymmetric wettability, the driving force that starts and sustains the entire mass transfer process can be acquired and regulated. The driving force can also be provided by the synergistic effect of capillary forces between layers with opposite wettabilities. In a unidirectional mass transfer process, the liquid first passes through the channels of the hydrophobic layer and then through those of the hydrophilic layer. The liquid is affected by the combined action of the capillary resistance of the hydrophobic layer and the capillary driving force of the hydrophilic layer.

Evidently, when the pore size on both sides of the Janus material is identical and the wettability remains unchanged, a thinner hydrophobic layer promotes the contact between the liquid and the hydrophilic layer. The capillary force provided by the hydrophilic layer further promotes the transfer of the liquid from the membrane, thus enabling the droplet transfer process to occur efficiently. Likewise, when the wettability and thickness configuration of the Janus material on both sides remain constant, an increase in the pore size of the hydrophobic layer reduces the capillary resistance of the fluid during transmembrane passage. At the same time, reducing the pore size in the hydrophilic layer can enhance the capillary attraction of the hydrophilic layer [70]. Cao and his colleagues developed a hydrophobic/hydrophilic integrated Janus fog-collection system (as shown in Figure 5.10c–e) by combining a dodecanethiol-modified copper mesh (with a pore size of 250 μm and a WCA greater than 140°) with commercial skimmed cotton [66]. In order to assess the fog-collection ability of this system, three different samples were prepared: the Janus fog-collection system, a hydrophobic copper mesh, and hydrophilic skimmed cotton.

In nature, fog doesn't always flow in a single direction. Cao et al. [66] carried out further optimization on the fog collection system to make it capable of performing the fog collection function under various fog-flow directions. The modified copper mesh was shaped into a hollow cylinder, and hydrophilic degreased cotton was closely arranged inside the cylinder. This design led to the fabrication of a cylindrical Janus fog collector (Figure 5.10f). During the movement of fog, this collector can simultaneously achieve efficient fog collection, directional transportation, and rapid preservation, thus implementing a continuous fog-collection process. Tiny droplets in the fog are caught by the outer copper mesh, where they then aggregate and grow. Once the collected droplets become larger and make contact with the inner hydrophilic skimmed cotton, they are captured and transferred within approximately 10 ms. Subsequently, the water within the hydrophilic cotton can move downward due to gravity for self-transportation (Figure 5.10g). Moreover, the Janus membrane has been optimized with a hydrophobic structure on both sides and a hydrophilic structure in the middle, allowing for fog collection regardless of the wind direction [71].

In foggy conditions, airborne water vapor typically condenses on the hydrophobic side of Janus materials. Without the need for any external driving force, the condensed water can transfer spontaneously from the hydrophobic side to the hydrophilic side. Although hydrophilic materials enable water to condense rapidly on their surfaces, the water has difficulty detaching from these surfaces, which inhibits further water condensation. In contrast, Janus materials possess a distinct edge in continuous condensation and fog collection processes. Beyond the applications already discussed, Janus materials demonstrate potential in numerous other fields. For example, they can be used in membrane distillation, sensors [72], oil/water separation [73], applications involving the unidirectional transport of air bubbles, and even in hemostasis applications, all thanks to the unique design featuring two surfaces with different wettabilities [73].

5.5 Microstructure of Other Biomimetic Fog-Water Harvesting Materials

The leaves of the dragon fruit plant exhibit excellent fog collection behavior through its 1D spines with a linear microstructure. Its spines, although not able to utilize their special structure and chemical gradient to provide sufficient power for fog droplet transport as cactus, exhibit efficient water supply, which greatly improves fog collection of droplets (Figure 5.11a–c) [74]. This structure is capable of remarkably lessening the deviation of the fog flow. Moreover, due to the innate hydrophilic property of the surface, the fog can amass in the form of droplets. As a result, its efficiency in water supply is 100 times higher than that of a 2D plane, effectively resolving the issue of fog water collection. Guo's team [74] was inspired by this to further develop a 3D fog catcher consisting of 1D microfine copper wires. To mimics the structure of dragon fruits, the copper wires distribute in regular rows for fog collection, while the substrate material serve as a path to discharge

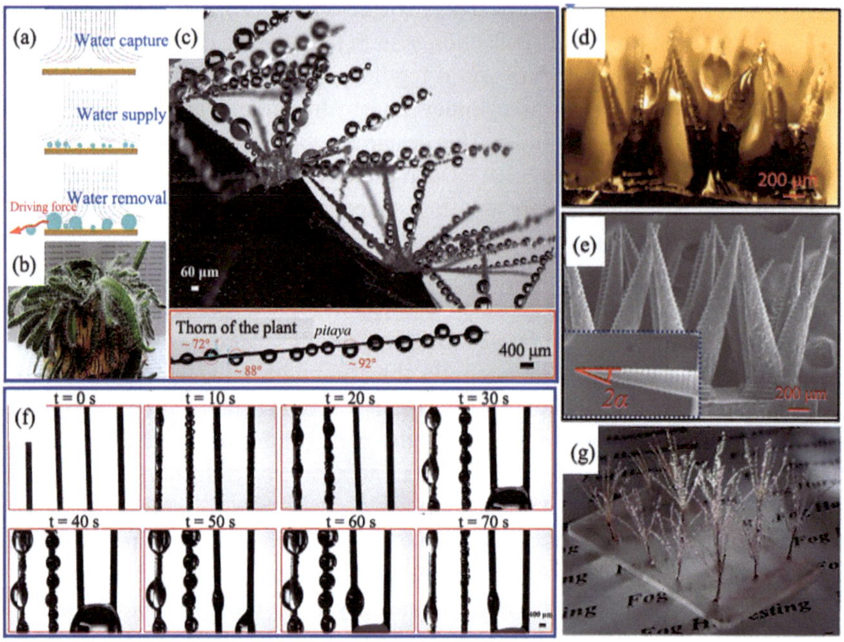

Figure 5.11 (a) The process of fog harvesting consists of three key steps: capturing water droplets from the fog, supplying the collected water for utilization, and removing any remaining or excess water. (b) An optical photograph of the pitaya is presented. (c) In the case of a pitaya covered with tiny thorns, it demonstrated outstanding fog collection performance. Within 20 s, with a fog speed of approximately 60 cm s^{-1}, it effectively harvested fog. The wettability of the thorn is indicated within the red box. Source: (a–c) Reproduced with permission [74] / American Chemical Society. (d,e) Water catchment structures resembling pinpricks, which are fabricated via 3D printing technology. Source: Reproduced with permission from [75] / John Wiley & Sons. (f) The time-lapse images of the water harvesting process on the original hydrophilic copper wire, SHB copper wire, superhydrophobic–superhydrophilic (SHB–SHL) copper wire, and superhydrophilic (SHL) copper wire are presented from left to right respectively. The scale bar represents 400 μm. (g) During the fog harvesting process, there is a fog collector that consists of thorns (using original copper wires) and a supporter made of PDMS. Source: (f,g) Reproduced with permission from [74] / American Chemical Society.

the collected water (Figure 5.11d,e). Porcupine has attracted much attention due to its unique structural features and surface properties. Chen's research group [76] observed the microstructure of the surface of the cage mouth of porcupine with the aid of SEM. It is concluded that the multi-scale structure of the cage mouth enhances the capillary rise in the direction of liquid transport and prevents reflux through the wedge-shaped angular structure. The surface micromorphology of the interior of the corkscrew and the hygroscopicity of the secreted nectar caused the surface to display superhydrophilicity and to form a stable water film in humid environments. This continuous, directional water transport phenomenon and water film dampening phenomenon inspired the design and preparation of various liquid infused surfaces.

Guo's and coworkers [4] designed liquid infused surfaces that are resistant to boiling water and self-repairable. By utilizing the intermolecular forces between PDMS molecular brushes and silicone oil, the liquid infused surfaces have stable hydrophobicity and sliding ability even when they are placed outdoors or subjected to cold/hot cycling tests. These liquid infused surfaces are characterized by strong droplet condensation and heat transfer, and the bottom of the nuclear energy base, and exhibit highly efficient fog collection when fog collection is performed. In two-dimensional bionic surfaces, surface chemistry is critical for water trapping. Compared with hydrophobic surfaces, hydrophilic surfaces have an energetic advantage in capturing water. These bionic surfaces, modeled after the back of desert beetles, were engineered to possess alternating surface wettability. Their effectiveness in capturing fog water has been verified. When a new SHL copper wire is exposed to fog, it is instantaneously covered with a water film and demonstrates a strong water-trapping capacity, as shown in Figure 5.11f,g. Also based on the principle that PDMS can be grafted onto metal oxide photocatalysts under UV illumination, Guo's and coworkers [45] further prepared tungsten trioxide (WO_3) nanoparticles, which were sprayed onto the surface of a copper substrate using an inorganic binder (aluminum phosphate). Acid-resistant, shear-resistant, and tarnish-resistant slip surfaces were obtained by infusing silicone oil after UV grafting PDMS molecular brushes. The fog collection efficiency was about 0.4605 g $(\text{h-cm}^2)^{-1}$ at a fog volume of 0.07 g s^{-1} and humidity of 85%. More importantly, the oil content in the collected water is extremely low, which saves trouble for further purification. In addition, a large number of bionic fog collection materials have been developed, which is of far-reaching significance for solving the problem of freshwater shortage.

In summary, fog collection technology presents a viable solution to global freshwater shortages, drawing inspiration from the microstructures of various organisms that have evolved over time. For instance, the elytra of desert beetles facilitate efficient water droplet capture and directional transport through the synergy of hydrophilic bumps and hydrophobic regions, with biomimetic designs leveraging 3D printing and nanocoating to optimize hydrophilic density and pattern geometry. Cactus spines utilize Laplace pressure differentials and surface roughness gradients to drive water transport, inspiring high-efficiency collection arrays fabricated via laser direct writing and MPAM. Similarly, the spindle-knotted structure of spider silk enables directional water movement through surface energy gradients and curvature differences, with ES technology allowing for large-scale production of analogous fibers to enhance water collection efficiency. The micro/nanoscale protrusions on lotus leaves have inspired Janus materials, enabling integrated "capture-transport-storage" systems. Additionally, bio-inspired designs based on dragon fruit spines and porcupine quills further expand the possibilities of fog-harvesting materials, as demonstrated by 1D copper wire arrays that mimic dragon fruit spines, achieving a hundredfold increase in water transport efficiency. Current research trends focus on multi-organism structural integration and multifunctional composites, such as beetle-cactus hybrid coatings and spider silk-beetle heterogeneous fibers, with advances in 3D printing and ES facilitating the controlled fabrication of complex microstructures. However, challenges remain in scaling up production and controlling costs. Future developments must balance efficiency, environmental adaptability, and sustainability to advance the practical application of biomimetic fog-harvesting materials, offering new strategies for sustainable water resource management in arid regions worldwide.

References

1 Wan, K., Gou, X., and Guo, Z. (2021). *Journal of Bionic Engineering* 18: 501.
2 (a) Yue, H., Zeng, Q., Huang, J. et al. (2022). *Advances in Colloid and Interface Science* 300. (b) Guo, J., Huang, W., Guo, Z., and Liu, W. (2022). *Nano Letters* 22: 3104.
3 Wang, X., Huang, J., and Guo, Z. (2022). *Advances in Colloid and Interface Science* 301.
4 Jing, X. and Guo, Z. (2019). *ACS Applied Materials & Interfaces* 11: 35949.
5 Wang, S., Feng, L., and Jiang, L. (2006). *Advanced Materials* 18: 767.
6 Parker, A.R. and Lawrence, C.R. (2001). *Nature* 414: 33.
7 Fu, Y., Wu, L., Ai, S. et al. (2023). *Nano Today* 52: 101979.
8 Guo, J., Guo, Z., and Liu, W. (2023). *Nanoscale* 15: 10567.
9 Bai, H., Wang, L., Ju, J. et al. (2014). *Advanced Materials* 26: 5025.
10 Pang, Y., Mei, G., Wang, F. et al. (2024). *Colloids and Surfaces A: Physicochemical and Engineering Aspects* 680: 132796.
11 Wang, X., Zeng, J., Yu, X. et al. (2019). *RSC Advances* 10: 282.
12 Hou, K., Li, X., Li, Q., and Chen, X. (2019). *Advanced Materials Interfaces* 7: 1901683.
13 Bai, H., Ju, J., Sun, R. et al. (2011). *Advanced Materials* 23: 3708.
14 Shahrokhian, A., Feng, J., and King, H. (2020). *Journal of the Royal Society Interface* 17: 20200038.
15 Ahmad, S., Tang, H., and Yao, H. (2019). *International Journal of Heat and Mass Transfer* 141: 1087.
16 Park, J.-H. and Kim, S. (2019). *MDPI Micromachines* 10: 201.
17 Peng, L., Chen, K., Chen, D. et al. (2021). *RSC Advances* 11: 14769.
18 Su, Y., Chen, L., Jiao, Y. et al. (2021). *ACS Applied Materials & Interfaces* 13: 26542.
19 Wang, S., Zhu, L., Yu, D. et al. (2023). *Advanced Functional Materials* 33: 2305244.
20 Lou, X., Huang, Y., Yang, X. et al. (2020). *Advanced Functional Materials* 30: 1901130.
21 Wang, G. and Guo, Z. (2019). *Nanoscale* 11: 22615.
22 Ogburn, R.M. and Edwards, E.J. (2009). *American Journal of Botany* 96: 391.
23 Yi, S., Wang, J., Chen, Z. et al. (2019). *Advanced Materials Technologies* 4.
24 Zhou, H., Jing, X., Li, S., and Guo, Z. (2021). *Chemical Engineering Journal* 417.
25 Dong, Z., Schumann, M.F., Hokkanen, M.J. et al. (2018). *Advanced Materials* 30: 7.

26 Xiao, Y., Huang, J., Guo, Z., and Liu, W. (2022). *Materials Today Physics* 28.

27 Sullivan, S.P., Murthy, N., and Prausnitz, M.R. (2008). *Advanced Materials* 20: 933.

28 Heng, X., Xiang, M., Lu, Z., and Luo, C. (2014). *ACS Applied Materials & Interfaces* 6: 8032.

29 Wang, J., Yi, S., Yang, Z. et al. (2020). *ACS Applied Materials & Interfaces* 12: 21080.

30 Zhou, H., Jing, X., and Guo, Z. (2020). *Langmuir* 36: 6801.

31 Peng, Z. and Guo, Z. (2022). *Biomaterials Science* 10: 5831.

32 Zhu, H., Guo, Z., and Liu, W. (2016). *Chemical Communications* 52: 3863.

33 Peng, Z., Fu, Y., and Guo, Z. (2023). *ACS Applied Materials & Interfaces* 15: 38110.

34 Song, D. and Bhushan, B. (2019). *Philosophical Transactions of the Royal Society A: Mathematical, Physical and Engineering Sciences* 377: 20180335.

35 Song, D. and Bhushan, B. (2019). *Philosophical Transactions of the Royal Society A: Mathematical, Physical and Engineering Sciences* 377: 20190128.

36 Wang, X., Zhang, W., Li, S. et al. (2023). *Chemical Engineering Journal* 472.

37 Wang, Q., Yang, F., and Guo, Z. (2023). *Chemical Communications* 59: 6048.

38 Tang, X. and Guo, Z. (2020). *New Journal of Chemistry* 44: 20495.

39 Tang, X., Huang, J., Guo, Z., and Liu, W. (2021). *Journal of Colloid and Interface Science* 604: 526.

40 (a) Chenxi, Y., Jian, W., Juan, L. et al. (2023). *ACS Sustainable Chemistry & Engineering* 11: 3147; (b) Compton, B.G. and Lewis, J.A. (2014). *Advanced Materials* 26: 5930.

41 Chen, W. and Guo, Z. (2019). *Nanoscale* 11: 15448.

42 Fu, Y., Ai, S., Guo, Z., and Liu, W. (2023). *Journal of Colloid and Interface Science* 649: 646.

43 Zheng, Y., Bai, H., Huang, Z. et al. (2010). *Nature* 463: 640.

44 Liang, X. and Guo, Z. (2020). *Chemical Engineering Journal* 395.

45 Fan, H. and Guo, Z. (2021). *Journal of Colloid and Interface Science* 591: 418.

46 Si, Y. and Guo, Z. (2016). *RSC Advances* 6: 30776.

47 Gou, X. and Guo, Z. (2021). *Chemical Engineering Journal* 411.

48 Bai, H., Tian, X., Zheng, Y. et al. (2010). *Advanced Materials* 22: 5521.

49 He, Z., Xie, H., Jamil, M.I. et al. (2022). *Advanced Materials Interfaces* 9.

50 Dong, H., Wang, N., Wang, L. et al. (2012). *ChemPhysChem* 13: 1153.

51 Lee, S., Kim, W., and Yong, K. (2011). *Advanced Materials* 23: 4398.

52 Tian, X., Chen, Y., Zheng, Y. et al. (2011). *Advanced Materials* 23: 5486.

53 Zhao, L., Song, C., Zhang, M., and Zheng, Y. (2014). *Chemical Communications* 50: 10651.

54 Thakur, N., Ranganath, A.S., Agarwal, K., and Baji, A. (2017). *Macromolecular Materials and Engineering* 302.

55 Liu, Y., Yang, N., Li, X. et al. (2020). *Small* 16: e1901819.

56 Shi, R., Tian, Y., and Wang, L. (2021). *ACS Nano* 15: 7907.

57 Tian, X., Bai, H., Zheng, Y., and Jiang, L. (2011). *Advanced Functional Materials* 21: 1398.

58 Chen, S., Yang, F., and Guo, Z. (2022). *Advances in Colloid and Interface Science* 309: 102779.

59 Venkatesan, H., Chen, J., Liu, H. et al. (2019). *Materials Chemistry Frontiers* 3: 2472.

60 Venkatesan, H., Chen, J., Liu, H. et al. (2020). *Advanced Functional Materials* 30.

61 (a) Wang, S., Feng, S., Hou, Y., and Zheng, Y. (2015). *Macromolecular Rapid Communications* 36: 459. (b) Algarni, S. *Water and Environment Journal* 2018 (32): 301.

62 G. Mei, Z. Guo, *Advanced Materials Interfaces* 2022, 9, 2102484.

63 Peng, H., Yang, H., Ma, X. et al. (2022). *Colloids and Surfaces A: Physicochemical and Engineering Aspects* 643.

64 Xu, S., Wang, Q., and Wang, N. (2020). *Advanced Engineering Materials* 23: 2001083.

65 Li, S., Zhu, J., Liu, C. et al. (2023). *Chemical Engineering Journal* 464: 142549.

66 Cao, M., Xiao, J., Yu, C. et al. (2015). *Small* 11: 4379.

67 Li, J., Li, W., Han, X., and Wang, L. (2021). *Journal of Colloid and Interface Science* 581: 545.

68 Feng, R., Song, F., Xu, C. et al. (2021). *Chemical Engineering Journal* 422: 130119.

69 Zhou, H. and Guo, Z. (2019). *Journal of Materials Chemistry A* 7: 12921.

70 Tian, X., Jin, H., Sainio, J. et al. (2014). *Advanced Functional Materials* 24: 6023.

71 Wang, X., Guo, Z., and Liu, W. (2023). *Advanced Materials Interfaces* 10: 2300115.

72 Tian, P., Gao, X., Wen, G. et al. (2018). *Journal of Colloid and Interface Science* 532: 517.

73 Long, Z., Zhao, Y., Zhang, C. et al. (2020). *Advanced Materials* 32: e1908099.

74 Zhong, L., Zhang, R., Li, J. et al. (2018). *Langmuir* 34: 15259.

75 Li, X., Yang, Y., Liu, L. et al. (2020). *Advanced Materials Interfaces* 7: 1901752.

76 Chen, H., Zhang, P., Zhang, L. et al. (2016). *Nature* 532: 85.

6

Surface Wettability of Fog-Harvesting Materials

Except the nano/micro-surface structures that affect the fog harvesting efficiency as discussed in Chapter 6, other impact factors such as the surface wettability also plays a vital role in assisting the fog collection. Besides, fog collection is currently under rapid development, and how to control the condensation, growth, and shedding of water droplets is the key to efficient water collection [1]. Individual superhydrophilic surfaces are extremely easy to capture water fog due to their high affinity for water, but small water droplets are difficult to aggregate on the surface and instead directly wet the skeleton to form a water film [2]. In contrast, superhydrophobic surfaces are capable of rapidly dropping water droplets during fog collection, but small fog beads tend to bounce back into the environment after collision, thus reducing collection efficiency [3]. In addition, researchers have combined the advantages of the special structure of the desert beetle by designing a surface with a mixed positive and negative pattern, where water droplets are captured and grown in the hydrophilic region, and then dislodged from the hydrophobic region for collection [4]. However, the water collected in the hydrophilic region needs to reach a certain mass before it can flow down by gravity, which reduces the fog collection efficiency. As the collection process advances and the superhydrophobic structure is placed in a high humidity environment for a long period of time, small water droplets gradually fill up the gaps between the rough structures on the superhydrophobic surface, destroying the superhydrophobic properties of the structure, thus forming a water film and preventing further fog collection. The frame has poor fog collection capability.

Common fog and water collection materials are usually designed as grids or planes placed perpendicular to the direction of fog flow. During fog collection, the water fog first collides with the water collection material, is captured by the microstructures and nanostructures on the surface of the water collection, and grows condensed on the surface of the material. When the weight of the droplets is greater than the adhesion of the fog collection material to the droplets, the droplets will slide off the surface. The entire fog collection process can be divided into two main parts: droplet capture and droplet transportation. Therefore, improving the droplet capture efficiency and accelerating the droplet transportation become the key elements for the preparation of efficient fog collection materials. Currently, the research on water-collecting surfaces mainly focuses on superhydrophobic surfaces, oil-filled surfaces, and quasi-liquid surfaces (QLSs) with special wettability. The fog collection

Bioinspired Materials for Fog Harvesting: Principles, Design, Preparation, and Applications, First Edition. Zhiguang Guo and Weimin Liu.

material will affect the fog collection when the droplet capture efficiency and fog flow in the material surface contact form, a good fog collection material, the fog flow in its surface contact surface large residence time is long, and the adhesion of condensed droplets is low.

6.1 SHB–SHL Fog Harvesting Materials

Desert beetle-inspired water harvesting materials have been shown to be effective in improving fog collection efficiency. However, the structure and distribution of surface wettability of such materials is often complex. Such materials must be fabricated through the use of rapid and high-resolution surface modification techniques. Using laser beam-based surface modification techniques, superhydrophobic two-dimensional surfaces can be directly and rapidly modified into high-resolution, large-area, and flexible hydrophilic patterns [5]. Wang et al. [5b] utilized laser direct writing to pattern tree-shaped laminar cones on superhydrophobic titanium dioxide films. The hydrophobicity of the surface was reduced after laser scanning, and the contact angle became 95.1°, effectively trapping fog. Due to the surface energy gradient at the boundary of the tree-shaped laminar cone pattern, tiny droplets on the superhydrophobic region are pushed into the tree-shaped region, resulting in faster formation of larger droplets. At the same time, since the cone pattern generates a Laplace pressure difference at each stage, this allows tiny droplets to push precisely and continuously toward the root of the tree pattern. Yu et al. [6] prepared superhydrophobic polyester composite fabrics by selectively depositing copper onto fabrics using a one-step copper mirror image reaction. The patterned fabric had a very high-water collection rate compared to other fabrics. In addition to the synergistic effect of superhydrophilic and superhydrophobic regions mentioned above, the curvature of the rough fabric increases the probability of droplet collision with the substrate. It reduces the kinetic energy of the droplets, making it easier to capture the tiny droplets carried in the air stream.

He et al. realized a composite hydrophilic and superhydrophobic surface by laser etching a unique pattern on the superhydrophobic surface [7]. The preparation process of SUSHB is shown in Figure 6.1a. First, copper sheets (OS) were sonicated in 0.1 M hydrochloric acid solution, anhydrous ethanol and deionized water for 5 min to obtain copper sheets with the surface oxide layer removed. Second, 1 g of (polydimethylsiloxane) PDMS (10 : 1 ratio of PDMS to curing agent) and 0.6 g of Aerosil R202 were added to 15 mL of hexane and stirred homogeneously to obtain the spray solution [8]. Third, the spraying solution was uniformly sprayed onto the copper sheet using a spray gun with the pressure maintained at 0.2 MPa and the spraying distance of 20 cm. The coated copper sheet was baked in an oven at 80 °C for 2 h to obtain the superhydrophobic copper sheet (SHB) . Then, the pattern was drawn on the laser marker and the etching pattern is shown in Figure 6.1b–f. After laser etching, we obtained superhydrophobic copper sheet (SSHP) with hydrophilic pattern. The SSHP was then plated in 0.5 M $CuSO_4$ plating solution as a cathode, with the cathode 5 cm away from the anode and the side with the hydrophilic pattern

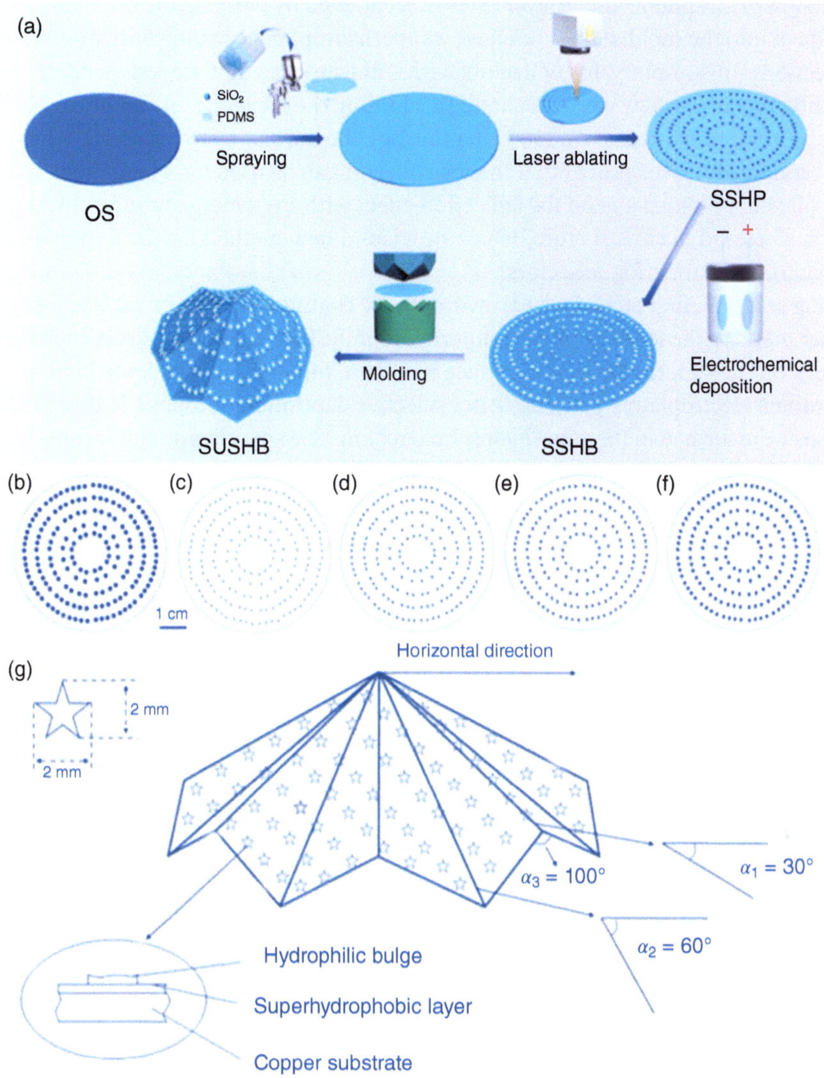

Figure 6.1 (a) Preparation diagram of fog collector. (b–f) are the CAD diagram of a circle, triangle, quadrangle, five-pointed star, and six-pointed star respectively. (g) Fog collector structure schematic. Source: Reproduced from [7] / with permission of Elsevier.

opposite to the anode. Superhydrophobic copper sheets with hydrophilic bumps (SSHB) were obtained by plating at a constant current pattern of 0.2 A and a plating time of 90 min. Finally, the mold was modeled using Solidworks software and fabricated using 3D printing. Then, the SSHB was placed in the center of the mold and pressurized to obtain superhydrophobic copper sheets (SUSHB) with corrugated structure and hydrophilic projections. Figure 6.1g schematically illustrates the structure of the SUSHB. The later mentioned SUSHP was obtained by placing the SSHP in a mold and extruding it; the SUSHP does not have a projecting structure.

The (superhydrophobic underlayer) SUS was obtained by pressing the original copper sheet into the mold; it does not have a superhydrophobic coating or hydrophilic projections. In summary, fog collectors with solid surfaces that are independent of the direction of fog flow were prepared. In addition, the contact angle of the fog collector surface remained above 150 ° after the fog collector was tested against UV light and sand impact. This proves that the fog collector can be used for a long time. The fog collector obtained a good fog collection effect with the synergistic effect of wetting surface and macrostructure, which provides a new method for the design and preparation of future fog collectors.

Pang et al. [9], first etched a superhydrophobic coating on a copper surface using a laser marking device to produce a superhydrophilic pattern. The superhydrophilic pattern on the superhydrophobic surface was then immersed as a cathode in a copper sulfate electroplating solution. After selective deposition of copper, hydrophilic humps were formed in the superhydrophilic region. The superhydrophobic samples containing hydrophilic hump arrays were found to be very effective in collecting water fog when compared to the prepared superhydrophilic, superhydrophobic and pristine copper sheets. The effect of different hump spacing on fog collection efficiency was investigated. The samples with the highest collection efficiency were identified and the reasons for the high collection efficiency were analyzed. In addition, multiple fog collection cycles were performed on the samples to verify the stability of the fog collection function. Excellent fog collection performance was achieved. A series of SHB-SHL copper surfaces with excellent fog collection efficiency were successfully prepared by means of facile etching, modification, laser scanning, and electroplating, etc. The SHB-SHL copper surfaces exhibit a typical characteristic of a large number of hydrophilic patterns arranged in an orderly manner in a nanorod structure and embedded in a superhydrophobic surface. The wettability distribution of the sample surface can be regulated by controlling the space between neighboring hydrophilic peaks. Compared with pristine copper surfaces, (superhydrophilic layer) SHL copper surfaces, SHB copper surfaces and (superhydrophilic pattern) SHP copper surfaces, SHB- SHL copper surfaces showed significantly higher fog collection efficiency due to their unique surface wettability and surface structure. Among the SHB-SHL samples, the L1.2 surface showed the highest recovery rate, which was 107% higher than the pristine copper surface. This enhancement mechanism can be attributed to the energy conversion of the droplets captured by the hydrophilic peaks, which makes them easily dislodged from the surface. The prepared surfaces have good fog collection properties and have promising applications in water harvesting and transportation materials [9].

Peng and Guo prepared hydrophilic-superhydrophobic three-dimensional copper sheets with efficient water-harvesting properties by incomplete combustion of PDMS [10]. Due to its excellent thermochemical stability and biocompatibility, as well as its simple and fast curing process, PDMS has become one of the most widely used materials in chemical production [9]. Using PDMS partially combusted by convective heat flow, surfaces with different wettability can be obtained by controlling the combustion time provides a basis for the preparation of superhydrophobic surfaces and alternating hydrophilic-superhydrophobic surfaces. Two kinds of three-dimensional hydrophilic-superhydrophobic hybridized copper sheets with different bumps, in which the bumps are hydrophilic and the flat areas are superhydrophobic, were prepared by spin-coating method and high-temperature calcination. In conclusion, a series of SHB-HL copper surfaces with excellent fog trapping efficiency were successfully prepared by simple etching, modification, laser scanning and electroplating methods. The SHB-HL copper surfaces exhibit the typical feature of a large number of hydrophilic patterns regularly arranged and embedded in a superhydrophobic surface with nanorod structure [11]. The wettability distribution of the sample surface can be adjusted by controlling the spacing between neighboring hydrophilic peaks [12]. Compared to pristine, SHL, SHB, and SHP copper surfaces, the SHB-HL copper surface shows significantly enhanced fog trapping efficiency due to its unique surface wettability and surface structure. The highest collection rate in the SHB-HL samples was 107% higher than that of the pristine copper surface. The enhancement mechanism can be attributed to the energy conversion of the droplets captured by the hydrophilic peaks, which makes them easily dislodged from the surface [13]. The excellent fog trapping properties of the prepared surfaces indicate the potential to realize a large number of applications in water harvesting and transportation materials. The results of this research provide an important reference for the preparation of three-dimensional water-harvesting structures with hydrophilic-hydrophobic hybrid surfaces.

Several species of desert beetles, cribripes Physasterna and Onymacris unguularis, are able to utilize their dorsal structures to collect water on foggy days [11]. The dorsal structures of the mudskipper all consist of hydrophilic projections and super sparse horizontal surfaces but there is a significant difference in the structure of the dorsal hydrophilic region The dorsal hydrophilic region of the mudskipper consists of hemispherical projections while the dorsal hydrophilic region of the mudskipper consists of linear projections [14]. Therefore, it is worth exploring which structure is more favorable for water collection. As shown in Figure 6.2, we mimicked the dorsal structure of nanobrachid beetles and prepared two copper

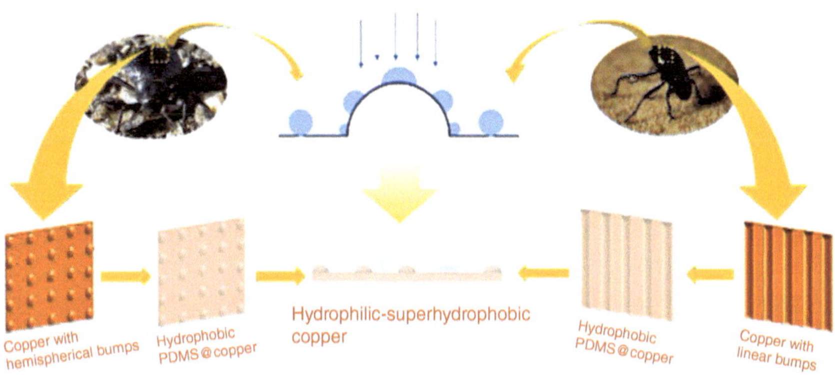

Figure 6.2 Schematic illustration of the preparation of hydrophilic–superhydrophobic copper sheets. Source: Reproduced with permission from [10] / The Royal Society of Chemistry.

sheets with different concave and convex structures. One is the original copper sheet with 5 × 5 hemispherical bumps (original – hb), and the other is the original copper plate with five linear bumps (original – lb). Incomplete combustion on the surface of PDMS oligomers can change the surface wettability and enhance the hydrophobic PDMS to superhydrophobic PDMS. with the increase of the calcination time, the surface of the PDMS appeared nano particles, increasing the surface roughness. With the increase of surface roughness, the hydrophobicity of the hydrophobic PDMS surface was enhanced and the contact angle gradually increased. However, when there are too many nanoparticles, the surface tends to be flat instead, which decreases the contact angle, thus leading to an increase and then a decrease in the contact angle. When the calcination time was 95 s, the water contact angle (WCA) reached a maximum value of about 153.0°. Therefore, two PDMS copper sheets with different concave and convex structures were calcined for 95 s to obtain hydrophilic-superhydrophobic samples (Hybrid-HB and Hybrid-LB). The WCA of the planar region of both Hybrid-HB and Hybrid-LB was above 150°, and the wettability of the bumps was different from that of the planar region. It is hypothesized that uneven surfaces lead to uneven heating of the surface, which results in different wettability of the raised surfaces from that of the flat surfaces.

In summary, hydrophilic-superhydrophobic hybridized copper sheets with two different concave-convex structures were prepared by spin-coating combined with high-temperature calcination. The method of obtaining superhydrophobicity is environmentally friendly [15]. In order to analyze the effect of different shapes of bulges on the water supply and water removal behaviors of fog collection, we compared and analyzed the water collection properties and mechanisms of superhydrophobic PDMS@copper and hydrophilic-superhydrophobic hybridized copper sheets with two different shapes of bulges. The experimental results show that the hydrophilic-superhydrophobic copper sheet with linear bulge has the highest water collection rate of $1.568\,g\,cm^{-2}\,h^{-1}$. Compared with the original hydrophilic copper sheet, its water collection efficiency is improved by about 151%, which proves that it has excellent fog collection performance. Fog collection mechanism analysis of the three surfaces also confirmed the advantages of the hydrophilic-superhydrophobic copper sheet with linear projections in water collection. However, since the hydrophilic-superhydrophobic hybridized surfaces are closely related to the bumps in the experiments, it is difficult to determine which factor, wettability or three-dimensional structure, plays a greater role in water collection [16]. This sample combines specific surface chefogry and surface morphology, which is promising for fog collection, and could provide inspiration for fog collection on surfaces with three-dimensional structures or surfaces with mixed wettability [17].

Inspired by the desert beetle, Guo et al. [4] report a wettability-integrated system consisting of a wettability hybrid coating and vein patterns, as in Figure 6.3a, which was successfully fabricated by a simple method. Enhanced water trapping, polymerization, and drainage were considered in our design, which allowed for the realization of optimal surface design for fog collection. Widely distributed hydrophilic points can easily capture fog droplets and trap them on the surface;

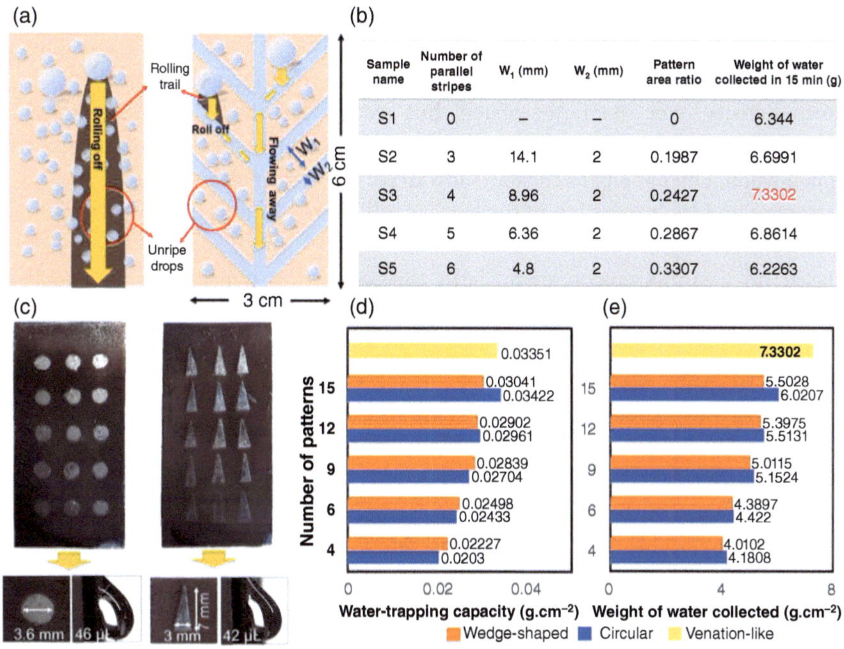

Sample name	Number of parallel stripes	W_1 (mm)	W_2 (mm)	Pattern area ratio	Weight of water collected in 15 min (g)
S1	0	–	–	0	6.344
S2	3	14.1	2	0.1987	6.6991
S3	4	8.96	2	0.2427	7.3302
S4	5	6.36	2	0.2867	6.8614
S5	6	4.8	2	0.3307	6.2263

Figure 6.3 (a) Comparison of drop motion trails on pattern less and patterned surfaces. (b) Weight of water collected in 15 min by each sample with increasing pattern density. (c) Optical photographs of circular and wedge-shaped patterned surfaces and the largest volume of a water drop trapped by a single pattern. Source: Reproduced with permission from [4] / American Chemical Society. (d) Water-trapping capacity of each sample with different pattern densities. (e) Weight of water collected in 15 min by different samples with different pattern densities.

the trapped water droplets form a three-dimensional topography that expands the contact area with the fog stream. At the same time, the vein-like pattern provides a fast and planned pathway method for water transport and removal. The results showed that the prepared wettable integrated surface could easily increase the water collection rate up to 225.6% compared to a single wettable surface [18].

In general, studies on superwettable patterned surfaces and their fog collection capabilities have shown that patterned surfaces can indeed improve water collection efficiency compared to single wettable surfaces, as shown in Figure 6.3b. For the mechanistic explanation, most believe that the hydrophilic zone is more favorable for droplet capture, while the superhydrophobic zone is favorable for droplet stripping [19]. In fact, the wettability contrast is the most crucial reason. Patterns with tight wettability contrasts create a wettability boundary and trap water in the pattern. The trapped water protrudes significantly from the surface, creating a three-dimensional topography with a large effective fog contact area. Circular and wedge-shaped surfaces were prepared using a similar approach, and these surfaces have been shown to be effective in improving water collection efficiency in other literature (Figure 6.3c). Experimental results showed that a single circular or wedge-shaped pattern could capture 46 or 42 μL of water when vertically displaced. The water collection capacity and water collection rate of each sample are shown in Figure 6.3d,e, respectively. The water collection rate increased with increasing pattern density over the range tested, but we believe there is a range in this increasing relationship. In addition, the size of the pattern affects the shape of the trapped water. Once the pattern size is too large, the deformation under gravity will be stronger and the projection of the surface will be reduced as the droplet volume increases [20].

Based on a similar mechanism, both the wettability hybrid coating and the wettability contrast coating by Guo and coworkers [21] significantly improved the fog collection capacity. However, the wettability-integrated system of this material has a higher water collection rate due to a well-designed drainage strategy. In summary, they designed a wettability-integrated fog collection system that consists of a wettability hybrid coating and a vein-like wettability contrast pattern. This wettability-integrated system showed efficient fog collection by constructing a three-dimensional surface topography that captured water droplets and drainage in a planned path [14]. The fog collection rate can be greatly increased compared to single wettability surfaces. The whole fabrication process is simple, low-cost, and environmentally friendly, which is a reliable solution for water shortage and industrial waste fog recycling in arid regions [22].

Li et al. [23] A deformable liquid repellent fabric surface inspired by lotus and hogweed was fabricated by a simple one-step method. The prepared PDMS@Fe_3O_4 fabric surface not only repels liquid droplets similar to lotus leaves, but also maintains an almost lubricant-infused smooth surface with a continuous film mimicking the edge of hogweed to a certain extent. In addition, the orientation of its micro-cilia can be changed by adjusting the direction of the magnetic field, suggesting that a reversible on-demand liquid rejection switching mode can be realized [24]. Utilizing the switching wetting state, the deformable surface can switch between the lotus state and the hogwash state, thereby maintaining the desired droplet mobility. In addition, deformable surfaces with the ability to dynamically change liquid repellency can be used to manipulate liquid repellency [25], anisotropic self-cleaning [26], unidirectional water manipulation [27], and programmable fog collection [27] and delivery [28].

6.2 SLIPS Fog Harvesting Materials

Inspired by hogwash [29], liquid-injected porous surfaces are capable of rapid continuous droplet nucleation growth. Some liquid-injected surfaces also exhibit excellent corrosion [30], self-healing [31] and anti-icing properties [32]. For liquid-injected surfaces, surface frame morphology, collection area, and lubricant properties (e.g. viscosity, chemical composition, etc.) play a key role in the fog collection process. The rough structure on the frame improves the contact area of the fog droplets with the surface, which also reduces the loss of lubricant and increases the service life. Currently, 3D frames injected with lubricant are used for fog collection due to their high specific area [33]. Micro/nanostructuring of the spine in a three-dimensional (3D) framework can further improve the fog collection performance [34]. Therefore, the preparation of liquid-infused surfaces using 3D frameworks is increasingly favored by researchers [35]. In addition, increasing the roughness of the structure can effectively increase the number of collisions of small droplets with the framework and provide good oil-locking ability through capillary action [36]. Exploring the reaction of long-chain alkanes with metals, different chain lengths produce different structures. Octadecanoic acid (stearic acid) reacts with copper to produce an irregular petal-like structure, whereas tetradecanoic acid leaches directly with copper in a daisy-shaped spherical micro- and nanostructure with good superhydrophobicity and a WCA of up to 160° [37].

Yue et al. [1] rapidly prepared stable superhydrophobic foam (SHBF) by in situ reaction and obtained three-dimensional micro-nano daisy-like clustered structure F superhydrophobic lubricating foam (SHBLF) covered by trimethylsilicone oil after oil injection. Due to the synergistic effect of the three-dimensional porous lattice system of the micro-nano-structure and the lubrication layer, the structure has excellent lubricant absorption and retention capabilities [38]. The presence of the lubrication layer greatly improves the ability of the structure to capture droplets during fog collection, accelerates droplet aggregation and facilitates droplet shedding. In addition, it exhibits excellent anti-fouling [32a, 39] and anti-icing [40] properties. The preparation method simplifies the preparation process and makes it easy to meet the reaction conditions. The prepared liquid-filled surfaces have good fog collection performance and stability, and the fog collection efficiency is improved by 98.7% compared with the original samples, which provides a new idea for the large-scale preparation of liquid-filled fog collection structures, and provides new ideas and insights for drought mitigation.

The collection process of the same fog structure is divided into three steps: droplet capture, droplet condensation and droplet removal. Therefore, these three steps can be optimized to improve the fog collection efficiency [4, 41]. When fog droplets collide with smooth surfaces, they tend to bounce back into the air, which reduces the collection efficiency. Therefore, increasing the surface roughness can increase the number of droplet collisions and improve the droplet capture efficiency. The specific wettability of the surface also plays a crucial role. The high WCA and low roll angle make it difficult for water droplets to form a relatively stable state on superhydrophobic surfaces, whereas superhydrophilic surfaces have a high affinity for water and can easily capture tiny water droplets. In addition, the special wettability of the surface also affects the growth and transport of water droplets by influencing the sliding angle, which feeds back into the fog collection behavior. However, the droplets on the SHBF surface are very small and some droplets fall directly from the surface, so other small droplets along the way are not cleaned, so a part of the surface is not released, some large droplets do not leave the surface, and a large number of microscopic droplets do not condense. And most of the water droplets on the surface of SHBF did not form a stable form [42].

The three-dimensional liquid injected into the framework obtained by oiling the stabilized superhydrophobic layer prepared by in-situ reaction has excellent lubricant absorption and locking ability due to the three-dimensional porous lattice system and chrysanthemum-like micro-nanoflower clusters, which greatly improves the structure's ability to capture droplets during the fog collection process, accelerates droplet condensation, and promotes droplet shedding [43]. It is an excellent fog collection material (98.7% higher fog collection efficiency than the original sample). It also exhibits excellent antifouling performance and anti-icing ability because the droplets cannot adhere to the surface for a long time.

For these (hyper)hydrophilic sites on the hydrophilic surface, the droplets are more inclined to form a water film on the surface, hindering the formation and movement of subsequent droplets. Although (super)hydrophilic sites show good droplet transport capacity at the beginning, with changes in external conditions (e.g., increase in humidity, increase in supercooling, etc.), droplet pinning also occurs, which is detrimental to fog collection [44]. Therefore, it is important to find new ways to achieve long-term fog collection. Inspired by neem plants, a lubricant-infused surface was prepared, which has better heat transfer properties and smaller contact angle hysteresis. Inspired by mussels, Li and coworkers designed a lubricant-infused surface with high transparency and excellent anti-adhesion properties. Lubricant-impregnated surface (LIS) can be applied to a variety of surfaces of any shaped objects with anti-fogging, anti-fouling, self-cleaning, and anti-icing. Zhang et al. [45] prepared a surface with layered coral-like micro-pumps and nano-pore structure as a substrate, and then used dimethyl silicone oil as a lubricant to make a low adhesion, anti-fouling LIS with excellent liquid repellency [46]. Unfortunately, the loss of lubricant from smooth surfaces over time leads to the exposure of rough substrates, which adversely affects the performance of the LIS.

Lee and coworkers [47] designed a solid smooth surface mimicking the structure of moth-eye nanopillars and coated it with a paraffin layer with good light transmission and self-healing ability. The liquid-injected surface made the droplets easy to dislodge with low contact angle (CA) hysteresis, but due to the low contact angle of the droplets on the surface the droplet dislodgement was slow. Deng and coworkers [48] prepared a smooth sage-like surface consisting of protrusions with sliding heads with low droplet adhesion and excellent compressive and impact stability on the liquid-injected surface, the performance of the lubricant (viscosity, surface tension, volatility, whether it reacts with the substrate surface chefogry, and whether it is miscible with the water droplets) plays a key role in the water collection efficiency (WCR) [49]. The contact area of the droplets with the substrate surface is also an important factor affecting the fog trapping efficiency. In order to improve the water collection efficiency when water droplets are slowly shed, three-dimensional (3D) frame substrates with high specific surface area are selected to be injected with a lubricant to capture fog droplets [33]. Lubricant-injected 3D frames can be fabricated by building micro- and nano-composite structures on the filament walls of the frames, and then injecting the lubricant [50]. To the best of our knowledge, relatively few studies have utilized liquid-injected 3D frame materials for water harvesting [32b].

Han et al. [49] prepared 3D frames with micro/nanocomposite structures on the surface of the filament wall through a two-step electrochemical reaction.

After low surface energy treatment and lubricant injection, superhydrophobic framework materials injected with lubricant were obtained. The lubricant-filled superhydrophobic framework has excellent lubricant retention capability due to the synergistic effect of the 3D grid system, the nano-spine cluster structure on the material surface and the lubricant layer. It also has efficient droplet trapping, agglomeration and removal capabilities. The lubricant-filled superhydrophobic framework maintains excellent collection properties after high-speed shear and long-term storage [51]. It also exhibits remarkable electrochemical corrosion resistance and anti-icing properties. The prepared lubricant-injected framework has good lubricant retention capability. It also has considerable corrosion resistance and anti-icing properties. After being subjected to shear at different speeds, the surface remains stable and smooth after long-term storage [3]. More critically, the lubricant-infused frame can be used to effectively capture fog droplets, accelerating their condensation and shedding. It is suitable for efficient water collection with high water collection efficiency. This study provides a promising strategy for the design of water-harvesting materials, which is expected to be used to ameliorate water scarcity in arid regions [2].

Jing and Guo [41b, 52] prepared biocompatible durable LISs by grafting polydimethylsiloxane (via Zn—O—Si bonds) onto ZnO nanorods under UV light and using the remaining unbound silicone oil as a lubricant. Although pdms impregnated surfaces have already been reported in articles, our article has many unique innovations and highlights in terms of preparation methods and applications. Through the high viscosity and strong intermolecular forces between the silicone oil and the PDMS brushes, the lubricant can be firmly retained on the smooth surface, which is further resistant to boiling water and hot liquids. Since water collection is related to surface structure and chefogry, surfaces with layered structures collect more water than single structures (nanostructures or microstructures). A simple hydrothermal method was used to grow ZnO nanowires on micropyramids, which have better water collection properties. Compared with pristine wafers, micropyramid wafers, and (superhydrophobic surface) SHS, LISs have stronger water trapping, water supply, and water removal capabilities, and thus they can collect more water than the three surfaces mentioned above [38]. In addition, based on the stable lubrication layer, LISs have good hot water vapor (105 °C) collection capability [52]. In addition, to explore the adaptability to harsh environments, the LISs were exposed to sunlight for 7 days and underwent several heating/cooling cycles, and the LISs still had good water collection efficiency and reusability.

A significant amount of research has focused on the development of low-carbon and energy systems [53]. Specifically, surfaces with specific wettability to promote gas–liquid condensation have been extensively studied, but current solutions have poor heat transfer performance due to low droplet removal efficiency. In this study, the condensation limit on a beetle-inspired bipolar QLSs in a vapor environment is breached, which provides a heat flux 100 times higher than atmospheric condensation. Unlike beet-inspired surfaces with viscous hydrophilic regions, the bipolar QLS consists of poly(ethylene glycolated) and siloxane polymers forming hydrophilic and hydrophobic level liquid patterns with contact angle hysteresis of

3° and 1°, respectively. More importantly, each hydrophilic smooth pattern acts as a smooth bridge, accelerating droplet aggregation and removal. As a result, the coagulated droplets grew rapidly and fell off. Studies have shown that compared to conventional beetle-shaped surfaces, the hydrophilic striped QLS has a 60% higher water collection rate in atmospheric condensation and a 170% higher heat transfer coefficient (HTC) in vapor condensation.

Durability challenges have led to the development of hydrophobic QLSs composed of chemically grafted flexible polymers. This hydrophobic QLS behaves as a liquid-like layer and shows strong full hydrophobicity, self-cleaning and anti-icing properties [44, 46]. It also shows sustainable high-performance heat transfer in vapor condensation, superior to superhydrophobic and liquid-injected surfaces [54]. However, hydrophobic QLS is not conducive to droplet nucleation, hindering its maximum potential for condensation. Hydrophilic QLS with low contact angle hysteresis can provide droplet condensation [49]. However, the heat transfer performance cannot be sustained at high heat flux due to slow droplet removal.

By utilizing the smooth interfaces and long-lasting durability of QLSs, Dai and coworkers reported an amphiphilic QLS consisting of smooth hydrophilic and hydrophobic chemical modes, which showed contact angle hysteresis of 3° and 1°, respectively. On the amphiphilic QLS, the droplets were smooth in both hydrophilic and hydrophobic modes. This surface overcomes the inherent limitation of droplet immobilization on the hydrophilic region and minimizes droplet immobilization on the hydrophilic region of conventional bipolar surfaces. In this work, we present the water collection rate limit under atmospheric condensation conditions. In addition, we have increased the HTC limit for vapor condensation, which provides a heat flux 100 times higher than atmospheric condensation. It is demonstrated that bipolar QLS pushes the limit of beetle-shaped condensation to unprecedentedly high heat transfer performance [55]. They investigated four surfaces [19], homogeneous silane, circular silane, circular QLS, and striped QLS under atmospheric and vapor condensation conditions, where amphiphilic QLS outperformed conventional beetle-shaped surfaces with viscous hydrophilic domains (Figure 6.4a–d). Two circular surfaces were designed with a hydrophilic circle diameter of 500 m and a circle spacing of 500 m, as this pattern size showed the best performance of the beetle-shaped surface [19, 56]. Circular silanes were made by coating silanes as a hydrophobic background to minimize pinning, and bare silica rings were used as hydrophilic domains for nucleation of parenchymal droplets (Figure 6.4b). Circular and strip QLSs were fabricated by chemically grafting hydrophobic QLSs and hydrophilic QLSs (i.e. polyethylene glycolyzed coatings), respectively (Figure 6.4c,d).

Atmospheric condensation experiments were performed on all four surfaces. Experimental macroscopic images were taken simultaneously for all surfaces at 10 and 50 min from the beginning under the same conditions, with a surface temperature of 2 °C and a humidity of 95%. The homogeneous silane had the lowest droplet growth at 40 min intervals. The hydrophobic contact angle of 119° provides a high nucleation barrier during condensation. Round silanes showed higher droplet growth compared to homogeneous silane surfaces due to the addition of

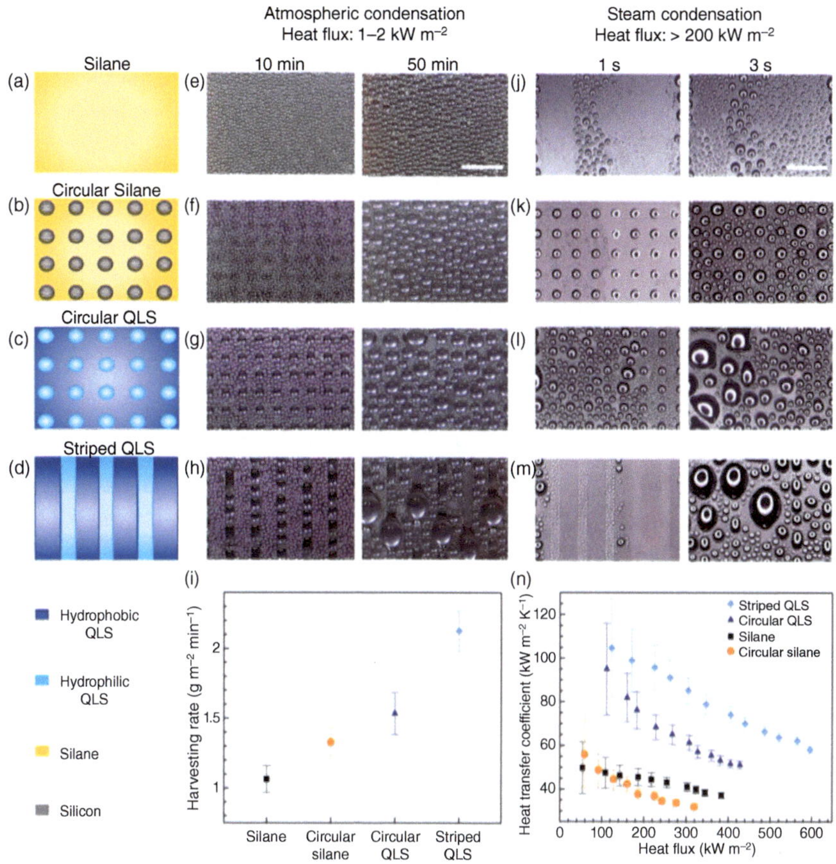

Figure 6.4 Pushing the limits of the water harvesting rate in atmospheric condensation and heat transfer coefficient in steam condensation on beetle-inspired surfaces with patterned wettability. Schematic for (a) silane (b) circular silane (c) circular QLS and (d) striped QLS. Macroscopic images of atmospheric water harvesting at a heat flux of 1–2 kW m^{-2} for 10 and 50 min on (e) silane (f) circular silane (g) circular QLS and (h) striped QLS. (i) Experimental water harvesting rate on the four surfaces. Macroscopic images of pure steam condensation at a heat flux >200 kW m^{-2} at 1 and 3 s in steady state on (j) silane (k) circular silane (l) circular QLS and (m) striped QLS. (n) Experimental heat transfer coefficient as a function of heat flux on the four surfaces. The scale bars are 2.5 mm. Source: Reproduced with permission from [53] / John Wiley & Sons.

hydrophilic regions to enhance droplet nucleation (Figure 6.4e,f). Greater droplet growth was achieved on circular QLS with smooth hydrophilic and hydrophobic patterns (Figure 6.4f,g). The hydrophobic QLS had an ultra-low 1° contact angle hysteresis compared to silanes containing 9°, greatly reducing droplet fixation and leading to faster droplet aggregation and growth. By changing from a circular to a striped design, the striped QLS achieved the highest droplet growth rate of any surface (Figure 6.4c,d). There are three reasons for this: (i) each smooth hydrophilic stripe acts as a bridge in which droplets expand laterally across the hydrophobic region to absorb droplets and rapidly increase in volume, (ii) the hydrophilic stripe improves overall nucleation, and (iii) the smooth nature allows for large droplets to be swept across for rapid droplet removal (Figure 6.4h). Water harvesting rates were quantified for all four surfaces. Striped QLS and circular QLS had 60% and 20% higher water harvesting rates than circular silanes, respectively. Homogeneous silanes have lower water harvesting rates than circular silanes. In addition, we show that oil-lubricated surfaces experience lubricant loss after 120 min of atmospheric condensation, but striped QLS remains smooth and trouble-free [57].

Thus, the limit of water harvesting rate of striped QLS under atmospheric condensation conditions has been broken. This study provides a new paradigm to push the limits of condensation heat transfer at high heat fluxes, which sheds light on the next generation of surface designs for water and energy sustainability [53].

6.3 Fog and Water Collection Materials Based on Gradient Wetting

Access to global freshwater resources has been a constant endeavor [12, 18, 58]. Currently, metal materials based on special wettability and special structure have become an effective way to collect fog [59]. However, how to further improve the efficiency of water collection by carefully designing the wettability and special structure of the materials is still a bottleneck. Therefore, we propose the use of temperature gradients to enhance fog collection on metallic surfaces. The temperature gradient enhancement mechanism resulted in a 54% increase in fog collection efficiency and about 541% increase in surface refresh rate. Further scattering measurements show that the capture time of fog particles on low-cooling surfaces is reduced by about 40%. Based on the theory of heat transfer and thermophoresis, the temperature gradient and the behavior of fog particles on metal surfaces were analyzed, and it was concluded that the temperature gradient induced by thermophoresis force can enhance the fog collection performance on metal surfaces [36, 53]. It provides ideas for designing better fog-collecting equipment and fog-collecting materials.

Inspired by the wetting surface of desert beetles, Wen et al. [9, 14, 60] prepared beetle-type hierarchical fog-collecting interfaces using antimicrobial needle arrays and hydrophilic–hydrophobic synergistic structures, which enabled the fog collector to have an efficient and continuous fog-collecting capability under harsh environments. Feng and coworkers designed several biomimetic surfaces combining the characteristics of various organisms, which effectively improved the fog-collecting performance of the materials [61]. Our team has also made continuous efforts in the past decade to design reasonable special wetting materials from the chemical gradient [61], to design more reasonable fog-collecting structures from the structural point of view [62], and to rationally utilize the synergistic effect of chemical and structural gradients to improve the fog-collecting efficiency [63]. Despite the rapid development of fog collection materials, the fog collection efficiency seems to encounter a bottleneck. In most of the past studies [13, 27, 64], the effect of the temperature state of the fog-collecting material on the water-collecting performance was usually ignored. Previously, including fog flow rate [18], wind speed [25] and relative humidity [22], have been shown to affect the water collection efficiency. However, in practice, it can be found that the collision of moist airflow with cold air leads to fog formation [16], and phase change water droplets may collect on low temperature surfaces such as metals, suggesting that temperature may affect fog accumulation. Leung and coworkers have discussed temperature gradients on glass substrates in their previous work, but there are many metal-based fog collection materials available. The underlying mechanism has not been explored, and the effect of temperature gradients on fog aggregation on metal surfaces has not been reported.

In their work, Fu et al. [35, 59, 65] demonstrated that temperature gradients on metal surfaces enhanced fog collection. The temperature gradient induced by low temperature can effectively trap fog particles and collect moisture, improving the collection efficiency. The relationship between surface. This temperature gradient water collection material demonstrates the temperature gradient-enhanced fog collection strategy on polymer metal surfaces, resulting in temperature gradient-enhanced MPS collection of moisture from fog. The fog collection efficiency increased by 54% to $1340 \pm 62\,mg\,cm^{-2}\,h^{-1}$ under the effect of a temperature gradient from 20 to 5 °C. The WCA and droplet sliding behavior remained essentially unchanged at different surface temperatures, suggesting that the surface wettability does not affect the water collection performance in our work, further validating the importance of the temperature gradient of the metal surface on the fog collection efficiency. Based on the scattering measurements of fog particles and the theory of heat transfer thermophoresis, the temperature gradient field formed in the space of the cooled surface is simulated, and the thermophoretic force acting on the fog particles is analyzed [14]. The fog particles are more captured by the cooling surface, resulting in higher fog collection efficiency. This provides a potential idea for the future design of a new generation of fog collection platforms.

The inherent fog collection mechanism of cacti has provided inspiration for the construction of energy-efficient and environmentally friendly water harvesting devices [15, 66]. However, related research has encountered a bottleneck in improving collection efficiency because it is difficult to replicate authentic natural branched spine clusters through conventional fabrication methods. 3D printing based on immersed surface stacking provides a tool to replicate branching cactus spines, enabling the study of water harvesting with various designs of artificial spines. Li and coworkers a cactus-inspired surface is decorated with artificial spines in multiple orientations for efficient water harvesting and transportation. Nanoscale hydrophobic coatings were sputtered on the surface of the 3D printed spines to accelerate water growth. The results showed that the hexagonally arranged clusters enhanced water airflow around the 3D printed spine, and the printed spine with a 10° tip angle and hydrophobic coating achieved the highest weight gain of $2\,mg\,min^{-1}\,mm^{-3}$.

Cacti have been found to be able to collect water continuously and efficiently using conical spines and banded trichomes on their fleshy stems [67]. Once the collected micronized water droplets grow to a critical size on the conical arrays, they can be directionally transported away from the original deposition site under

Figure 6.5 Biomimetic water collection structure: (a) natural cactus with the scanning electron microscope (SEM) images showing; (b) the spines and water collection capability; (c) schematic diagram of the (immersed surface accumulation) ISA-based 3D printing (ISA-3DP) process; the optical system in the ISA-3DP process (inset shows the magnification of the light guide tool and the optical fiber with projected 2D micropatterns) with the designed models; (d) the water collection capability and (e) the SEM image of the 3D-printed cactus-mimetic spine arrays. Source: Reproduced with permission from [15] / John Wiley & Sons.

a Laplace pressure gradient [18, 64b]. The rapid regeneration of the deposition site ensures efficient fog collection. A typical cactus stem is shown in Figure 6.5a. Clusters of spines are uniformly distributed on the stem, and the apex angle 2α of each spine is about 10°. The different orientations of the spines ensure the efficiency of water collection for fog flow in different directions (Figure 6.5b). However, the collected water is not easily transported from the tip to the bottom of the cactus. Inspired by the principle of water collection capacity of multiple spines, they digitally designed artificial spine arrays in different arrangements and fabricated them using 3D printing to accelerate water collection and transportation. During the 3D printing process, a set of tapered spines with tip angles ranging from 10°, 20°, 30°, 40° to 50° were obtained. The water collection capacity of the 3D printed spine arrays was investigated under a saturated water flow at 95% relative humidity at room temperature. The layer-based immersed surface accumulation 3D printing (ISA-3DP) process, which enables one to build microscale bionic structures around inserted objects, was used to fabricate artificial multibranched spines (Figure 6.5c) [68]. A step effect is created due to the layer-by-layer accumulation of material. This step effect is beneficial because, in addition to the top of the spine, water droplets are collected at the corners of the adjacent staircases due to the curvature change, enabling the collection of more water compared to smooth surfaces. In ISA-3DP, a light-guiding tool consisting of an optical rod and an objective lens was inserted into a resin canister filled with a light-cured liquid resin (E-glass/multi-walled carbon nanotubes (MWCNTs). 2D patterned beam was focused on the top surface of the light-guide tool, and a series of mask images were sequentially projected as the light-guide tool moved. Under the irradiation of the 2D patterning beam, the light-curing composite resin selectively cures to a solid. The 3D printed artificial spine is flexible, and the shape of the 3D printed spine recovers after bending under a compressive load of up to 6 N. The 3D printed artificial spine is flexible, and the shape of the 3D printed spine recovers after bending under a compressive load of up to 6 N. The recycled 3D printed artificial spine still showed similar water collection capacity. As shown in Figure 6.5e, a 3D printed multi-branched spine array with a 10° top angle was designed. When it was placed in the same environment as a natural cactus spine (Figure 6.5b), water droplets grew on each 3D printed spine (Figure 6.5d).

This study opens up interesting prospects for designing next-generation structural materials with special spatial distribution of bionic features for energy-free and efficient water harvesting. The results reported here are believed to contribute to the development of environmentally friendly water harvesting, water transportation and water separation devices.

Spider webs collect water through their capture filaments, which recover their deformed daytime shape at night through a water-sensitive shape memory effect [41a, 63a]. The unique smart functionality and geometry of spider silk has inspired the development of man-made fibers with periodic knots for targeted water harvesting, with great potential applications in water-scarce areas. Existing such fibers are mainly based on nylon filaments coated with synthetic polymer solutions derived from petroleum. Unlike the use of synthetics, all-silk protein fibers (ASPF) with periodic knots have an extremely high volume-to-mass ratio water-harvesting capacity [69]. The main body of the fibers prepared by Hu and coworkers consisted of degummedsilk of the mulberry silkworm coated with recombinantly modified spider trailing-silk macropotentiary glandular arachnid ampullate spidroin 2. It was 252 times lighter than synthetic polymer-coated nylon fibers, which have been reported to have the highest water absorption properties. At the shortest length (0.8 mm) of the three-phase contact line, the ASPF collected a maximum of 6.6L of water, which in terms of volumetric mass index is 100 times more efficient for water collection than the best available water-collecting synthetic fiber.

Directional water-harvesting fibers with spider-trapping ability were prepared using protein-based silk and coated materials. The results show a significant increase in water harvesting efficiency compared to non-protein-based man-made fibers. Uncoated silk fibers could collect up to 2.2 L of water while coated silk was 6.6 L, which is three times higher due to the arachnid protein knots on the fiber surface. The fibers are 252 times lighter than synthetic polymer-coated nylon, which has been reported to have the highest water-harvesting performance and 100 times higher water-harvesting efficiency according to (volume to mass index) VTMI. This extremely high-water harvesting efficiency is achieved through the synergistic effect of the geometry and hydrophilicity of the silk protein material.

Their work provides a high-performance solution for sustainable water harvesting because this new biodegradable material is so portable and disposable. Because the silkworm silk used here is available in large quantities and the effective tandem use of recombinant spider silk proteins shows great potential for scalability. The use of recombinant spider silk proteins as a coating material also addresses the issue of high cost and broadens the application radar. Effective utilization of recombinant arachnid proteins shows great potential for scalability due to the abundance of silkworm silk resources.

6.4 Janus Effect Fog Collection Material

It is well known that fog collection can be divided into three main stages: (i) the fog capture process. The basic research is to promote rapid droplet capture by different wettability matrices, including hydrophilic matrix, hydrophobic matrix, and hybrid hydrophilic-hydrophobic matrix. (ii) Droplet transportation process. Captured droplets are transported to the storage area driven by chemical and structural forces. On the one hand, it can provide a regeneration surface for droplet capture, and on the other hand, it can accelerate the fog collection cycle. (iii) Drainage process. The water stored on the sample is not really collected, and how to drain the water from the storage area into the container is still a factor to be considered [70]. For traditional fog collection devices [35, 65], inspired by desert beetles, hydrophilic patterns on hydrophobic substrates and hydrophilic-hydrophobic hybrids of wettable substrates are widely used for fog collection. In our past work, we prepared hydrophilic bulks on superhydrophobic substrates. By adjusting the arrangement of the bulge, the fog collection efficiency was effectively improved. In this mode of operation, droplets are captured on the hydrophilic bulge and form large droplets as they continue to condense and combine with neighboring droplets. They are then transported and discharged along the superhydrophobic substrate. In addition, tang et al. prepared asymmetric wettable Janus surfaces and explored fog collection patterns for hydrophobic-hydrophilic/superhydrophilic (HB-HL/SHL) surfaces Fog condenses and grows at the hydrophilic site on the (hydrophobic-hydrophilic) HB-HL side, and is then transported to the SHL side, where the final collected water is discharged from the fog collection device. However, continuous fog collection eventually leads to the accumulation of water droplets on the HB-HL side and the eventual formation of a liquid film that slows down the fog collection process. Li and coworkers used electrostatic spinning to prepare Janus porous membranes (JPMs) with asymmetric wettability and water transport channels on copper meshes.

In the fog collection process, there are mainly film condensation and drop condensation. It presents as a membrane condensation pattern. The captured fog accumulates on the sample surface to form a liquid film, which severely hinders the heat transfer of condensation. The fog collection method mainly involves small water droplets colliding and adsorbing on the surface of the water film. Therefore, all water droplets colliding with the water film are captured. When the captured water increases, they fall into the collection vessel under the effect of gravity.

Superhydrophobic surfaces have few nucleation sites and weak surface adhesion. As a result, a large number of small water droplets could not be pinned to the surface while affecting the subsequent successive condensation and growth processes. Ultimately, only a small amount of fog is captured and collected, which greatly reduces the fog collection efficiency. The droplets condensed as droplets on the superhydrophobic side. When the droplets contacted the micro- and nano-wires, they were rapidly transferred to the superhydrophilic side under the action of capillary driving force, which provided a new site for subsequent droplet condensation. At the same time, the transported droplets wet the micro- and nano-wires, and the subsequent smaller capillary force can drive the droplets to unidirectional transport. The captured droplets were unidirectionally transported to the superhydrophilic side, stored in the superhydrophilic side, and finally collected.

An in-depth study of the enhanced fog collector of the Janus system will provide a great deal of insight into the optimal design of the collector [17]. To this end, a series of experiments were conducted to reveal it. During atomization, if there is no hydrophilic backside, the fog water will gradually agglomerate and cover the upper surface of the hydrophobic copper screen (Figure 6.6a, b). Although the fog stream can pass through the mesh, the fog droplets do not condense on the lower surface. After passing through the mesh, the fog flow does suffer a sharp velocity loss from 50 to 1–3 cm. The deceleration of the fog mitigates collisions between the droplets and the substrate, resulting in no condensation on the lower surface of the mesh [71].

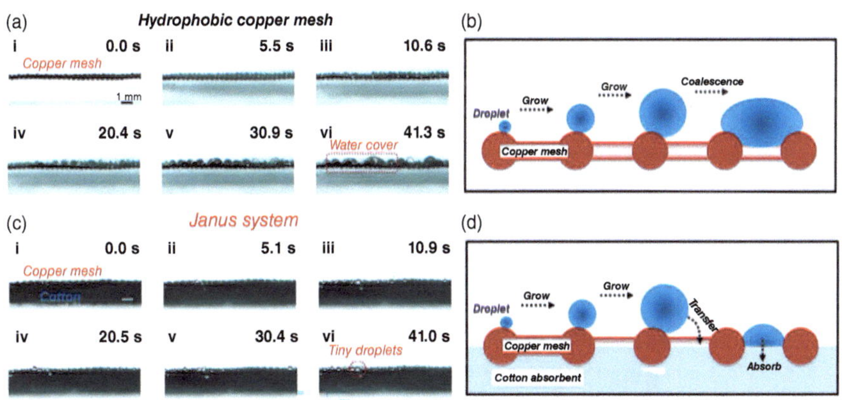

Figure 6.6 (a) The fog harvesting behaviors on a horizontal hydrophobic copper mesh. (b) The fog was first condensed on the hump of the copper mesh and started to gradually grow. The collected droplets coalesced with each other, and finally covered the surface of mesh. (c) In contrast, the Janus system can rapidly deliver the water droplets on the surface into the hydrophilic region on the basis of wettability difference. Even after 40 s fogging, the upper surface of Janus system remained dry, and only tiny droplets (diameter smaller than 500 μm) were observed. (d) Once the collected droplet is in contact with the hydrophilic backside, it would be simultaneously absorbed and transported. This rapid transportation process of collected water can regenerate a fresh surface for fog collection and reduce the re-evaporation rate of surface water, indicating a potential strategy for the enhancement of fog harvesting. The scale bars are 1 mm. Source: Reproduced with permission from [17] / John Wiley & Sons.

In contrast, the droplets collected in the Janus system are small in size, similar to the aperture of the mesh. The droplets condense before the grid rises. Once a growing droplet contacts the hydrophilic layer, it spontaneously transfers unidirectionally to the backside driven by wettability differences (Figure 6.6c,d). This directional water transfer phenomenon can also be found in composite membranes with gradient or Janus wettability. However, the droplet permeability of our Janus system (50 ms) was significantly higher than that of the previous composite membranes due to the lower resistance caused by the hydrophobic side with large pores (Figure 6.7a–c). Note that the hydrophobic modification of the copper mesh is indispensable for the directional penetration of droplets. After the droplets penetrate the pristine copper mesh, the water film is retained on the upper surface due to its high contact angle hysteresis. According to previous reports, wetting of the surface may not be preferred during fog collection. Fortunately, the hydrophobic mesh avoids water retention and contributes to the complete removal of moisture.

(Atmospheric water harvesting) AWH uses atmospheric water generators, which are devices that require only simple equipment to absorb moisture from the air, thus providing a solution that is not limited by geographical conditions [72]. Water vapor adsorbents are the core media of (atmospheric water generators) AWGs and are mainly classified as hygroscopic salts, silicones, zeolites, polymer gels, COFs, activated carbon felts, fibrous membranes, and hygroscopic (metal-organic frameworks) MOFs [73]. Among these adsorbents, hygroscopic MOFs offer several advantages, including a certain amount of water vapor adsorption capacity, excellent porosity and specific surface area due to their high hydrothermal stability, and abundant hydrophilic active sites. At the same time, the relatively low desorption activation energy of MOFs facilitates their water desorption in sunlight. Therefore, their

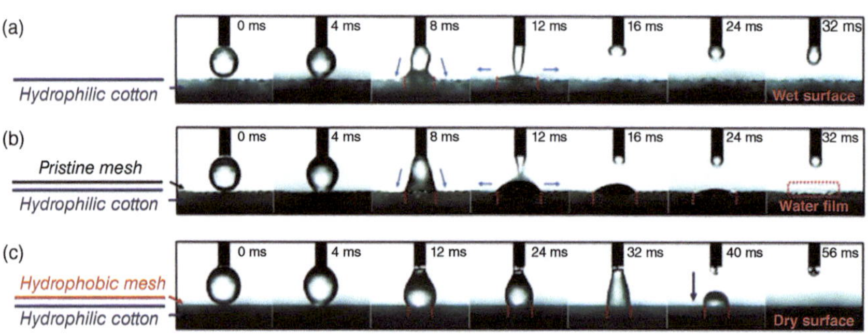

Figure 6.7 The transportation behaviors of different integrated meshes were recorded with a high-speed video camera. (a) A droplet can be rapidly absorbed by hydrophilic cotton absorbent, creating a wetted cotton surface. (b) The integrated system constructed by an unmodified copper mesh can also achieve rapid water transfer; however, the water film would remain on the surface due to the high contact angle hysteresis of the pristine mesh. (c) In comparison, the Janus system can directionally transfer the droplet into the hydrophilic site in ~50 ms, and still have a dry and fresh surface for the next cycle of fog collection. Therefore, the hydrophobic modification of copper mesh was indispensable for fabrication of the efficient Janus fog collector. Source: Reproduced with permission from [17, 72] / John Wiley & Sons.

desorption is hindered by the slow mass transfer due to low ambient temperature. Although MOF-based AWH shows great promise, it is challenging to effectively implement and realize commercial applications in arid regions due to multiple factors, including ineffective heat capture and transfer. Therefore, improving these factors is essential to overcome the limitations of MOF-based AWGs and maximize their potential for application in arid regions.

The first thing to note is that the heat transfer efficiency of powdered MOFs is low due to the presence of air gaps between MOF crystals. Therefore, some researchers have attempted to process powdered MOFs into blocks or pellets on a filled bed by physical compression or by using adhesives [74]. This method can enhance the heat transfer capacity and water storage density to some extent [33]. Alternatively, coating techniques using adhesives can load the MOF blocks onto metals with higher thermal conductivity, thereby increasing heat transfer efficiency. However, the use of an adhesive can cover the gaps between the pores, thereby clogging the pores and thus reducing their adsorption capacity. In-situ growth of MOF coatings on metal substrates without the use of adhesives can solve this problem, but it requires overcoming the low affinity between the substrate and the MOF. Several studies have successfully grown (MOF) coatings in situ on metal substrates, such as cau 10 (Al) on aluminum foil, hybrid (materials of institut lavoisier) MIL-96 (Al) and MIL-100 (Al) MOFs on aluminum foils, (self-assembled monolayer) SAM on [Pd @ MOF-5] @ Au (111) foils, and copper foils on (Hong Kong University of Science and Technology-1) HKUST-1. Among them, aluminum foils are promising substrates for industrial applications due to their low cost and low density, and aluminum-based MOFs also exhibit good hydrothermal stability and suitable adsorption behavior. It is theoretically feasible to grow Al-based MOFs on aluminum foil, which can be used as an aluminum ion source and nucleation site for Al-MOF crystals, and to improve the heat transfer performance of integral MOFs. But even then, the desorption of the integral MOF is limited at ambient temperature because the heat of conversion from solar radiation is quickly exchanged with the surroundings. When the converted heat is much larger than the exchanged heat, the temperature of the integral MOF will increase. This problem can be overcome if the monolithic MOF has an increased photothermal conversion capacity, i.e. if a photothermal adsorbent is prepared by compounding the adsorbent with a photothermal material [75]. Some studies have used them in direct contact with each other, but we witnessed several problems in the experiments we conducted [76]. First, the photothermal materials may clog the pores of the adsorbent, leading to a decrease in water vapor adsorption capacity. Second, their direct contact may affect their chemical stability, especially after several cycles. Third, the photothermal materials may be partially covered by the adsorbent, but may only realize a moderate photothermal conversion capacity.

"Indirect contact" structures would therefore be more rational, but there are some challenges, especially because it is difficult to establish effective heat transfer between the sorbent and the photothermal material. Considering that only one side of the metal substrate is used for MOF in-situ growth, the concept of a "Janus wet photothermal hybrid" with dual in-situ growth – a photothermal adsorbent with two sides having different functions – is proposed here [8a]. First, Cu is deposited on the aluminum foil side by arc spraying to form a Cu/Al bimetallic substrate. Second, the aluminum foil side was used for in situ growth of aluminum fumarate MOF, which is known to have excellent hydrothermal stability and high-water absorption above 20% relative humidity (RH) [77]. Third, the Cu side was used for in-situ growth of a CuxS photothermal semiconductor layer with high photothermal conversion efficiency for rapid heat generation from sunlight [78]. Dual in-situ growth enabled precise and controlled growth of the MOF coating and CuxS layer on opposite sides, resulting in a CuxS-Cu/Al-MOF wet photothermal hybrid with a Janus structure, where the two sides of the Janus structure acted as two independent "factories" performing the contrasting functions of photothermal conversion and water adsorption-desorption, respectively, leading to a functionality that was not only a high efficiency photothermal semiconductor, but also a high efficiency photothermal semiconductor. desorption, leading to functional asymmetry. At the same time, the Cu/Al substrate with high thermal conductivity creates a "heat conveyor belt" from CuxS to MOF, which can be heated up to 71.5 °C in 20 min and desorb 97% of the adsorbed water in 40 min under 1 simulated solar irradiation (0.1 $\mathrm{w\,cm^2}$). In the self-designed AWG, the water desorption of CuxS-Cu/Al-MOF was 3.21 times higher than that of the same mass of powdered MOF. CuxS-Cu/Al-MOF is the first Janus hybrid, which endows the adsorbent with photo-thermal conversion capability and can be applied in arid regions [72].

In brief, the collection of microdroplets from fog is a promising solution to the global freshwater crisis. Different types of fog collectors have been widely reported in the last decade. The improvement in fog collection can be attributed to the immediate transportation of harvested water, efficient regeneration of fog collection surfaces, etc. [17]. The hydrophobic/hydrophilic synergistic Janus system has enhanced fog collection capabilities by learning from nature's water retention strategies. Directional transportation of surface water, reduced reevaporation rates of harvested water, and a thinner boundary layer on the collection surface contribute to the improved collection efficiency.

References

1 Yue, H., Fu, Y., and Guo, Z. (2024). *Colloids and Surfaces A: Physicochemical and Engineering Aspects* 683: 132994.

2 Wang, L., Jiang, G., Zhu, D. et al. (2023). *Small* 19: 2302339.

3 Wu, S., Liang, Z., Li, Y. et al. (2022). *Advanced Science* 9: 2105986.

4 Guo, J., Huang, W., Guo, Z., and Liu, W. (2022). *Nano Letters* 22: 3104.

5 (a) Wang, Q., Yang, F., and Guo, Z. (2021). *Journal of Materials Chefogry A* 9: 22729. (b) Wang, B., Zhou, X., Guo, Z., and Liu, W. (2021). *Nano Today* 40: 101283.

6 Yu, Z., Zhang, H., Huang, J. et al. (2021). *Journal of Materials Science & Technology* 61: 85.

7 He, S., Wang, X., Li, S. et al. (2023). *Surfaces and Interfaces* 43: 103545.

8 (a) Wang, X., Guo, Z., and Liu, W. (2023). *Advanced Materials Interfaces* 10: 2202123. (b) Wang, X., Guo, Z., and Liu, W. (2023). *Langmuir* 39: 8926.
9 Pang, Y., Mei, G., Wang, F. et al. (2024). *Colloids and Surfaces A: Physicochemical and Engineering Aspects* 680: 132641.
10 Peng, Z. and Guo, Z. (2022). *Biomaterials Science* 10: 5831.
11 Peng, Z., Fu, Y., and Guo, Z. (2023). *ACS Applied Materials and Interfaces* 15: 38110.
12 Bai, H., Wang, L., Ju, J. et al. (2014). *Advanced Materials* 26: 5025.
13 Mishchenko, L., Khan, M., Aizenberg, J., and Hatton, B.D. (2013). *Advanced Functional Materials* 23: 4577.
14 Zhou, H., Zhang, M., Li, C. et al. (2018). *Small* 14: e1801335.
15 Li, X., Yang, Y., Liu, L. et al. (2019). *Advanced Materials Interfaces* 7: 1901752.
16 Zhong, L., Zhu, H., Wu, Y., and Guo, Z. (2018). *Journal of Colloid and Interface Science* 525: 234.
17 Cao, M., Xiao, J., Yu, C. et al. (2015). *Small* 11: 4379.
18 Bai, H., Ju, J., Zheng, Y., and Jiang, L. (2012). *Advanced Materials* 24: 2786.
19 Ju, J., Yao, X., Yang, S. et al. (2014). *Advanced Functional Materials* 24: 6933.
20 Zhang, S., Huang, J., Chen, Z., and Lai, Y. (2017). *Small* 13: 1602992.
21 Zhong, L., Feng, J., and Guo, Z. (2019). *Journal of Materials Chefogry A* 7: 8405.
22 Xue, P., Nan, J., Wang, T. et al. (2017). *Small* 13: 1601807.
23 Li, D., Huang, J., Han, G., and Guo, Z. (2018). *Journal of Materials Chefogry A* 6: 22741.
24 Wang, B., Zhang, Y., Shi, L. et al. (2012). *Journal of Materials Chemistry* 22: 20112.
25 Tian, X., Shaw, S., Lind, K.R., and Cademartiri, L. (2016). *Advanced Materials* 28: 3677.
26 Tian, Y., Su, B., and Jiang, L. (2014). *Advanced Materials* 26: 6872.
27 Manna, U., Raman, N., Welsh, M.A. et al. (2016). *Advanced Functional Materials* 26: 3599.
28 Zhao, Y., Wang, H., Zhou, H., and Lin, T. (2017). *Small* 13: 1601070.
29 Zhou, H., Li, Q., Zhang, Z. et al. (2023). *Advanced Fiber Materials* 5: 1555.
30 Zhang, S., Huang, J., Cheng, Y. et al. (2017). *Small* 13: 1701867.
31 Yu, M., Liu, M., and Fu, S. (2021). *ACS Applied Materials and Interfaces* 13: 32149.
32 (a) Wang, C. and Guo, Z. (2020). *Nanoscale* 12: 22398. (b) Gou, X. and Guo, Z. (2021). *Chemical Engineering Journal* 411: 128495.
33 Kalmutzki, M.J., Diercks, C.S., and Yaghi, O.M. (2018). *Advanced Materials* 30: e1704304.
34 Compton, B.G. and Lewis, J.A. (2014). *Advanced Materials* 26: 5930.
35 Fu, Y., Ai, S., Guo, Z., and Liu, W. (2023). *Journal of Colloid and Interface Science* 649: 646.
36 Borbora, A., Xu, Y., Dey, S. et al. (2023). *Advanced Materials* 35: 2302264.
37 Wang, S., Feng, L., and Jiang, L. (2006). *Advanced Materials* 18: 767.
38 Dai, X., Sun, N., Nielsen, S.O. et al. (2018). *Science Advances* 4: eaaq0919.
39 Guo, P., Zheng, Y., Wen, M. et al. (2012). *Advanced Materials* 24: 2642.

40 Yuan, Y., Xiang, H., Liu, G. et al. (2022). *Advanced Materials Interfaces* 9: 2101968.

41 (a) Wang, S., Zhu, L., Yu, D. et al. (2023). *Advanced Functional Materials* 33: 2305244. https://doi.org/10.1002/adfm.202305244. (b) Jing, X. and Guo, Z. (2019). *ACS Applied Materials and Interfaces* 11: 35949.

42 Greyson, E.C., Babayan, Y., and Odom, T.W. (2004). *Advanced Materials* 16: 1348.

43 He, Z., Xie, H., Jamil, M.I. et al. (2022). *Advanced Materials Interfaces* 9: 2200275.

44 Verho, T., Bower, C., Andrew, P. et al. (2011). *Advanced Materials* 23: 673.

45 Zhang, K., Xu, S., Tan, J. et al. (2023). *Colloids and Surfaces A: Physicochemical and Engineering Aspects* 669: 131495, https://doi.org/10.1016/j.colsurfa.2023.131495.

46 Chen, X., Wen, G., and Guo, Z. (2020). *Materials Horizons* 7: 1697.

47 Han, G. et al. (2020). *ACS Nano* 14: 10198–10209, https://doi.org/10.1021/acsnano.0c03463.

48 Li, X. et al. (2020). *National Science Review* 8, https://doi.org/10.1093/nsr/nwaa153.

49 Han, X. and Guo, Z. (2021). *ACS Applied Materials and Interfaces* 13: 46192.

50 Zhou, H., Jing, X., Li, S., and Guo, Z. (2021). *Chemical Engineering Journal* 417: 129215.

51 Xu, C.L. and Wang, Y.Z. (2017). *Advanced Materials Interfaces* 4: 1700550.

52 Peng, Y., He, Y., Yang, S. et al. (2015). *Advanced Functional Materials* 25: 5967.

53 Boylan, D., Monga, D., Shan, L. et al. (2023). *Advanced Functional Materials* 33: 2211113.

54 Zhang, P., Zhang, L., Chen, H. et al. (2017). *Advanced Materials* 29: 1702995.

55 Hoque, M.J., Sett, S., Yan, X. et al. (2022). *ACS Applied Materials & Interfaces* 14: 4598.

56 (a) Ghosh, A., Beaini, S., Zhang, B.J. et al. (2014). *Langmuir* 30: 13103. (b) Restagno, F., Bocquet, L., and Biben, T. (2000). *Physical Review Letters* 84: 2433.

57 Sarma, J., Zhang, L., Guo, Z., and Dai, X. (2022). *Chemical Engineering Journal* 431: 133475.

58 (a) Bai, H., Ju, J., Sun, R. et al. (2011). *Advanced Materials* 23: 3708. (b) Bai, H., Tian, X., Zheng, Y. et al. (2010). *Advanced Materials* 22: 5521.

59 Fu, Y., Li, L., Ai, S. et al. (2024). *Materials Today Sustainability* 25: 100654.

60 Wen, C. et al. (2019). *ACS Applied Materials & Interfaces* 11: 34330–34337, https://doi.org/10.1021/acsami.9b11862.

61 Wang, S., Yang, H., Hu, G., and Feng, W. (2018). *Materials Technology* 33: 513.

62 Peng, F., Zhang, D., Liu, X., and Zhang, Y. (2021). *Journal of Magnesium and Alloys* 9: 1471.

63 (a) Sullivan, S.P., Murthy, N., and Prausnitz, M.R. (2008). *Advanced Materials* 20: 933. (b) Tang, X. and Guo, Z. (2020). *New Journal of Chefogry* 44: 20495.

64 (a) Long, Z., Zhao, Y., Zhang, C. et al. (2020). *Advanced Materials* 32: e1908099. (b) Tian, X., Bai, H., Zheng, Y., and Jiang, L. (2011). *Advanced Functional Materials* 21: 1398.

65 Fu, Y., Guo, J., Pei, K. et al. (2023). *Surfaces and Interfaces* 42: 103356.

66 Cao, M., Ju, J., Li, K. et al. (2014). *Advanced Functional Materials* 24: 3235.

67 Ju, J., Zheng, Y., and Jiang, L. (2014). *Accounts of Chemical Research* 47: 2342.

68 Yang, Y., Li, X., Zheng, X. et al. (2018). *Advanced Materials* 30: 1704912.

69 Yang, Y., Song, X., Li, X. et al. (2018). *Advanced Materials* 30: 1706539. https://doi.org/10.1002/adma.201706539.

70 (a) Li, S., Zhu, J., Liu, C. et al. (2023). *Chemical Engineering Journal* 464: 142549. (b) Li, S., Lai, S., Zhang, R., and Guo, Z. (2023). *Chemical Engineering Journal* 470: 143659.

71 Tian, P., Gao, X., Wen, G. et al. (2018). *Journal of Colloid and Interface Science* 532: 517.

72 W. Chen, Y. Liu, B. Xu, M. Ganesan, B. Tan, Y. Tan, F. Luo, X. Liang, S. Wang, X. Gao, Z. Zhang, R. Ye, D. Y. C. Leung, S. K. Ravi, Y. Fang, *Small* 2024 20, 2306521. https://doi.org/10.1002/smll.202306521.

73 (a) Li, B., Wen, H.M., Cui, Y. et al. (2016). *Advanced Materials* 28: 8819. (b) Xu, J., Li, T., Chao, J. et al. (2020). *Angewandte Chemie (International Ed. in English)* 59: 5202. (c) Zhang, Y.S. and Khademhosseini, A. (2017). *Science* 356: eaaf3627.

74 Lin, C., Ma, W., Zhang, Y. et al. (2023). *Small* 19: 2301723.

75 Xu, S., Wang, Q., and Wang, N. (2020). *Advanced Engineering Materials* 23: 2001083.

76 Yang, Y., Ren, Z., Chen, W. et al. (2023). *Separation and Purification Technology* 320: 124175.

77 He, H., Huang, W., and Guo, Z. (2023). *Materials Today Physics* 30: 100927.

78 He, H. and Guo, Z. (2023). *Journal of Colloid and Interface Science* 629: 508.

7

Preparation of Fog-Collecting Materials

The previous sections introduced various fog-harvesting materials, such as those inspired by desert beetles and others based on unique principles. Each material possesses distinctive features, offering diverse approaches to fog collection. Building on this, this book will further explore the fabrication methods for fog-harvesting materials, including one-dimensional, two-dimensional, and three-dimensional materials. The preparation techniques for materials of different dimensions will provide crucial theoretical and practical support for advancing fog-harvesting technologies.

7.1 Preparation of 1-Dimensional (1D) Fog-Collecting Materials

One-dimensional materials are linear and consist of an infinite number of points. The common one-dimensional fog-collecting materials are often inspired by spider silk and cactus spines. The essence of fog collecting lies in the directional driving of droplets by roughness gradient (wettability gradient) and structure gradient (Laplacian pressure difference) [1]. These two driving forces inspired the subsequent production of a series of water collection surfaces. Through different chemical and physical methods, wettability driving and structural gradient driving forces can be assigned to the surface of the water-collecting material, which can effectively improve the water transport mechanism on the surface and thus improve the collection effect of fog.

Zheng et al. [2] first reported the fog-collecting phenomenon of spider silk in 2010, and they found that spider silk can effectively collect water from the air. Spider silk is composed of the conical spindle junction with a larger diameter and the connection points with a smaller diameter alternately. The spindle junction part is composed of nanofibers with a high density of cross arrangement, which has a higher roughness. The roughness of the joint is less. In addition, each repeating unit of spider silk can be regarded as a spindle-shaped structure, so its diameter varies in a gradient (Figure 7.1a–d). This set of synthetic patterns gives spider silk a unique surface energy structure gradient and geometric structure gradient. These two gradients provide the driving force for the droplet's directional transport. The surface energy gradient provides the driving force for surface wetting, which makes the water droplets

Bioinspired Materials for Fog Harvesting: Principles, Design, Preparation, and Applications,
First Edition. Zhiguang Guo and Weimin Liu.

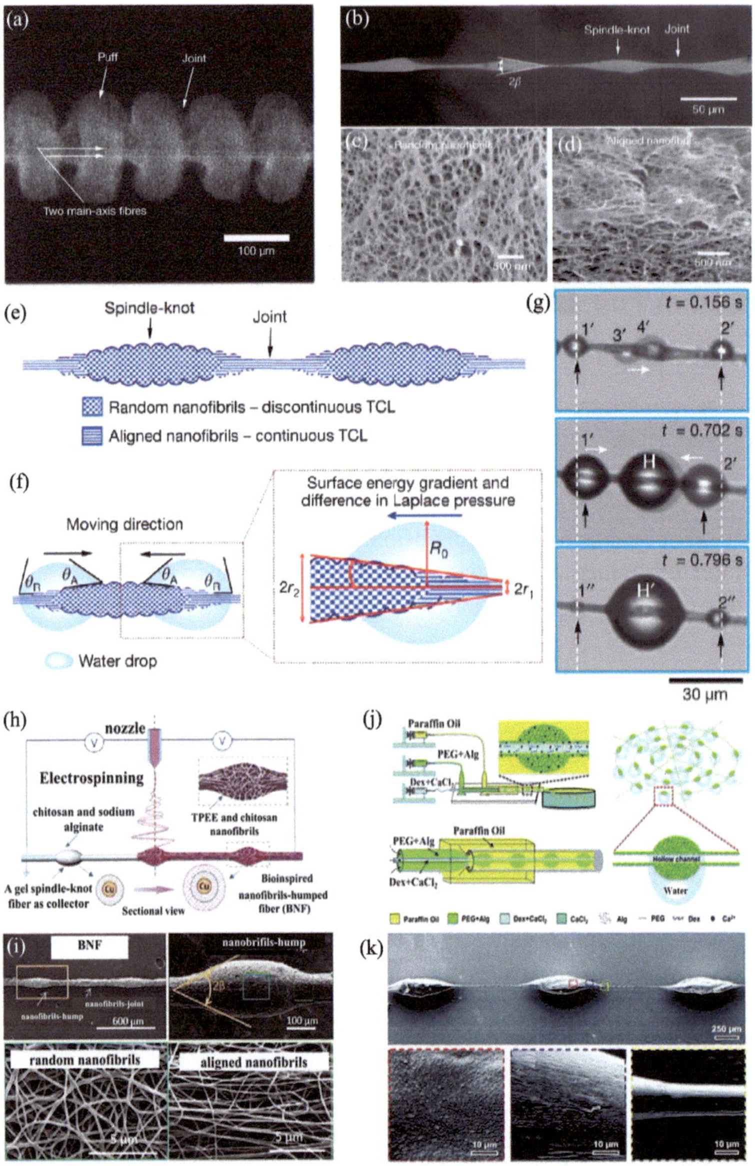

Figure 7.1 (a) The structure of dried spider silk; (b) SEM images of spider silk after wet reconstruction; (c) magnified images of spindle junctions, from which it can be seen that periodic spindle junctions are randomly interwoven by nanofibers; (d) the joint image consists of nanofibers arranged relatively parallel to the filament axis; (e) spindle junctions are made of highly randomly interwoven nanofibers, and joints are made of relatively aligned nanofibers; (f) schematic diagram of surface energy gradient generation; (g) optical image of droplet aggregation at the spindle junction, arrow indicating the direction of droplet movement; Source: Reproduced with permission from [2] / Springer Nature. (h) Schematic diagram of preparation of BNF by electrospinning of TPEE and CS nanofibers; (i) SEM photos of BNF with two nanofiber humps; Source: Reproduced with permission from [3] / American Chemical Society. (j) Schematic diagram of microfluidic systems for the preparation of HSF and water collection; Source: Reproduced from [4] / with permission of Elsevier. (k) SEM images showing dehydrated microfibers. Source: Reproduced with permission from [5] / The Royal Society of Chemistry.

more inclined to move toward the more hydrophilic region. The geometric structure gradient can generate additional pressure on the liquid droplets to drive the droplet to move in the direction of about a larger diameter (Figure 7.1e,g). The hydrophilicity of spider silk and the nanoscale fiber structure on its surface further enhance the ability of this droplet to be transported directionally. This excellent directional droplet transport enables spider silk to exhibit efficient fog-collecting behavior. The study of spider silk's water collection behavior further enriched the mechanism of bionic fog collection, and the new mechanism is helpful in designing and preparing a new generation of fog collection materials. In 2016, Zheng's team found that during the fiber preparation process, by adjusting the fluid flow rate and the content of each component, the structure and morphology of the fiber can be effectively adjusted to obtain the spindle fiber with the best roughness and radius of curvature [6]. When applied to fog collection, the surface of the spindle junction microfibers traps tiny droplets, which then grow and combine with other droplets. The condensed droplets then move away from the center of a knot like the spindle junction. After continuous directional movement of condensed droplets, the largest volume of droplets can overcome the adhesion of the spindle junction and fall off from the spindle junction. In addition, the results also show that the higher the height of the spindle junction, the larger the suspension volume at the time of droplet shedding. In the case of the same length, because of the longer contact line of the three-phase line, the larger spindle junction has a larger droplet suspension volume. However, given the size of the spindle junction, the droplet size is too small to move through the small Laplacian pressure difference in time, resulting in a surface-trapping blockage [7].

In 2020, Zheng's team [3] designed bioinspired nanofibril-humped fibers (BNFs) with a hierarchical structure by using a fluid coating method and electrospinning techniques, in which periodic spindle junctions are composed of staggered fluffy-like nanofibers and joints are composed of aligned nanofibrils (parallel to the axis). The design of the hierarchical structure mainly involves two dimensions: On the one hand, the staggered nanofibril has two different states, which are distributed inside the spindle node and the joint. The design not only realizes the surface roughness gradient but also constructs a channel inside the joint to transport extremely tiny droplets into the anisotropic capillary, which solves the above problem, namely internal transportation. On the other hand, the outside of the fiber can accomplish fog collection similar to spider silk, that is, external transport. When dry man-made fibers are placed in a humid environment, the hydrophilic fibers can trap tiny droplets because one component of the fiber is chitosan (CS), which has a hydrophilic character with a rich hydroxyl group in its structure. Next, the captured tiny droplets can be transported into a capillary-driven spindle junction. Droplets of different radii can form on the outside of joints and spindle junctions. Driven by the Laplacian pressure difference, the small droplets can be transported outside the spindle junction until they are combined with the larger droplets. Both external and internal transport modes are closely related to the hierarchical structure. Therefore, the synergistic effect of these two modes of transport results in a significant increase in fog collection efficiency. Under the condition of 150 $g\,m^{-2}$, the fog collecting efficiency of the spindle junction is about

$0.83\,mL\,s^{-1}\,mm^{-2}$. In 2022, Meng's team [5] proposed a novel hierarchical structural design, namely a spindle junction microfiber with continuous hollow channels, with greater capillary forces compared to the microchannels designed by Zheng, building the hollow structure through a multiphase laminar flow microfluidic strategy. In the spinning process, two aqueous phase system (ATPS) is introduced, and the calcium-containing dextran (inner phase) is injected into the polyethylene glycol containing sodium alginate (middle phase) to form a stable laminar flow. At the interface, Ca^{2+} crosslinks with sodium alginate to form microfibers. At the same time, due to the high viscosity of ATPS, alginate is prevented from spreading inward, resulting in a hierarchical structure, that is, a hollow structure. This is similar to the intermediate hollow structure, which has a strong shell to improve the stability of the microfibers. In addition, there is another hierarchical structure, namely the surface roughness of the spindle junction, the surface roughness of the node, and the surface roughness of the connecting area between the spindle junction and the node, which have the characteristics of small particles, folds parallel to the axis, and smooth, respectively. From their experimental results, the ability to hang water and the efficiency of collecting fog has been improved.

To achieve efficient fog collection, Guo et al. [4] used 3D printing technology to construct a bionic spider web. Using 3D printing technology to print micron-scale bionic spider webs, spider web spindle knots are periodically arranged. The size of the spindle junction and the joint curvature on spider silk were optimized by finite element analysis combined with experimental verification. Compared with the rest of the water collection surface, the artificial spider web water collection effect is better, and the mechanism has three aspects: First, the flow field around the optimized structure increases the effective deposition area more obviously, which is conducive to condensation of moist air on the surface of spider silk. Second, because the Laplacian pressure at both ends of the droplets is different, the condensed water droplets move along the conical surface at a faster speed. Third, the hydrophilic surface has greater adhesion to small droplets, the three-phase contact line of the droplets becomes longer, and the water film formed on the surface can accelerate the movement of the droplets and finally improve the collection ability of fog. Liu et al. [5] used a simple and flexible multiphase laminar flow microfluidic method to accurately control the structure and morphology to prepare bionic spindle junction microfibers with hollow channels. Notably, the compartment effect of the two-phase aqueous system (ATPS) is introduced to form stable hollow spindle junction microfibers (HSFs). Compared with solid spindle junction microfibers, HSFs exhibit better-hanging capacity, droplet growth rate, and water collection efficiency. The study found that the unique hollow structure brings additional capillary forces and provides a longer length for the water collection behavior of the three-phase contact line, enhancing the fog collection capability of HSFs. Zhang et al. [8] proposed a new dynamic interface spinning technique for controllable and flexible preparation of biomimetic fog-collected microfibers. The design combines the knot-node microstructure obtained from spider silk and the groove microstructure obtained from the spine of cacti. The surface of the cactus spine is covered with gradient microgrooves that facilitate one-way transport of water droplets during fog collection, while the wet-reconstructed spider silk, consisting of multiple spindle knots, is an effective water collection point. Through the clever combination of these two advantageous water collection structures, the water collection capacity of the obtained bionic microfibers has been significantly improved. Liu et al. [9] prepared a heterogeneous rough spindle junction microfiber based on calcium chloride and chitosan by electrospinning. The surface morphology of the spindle junction and the size of the spindle junction and joint can be controlled by adjusting the content and flow of calcium chloride. The results show that the surface morphology of the spindle junction directly affects the fog-trapping efficiency. Due to the hydrophilicity of chitosan gel, the increase in roughness will enhance the hydrophilicity of the spindle junction, which is conducive to increasing the surface energy gradient between the joint and the cross section of the spindle junction. The distance between the spindle junction and the nodule also affects the fog-trapping efficiency, and an increase in the distance will cause the surface curvature gradient to decrease, impeding the water transport. They then used thermoplastic polyester elastomer (TPEE) and CS to prepare nanofiber-raised fibers (BNFs). Combined with electrospinning technology and fluid coating method, they achieved periodic bumps composed of randomly interwoven nanofibers and joints composed of directional nanofibers. Its structure is highly similar to the micro/nanostructure of wet spider silk. In particular, the interlaced nanofibers have a large specific surface area and can effectively trap fog droplets and transport the captured water into the channel between the nanofibers under wet conditions, merging into large droplets at the peak through internal transport. This efficient water collection capacity is due to the synergistic effect of internal and external transport modes on BNFs. The external transport mode is induced by both the Laplacian pressure difference and roughness gradient. Anisotropic capillary channels between nanofibers induced internal transport patterns. It is this unique transport mechanism that makes BNFs the perfect solution to the problem of rapid transport of newly captured tiny water droplets, fundamentally enabling efficient fog collection [3].

7.2 2D Fog Collection Material Design

7.2.1 Imitation Cactus Spines

In the bionic design of two-dimensional fog-collecting materials, the main bionic objects are cacti, Nepenthes, Namib Desert beetles, etc. In the preparation process, surface wetting gradient, Laplacian pressure difference and capillary force effects are constructed to improve the fog collecting efficiency. However, two-dimensional design requires a more imaginative approach to transform the three-dimensional structure of bionic objects into two-dimensional structures, and two-dimensional materials can achieve efficient fog collection while integrating some special functions, such as antibacterial and reducing lubrication and loss [10]. Cone-shaped cactus spines allow for efficient fog collection, but the manufacture of

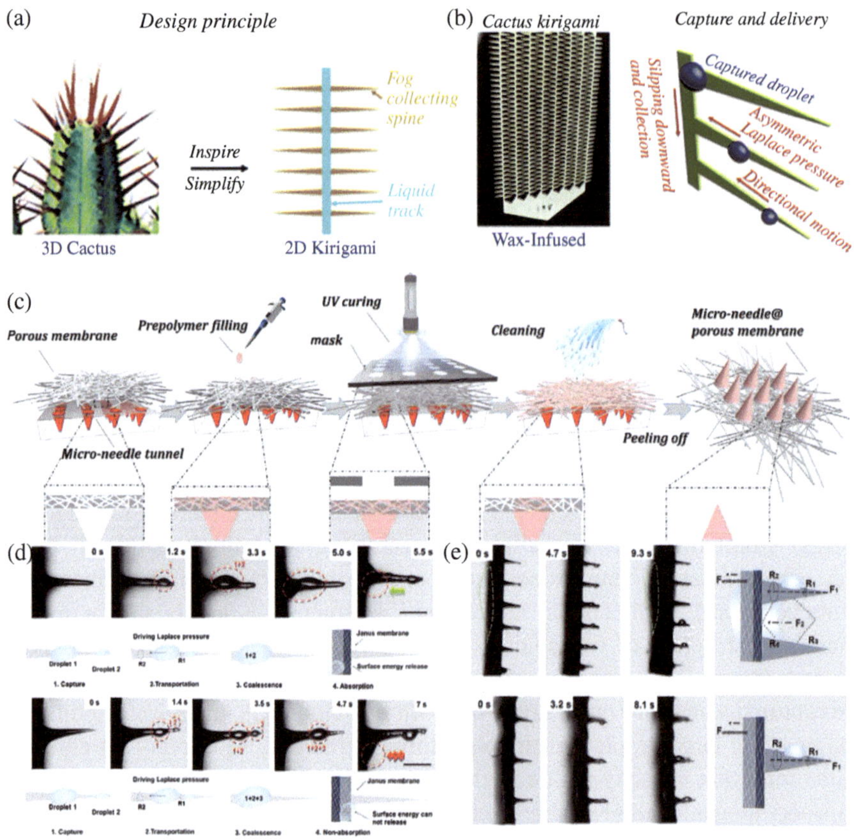

Figure 7.2 (a,b) Design of cactus cutouts, and optical images of typical cactus cutouts with dense spines; Source: Reproduced with permission from [11] / The Royal Society of Chemistry. (c) Schematic diagram of the preparation method of the microneedle porous film; (d) fog collection process of the microneedle array; (e) MN-HL/HB fog collection of different microneedle arrays: 16 × 16, 8 × 8. Source: Reproduced with permission from [12] / John Wiley & Sons.

such 3D structures is complex and cumbersome. As shown in Figure 7.2a,b, Cao and coworkers [11] designed a cactus-like kirigami structure, which simplifies the three-dimensional spiniform structure into a two-dimensional triangle. This design not only improves the efficiency of the material to capture the droplets but also self-propels the droplets, thus improving the efficiency of droplet polymerization and separation. It can be seen that the preparation idea of the two-dimensional fog-collecting material designed like cacti is to insert plane triangles on a macroscopic two-dimensional plane base, and there are wettability differences between the triangle area and the base. The triangular hydrophobic ridge creates a Laplacian pressure that can drive droplets from the tip to the root. The droplets collected at the roots are then pushed by gravity along a waxy track into the collection container. For millimeter-scale droplet self-propulsion, asymmetric structures that can generate anisotropic Laplacian pressures, such as triangles and cones, are very useful.

In addition, self-propulsion under Laplacian pressure requires sufficient droplet size, which requires droplet agglomeration. In addition, this process often involves the release of condensation energy, which allows the droplet to gain kinetic energy to overcome the droplet contact angle lag on the spine surface. The macroscopic two-dimensional plane is an amplification of the two-dimensional triangle action to achieve efficient fog collection. The construction of this hierarchical structure saves the tedious preparation process. At the same time, the cost is greatly reduced. When the fog flow velocity is 220 $cm\,s^{-1}$, the fog collection efficiency is as high as 4000 $mg\,cm^{-2}\,h^{-1}$. Zhou et al. [12] developed a mixed porous Janus film with a microneedle array (MN-HB/HL) using a straightforward and controllable top–down microforming technique. This innovative film leverages the Laplacian pressure from the microneedles along with the wettability gradient of Janus to facilitate effective fog collection, directed droplet movement, and rapid absorption. In comparison to conventional mesh and fabric materials, MN-HB/HL demonstrates an impressive fog collection efficiency of up to $kg\,m^{-2}\,h^{-1}$ (Figure 7.2c–e). Inspired by the structures of cactus and pine needles, He et al. [13] synthesized a CuO nanogrid layered structure on copper foam, which consists of a mesh tip structure with synergically enhanced fog collection properties. Conical structures can quickly trap droplets, forming a rough surface for building hydrophobicity, while the strong ability to restrict air between grid structures can make droplets fall off quickly. Further analysis shows that the directional gap between the grid structures can provide multiple channels for the directional transport of droplets. Driven by the Laplacian pressure difference and the surface energy gradient, the liquid can be discharged along the unidirectional microchannel formed by the grid gap. In addition, copper foam greatly increases the actual contact area with the droplet, and a good combination of surface structure and copper foam significantly improves the collection efficiency of fog. As an efficient method of collecting invisible fog in a humid environment, the Janus membrane still has some inherent limitations: (i) Due to the extremely low adhesion between fog droplets and the substrate, fog droplets condense poorly on the superhydrophobic side; (ii) the droplets cannot leave the superhydrophilic side in time, which hinders the continuous transportation of water in the micropores. Inspired by the alternating hydrophobic/hydrophilic bump on the back of a desert beetle and the asymmetric geometry of cactus spines, Su et al. [14] designed and prepared a layered hydrophilic/hydrophobic/uneven Janus film by ablating the aluminum film with a femtosecond laser to achieve self-driven fog collection. Compared with the traditional fog collecting device, the film has the following advantages: (i) The condensation performance of the fog drops is good; (ii) the transfer speed of the droplets is fast, and the evaporation of the collected droplets is low; (iii) the bottom surface has very low adhesion to water, making it easier for droplets to fall. The potential mechanism of efficient fog collection is the synergistic effect of stratified hydrophilic/hydrophobic/convex structures. The bulges reduce the thickness of the boundary layer so that the fog flow can effectively impact the surface of the structure. The existence of a hydrophilic region makes the water droplets grow faster, and the new hydrophobic region is updated faster, which is conducive to the condensation and condensation of the subsequent fog. The water film of the hydrophilic square diagram gradually increases under the synergistic action of the chemical wettability gradient and the Laplacian pressure difference until it contacts the micropores and is absorbed to the bottom surface [10b].

7.2.2 Nepenthes Imitate

The efficiency of fog collection can be improved in two aspects: (i) improving the surface fog capture capacity; (ii) accelerating the transport speed of the droplets on the surface so that the droplets can fall off the surface as soon as possible and make room for the subsequent water collection. However, it was found in the experiment that with the progress of the fog collection process, droplets would be nailed in the area originally with superhydrophobic properties, and the state of droplets on the surface would change from the Cassie state to Wenzel state. The final result was that droplets could not slide off the surface in time, and the efficiency of the surface fog collection would be affected. The lubricant injection surface inspired by Nepenthes has garnered significant interest due to its excellent sliding characteristics, as illustrated in Figure 7.3a–d. Based on the inherent design principles of this lubricant injection surface, water droplets do not nucleate within the micro/nanostructures of the substrate during fog collection, which prevents droplet pinning. The remarkable mobility of droplets on smooth lubricant injection surfaces (SLISs) makes them highly effective for liquid collection and manipulation applications. SLISs are employed for harvesting water, and their performance can be enhanced by modifying both the micro/nanostructure and surface chefogry of these smooth surfaces [17]. For example, Ahn et al. [16] have proposed a method for realizing stretchable smooth materials with surfaces with tri-Wiener structures. Specifically, 3D micro-wrinkles are programmatically formed on the surface of a polydimethylsiloxane (PDMS) elastomer through chemical oxidation, and a lubricant is then injected into the structure. The chemical oxidation of PDMS with acidic solution can produce a large area of various shapes of fold structures (Figure 7.3e,f). Due to the existence of a three-dimensional structure, the sliding angle of water droplets on the smooth surface of random folds is less than 5°, and the water collection is greatly improved. The neatly arranged slivers can not only effectively capture small droplets such as fog but also accelerate the agglomeration and drainage of droplets, thereby improving the water collection efficiency (Figure 7.3g). Gou et al. [18] reported the effects of two different forms of SLISs on water collection behavior. The modified SLISs were prepared by etching copper foil, controlling the morphology of cationic or anionic surfactants, surface hydrophobic modifications, and injection of a hydrophobic lubricant. SLISs with nanocrystalline layers showed better water collection performance (782.4) compared to the needle assembly surface (580.8) because of good lubrication layer retention in the nanocrystalline layer structure. The prepared SLISs exhibit chemical and physical stability under harsh environments such as corrosive solutions, long-term storage, and UV exposure, which is attributed to the surface texture of SLISs. Compared with flat SLIS, porous 3D frames have a more significant water collection capacity. For example, Pei et al. [19] studied the water-trapping performance of three-dimensional copper foam with

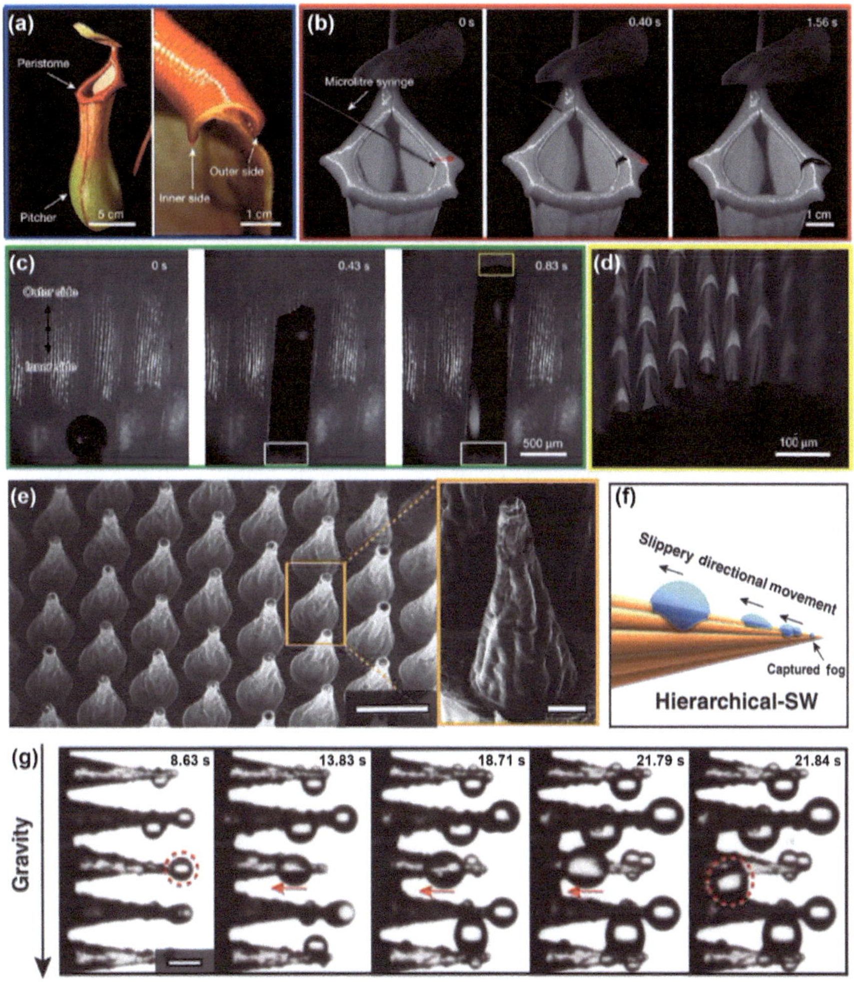

Figure 7.3 (a) Optical images depicting the mouth margin (left) alongside cross-sectional views of the peribody (right); (b) water droplets are applied to the mouth edge using a microliter syringe, with the red arrow indicating the direction of water movement; (c) optical images showing water transitioning from inside to outside; (d) magnified view of the water boundary. Source: Reproduced with permission from Chen et al., (2016) / Springer Nature. (e) ESEM images (viewed at a 45°) of hierarchical-SW featuring hexagonally arranged cones with aligned wrinkled surfaces. The dimensions for hierarchical-SW include tip diameter, base diameter, height, and tip-to-tip spacing measuring 70, 300, 700, and 400 μm, respectively. (f) A schematic representation illustrating the fog capture and removal process involving hierarchical-SW. (g) Time-lapse images demonstrating how fog is captured and directed toward the bases of cone structures in hierarchical-SW. Scale bars: (e) 500 μm (left), 100 μm (right); (g) 200 μm. Source: (e–g) Reproduced with permission from [16] / John Wiley & Sons; (f) [15] / with permission of Elsevier.

superhydrophobic surface and hydrophobic SLIS. The superhydrophilic copper foam surface covered with dense nano-ridges was prepared by the etching method with NaOH and $(NH_4)_2S_2O_8$ mixed solution. The modified surface was modified by chemical vapor deposition with fluoroalkyl silane. The modified surface showed superhydrophobicity. A hydrophobic SLIS is obtained by injecting a hydrophobic lubricant (PMX200/5cs) into the superhydrophobic surface (SHS). The modified smooth surface (~800 mg cm^{-3} h^{-1}) has better water collection performance than the SHS (~500 mg cm^{-3} h^{-1}). This is because the tiny droplets are trapped by the nanocanths, grow larger, and eventually form a film of water on the SHS. In this case, the formation of a water layer prevents the removal of water and the adsorption of new water droplets on the superhydrophobic foam, resulting in its unsatisfactory water collection performance. The smooth interface, under the synergistic effect of nano-spines and low SAs, can quickly capture tiny droplets and continuously promote the transport and removal of water, thus improving the efficiency of water collection. In addition, these prepared smooth foams maintained stable fog-trapping efficiency (~25%) and fog-trapping rate (15,200) for 30 consecutive cycles and also maintained low lubricant loss (~6.7%). This stability is mainly due to the action of the nano-spines on the modified smooth foam, which imparts a strong capillary force, thus trapping the lubricant on the foam. In addition to hydrophobic SLISs, hydrophilic SLISs have also been developed and have shown excellent water collection performance. For example, Dai et al. [20] studied water collection rates on different surfaces and found that hydrophilic smooth rough surfaces (SRSs) showed faster nucleation and removal rates of water droplets than hydrophilic or hydrophobic SLISs and graded SHSs (without lubrication). Combining the characteristics of pitcher plants (SLIS) and rice leaves (oriented and graded structure), a hydrophilic SRS layered structure was prepared to facilitate the collection and flow of water droplets. Hydrophilic SRSs, hydrophilic or hydrophobic SLISs, and graded SHSs have water-trapping properties of approximately 500, 350, 200, and 175, respectively. By adjusting the surface chemical composition and increasing the surface roughness, the water collection performance can be significantly improved. After 1000 consecutive drops fall off, the liquid repellency is still good because the lubricant remains on the nanotextured surface. This is because the capillary forces in the nanostructure protect the lubricant from being damaged. In addition, the use of higher viscosity lubricants can further reduce lubricant detachment.

7.2.3 Imitation Desert Beetle

Desert beetles effectively capture moisture from the fog using uneven surfaces randomly distributed between wax-coated hydrophobic zones and wax-free hydrophilic zones. Fog condenses on hydrophilic ridges and grows rapidly. When the volume of the droplet reaches a certain level, it falls off the hydrophilic bulge under the action of gravity, passes through the hydrophobic area, and enters the mouth of the beetle [21]. Based on this method of collecting fog, the surface with intimate and sparse patterns has attracted the attention of many scholars. Based on this method of collecting fog, the surface with intimate and sparse patterns has attracted the attention of many scholars. Inspired by the fog-collecting ability of the wettability gradient pattern on the back of the desert beetle, Sun et al. [15] used a smooth aluminum alloy substrate as a hydrophilic component and a superhydrophobic adhesive paper of different pore sizes as a superhydrophobic component. The fog-water collection efficiency was evaluated by adjusting and combining hole sizes on surfaces of superhydrophobic–hydrophilic hybrid patterns. Compared with homogeneous superhydrophobic and hydrophilic surfaces, the prepared single-aperture mixed superhydrophobic and hydrophilic patterned surfaces have higher fog-trapping efficiency. This is because mixing the superhydrophobic–hydrophilic surface can make the droplet reach the critical size of fall in the fastest time, accelerate the fall of the droplet, and greatly improve the fog-trapping efficiency. Zhang et al. [22] used electrospinning and ultrasonic techniques to successfully prepare a Janus membrane material with multiple bioinspired properties for fog collection and seawater desalination. The membrane consists of polyurethane nanofibers and surfacing anchored hydrophilic carbon nanospheres and carbon nanotubes, the introduction of which mimics the hydrophilic bumps of desert beetles. In the process of fog collection, the tiny droplets on a single nanofiber are first captured by the hydrophilic carbon nanosphere (CNS) process. When the droplets condense to the critical size, the high dynamic pressure generated by the process stimulates the droplets to roll from the junction of hydrophilic and hydrophobic parts to the hydrophobic region and fall off from the surface. And as the droplets move, they combine with neighboring water droplets to shorten the fog-trapping and coalescence process. By mimicking the wettability pattern of the Namib Desert beetle and the layered microchannel structure of the Sarracenia trichome, Huan et al. [23] proposed a novel integrated bioinspired Janus grid with a bi-directional wettability gradient and a microchannel connection pattern gradient for fog collection. Compared with previous surfaces, the obtained surfaces not only maintain effective droplet capture through spontaneous penetration of collected droplets, but also show excellent droplet transport capacity through the synergistic action of multiple driving forces. More importantly, the surface also accelerates the regeneration rate of the surface through the connectivity of the microchannels, improves the coalescence frequency of the droplets, and achieves a continuous and efficient fog-trapping cycle. As illustrated in Figure 7.4b, Zhou et al. [29] introduced an effective approach that integrates extrusion compression molding with surface modification for the large-scale fabrication of superhydrophobic–hydrophilic polypropylene/graphene nanosheet (UPP/GNS) films aimed at quantitative fog collection. The microcolumns inspired by desert beetles on the UPP/GNS surface feature hydrophilic tops and superhydrophobic bottoms and sidewalls, resulting in a mixed wettability due to the micro/nanostructures present. This mixed wettability significantly enhances the condensation and aggregation of droplets, facilitating mass transfer within the fog collection system. Notably, UPP/GNS films exhibit excellent resistance to chemical wear, passive anti-icing properties, active photothermal anti-icing capabilities, and sufficient durability. These attributes allow them to sustain stable fog collection efficiency even when exposed to various external factors such as salt solutions, acid/alkali environments, and high temperatures. Ultimately, this

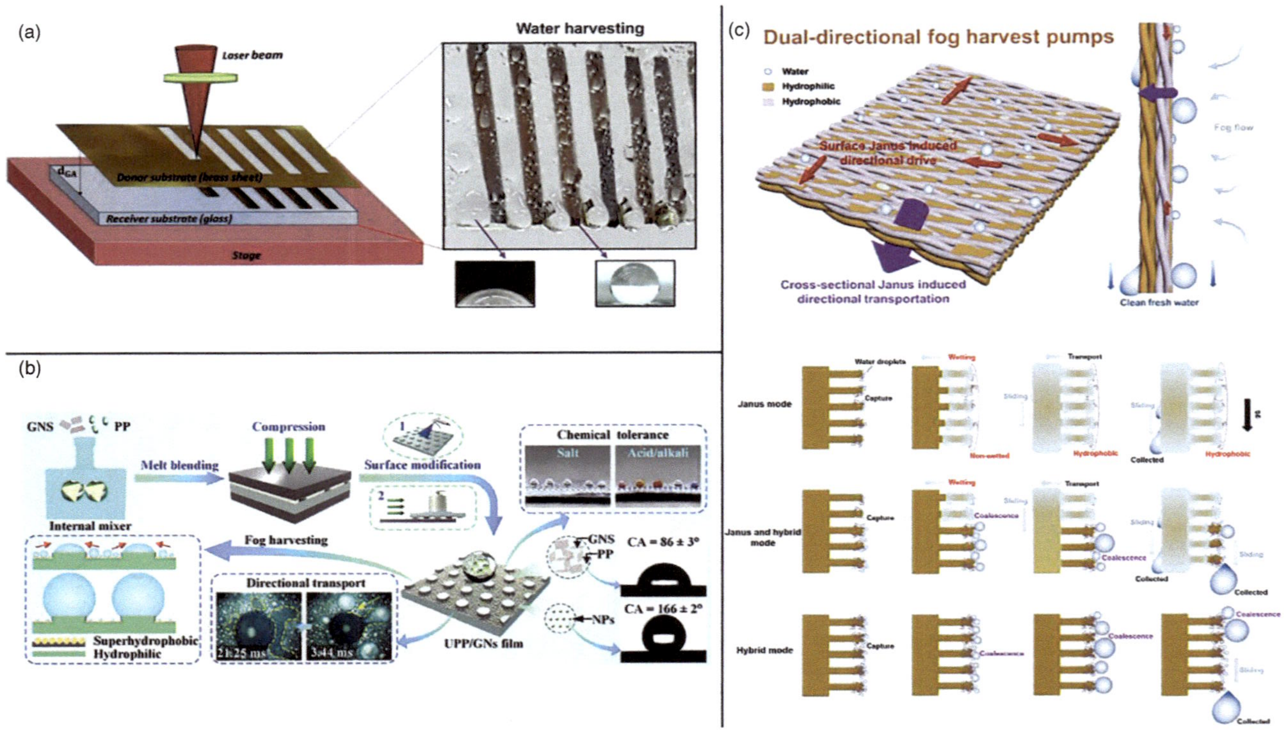

Figure 7.4 (a) A schematic illustration depicting the method for creating a superhydrophobic–hydrophilic mixed pattern surface on glass. Source: Reproduced with permission from [24, 25] / Elsevier. (b) A flowchart outlining the preparation process of UPP/GNS films. Source: [26] / with permission of Elsevier. (c) A schematic representation of hydrophilic/superhydrophilic asymmetrical grids along with three distinct operational modes for water collection. Source: Reproduced from [27] / with permission of Elsevier; Reproduced with permission from [28] / American Chemical Society.

combination of hydrophilic and superhydrophobic characteristics leads to a water collection efficiency of 1251 $mg\,cm^{-2}\,h^{-1}$ for UPP/GNS films. In another study depicted in Figure 7.4a, Bakhtiari et al. [25] utilized a straightforward physicochemical mixing technique to create glass surfaces with varying wettability profiles. Initially, integrated geometric patterns were applied onto the glass substrate using a laser-induced forward transfer method. Following this step, stearic acid (SA) treatment was employed to alter the wettability of the superhydrophilic deposition pattern into a superhydrophobic one. The resultant surfaces exhibited both superhydrophobic and hydrophilic features through different regular geometric patterns as well as fractal designs; their effectiveness in fog collection was evaluated while also examining how these design shapes influenced water accumulation on glass surfaces. In samples featuring classification patterns where hydrophilic regions are bordered by superhydrophobic areas, only minor increases in water collection efficiency were observed since droplets formed primarily within the hydrophilic zones before moving along the surrounding superhydrophobic structures once they reached a critical size. Due to the discrete nature of the hydrophilic region, water droplets growing on these regions can only leave the hydrophilic region when they reach a critical size, which reduces the water collection speed. However, when the sample has interconnected hydrophilic tubes, the growing droplets can move in the hydrophilic tube, and the rolling water droplets contact other captured water droplets on their path and carry them away. This process can lead to an increase in the size of the rolling water droplets, which are removed from the surface more quickly with the help of a superhydrophobic pattern. The parent-sparse surfaces of desert beetles can be thought of as surface Janus (S-Janus) and common Janus materials are thought of as interface Janus (CS-Janus). Wu et al. [28] drew inspiration from the surface of beetles to successfully create an asymmetrical mesh with hydrophilic and superhydrophilic properties, utilizing the water-directional transport capabilities inherent in the interfacial Janus (CS-Janus) structure. A stable superhydrophobic coating made from SA was partially applied onto zinc oxide nanorods that were supported by a stainless-steel substrate through a straightforward unilateral spray drying technique. By varying the concentration of the SA solution, it is possible to manipulate surface wettability as well as fog-trapping and transportation behaviors on the mesh. The process of fog collection was thoroughly investigated, leading to the development of an optimal Janus pump exhibiting bidirectional water collection performance, which significantly enhanced the material's overall fog collecting efficiency (Figure 7.4c).

7.2.4 Imitation Lotus Leaf-Janus

Two-dimensional design usually involves the construction of a membrane-like material. The name Janus is derived from the "two-faced god" in ancient Roman mythology, and the design was inspired by the wettability difference between the two sides of the lotus leaf [30]. One of the most important properties of these membranes is their ability to manipulate and control fluid transport. In recent years, there have been several studies using Janus to build fog collection materials. The reason is that Janus can facilitate the transport of the collected droplets without the input of external field energy. However, commonly used membrane substrates require complex top–down manufacturing processes or poor mechanical properties, which limits their practical application. Ding et al. [31] reported the preparation of a wooden Janus film. Unlike other Janus membranes, the graded structure and surface wetting gradient in this work (necessary for Janus membranes) are inherent to the wood itself. And the structure of the wood itself is biobased, renewable, and mechanically robust. The general method of making such films is to first vulcanize them with fluorosilane/TiO_2 nanoparticles and then irradiate one side of them with ultraviolet light to produce the desired surface wettability gradient. The laser cut then obtains a thin wood film and a surface with a smooth hierarchical structure on both sides. The experimental results showed that the water droplets adhered to the lower surface of the horizontally placed film. Although the hydrophobic surface is at the bottom, the water droplets are still transported to the hydrophilic surface at the top. This is driven by the Laplacian pressure on the asymmetric wettable Janus. The use of this natural internal hierarchical structure offers great potential for future smart buildings in arid regions. The above method of constructing Janus membranes directly takes advantage of the unique structural advantages of wood. In addition, there are many other methods for constructing Janus films using copper mesh (CM). Alnaser and coworkers [32] used two materials, copper foam and CM, to construct two types of Janus membranes using femtosecond laser technology. One side of the femtosecond laser-treated membrane produced a superhydrophilic surface, while the untreated side remained hydrophobic. Both membranes facilitate the easy transport of collected droplets from the hydrophobic side to the hydrophilic side, and vice versa. The ability to collect fog is a result of the different wettability of both sides and the surface micro-nanostructure constructed by the femtosecond laser. This superhydrophilicity of the surface is due to the increase in surface roughness. In addition, this is also due to the appearance of a large number of polar oxidation sites on the surface after femtosecond laser treatment, which has a strong affinity for hydroxylation. The structure of the copper Janus membrane is relatively simple, and a drop of water adsorbed on the hydrophobic surface flows directly through the lower side of the membrane. Moreover, in a foam, its internal winding structure allows water droplets to diffuse to the inner hydrophilic part before being transported to the bottom. The experimental results show that the fog collection efficiency of the mesh Janus membrane is higher than that of the copper foam-based Janus membrane. This femtosecond laser modification cannot be maintained for a long time and is easily damaged, so it requires a certain hydrophilic coating protection.

The reason for the high fog-trapping efficiency of the Janus membrane is its hydrophobic upper surface, which reduces the flow resistance, promotes the collision of water droplets and the accumulation of large water droplets, and quickly transports to the hydrophilic area [12, 33]. Song et al. [34] designed and fabricated a highly porous, asymmetric, ultra-wettable Janus membrane that can be used for freshwater harvesting. The CM is chemically etched and then modified by coating as needed to obtain a black copper mesh (SPCM) with a photothermal effect. Then, a chemical cross-linking reaction is performed after the coating to obtain compatibility between the asymmetric hyperwettability layers (CSPCM). The final surface obtained by electrospinning on the CSPCM surface is called ECSPCM, and its structure is designed with different wettability and acicular hydrophilic $Cu(OH)_2$ nanowires on both sides of the mesh on a CM substrate. The migration and collection of droplets on the surface of the SPCM and ECSPCM Janus membrane were compared. As can be seen from the results, ECSPCM's Janus can capture smaller droplets. This is because the nanofibers with superhydrophilic backsides in Janus produce a larger wetting gradient. In addition, the hydrophilic nanowire structure on the Janus membrane increases the surface nucleation sites and the specific surface area in contact with the fog. Popular in this work is the design of the structure, which can capture fog at night while desalting during the day using a coating with a photothermal effect. This provides more ideas for the design of fog collectors (FC), and the combination of certain chemical properties on top of hydrophobic surface (HS) can make FC more efficient in meeting the daily water needs of humans.

7.3 Three-Dimensional (3D) Preparation of Fog-Collecting Materials

Three-dimensional materials have unique advantages in fog collection. The combination of a small-scale 3D structure and 2D plane to collect water can greatly improve the water collection efficiency of the 2D plane. At the same time, the large-sized 3D structure can be directly used for fog collection after special wettability. A typical example of a small 3D structure combined with a 2D plane for water collection is the back structure of the Namib Desert beetle. The Namib Desert beetle has excellent water collecting capacity due to its hydrophilic ridges and hydrophobic depressions on its back. It is found that apart from the influence of surface wettability, the influence of concave and convex structures on the improvement of water collection capacity cannot be ignored. In recent years, a variety of raised 3D structures have been combined with 2D planes for fog collection [14, 35]. For example, Zhong et al. [26] modeled the raised structure on the back of the Namib Desert beetle and prepared SHSs with hemispherical bulges for fog collection. To explore the mechanism of improving the surface water collection efficiency of the convex structure, they simulated the droplet collision at the convex and plane of the fog flow, as well as the transient velocity vector and velocity profile distribution of the tiny droplet in the fog flow in and around the single convex. It can be seen from the simulation results that the hemispherical convex structure affects the fluid velocity contour composed of droplets and air. The convex structure makes the fluid colliding with it more concentrated, resulting in a higher density of velocity vectors at the convex structure. Therefore, the raised structure captures more droplets than the flat structure. In addition, the simulation diagram shows that the droplet velocity on the raised top is faster than that on the plane, so the raised structure can accelerate the droplet growth rate on the top. The presence of hemispherical bulges greatly improves the water collection efficiency of the surface. Inspired by the raised structure, hydrophilic and hydrophobic alternating pattern, and vein pattern of the desert beetle Elytra, Pang et al. [24] designed a hydrophilic and hydrophobic patterning material with raised for fog collection through chemical etching and template method. Compared with the original flat zinc sheet and the superhydrophobic zinc sheet with the bulge, the fog collecting efficiency of the hydrophilic and hydrophobic patterned material with the bulge is increased by 75% and 41%, respectively. Inspired by desert beetles, Wang and coworkers [27] used a laser marking machine to remove the superhydrophobic coating on the surface of the copper sheet. After selective deposition of copper, hydrophilic bumps were grown in the hydrophilic region, and superhydrophobic materials with an array of hydrophilic bumps were prepared. The surface morphology and chemical composition of each sample were characterized and analyzed. Finally, the fog-collecting efficiency of each sample was tested. Compared with the prepared superhydrophilic, superhydrophobic, and original copper sheet, the superhydrophobic sample with the array hydrophilic bump has not only the characteristics of efficient fog capture and rapid removal of water droplets, but the surface structure is also reconstructed by using liquid droplets hanging on the surface, which greatly increases the effective fog-trapping area, so the fog collection efficiency reaches 207% of the original copper sheet.

Large-scale 3D structures that can be used for fog collection are always inspired by our lives, such as paper-cut art and corrugated paper. Inspired by the art of paper cutting, Li et al. [36] prepared an efficient kirigami structural fog collector with aerodynamic assistance (Figure 7.5a). The tubular fog vortex provided by the 3D structure based on paper-cut art is the source of its fast fog-capture ability. The velocity vector diagram of fog flow on a 3D structure surface with different angle θ (θ is the bending angle) is simulated. Through the flow line, it can be seen that there is a counter-rotating vortex at all folds, and with the decrease of θ, the vortex becomes smaller and closer to the substrate. Because of the presence of this counter-rotating vortex, the fog flow forms additional areas of stagnation before bending the base. These stagnation zones change the trajectory of the droplet and attract the droplet to aggregate. Fog collectors with corrugated structures inspired by corrugated paper are also suitable for large-size preparation and have unique advantages in the fog collection process. The ripple structure can increase the number of collisions of fog flow on the surface of the fog collector, and the increase of collision frequency increases the probability of tiny droplets in fog flow being captured by the hydrophilic region on the surface of the fog collector, thus improving the fog collection efficiency of fog collector. Wang et al. [37] prepared three fog collectors with both kirigami structure and wetting gradient. Hexagonal ZnO nanowires were grown by anodic oxidation on the fog-facing surface, and superhydrophobic SiO_2 nanoparticles were sprayed on the back. The droplets nucleate and grow on the superhydrophilic surface, and are then directionally transported to the SHS, where they gradually grow and are collected. Among them, sample 3 has the best fog collection effect. The fog collection efficiency can reach about $3.5\,g\,h^{-1}\,cm^{-2}$, which is 1.82 times that of an unmodified zinc sheet (Figure 7.5b).

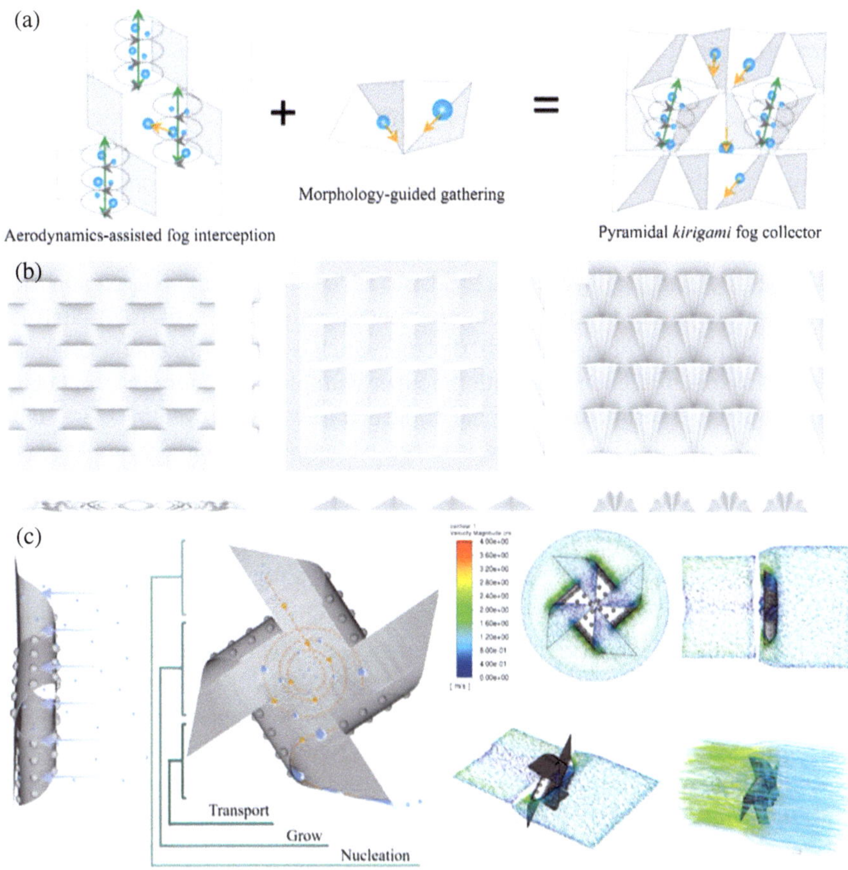

Figure 7.5 (a) Kirigami structure based efficient fog collector. Source: [36] / Springer Nature / CC BY 4.0. (b) Front view and side view of fog collectors with wavy, V-shaped and W-shaped structures. Source: Reproduced with permission from [37] / American Chemical Society. (c) Schematic diagram of droplet nucleation, growth and transport in windmill and flow field simulation of fog collector. Source: [38] / American Chemical Society.

The key to improving the efficiency of the 3D fog collector, which is inspired by the corrugated structure of corrugated paper, is that the corrugated structure can increase the number of collisions of fog flow on the surface of the fog collector. Peng et al. [39] prepared superhydrophilic zinc tablets by electrochemical anodizing. The SHS was prepared by grafting PDMS onto ZnO by photocatalytic reaction combined with the spraying method. The 3D structure of the corrugated paper was prepared by the template method, and the superhydrophilic and superhydrophobic 3D structure (TSHL-SHB) was prepared by the mask spraying method. The combination of hybrid wettability and a 3D structure gives TSHL-SHB excellent water collection performance. First, the hybrid wettability speeds up the fog collection cycle, the superhydrophilic tip accelerates the capture of droplets, and the superhydrophobic region promotes the rapid transport and separation of droplets. The 3D structure of corrugated paper improves the efficiency of droplet interception. One is that the 3D structure increases the surface area in contact with the fog stream. Second, it is because the concave of the imitation corrugated paper structure forms a tubular vortex connected up and down. This tubular vortex is a vortex from the outflow to the center, which can effectively improve the efficiency of droplet trapping. Most of the droplets in the fog stream will gather near the vortex core along the direction of the vortex. Then small droplets gather near the vortex core and grow into large droplets. Finally, large droplets break out of the vortex and are ejected to the surface in an opposite direction. The water collection efficiency of TSHL-SHB is much higher than that of the original 2D plane, which is 418% of that of the original zinc sheet. Meanwhile, compared with the superhydrophilic and superhydrophobic hybrid 2D plane, the superhydrophobic 3D structure was improved by 168%, and the superhydrophobic 3D structure was improved by 150%. Inspired by umbrellas, corrugated paper, and beetles, He et al. [40] prepared umbrella-like corrugated structures with pentagonal star-like hydrophilic bumps. The circular umbrella structure can ignore the direction of the fog flow, realize 360° fog collection, and ignore the influence of the direction of the fog flow. The surface of the umbrella is distributed with corrugated structures, which can increase the collision times of fog flow in the structure and improve the capturing ability of fog flow. Hydrophilic bumps are uniformly distributed on the surface of the lattice structure, and the hydrophilic bumps and SHSs form a beetle-like surface, which accelerates the trapping and separation of water droplets. Compared with ordinary circular bumps, the hydrophilic bumps shaped like pentagonal stars have more nucleation sites and better trapping ability for fog flow. Under the multi-bionic structure, the fog collecting efficiency of the fog collector is greatly enhanced, which is 232.6% of the original copper sheet.

The idea of combining geometric gradient conical structures with special wettability bases inspired by cactus spines has also been widely used in the preparation of 3D fog collectors. For example, Zhou et al. [41] designed an efficient fog collection device consisting of a one-dimensional (1D) conical copper needle with gradient wettability and a hydrophilic sliding surface, which can simultaneously satisfy the capture, condensation, and transport of water droplets. Besides, the surface of the copper needle with a wettability gradient is covered with two micro-nanostructures with different roughness, showing the wettability difference between the superhydrophobic tip and the hydrophilic tail. Therefore, the tiny droplets can be quickly captured by the copper needle, and under the action of the Laplacian pressure difference and wettability gradient, the small droplets move directionally, condense into large droplets, and accelerate droplet transport and removal. Further combined with the hydrophilic sliding surface, the droplets moving to the end of the copper needle can be quickly removed under the action of the sliding surface. This combined system can realize continuous and efficient circulation in the fog collection and effectively improve the water collection rate. On this basis, Zhou et al. [42] also designed a fog collection system consisting of a superhydrophilic copper needle with a zinc substrate surrounded by regular hydrophilic bumps and hydrophobic slippage. Among them, the regularly raised top has regular circular holes into which a superhydrophilic copper needle is vertically inserted and fixed with a sponge base. The system exhibits an efficient droplet capture-coagulation-transport cycle during fog collection. A large number of captured droplets move quickly to the top of the bulge due to the meniscus effect of the oil produced by the bulge, greatly increasing the movement speed of the droplets. Inspired by the way seeds move toward aquatic vegetation in nature, the oil meniscus face around the hydrophilic bulge on the hydrophobic slip surface generates a capillary driving force for the droplet to move toward the bulge in all directions. Under the synergistic action of the whole fog collection system, the fog collection efficiency is greatly increased, up to $12.52\,g\,h^{-1}$. Based on the above work, Fu et al. [43] further developed a combined fog collector that can collect both fog and dew. The collection part of the system has an alternating concave and convex gradient structure, giving it different chemical wettability. Hydrophilic sponges will be used as storage sections and mechanically assembled to form a fog/dew collection system. This combined approach can improve collection efficiency by adjusting the liquid behavior throughout the fog collection process, including droplet nucleation during capture, high-speed transfer over conical channels, and droplet separation and storage from the Janus sponge. The synergistic effect of cone capture and sponge separation collection improves the efficiency of fog collection. Compared with the traditional fog collection structure, the collection system greatly improves the efficiency of fog collection. At the same time, the collection system can effectively collect dew under traditional supercooling conditions.

At present, most of the 3D fog collectors prepared are static fog collectors, and researchers have developed a new 3D dynamic fog collector system based on the static 3D fog collector inspired by the windmill and have the power generation capacity at the same time. Inspired by beetles and origami windmills, Wang and coworkers [38] designed a fog collector with a windmill structure capable of both fog collection and wind power generation (Figure 7.5c). The fog collector consists of a metal origami windmill made of aluminum sheets and a shell. A mold is used to press out a round bulge at the tip of the inner windmill. The rough nanostructure of alumina was grown on the surface of the windmill by anodizing, and some parts were modified by stearic acid to be superhydrophobic. The shell is prepared by 3D printing technology, and the spraying method is used to make the surface of the shell uniformly covered with a layer of superhydrophobic nano silica particles. When the fog collector works in the fog flow, the droplets nucleate in the hydrophilic bulge, move and grow on the surface with the rotation of the windmill, and finally overcome the pinning effect under the combined action of centrifugal force and gravity, break away from the surface of the windmill, and are intercepted and collected in the inner wall of the shell. This fog collector not only can realize efficient water collection but also has certain potential for practical application. Zhong et al. [44] produced a large-scale Gradient-Janus wire (GJW) with a diameter of 0.3 mm through liquid confinement modification technology. The opposite conical wettability phenomenon appeared along the same direction of the GJW, which was characterized by an increase in the superhydrophilic zone and a decrease in the hydrophobic zone. This feature allows the GJW to hover 0.6 mg of fog in 30 s during fog collection. Moreover, when the collected water droplets are fixed at the end of the GJW, effective torque is generated as the driving force for rotation. On this basis, a GJW wheel was designed, which arranged 60 GJWs together in a radial direction to form an ultra-light structure weighing only 1.9 g. When the coil is loaded in a magnetic field, the peak output power is about 0.25 μW, while maintaining a high-water collection efficiency, averaging about $38 \pm 2.12\,\mathrm{mg\,min^{-1}}$.

In summary, the preparation of fog collection materials encompasses one-, two-, and three-dimensional structures. One-dimensional materials utilize structural gradients, such as the unique configuration of spider silk, to enable the directional transport of fog droplets. Two-dimensional materials, often biomimetic and diverse, mimic structures like cactus spines to harness the Laplace pressure for fog droplet collection. Three-dimensional materials leverage their volumetric structural advantages to enhance collection efficiency. Moving forward, optimizing material properties and broadening application areas, such as industrial water recovery and self-sustaining water systems in buildings, will be essential for advancing fog collection technology.

References

1 Daniel, S., Chaudhury, M.K., and Chen, J.C. (2001). *Science* 291: 633.
2 Zheng, Y., Bai, H., Huang, Z. et al. (2010). *Nature* 463: 640.
3 Liu, Y., Yang, N., Gao, C. et al. (2020). *ACS Applied Materials and Interfaces* 12: 28876.
4 Guo, C., Wang, C., Huang, Q. et al. (2022). *Heliyon* 8: e10007.
5 Liu, H., Wang, Y., Yin, W. et al. (2022). *Journal of Materials Chefogry A* 10: 7130.
6 Chen, Y., Li, D., Wang, T., and Zheng, Y. (2016). *Scientific Reports* 6: 19978.
7 (a) Nguyen, L.T., Bai, Z., Zhu, J. et al. (2022). *ACS Sustainable Chefogry and Engineering* 10: 11176. (b) Kowalski, N.G. and Boreyko, J.B. (2022). *Soft Matter* 18: 7148.
8 Zhang, M., Zheng, Z., Zhu, Y. et al. (2022). *Chemical Engineering Journal* 433: 134495.
9 Liu, Y., Yang, N., Li, X. et al. (2020). *Small* 16: 1901819.
10 (a) Gürsoy, M., Harris, M.T., Downing, J.O. et al. (2017). *Colloids and Surfaces A: Physicochemical and Engineering Aspects* 529: 195. (b) Park, K.-C., Kim, P., Grinthal, A. et al. (2016). *Nature* 531: 78.
11 Bai, H., Zhao, T., Wang, X. et al. (2020). *Journal of Materials Chefogry A* 8: 13452.
12 W. Zhou, C. Zhou, C. Deng et al. (2022). *Advanced Functional Materials* 32: 2113264.
13 He, W.-X., Wang, X.-W., Chu, Z.-W. et al. (2023). *Colloids and Surfaces A: Physicochemical and Engineering Aspects* 657: 130512.
14 Su, Y., Chen, L., Jiao, Y. et al. (2021). *ACS Applied Materials and Interfaces* 13: 26542.
15 Sun, R., Zhao, J., Liu, C. et al. (2022). *Progress in Organic Coatings* 171: 107016.

16 Ahn, J., Jeon, J., Heu, C.S., and Kim, D.R. (2018). *Advanced Materials Interfaces* 5: 1800980.
17 (a) Li, J. and Guo, Z. (2019). *Research* . (b) Gou, X. and Guo, Z. (2021). *Chemical Engineering Journal* 411: 128495.
18 Gou, X. and Guo, Z. (2020). *Langmuir* 36: 8983.
19 Pei, W., Li, J., Guo, Z. et al. (2021). *Chemical Engineering Journal* 409: 128180.
20 Dai, X., Sun, N., Nielsen, S.O. et al. *Science Advances* 4: eaaq0919.
21 (a) Uddin, M.N., Desai, F.J., Rahman, M.M., and Asmatulu, R. (2020). *Nanoscale Advances* 2: 4627. (b) Knapczyk-Korczak, J., Ura, D.P., Gajek, M. et al. (2020). *ACS Applied Materials and Interfaces* 12: 1665.
22 Zhang, Y., Cai, Y., Shi, J. et al. (2022). *Desalination* 540: 115975.
23 Huan, J., He, Y., Geng, X. et al. (2023). *Colloids and Surfaces A: Physicochemical and Engineering Aspects* 659: 130780.
24 Pang, Y., Wang, F., Mei, G. et al. (2024). *Surfaces and Interfaces* 46: 104155.
25 Bakhtiari, N., Azizian, S., and Jaleh, B. (2022). *Journal of Colloid and Interface Science* 625: 383–396.
26 Zhong, L., Zhu, H., Wu, Y., and Guo, Z. (2018). *Journal of Colloid and Interface Science* 525: 234.
27 Pang, Y., Mei, G., Wang, F. et al. (2024). *Colloids and Surfaces A: Physicochemical and Engineering Aspects* 680: 132641.
28 Wu, J., Yan, Z., Yan, Y. et al. (2022). *ACS Applied Materials and Interfaces* 14: 49338–49351.
29 Zhou, W., Wu, T., Du, Y. et al. (2023). *Chemical Engineering Journal* 453: 139784.
30 Wang, X., Zhang, W., Li, S. et al. (2023). *Chemical Engineering Journal* 472: 145042.
31 Ding, Y., Tu, K., Burgert, I., and Keplinger, T. (2020). *Journal of Materials Chefogry A* 8: 22001.
32 Yalishev, V.S., Abbasi, N.A., Iqbal, M., and Alnaser, A.S. (2023). *Langmuir* 39: 1815.
33 Tang, J., Peng, L., Chen, D. et al. (2022). *ACS Applied Materials and Interfaces* 14: 2202.
34 Song, M., Zhu, Z., Qi, J. et al. (2023). *Environmental Science: Nano* 10: 996.
35 Park, H., Hwang, J., Lee, T.H. et al. (2021). *ACS Applied Materials and Interfaces* 13: 27575.
36 Li, J., Ran, R., Wang, H. et al. (2021). *Nature Communications* 12: 5484.
37 Wang, X., Guo, Z., and Liu, W. (2023). *Langmuir* 39: 8926.
38 Li, L., Wang, X., Xin, Y. et al. (2024). *ACS Applied Materials and Interfaces* 16: 15496.
39 Peng, Z., Fu, Y., and Guo, Z. (2023). *ACS Applied Materials and Interfaces* 15: 38110.
40 He, S., Wang, X., Li, S. et al. (2023). *Surfaces and Interfaces* 43: 103545.
41 Zhou, H., Jing, X., and Guo, Z. (2020). *Langmuir* 36: 6801.
42 Zhou, H., Jing, X., Li, S., and Guo, Z. (2021). *Chemical Engineering Journal* 417: 129215.
43 Fu, Y., Wu, L., Ai, S. et al. (2023). *Nano Today* 52: 101979.
44 Zhong, L., Chen, H., Zhu, L. et al. (2024). *ACS Nano* 18: 10279.

8

Sustainable Use of Fog Collectors

In the natural environment, fog collectors face numerous severe challenges. The water collection surface may be contaminated by windblown dust, which can affect its wettability and microstructure. Long-term impacts from sand particles, such as abrasion and friction, can lead to surface wear, reducing the efficiency of droplet capture and transfer. The erosion caused by acid rain can damage the chemical stability of the surface, altering its surface energy and chemical composition. The proliferation of bacteria can result in microbial contamination, clogging the collection channels and obstructing the aggregation and transfer of droplets. Prolonged exposure to ultraviolet radiation may cause material aging, thereby reducing its mechanical properties and surface stability. To effectively resist these threats and maintain the efficient operation of fog collectors, the sustainability of the water collection surface is critical. In this context, the present study thoroughly investigates various strategies to enhance the mechanical stability, chemical stability, and antibacterial properties of the collection surface.

8.1 Mechanical Stability

In water collection systems, hydrophobic regions serve as a core component and play a critical role. Their primary function is to facilitate the formation and maintenance of nearly spherical droplets on the surface. This spherical configuration not only promotes droplet growth but also enables their rapid removal from the surface, thereby significantly enhancing water collection efficiency. However, despite the numerous advantages of hydrophobic surfaces in the water collection process, most hydrophobic materials with microstructured surfaces suffer from a notable lack of mechanical stability. The surface textures of these materials are fragile and susceptible to damage under external forces; even simple friction from a fingertip can cause the surface structure to degrade [1]. The destruction of such surface textures severely impairs the water collection process. Once the microstructures and nanostructures of the hydrophobic surface are damaged, its hydrophobicity is lost, and the wettability of the surface may shift from hydrophobic to hydrophilic [2].

Bioinspired Materials for Fog Harvesting: Principles, Design, Preparation, and Applications, First Edition. Zhiguang Guo and Weimin Liu.

This transition results in a significant decline in water collection efficiency, thereby compromising the performance of the entire water collection system. Consequently, developing hydrophobic surfaces with enhanced mechanical durability is crucial for extending the operational lifespan of water collection devices. Currently, although a few superhydrophobic surfaces have demonstrated some degree of mechanical durability – retaining high water contact angles (WCAs) even after tests involving sand impact [3], tape peeling [4], soft cloth wiping [5], and sandpaper abrasion [6] – the stability of these surfaces is difficult to maintain over the long term. After multiple cyclic tests, their hydrophobicity tends to degrade. This indicates that while these superhydrophobic surfaces exhibit short-term mechanical durability, their long-term stability remains a challenge. Therefore, future research must further explore and develop hydrophobic surfaces with higher mechanical stability and long-term durability to meet the performance and lifespan requirements of water collection devices in practical applications.

Wang et al. [7] employed meticulous structural engineering to develop a super-reliable surface with an armored structure. In this design, nanoscale structures endowed the surface with supercritical performance, while specific structural components provided notable mechanical stability. Specifically, the microstructures were fabricated using photolithography to create inverted pyramidal cavities on a silicon substrate, and these were encapsulated with fluorinated nanoclusters to achieve super-silent properties. When subjected to external forces, the interconnected microstructures effectively shielded objects with dimensions exceeding the maximum framework size, thereby protecting the internal nanostructures (Figure 8.1a). The mechanical stability of this structure was primarily determined by its geometric configuration. Research demonstrated that increasing the angle α between the sidewalls of the localized structures and the substrate is an effective strategy for improving structural stability. However, this modification also increases the liquid–solid contact fraction (f_{mirco}), which significantly affects the focal performance of the surface. Experimental findings revealed that when α reached 120°, the optimal balance between mechanical stability and hydrophobic performance was achieved. Even after repeated scraping with a steel blade, the nanostructures maintained a high degree of integrity. However, wear caused the fluorosilane layer on the tops of the microstructures to be removed, leading to a localized transition in wettability from hydrophobic to hydrophilic (Figure 8.1b). Overall, the gas–liquid–solid composite surface demonstrated excellent stability, primarily due to the outstanding anti-friction properties and mechanical robustness provided by the microstructures (Figure 8.1c). To more comprehensively evaluate the effects of mechanical fatigue on the superhydrophobic surface, the researchers measured the static contact angle (θ^*) and roll-off angle (θ_{roll}) under varying aperture widths and liquid–solid contact fractions (f_{micro}) (Figure 8.1d,e). The results indicated that although static contact angles slightly decreased and roll-off angles increased following fatigue, the surface retained its previous mobility levels. Compared with the Cassie–Baxter model, the inverted pyramidal structure's hydrophobic performance was found to be independent of its size, and the armored structure effectively protected the nanostructures.

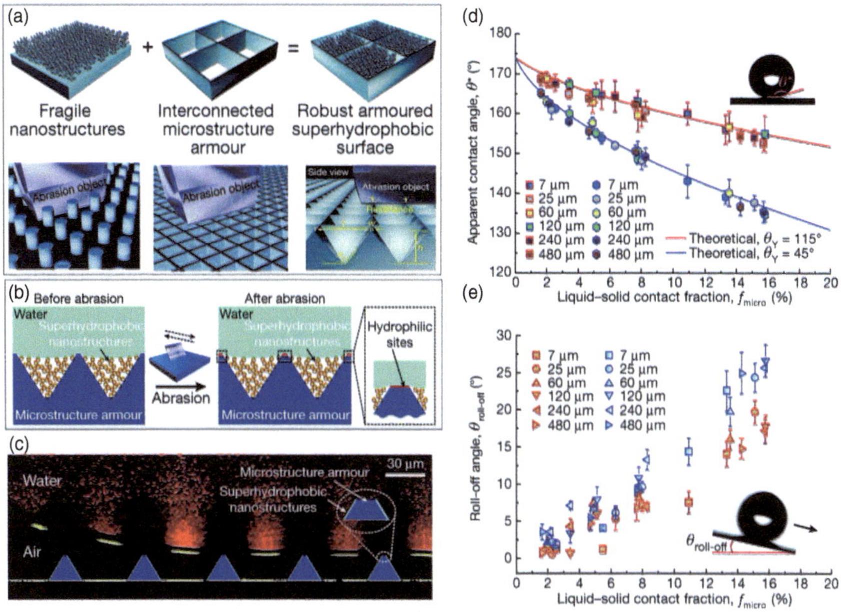

Figure 8.1 (a) This figure illustrates a strategy to enhance the mechanical stability of superhydrophobic surfaces by incorporating hydrophobic nanostructures into a protective microstructural "armor." The interconnecting microstructural topology serves to block abrasive materials larger than the framework dimensions. (b) It presents the water-repellent mechanism of the armored superhydrophobic surface before and after wear. (c) This diagram demonstrates that even after wear, the gas–liquid–solid composite interface can maintain stability. (d) It compares the contact angles before wear (represented by square symbols) and after wear (represented by hexagonal symbols) with theoretical models. (e) It provides the rolling angle measurement data before wear (indicated in red) and after wear (indicated in blue). Source: Reproduced with permission from [7] / Springer Nature.

In the literature related to fog collection, some studies have included tests on mechanical durability. Most fog-collecting surfaces fabricated through methods such as chemical etching, spraying, and 3D printing exhibit relatively fragile properties, making them difficult to withstand even mild friction tests. Hydrophobic armored structures with high wear resistance hold promising application potential in water collection systems, as they can effectively shield the equipment from outdoor sand and dust erosion, thereby extending its service life and enhancing operational stability. Moreover, the use of substrates with superior wear resistance and the optimization of spraying techniques are also regarded as effective approaches to improving mechanical stability. Copyright 2020. Reproduced with permission from *Nature*.

8.2 Chemical Stability

Water-harvesting interfaces operate under stochastic environmental stressors that compromise their functional durability. Acidic rainwater (pH < 5.6) induced by atmospheric contaminants initiates surface chemical corrosion processes, leading to progressive degradation of hydrophilic constituents and subsequent performance deterioration. Furthermore, sustained thermal extremes (−40 to 80 °C) and high-energy UV exposure (λ = 300–400 nm) induce photo-oxidative chain scission mechanisms, significantly destabilizing interfacial polymer matrices [8]. To ensure operational robustness, engineered water-capturing surfaces necessitate multi-phased stabilization protocols: (i) pH-responsive buffering networks to neutralize ionic imbalances, (ii) thermally adaptive supramolecular architectures maintaining structural integrity across cryogenic to hyperthermal regimes, and (iii) UV-absorbing nano-additives for selective photonic energy attenuation. This synergistic stabilization paradigm enables sustainable hydrologic functionality in heterogeneous outdoor environments [8].

Song et al. [9] engineered a hybrid wettability interface by integrating polydimethylsiloxane (PDMS) and graphene (PDMS/G, 3.3 wt%) onto copper mesh substrates. Laser ablation facilitated PDMS/G film deposition, followed by ultrasonic vibration to achieve graded wettability patterning. Corrosion resistance evaluations involved immersion in aggressive media (1 M HCl, NaOH, and NaCl; 2 h; Figure 8.2a), with post-exposure analyses revealing retained hydrophobicity (Figure 8.2b) and stable water capture rates (WCR: 5.26 g cm^{-2} h^{-1}). These findings confirm the graphene-polymer composite's exceptional chemical resilience, validating its applicability in corrosive environments. Concurrently, Zhou et al. [10] devised a Janus copper foam system with alternating hydrophobic–hydrophilic nanoscale domains via hydrothermal synthesis and UV-assisted photocatalysis. ZnO nanowire arrays were epitaxially grown on $Cu(OH)_2$ substrates, followed by low-concentration PDMS grafting to modulate surface energetics. The hierarchical architecture demonstrated enhanced fog-harvesting efficiency. Accelerated aging protocols – including thermal cycling (120 °C/20 °C, 5 min/2 min intervals, 10 cycles) and prolonged UV irradiation (36 h, 300–400 nm) – revealed minimal degradation in WCA and collection capacity (8.15% mass loss). Robust Zn–O–Si covalent networks were identified as the mechanistic basis for sustained interfacial stability, underscoring the system's ultraviolet resistance and thermal durability.

Generally, employing coating materials with strong chemical stability can significantly improve the corrosion and weather resistance of the fog-collection surface. Furthermore, during the processes of chemical etching or deposition, it is essential to ensure that the coating fully encapsulates the substrate to minimize the risk of substrate exposure to the surrounding environment.

8.3 Antibacterial

Prolonged operation of water-harvesting systems in hygroscopic environments poses significant microbial colonization risks, wherein pathogen proliferation on interfacial surfaces compromises hydrologic safety [12]. To address this challenge, diverse antimicrobial paradigms have emerged, with superhydrophilic zwitterionic coatings demonstrating exceptional efficacy in suppressing bacterial adhesion

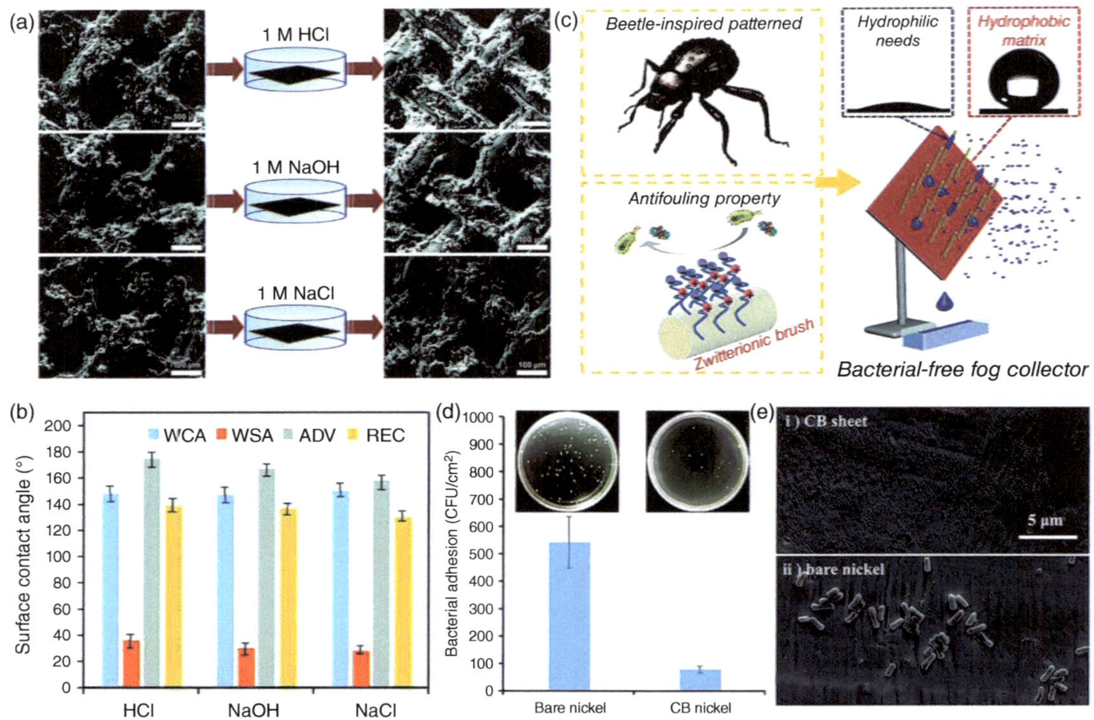

Figure 8.2 (a) Comparative SEM micrographs illustrating surface morphology evolution following 2 h immersion in corrosive media (1 M HCl, NaOH, NaCl). (b) Wettability characterization of engineered heterogeneous wettability interfaces post-chemical exposure, demonstrating retained hydrophobic functionality. Source: Reproduced with permission from [9] / The Royal Society of Chemistry. (c) Schematic representation of a bioinspired antibacterial fog-harvesting architecture featuring alternating hydrophilic–hydrophobic domains. (d) Quantitative bacterial adhesion analysis: Optical micrographs contrasting *E. coli* colonization densities (107 cells mL^{-1}, 2 h) on pristine versus carboxybetaine-functionalized nickel substrates. (e) SEM validation of long-term antimicrobial efficacy (109 cells mL^{-1}, 72 h), revealing minimal biofilm formation on zwitterionic surfaces versus severe contamination on untreated controls. Source: Reproduced with permission from [10, 11] / American Chemical Society.

through interfacial hydration mechanisms [13]. These coatings establish a tightly bound aqueous boundary layer that sterically hinders direct microbe-surface interactions, thereby disrupting biofilm formation. Wen et al. [11] innovatively engineered a bioinspired fog-harvesting architecture combining antibacterial needle arrays (ABN) with desert beetle-mimetic wettability gradients (Figure 8.2c). The ABN substrate was functionalized with zwitterionic carboxybetaine (CB) polymer brushes, achieving dual functionality: enhanced fog droplet nucleation via polar interactions and antimicrobial protection through electrostatic hydration barriers. Gravitational forces facilitated rapid droplet detachment from hydrophobic regions, while system performance optimization was achieved through parametric adjustments. At a 60° inclination angle, peak water capture rates (WCRs) were attained, with further enhancements observed through CB needle density modulation – a direct correlation between increased specific surface area and elevated WCR. Antimicrobial validation via in vitro bacterial adhesion assays revealed stark contrasts between unmodified and CB-functionalized nickel substrates. At *E. coli* inoculum concentrations of 107 cells mL^{-1} (2 h incubation), quantitative analysis demonstrated an 85.6% reduction in surface-adhered pathogens on CB-coated surfaces (78 CFU) versus bare nickel controls (542 CFU; Figure 8.2d). Extended exposure (72 h, 109 cells mL^{-1}) corroborated these findings, with scanning electron microscopy confirming negligible biofilm formation on CB interfaces compared to heavily contaminated bare substrates (Figure 8.2e). The CB coating's hyper-hydration capacity creates a steric exclusion zone, effectively decoupling microbial cells from the underlying material – a critical advancement for durable antibacterial hydrologic systems.

To enhance the durability of water collection surfaces, it is essential to consider factors such as structural design, material selection, and the preparation process. The armored structure stands out as a promising approach to improving the wear resistance of these surfaces, despite the limited research in this area, making it a valuable avenue for exploration. In addition, selecting appropriate coating materials is a direct and effective method to enhance both chemical stability and antibacterial properties. Finally, while the preparation process is often underestimated, it plays a crucial role. Parameters involved in processes like electrochemical corrosion, chemical etching, and spraying determine the coating's density and thickness, significantly influencing the overall performance of the water collection surface.

References

1 Verho, T., Bower, C., Andrew, P. et al. (2011). *Advanced Materials* 23: 673.

2 Milionis, A., Loth, E., and Bayer, I.S. (2016). *Advances in Colloid and Interface Science* 229: 57.

3 Li, Y., Chen, S., Wu, M., and Sun, J. (2014). *Advanced Materials* 26: 3344.

4 Maitra, T., Antonini, C., Auf der Mauer, M. et al. (2014). *Nanoscale* 6: 8710.

5 Wang, H., Liu, Z., Zhang, X. et al. (2016). *Advanced Materials Interfaces* 3: 1600040.

6 Lu, Y., Sathasivam, S., Song, J. et al. (2015). *Science* 347: 1132.
7 Wang, D., Sun, Q., Hokkanen, M.J. et al. (2020). *Nature* 582: 55.
8 Jing, X. and Guo, Z. (2019). *ACS Applied Materials and Interfaces* 11: 35949.
9 Song, Y.-y., Liu, Y., Jiang, H.-b. et al. (2018). *Nanoscale* 10: 16127.
10 Zhou, H., Jing, X., and Guo, Z. (2020). *Journal of Colloid and Interface Science* 561: 730.
11 Wen, C., Guo, H., Bai, H. et al. (2019). *ACS Applied Materials and Interfaces* 11: 34330.
12 Fan, H. and Guo, Z. (2020). *Biomaterials Science* 8: 1502.
13 Cheng, G., Zhang, Z., Chen, S. et al. (2007). *Biomaterials* 28: 4192.

9

Non-phase Fog Collection

In field of biomimetic fog water collection, the development of mechanical properties plays a crucial role in creating efficient water-harvesting systems. The choice of materials with excellent mechanical performance enables researchers to refine the design and structure of fog collectors, enhancing their ability to capture tiny water droplets effectively. The mechanical properties not only govern the durability and stability of the collectors but also directly impact their performance under varying environmental conditions, such as wind speed, humidity, and temperature fluctuations. As we delve deeper into the study of mechanical properties, it becomes evident that transitioning to non-phase-change water collection techniques opens up new avenues for maximizing water resource efficiency. These innovative methods draw inspiration from nature, mimicking the processes involved in water droplet formation and aggregation. Unlike traditional phase-change systems, non-phase-change water collection techniques focus on gentle mechanisms that facilitate the harvesting of moisture from the air without requiring significant energy input. This transition is significantly supported by advancements in material science and a comprehensive understanding of water condensation mechanisms. By integrating progressive material fabrication techniques with optimized design strategies, non-phase-change water collection systems present higher capture efficiency and greater adaptability to diverse environments. This positions them as a vital solution to address water scarcity issues, particularly in arid regions.

9.1 Overview of the Development of Non-phase Bionic Fog Collection

In regions with high relative humidity and a stable atmosphere, water vapor typically condenses into small droplets (1–120 μm in diameter) suspended in the air. Over the past decade, researchers have mainly used bionic functional surfaces to collect small water droplets from the atmosphere. With advancements of science and technology, there are now numerous studies have demonstrated that the collection of liquid droplets in the air can be achieved. We refer to this class of method of collecting

Bioinspired Materials for Fog Harvesting: Principles, Design, Preparation, and Applications, First Edition. Zhiguang Guo and Weimin Liu.

liquid-phase small droplets as non-phase fog collection, in contrast to phase fog collection, in which involves a gas–liquid phase process during the collection process of water vapor when it exists in a gaseous state. Capturing water vapor significantly more challenging and requires specialized material structures and compositional designs. This chapter focuses on non-phase fog collection.

The process of complete non-phase fog collection consists of three main aspects, i.e. fog capture, fog supply, and water transportation (discharge) [1]. Fog capture refers to the process of collecting small airborne fog droplets on the surface of a material, followed by the accumulation of the captured water on that surface. Fog supply is the fusion process between the airborne fog and the water droplets on the surface of the material, which occurs after a significant number of droplets have been collected. This process results in the gradual growth of the water droplets on the surface. Water transportation (expulsion) refers to the removal of the captured water from the surface, facilitated by a driving force, which may be generated by a chemical gradient in the material's specialized structure or by gravity. Therefore, achieving efficient fog collection fundamentally relies on enhancing these three stages of the fog collection process.

Considering these three processes of fog collection, it is evident that the removal of already condensed droplets allows the condensation of new droplets at the capture site, i.e. the formation of a continuous, efficient collection cycle requires the ability of the surface to transport droplets. Hydrophilic surfaces are effective for capturing droplets but are less efficient for their removal. In contrast, hydrophobic surfaces are more suitable for droplet removal. This raises the question of how to design capture surfaces for efficient fog collection. Surface wettability design is one of the key aspects addressed in this chapter. It has been suggested in a related report that, in addition to the wettability (surface chemistry) design, the impact of the fluid properties of the fog (consisting of tiny water droplets and flowing air) on the fog collection cannot be overlooked. The air flow can be disturbed by different surface geometries, thus significantly affecting the movement of the droplets. How can a comprehensive fog collection system be designed to address all three aspects of fog collection?

The evolution of plants and animals over millions of years has resulted in the development of sophisticated adaptations that reflect nature's intelligence. By studying organisms capable of collecting fog, scientists have discovered that these plants and animals possess a comprehensive fog collection system for capturing, condensing, and removing water droplets. For example, in Chapter 2, it was mentioned that nano-fabric desert beetles achieve fog capture and removal through a structure composed of hydrophilic projections surrounded by hydrophobic areas. This dorsal structure, characterized by its hydrophobic and hydrophilic components, coordinates the process of fog capture and removal. Inspired by the combination of hydrophilic bumps and a hydrophobic periphery, several desert beetle-inspired fog collectors have been developed, demonstrating excellent fog collection capabilities. The primary biomimetic concept is to enhance the fog collection efficiency by creating a wettability-heterogeneous surface. This approach can be divided into two main aspects: the first involves regulating surface wettability to achieve a disordered affine-heterogeneous surface, while the second focuses on designing a patterned wetting surface. For example, surfaces with varying concentrations of hydrophilic particles can be created by controlling the amounts of iron and cobalt particles on fiber fabrics using selective chemohydrophobic modifications belonging to conformationally disordered interphase surfaces [2]. It was found that increasing the hydrophobic proportion would shorten the time for the first drop of water to roll off the fiber surface, while increasing the hydrophilic proportion extends this time. The superhydrophobic–superhydrophilic surfaces prepared in this manner offer advantages such a slow cost and convenient preparation procedures; however, the distribution of hydrophobic and hydrophilic particles on the surface is disordered. In our subsequent work, we found that the arrangement of different hydrophilic and hydrophobic patterns has an important influence on fog collection [3]. Therefore, we will also present the design of patterned wetting surfaces in the same section.

The upper and lower surfaces of lotus leaves exhibit distinct wettability characteristics, with the upper surface exhibiting superhydrophobicity and the surface demonstrating superhydrophilicity. Inspired by the asymmetric wettability of lotus leaves, a series of Janus materials with asymmetric wettability have been designed and prepared. In Janus materials, the asymmetric wettability on both sides generates a driving force that propels droplets from the hydrophobic side to the hydrophilic side, while preventing movement in the reverse direction. In fact, this difference in wettability corresponds to the difference in surface energy of the droplets on both sides, with the droplets tending to move toward the area of lower surface energy. We rationalize the difference in wettability between the two sides of the Janus fog-collecting material by categorizing it in the section on the design of wettable surfaces.

During fog collection, Janus materials can accelerate the unidirectional osmotic removal of captured droplets, addressing the issues of slow regeneration of capture sites and the evaporative loss of large droplets caused by gravity-dependent shedding. Therefore, many research teams have rationalized the integration of Janus materials with other biomimetic devices into a fog collection system to achieve higher fog collection efficiency compared to a homogeneous wettable surface. For instance, the Janus material was combined with the wettability characteristics of the hydrophilic-hydrophobic interphase of the desert beetle, resulting in the successful preparation of hydrophilic-hydrophobic interphase/superhydrophilic copper foams by hydrothermal growth and local modification [4]. This Janus copper foam integrally balances the three basic processes of fog collection. Its hydrophilic-hydrophobic surface enhances fog capture and condensation growth, while its asymmetric wettability that further facilitates droplet transportation and removal. As a result, Janus copper foam shows better fog collection than copper foams with homogeneous wettability that rely mainly on gravity-induced droplet removal. Similarly, combining the conical structure of a cactus with the Janus property presents an innovative design concept for a fog collection device [5].

The conical spindle knot structure formed by spider silk in humid environments plays a key role in directed water collection [6]. This unique structure creates both a surface energy gradient and a geometric structure gradient, which together provide the driving force for the directional transport of droplets. The surface energy gradient encourages the droplets to move from the junction to the more hydrophilic spindle junction, and the geometric structure gradient generates a Laplace pressure difference on the droplets, propelling them from the junction (a high curvature region) to the tapered spindle junction (a low curvature region). The cactus is regarded as an integrated fog fog collection system with structural and functional diversity. Its needles consist of three parts with different structural characteristics: a directional barb at the tip, a gradient groove in the middle, and a banded trichome at the root. The three parts of the structure act synergistically to provide the conditions and driving force for droplet capture and directed transport [7]. In addition to the previously described desert beetle-inspired hydrophilic-hydrophobic two-dimensional surfaces, researchers have developed one-dimensional fog-collecting devices inspired by cactus thorns and spider silks [6, 7], which have also demonstrated satisfactory water-collection results. For instance, the periodically roughened gradient-tapered copper wire prepared by Zheng and coworkers [5, 8] is capable of generating power for remote unidirectional droplet transport through the surface energy released from droplet merging, the Laplace pressure caused by the gradient of the geometric curve, and the difference in surface energy caused by the periodic roughness. This design aims to enhance the water discharge (transportation) process thereby improving the overall effectiveness of fog collection. Using surface energy gradients and geometric structure gradients as the main drivers for the design of geometrical structures for bionic fog collectors, we will introduce several basic fog collection structures, such as linear, conical, and spindle structures., The chapter will conclude with an overview of the design of composite structures, or multifunctional bionic fog collection materials, which address multiple fog collection processes.

In addition to the aforementioned species of plants and animals, pitcher plants exhibit a similar continuous, directional transportation of fog, as well as the establishment of a complete fog collection system [9]. In humid environments, the pitcher plant relies on its highly hygroscopic nectar and microstructure to form a smooth film of water on its surface, reducing the friction between insects and their surface, which in turn is used to capture insects [10] Inspired by pitcher plants, many slippery-infused surfaces have been prepared. Due to the characteristics of low surface energy and high nucleation density, fog droplets can condense rapidly when slippery surfaces are applied to fog collection [11]. Additionally, the infused lubricant layer r creates a continuous, flat surface with a uniform atomic distribution. This not only facilitates the condensation of water droplets but also results in droplets that exhibit very low contact angle hysteresis and high sliding properties. These characteristics ensure the rapid removal of droplets and the formation of new nucleation sites [12]. The section on slippery surfaces for fog collector preparation will not be treated as a standalone chapter. Instead, the advantages of slip surfaces in fog collection will be integrated into other sections.

Based on several typical bionic designs, several reports have proposed the use of external field stimuli to enhance the efficiency of fog collection. Also mimicking the conical needle-prick structure of a cactus, Liu and coworkers [13] prepared a magnetically responsive flexible needle-prick array based on the magnetic response, which improves the ability of this fog-harvesting system to capture and collect fog droplets in environments with low fog flow rates. The possibility of collision between the oscillating conical needles and the fog droplets increased when an external magnetic field was applied. Additionally, the effective fog collection area of the needles expanded due to the periodic oscillation of the conical needles. As a result, the magnetically controlled flexible conical needling array exhibits good fog collection capability in low fog flow environments. In such conditions, the low fog velocity makes it challenging for the conical spikes to capture fog without the assistance of an external magnetic field, resulting in negligible fog-capturing efficiency. The design of the flexible conical needle stinger acknowledges the significant impact of fog flow rate on fog collection efficiency, which researchers have considered during the fog collection process through an intelligent design strategy. Intelligent water fog materials can adapt to their environment, enhancing their sensitivity to fog. In addition to magnetic fields, common external stimuli include stress, temperature, and light, and so on. In the fourth part of this chapter, we will describe the responses to different external stimuli accordingly.

From the perspective of the fluid properties of fog, various surface geometries will affect the movement of the droplets. For example, Aizenberg and coworkers [1b] found in their experiments investigating water vapor condensation on slippery bumps that discontinuities in the surface enhance the diffusion flux of water vapor, thereby increasing the water supply. Surface geometry significantly influences a crucial factor in fog collection: the velocity of fog flow reaching the capture surface. In our earlier discussion of the bionic fog collection concept inspired by the Namib Desert beetle, we noted that the geometry of the surface bumps can impact the distribution of fog flow, which in turn affects the availability of fog during the collection process. Therefore, for two-dimensional fog collection surfaces, the pattern of hydrophilic projections and hydrophobic surroundings is generally recognized as having enhanced fog collection capability. However, we found that the effect of one-dimensional copper wires with their diameters on the fog collection effect is more significant than the effect of surface wettability [14]. In other words, we should realize that one- and two-dimensional materials are placed in the fog stream and their interference with the fog stream is quite different. Airflow through a two-dimensional plane undergoes a large degree of bypassing, but for one-dimensional wire materials, this degree of deviation is greatly reduced, which results in fog droplets being captured by the surface with higher energy. Thus, superhydrophobic copper wires, superhydrophilic copper wires, pristine copper wires, and hydrophilic and hydrophobic copper wires show only small differences in their ability to fog capture fog. This factor of fog flow interference should be consciously taken into account when designing fog-harvesting devices. For example, when designing a conical array on a two-dimensional plane, the interference of the array's arrangement (quadrilateral, hexagonal) on the fog flow should be taken into account [15]. Therefore, it is essential to consider, from an aerodynamic perspective, what additional aspects can be designed and adapted to enhance fog collection efficiency. We will address this topic in the final section of this chapter.

9.2 Surface Wettability Design

9.2.1 (Super)Hydrophilic Surface

In the initial stage of fog collection, fog droplets with a certain velocity are captured by the surface of the material serve as nucleation sites for condensation. By increasing the number of nucleation sites, the ability of the material surface to capture droplets can be increased, thereby increasing the rate of droplet nucleation and collection [16]. According to the classical nucleation theory (Volmer) , the free energy barrier for the formation of liquid nuclei on flat surfaces strongly depends on the intrinsic wettability of the material surface and increases with a rising contact angle (CA). Under the same conditions, hydrophobic surfaces have higher nucleation energy barriers compared to hydrophilic surfaces, and nucleation of water vapor and small water droplets on hydrophilic surfaces is energetically more favorable [16, 17]. The data show that the nucleation rate on hydrophilic surfaces with CA ~ 25° is 10 times higher than that on hydrophobic surfaces with CA ~ 110°. Xu et al. found that the nucleation rate increased from 105 m^{-1} s^{-1} to 1015 m^{-1} s^{-1} when the contact angle was reduced from 80° to 60° [18].

The hydrophilic surface exhibits a significantly higher coalescence rate compared to the hydrophobic surface, as evidenced by the rapid growth rate of the droplets [19]. There are three reasons for this. First, due to the low contact angle, droplets exhibit a larger contact area on hydrophilic surfaces compared to hydrophobic one, which reduces the effective distance between droplets and promotes droplet merging. Second, a higher nucleation density leads to faster coalescence of droplets Third, on some hydrophobic surfaces that utilize lubricants, small water droplets can easily form a water film on the hydrophilic surface if they are not removed in time. Due to the low heat transfer of the water film, it is difficult for droplets in the fog to continue to condensing, and they may even rebound from the surface of the water film [20].

Azad et al. [21] investigated the effect of the wettability of polypropylene hydrocarbon mesh surfaces on fog collection efficiency and showed that the superhydrophilic mesh exhibited the highest fog collection efficiency (Figure 9.1a–d). The amount of water captured by the hydrophilic, hydrophobic, and superhydrophilic mesh samples was 443 ± 217 μL, 1284 ± 406 μL, and 2384 ± 328 μL, respectively. The fog droplets had a significantly lesser impact on the hydrophobic mesh surface as on the superhydrophilic mesh. Additionally, the tiny droplets that impacted the hydrophobic surface did not merge together as quickly as those on the superhydrophilic surface. The faster droplet fall on the superhydrophilic grid can be attributed to two factors: the high number of fog droplets impacting the superhydrophilic surface and the formation of a continuous thin film of liquid on that surface, which prevents blockages between the grids. Hydrophobic surfaces exhibit a lower affinity for water, causing droplets to roll off more easily. In contrast, hydrophilic mesh surfaces possess a higher affinity for water than hydrophobic mesh surfaces; however, they exhibit the lowest mobility for water droplets [24]. In addition, for a droplet to move gravitationally over a solid surface, it must have a critical size greater than the length of the capillary [25]. The disadvantage of large

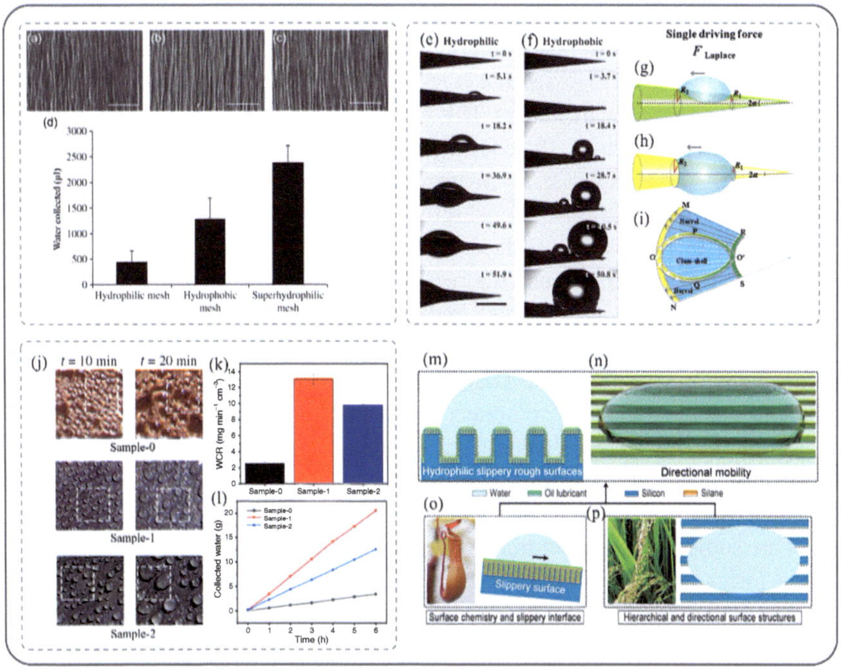

Figure 9.1 (a–d) SEM images of different wetted surfaces and corresponding fog collection columns. SEMs of a ribbon surface of (a) hydrophilic mesh, (b) hydrophobic mesh, and (c) superhydrophilic mesh; scale bar = 10 μm. (d) Amount of water collected from the fog stream over 30 min by polyolefin mesh (surface area approximately 24 cm^2) with different degrees of wettability. $p < 0.001$ indicated that the differences among all sample types were highly significant. Source: Reproduced with permission from [21] / IOP Publishing. (e–i) Hydrophilic and hydrophobic cones are used for fog collection. (e–f) Microscopic observations of typical fog-collection processes on three kinds of conical copper wires (CCWs) with different wettability. (g–i) Mechanism of the distinct velocity of water drop motion on CCWs with different wettability. A water drop is apt to be in (g) clam-shell conformation and (h) barrel conformation on a hydrophobic and hydrophilic CCW, respectively. (i) Comparison between the single driving force F_{Laplace} on clam-shell and barrel droplets. Source: [22] / John Wiley & Sons. (j–l) The fog collection performance of the original copper sheet (sample 0) and CuO Lubricant-Infused surfaces with different morphologies (layered CuO named sample-1 and needle-like CuO named sample-2). (j) Optical images of the water-harvesting procedure at 10 and 20 min for Sample-0, Sample-1, and Sample-2. (k) Water collection rate (WCR) for Sample-0, Sample-1, and Sample-2. (l) Weight of collected water in 6 h for Sample-0, Sample-1, and Sample-2. Source: Reproduced with permission from [23] / American Chemical Society. (m–p) Hydrophilic directional SRS inspired by pitcher plants and rice leaves. Side view (m) and three-dimensional view (n) of the hydrophilic directional SRS. Photos and schematics showing the pitcher plant–inspired slippery surface (o) and rice leaf–inspired directional structured surface. (p) The photography of rice leaf was reprinted with permission from Bixler and Bhushan. Source: Reproduced from [19] / with permission of American Association for the Advancement of Science – AAAS / CC BY-NC 4.0.

droplets during fog collection is that they cannot be quickly removed from the surface, which prevents the regeneration of nucleation sites blocks the voids, the voids, and may even lead to losses due to re-evaporation [26]. As a result, the overall efficiency of fog collection decreases. This explains why the superhydrophilic mesh has the highest fog collection efficiency and the hydrophilic mesh has the lowest fog collection efficiency. In contrast, for hydrophobic and hydrophilic surfaces, droplets continue to grow until the gravitational force is sufficient to overcome the adhesion, allowing them to roll off the surface. This phenomenon causes the droplets to adhere to the surface and between the two neighboring grids, obstructing a certain area and hindering fog flow, ultimately reducing fog collection efficiency.

Park et al. [24] investigated the effect of surface wettability properties on the fog trapping ability of woven meshes. They developed a model that combines hydrodynamics and surface wettability to predict the overall fog collection efficiency of the mesh. Droplet re-entrainment is induced aerodynamically, with deposited droplets breaking away from the air flow before reaching a critical volume, where gravity drainage plays a dominant role. As deposited droplets aggregate, the growing droplets are subject to a competition between aerodynamic drag forces and surface adhesion forces. When the drag forces are greater than the adhesion forces, the droplets are re-entrained in the fog stream, resulting in a decrease in fog collection efficiency. Surface wettability plays a crucial role in limiting the re-entrainment of collected droplets and preventing mesh clogging. That is, the surface must have sufficient adhesion to capture droplets from the fog stream, while simultaneously ensuring that the captured droplets are rapidly transported to the bottom of the mesh. From this perspective, superhydrophilicity offers the highest affinity for tiny droplets impacting the surface and facilitates their rapid diffusion and transport across the surface [27].

Ju et al. [22] showed that droplets grow faster on hydrophobic taper wires than on hydrophilic ones; however, the droplet transport velocity rate is much lower (about 2.5 times lower) than that on hydrophilic taper wires (Figure 9.1e,f). The more favorable droplet transport rate on the hydrophilic surface can be attributed to two main factors. On the one hand, the droplets have a longer pinned contact line on the hydrophilic taper wire than on the hydrophobic taper wire, which have a bucket shape (Figure 9.1h), compared to a scallop shape on the hydrophobic taper wire (Figure 9.1i). This extended pinned contact line on the hydrophilic surface generates a greater Laplace driving force on the tapered surface of the water droplet. On the other hand, the liquid film formed by the droplets on the hydrophilic surface facilitates "lubrication" and reduces the resistance to the transport of the droplets.

Under large subcooling conditions, a Wenzel wetting state is inevitably formed due to the nucleation of droplets that become pinned to the surface, which greatly reduces the fluidity of the droplets [28]. In a highly humid environment, hydrophilic surfaces are highly susceptible to the formation of a water film. The strong adhesion between the water film and the material surface can severely impede the growth and shedding of droplets. Gou et al. investigated the fog collection performance of slippery surfaces obtained by infusing lubricating oil with copper oxides of different morphologies, with lamellar CuO, referred to as Sample-1, and acicular CuO, referred to as Sample-2. The results show that the original hydrophilic copper sheet (Sample 0) forms a water film covered with droplets on its surface after a fog collection period of 10 min (Figure 9.1j). After 20 min, the fog collection efficiency of the Sample 0 surface (2.54 mg min^{-1} cm^{-2}) was much lower than that of the slippery surfaces of the same specifications (Sample-2 : 9.68, Sample-3 : 13.04 mg min^{-1} cm^{-2}), and the amount of water collected by it was much smaller than that of the slippery surface in 6 h (Figure 9.1l) [23]. This demonstrates the unique advantages of slippery surfaces in the field of fog collection.

Inspired by the surface of hogweed and the leaves of rice, Feng et al. [29] used a hydrophilic liquid lubricant to modulate the properties of solid surfaces, which can be made hydrophilic and not water-repellent. The liquid lubricant allows the droplets to slip off quickly due to its good fluidity. This liquid-lubricated surface (SRS) is characterized by a hydrophilic surface, low sliding resistance, an oriented structure, and a large surface area, all of which maximize droplet fluidity (Figure 9.1m–p) [19]. Specifically, the SRS specifically features oriented microgrooves infused with liquid to enhance droplet nucleation and transport. Notably, the presence of nanoscale textures on the microgrooves aids in retaining a thin layer of hydrophilic lubricant, providing a smooth interface for droplet removal. Such a slippery surface provides high droplet nucleation density, as well as fast droplet aggregation and removal under a high-humidity environment, which is not possible on conventional hydrophilic/hydrophobic surfaces [30].

To date, commonly used methods for wettable surface modification methods include the sol–gel method, layer-by-layer self-assembly technique, and chemical vapor deposition (CVD), which involves chemical reactions such as etching, polymerization, and molecular grafting, and it offers the advantages of simplicity, speed, and efficiency [31]. For (super)hydrophilic surfaces hydrophilic inorganic meso- or nanoparticles are often used to coat the material surface by inkjet printing [32], spin-coating [33], sputtering, sol–gel [34], solution-growth [35], photolithography [36], electrochemical deposition [37] and other methods. Dai et al. showed through molecular dynamics simulations that the efficient nucleation of droplets is attributed to hydrophilic surface functional groups. TiO_2 [34, 38], SiO_2 [39], and ZnO [33] nanoparticles have been widely studied due to their excellent hydrophilicity and accessibility. In addition to this, the grafting of hydrophilic groups onto the material surface by ion radiation [40], electron beam [41], and plasma techniques [42] are also commonly used modification methods.

9.2.2 (Super)Hydrophobic Surfaces

The efficiency of fog collection on a material surface depends on both the droplet nucleation rate and the migration rate. The droplet transportation process is a critical factor that affects the fog collection. If the droplets captured on the surface of the material are not removed in time, it can lead to a decrease in the wettability of the material surface, which in turn affects the overall fog collection cycle. When tiny droplets agglomerate to form larger droplets, the reduction in the specific surface area of the droplet releases surface energy E_S. A portion of this energy is dissipated by viscosity (E_{vis}), another portion is used to overcome the interfacial adhesion force to do work (E_W), and the remaining energy is converted into kinetic energy (E_K) to drive the droplet removal [43]. Hydrophilic regions tend to nucleate more droplets and increase surface adhesion, which severely hinders droplet shedding. Hydrophobic surfaces facilitate rapid droplet roll-off. Zhang and coworkers [44] prepared a super bi-phobic coating by proportionally adding silicon carbide particles to the nano-SiO_2 superhydrophobic coating. The coating has excellent superhydrophobic properties, a high nucleation rate, high droplet removal efficiency, and a good fog-collecting effect. The collision of water droplets between the arrays is conducive to the removal of water droplets. In particular, when the prepared samples are placed in parallel at a distance of 2 mm, the collision of water droplets between the two sample surfaces can promote the removal of water droplets, and the water collection rate of the collision surface is 217% higher than that of the non-collision surface. In addition, the "rolling angle" is generally used to evaluate the ability of the hydrophobic meter to remove water droplets. The smaller the inclination angle of water droplets rolling on the hydrophobic surface, the smaller the rolling angle of the corresponding surface, the easier it is to quickly remove water droplets.

As the solid surface begins to tilt, contact angle hysteresis (CAH) occurs due to the inconsistency of the contact angle between the front and back ends of the water droplet. At this time, the contact angle of the droplet at both ends of the forward contact angle θ_{ad} and backward contact angle θ_{re}, respectively, and the difference between them is called the contact angle hysteresis (CAH = $\theta_{ad} - \theta_{re}$). The resistance due to contact angle hysteresis is F_R (mN), $F_R = \pi R_0 \gamma(\cos\theta_{re} - \cos\theta_{ad})$, R_0 (mm) denotes the contact radius of the three-phase contact line, and γ is the surface tension [45]. The larger the contact angle hysteresis, the less likely a droplet will roll off a solid surface. On the contrary, if the contact angle hysteresis is smaller, the droplets roll off the solid surface more easily. In the case of materials with hydrophilic and hydrophobic sites on the surface used for fog collection, the captured droplets gradually increase in size and form asymmetric contact angles on surfaces with different wettability. At this point, droplets tend to migrate e from hydrophobic to hydrophilic regions, driven by chemical gradient forces (Figure 9.2). The chemical gradient force can be expressed as $F_{Chemical} \approx \gamma \times (\cos\theta_B - \cos\theta_A) \times l$, θ_B and θ_A denote the contact angles of the droplets in the high and low wetting surface regions, respectively, and l (mm) denotes the length of the three-phase contact line enclosure [6, 45].

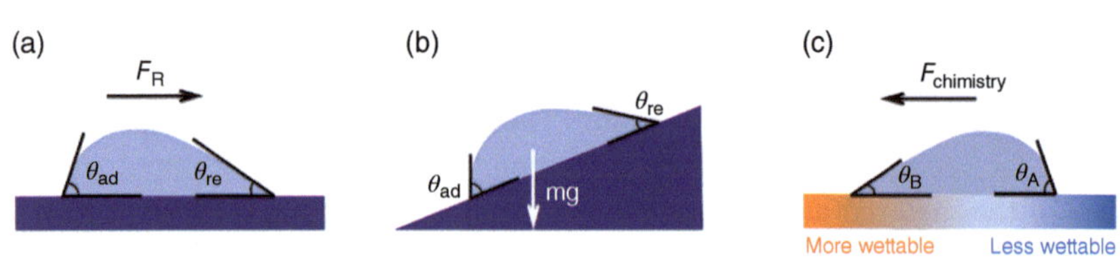

Figure 9.2 (a–b) Trend of droplet motion on horizontal and inclined surfaces; (c) Trend of droplet motion on a surface with a wetting gradient.

The fog collection efficiency of hydrophobic surfaces is mainly limited by two aspects: on the one hand, hydrophobic surfaces struggle to nucleate the captured droplets due to high nucleation energy barriers, causing smaller droplets to detach rapidly from the surfaces. This detachment results in a decrease in the fog collection efficiency. On the other hand, the air pockets captured with the microstructure of the lotus leaf-inspired superhydrophobic surfaces are easily destroyed under pressure or excessively high humidity conditions [46], resulting in a transition of the droplet contact state with the solid surface transitions to a fully wetted Wenzel state [47].

In summary, optimizing the surface wettability of hydrophilic and hydrophobic phases represents a more effective strategy for designing fog collection surfaces. Hydrophilic regions maximize droplet nucleation density, while hydrophobic regions areas facilitate droplet removal. Numerous experiments have shown that hydrophilic/hydrophobic surfaces obtain greater droplet nucleation density due to the nucleation sites in the hydrophilic region of their surfaces. Meanwhile, the hydrophobic regions ensure rapid droplet removal, thus displaying higher fog collection efficiency. We broadly categorize the currently reported hydrophilic and hydrophobic fog-harvesting materials into disordered hydrophilic/hydrophobic surfaces and patterned hydrophilic/hydrophobic surfaces. We will summarize these two types of surface wettability designs in the next subsection.

9.2.3 Heterogeneous Design Between Affinities

Inspired by desert beetles, hydrophilic wettable surfaces that integrate hydrophobic and hydrophilic regions on the same two-dimensional surface have been designed for efficient fog collection. However, surface wettability, surface structure, and the distribution ratio of hydrophilic/hydrophobic regions are three critical factors that affect water collection efficiency. To date, hydrophilic-hydrophobic hybrid surfaces have been intensively investigated in terms of both surface structure design and surface chemical modulation [48]. Concerning substrates, hybridized surfaces with hydrophilic-hydrophobic domains have been applied to a wide range of substrates for water-harvesting applications, such as copper mesh [49], copper sheet [50], stainless steel mesh [51], textile [52], cellulose film [53], etc. In summary, the wettability of the hydrophobic-hydrophilic interphase on the back of a bionic desert beetle has extensive and profound implications for the design of fog-water-harvesting surfaces [54]. Concerning the construction methodology of the hydrophilic/hydrophobic interphase, the preparation techniques are diverse and may potentially encompass electrostatic spinning, braiding, spraying, impregnation, and laser etching.

Joanna et al. [55] successfully prepared a composite web consisting of hydrophobic and hydrophilic fibers using a dual-nozzle electrostatic spinning equipment. The hydrophobic and hydrophilic materials were hydrophobic polystyrene and hydrophilic polyamide, respectively. Combining these two polymers and the electrostatic spinning device, both polymer filaments were generated simultaneously from the nozzle and stacked layer by layer to form a composite material. This process

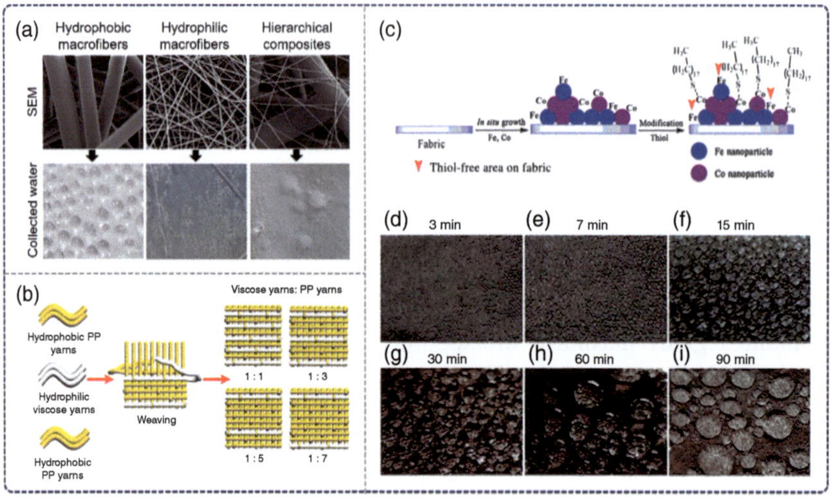

Figure 9.3 (a) The SEM images and fog collection images of hydrophobic microfibers, hydrophilic microfibers, hydrophilic and hydrophobic hierarchical composites. Source: Reproduced with permission from [55] / American Chemical Society. (b) Weaving process of the freely-regulated hydrophilic-superhydrophobic patterned weft-backed woven fabric. Source: Adapted from [26]. (c–i) the preparation procedure of the superhydrophobic–superhydrophilic hybrid (SSH) fabric and their fog collection performances. Source: Adapted from [56]. (c) Schematic illustration of the preparation procedure of the SSH fabric from in situ growth of Fe and Co NPs with thiol modification. (d–i) Optical images of harvested droplets with various times of (a) 3 min, (b) 7 min, (c) 15 min, (d) 30 min, (e) 60 min and (f) 90 min. Source: Reproduced with permission from [2] / The Royal Society of Chemistry.

resulted in a wettable, layered structure that exhibited a combination of hydrophobic and hydrophilic properties (Figure 9.3a). This combination of hydrophobic and hydrophilic fibers plays a vital role in improving water collection efficiency [57]. Therefore, the key to designing efficient water-harvesting materials lies in the combination of hydrophobic and hydrophilic materials and the balance between the two. Gao et al. [58] prepared a hydrophilic-superhydrophobic patterned fabric surface by using a braiding method. This hybrid wettable surface was made from hydrophilic viscose yarns and hydrophobic polypropylene yarns. As shown in Figure 9.3b, the hydrophobic and hydrophilic yarns were treated as warp and weft yarns, respectively, and the hydrophilic-superhydrophobic patterned surface was formed by interweaving the warp and weft yarns. Different ratios were compared in the experiment, and several ratios were 1 : 1, 1 : 3, 1 : 5, and 1 : 7 (viscose yarn: polypropylene yarn). This experiment proved that the hydrophilic-superhydrophobic fabrics prepared by this method were 59.2 times more efficient in water collection than beetles at the 1 : 1 ratio. This finding further confirms the effectiveness of the bionic strategy for creating hybrid wettable surfaces.

Wang et al. [56] devised fabrics featuring a superhydrophobic–superhydrophilic interphase, by selectively modifying iron and cobalt metal nanoparticles with *n*-octadecanethiol, as illustrated in Figure 9.3c. The pristine fabrics were immersed in $Co(NO_3)_{two}$ and $FeSO_4$ solutions at ambient temperature, Subsequently, the reduced properties of sodium borohydride were utilized to promote the growth of fabrics with iron and cobalt nanoparticles adhering to their surfaces. Following this, a selective hydrophobic modification was executed using an ethanol solution of *n*-octadecyl mercaptan. Fe and Co were randomly dispersed on the fabric surface. The cobalt nanoparticles on the fabric surface could be modified using *n*-octadecanethiol, whereas the Fe particles on the surface could not be modified by *n*-octadecanethiol. This process results in the creation of a super-wettable surface similar to that of a desert beetle, exhibiting a unique combination of affinity and hydrophobicity. The surface displayed superior water fog collection efficiency. Initially, droplet nucleation on the fabric did not interact with each other, resulting in minimal initial coverage (Figure 9.3d,e). As the surface coverage of the droplets becomes substantial, the droplets frequently amalgamate with neighboring droplets. The center of mass of the merged droplet remains constant, as depicted in Figure 9.3f,g. Ultimately, when the droplet diameter exceeds a certain threshold, the large droplet detaches from the surface and is eliminated by gravity. Similarly, Tian et al. [59] employed layered nanostructures to emulate the back of a beetle. ZnO nanoparticles vertically aligned on a stainless-steel mesh were initially synthesized by hydrothermal method. Subsequently, Fe_2O_3 nanoparticles were deposited on the ZnO-coated stainless steel (ZnO/SSM) surface via impregnation process. Lastly, the superhydrophobic–superhydrophilic surface was prepared using selective modification with octadecylthiol. This experiment ingeniously utilizes the varying strengths of interactions between alkyl mercaptans and excess metal oxides to attain heterogeneous structured surfaces. The hydrophilic-hydrophobic surface exhibited the highest fog collection efficiency (977.0 $mg\,cm^{-2}\,h^{-1}$) compared to the single wettable surface.

Park et al. [1b] developed superior hydrophobic-superhydrophilic material surfaces with hydrophilic–hydrophilic wetting states by integrating targeted thiol modification with spraying techniques. A blend of Cu_2O microspheres, which can be readily modified by octadecanethiol, and ZrO_2 nanospheres, which are not modifiable by octadecanethiol, together with an inorganic aluminum phosphate binder formed the spraying solution, and a hybrid wettable surface with a micro- and nano-structured surface was constructed on a stainless-steel mesh. The surface was capable of attaining the highest fog collection efficiency of 1707.25 $mg\,cm^{-2}\,h^{-1}$. The incorporation of inorganic aluminum phosphate significantly enhanced the stability of the material surface, ultimately enabling the fabricated material to successfully accumulate 15 L of water. The team discussed the impact of the ratios of Cu_2O and ZrO_2 on the effectiveness of fog collection. Through numerous experiments, it has been confirmed that the material surface exhibits hybrid superhydrophilic-superhydrophobic wettability solely when the ratio of Cu_2O to ZrO_2 is 8 : 1, 7 : 1, 6 : 1, and 5 : 1. Among them, the 8 : 1 sample exhibited the best water collection within 30 min, which was twice as effective as the superhydrophilic material (Figure 9.4a). The team also engaged in an in-depth discussion regarding the synergistic effects of surface micro-nano-structure and mixed wettability on fog collection. They concluded that the larger droplets on

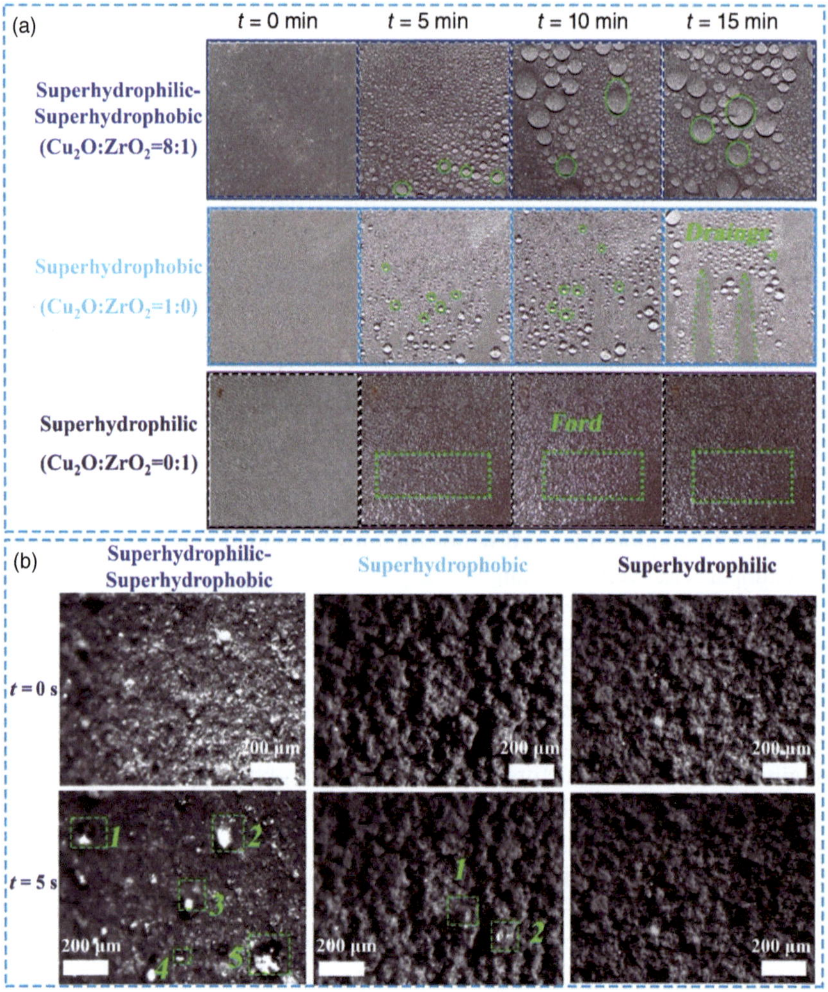

Figure 9.4 (a) Optical photographs of the fog-harvesting progress of the as-prepared hybrid superhydrophilic-superhydrophobic surface (Cu_2O: ZrO_2 = 8 : 1), superhydrophobic surface (Cu_2O: ZrO_2 = 1 : 0) and superhydrophilic surface (Cu_2O: ZrO_2 = 0 : 1) at 0, 5, 10 and 15 min. The water fog sprayed on the superhydrophobic surface first appears in the shape of tiny droplets, grows slowly, and rolls off quickly. The droplet diameter on the hybrid superhydrophilic-superhydrophobic surface is larger than that on the superhydrophobic surface. Nevertheless, the water fog sprayed on the superhydrophilic surface always appears in the form of a water film. (b) Comparison of fog capture capacity of the as-prepared hybrid superhydrophilic-superhydrophobic surface, superhydrophobic surface and superhydrophilic surface in 5 s (Scale bar: 200 μm). The hybrid superhydrophilic-superhydrophobic surface captures five droplets within 5 s, and the superhydrophobic surface captures only two droplets at the same time. No droplets are captured on the superhydrophilic surface in 5 s. Source: Reproduced with permission from [29] / Elsevier.

the sample surface can be perceived as a series of patterned bumps, which can capture more water droplets, expedite the water supply process, and ultimately achieve exceptional fog collection. The results of the flow field analysis indicated that these patterned bumps can effectively increase the growth rate of droplets on the surface. Droplets concentrated around the apex of the bumps have a higher density and move faster than flat droplets, which makes them more susceptible to surface capture. Figure 9.4b illustrates a comparison of the fog retention capacity of the prepared superhydrophilic-superhydrophobic surface, superhydrophobic surface, and the superhydrophilic surfaces over a 5 s period. The superhydrophilic-superhydrophobic hybrid surface captured approximately five droplets in 5 s, whereas the superhydrophobic surface only captured about two droplets simultaneously. No droplets were captured by the superhydrophilic surface in 5 s. The superhydrophilic surface did not capture any droplets simultaneously. Moreover, a specialized structure composed of hydrophobic/hydrophilic nanoparticles can stimulate water collection behavior. The hydrophobic particles are larger, while the hydrophilic nanoparticles are small, resulting in a significantly smaller hydrophilic area on the material's surface compared to the hydrophobic area. This unique secondary structure can effectively promote the removal of tiny water droplets from the material surface. In conclusion, the prepared superhydrophilic-superhydrophobic hybrid surface possesses excellent fog collection efficiency.

Lin et al. [60] utilized the photoresponsive wettability of TiO_2 to synthesize $CuO@TiO_2$ with a layered architecture on the copper surface (Figure 9.5). Specifically, highly hydrophilic acicular copper hydroxide was initially prepared through alkaline ammonia etching. This was followed by multiple rinses with a tetrabutyl titanate/ethanol solution, and additional rinsing with ethanol to produce $Cu(OH)_2@TiO_2$-coated surfaces. The $CuO@TiO_2$ surface was obtained after the sample was calcined. The samples were modified to superhydrophobic surfaces using a cetyltrimethoxysilane/ethanol mixture solution as a hydrophobic modifier. Following UV irradiation for various durations, the cetyltrimethoxysilane on the $CuO@TiO_2$ surface undergoes partial degradation. Consequently, the superhydrophobic surface can be engineered to form a hydrophobic-hydrophilic alternating surface by judiciously designing the irradiation duration. When droplets encounter such a surface with hydrophobic domains, they initially agglomerate in the hydrophilic regions, independently forming droplets that grow in situ. Subsequently, two adjacent droplets collide and coalesce into a larger droplet. Ultimately, the droplet is dislodged from the surface by gravity and aggregation forces. It is demonstrated that the $CuO@TiO_2$-coated surface with hierarchical structure and hydrophilic domains plays a pivotal role in droplet transportation. Furthermore, the synergistically designed surfaces of the two types can enhance the efficiency of fog collection.

Contrary to previous studies, Kang et al. [61] isolated oil and water by affixing a copper mesh post ammonia-based etching within an oil–water separator and subsequently discharging the *n*-octadecanethiol/ethanol blend into the separator. After 30 s, the blend was entirely partitioned by rinsing the specimens with

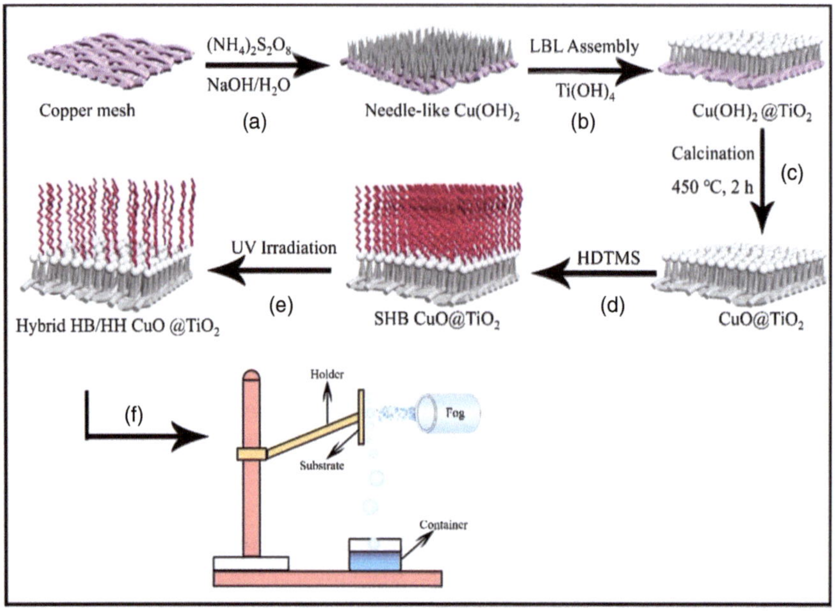

Figure 9.5 Schematic illustration of the preparation process of CuO@TiO_2 surface with alternating hydrophobic–hydrophilic wettability: (a) Chemical etching method to produce $Cu(OH)_2$ on copper mesh, (b) production of $Cu(OH)_2$@TiO_2 by LBL self-assembly deposition, (c) calcination treatment of $Cu(OH)_2$@TiO_2 to obtain CuO@TiO_2, (d) chemical modification to obtain superhydrophobic CuO@TiO_2 surface, (e) UV irradiation to result in CuO@TiO_2 surface with alternating hydrophobic–hydrophilic wettability, (f) homemade testing system to evaluate water-harvesting performance. Source: Adapted from [60].

deionized water and ethanol, it was discovered that the portion of the sample surface in contact with the blend displayed superhydrophobicity, with a water contact angle of approximately 155.2° and a rolling angle of approximately 8.4°. In contrast, the areas that were not in contact with the blend still displayed superhydrophilicity. In the macroscopic state, a circular pattern of superhydrophobicity was left on the superhydrophilic copper mesh. By reiterating this process, the patterning design of the hydrophilic sites on the back of the bionic desert beetle was macroscopically magnified, enabling more efficient fog collection. During the fog collection test, a multitude of minute droplets adhered to the superhydrophobic region, and a water film formed on the superhydrophilic region could be observed on the surface with patterning. It was demonstrated that the material with the patterned design of the hydrophilic sites displayed the highest fog collection efficiency (approximately 1316.9 mg h^{-1} cm^{-2}) when the sample was positioned at an inclination of 90°. Furthermore, there was no significant change in the collection efficiency after 10 cycles of fog collection.

To address the issue that droplets on the hydrophilic surface could not be evacuated promptly, Zhan et al. [62] deposited polytetrafluoroethylene (PTFE) nanoparticles on the surface of the copper mesh through a femtosecond laser direct

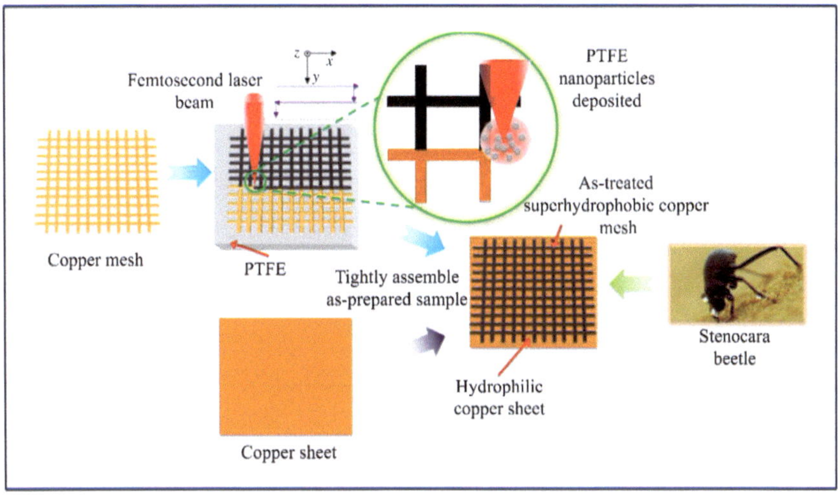

Figure 9.6 The preparation procedure for the hybrid superhydrophobic-hydrophilic micro/nanopatterned surfaces. Source: Reproduced with permission from [74] / The Royal Society of Chemistry.

writing technique, and then securely bonded the resultant superhydrophobic copper mesh with the hydrophilic copper sheet to fabricate a composite copper mesh-copper sheet with superhydrophobic-hydrophilic properties (Figure 9.6). The composite copper mesh-copper sheet with superhydrophobic-hydrophilic nature exhibited a higher water collection rate compared to other samples pristine copper mesh (hydrophobic), pristine copper sheet (hydrophilic), treated copper mesh (superhydrophobic) and pristine copper mesh-copper sheet (hydrophobic-hydrophilic). Moreover, altering the mesh size of the copper mesh and the inclination angle of the collection fog, it was found that the samples utilizing 40 mesh copper mesh at 90° inclination angle exhibited the highest water collection rate, and the water collection rate diminished with increasing mesh size. Propelled by the wettability gradient, droplets deposited on the superhydrophobic copper mesh spontaneously migrate to the hydrophilic copper plate and proliferate. Additionally, due to the high contact angle of the droplets on the copper mesh, droplets with diameters surpassing that of the copper mesh can effortlessly fall under gravity. The superhydrophobic-hydrophilic hybrid surface effectively balances the processes of droplet nucleation, growth, and removal., and the work provides a strategy to optimize the wettability by modifying the contact angle disparity between the hydrophobic copper mesh and the hydrophilic surface.

Rather than adopting a novel methodology, Zhang et al. [63] fabricated hydrophilic-superhydrophobic interphase wettability on the surface of a three-dimensional convex-structured copper plate through the imperfect calcination attributes of PDMS. As illustrated in Figure 9.7a, copper substrates (2.5 × 2.5 cm) were engineered into two categories of convex configurations through molding techniques. One of the copper plates possesses a 5 × 5 hemispherical bump (HB) structure with a diameter of 1 mm on the surface, while the other plate features a

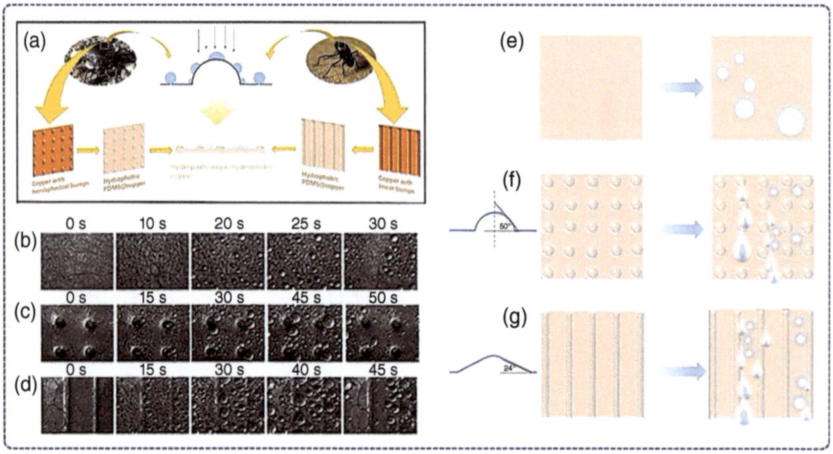

Figure 9.7 (a) Schematic illustration of the preparation of hydrophilic–superhydrophobic (hybrid) copper sheets. (b–d) Optical images showing the progress of water harvesting on PDMS@copper after calcination. The growth and aggregation of tiny droplets on SHB (b), Hybrid-hemispherical bumps (HB) (c), and Hybrid-linear bumps (LB) (d). (e–g) Schematic illustrations of water removal from SHB (e), Hybrid-HB (f), and Hybrid-LB (g). Source: Reproduced from [63] / with permission of Elsevier; Reproduced with permission from [3a] / The Royal Society of Chemistry.

linear bump (LB) structure with a diameter of 1 mm. PDMS films were fabricated by spin coating on these two copper sheets with distinct surface morphologies. Subsequently, the samples were vacuum-cured and calcined to yield copper flakes with mixed hydrophilic-superhydrophobic wettability. It has been demonstrated that the imperfect combustion of the PDMS oligomer alters the surface texture and, consequently, its wettability, transforming a hydrophobic surface into a superhydrophobic one. Simultaneously, the uneven surface results in non-uniform surface heating, thereby achieving the disparity in wettability between elevated and planar surfaces. The construction of a hybrid wettability system featuring hydrophilic raised regions and superhydrophobic flat areas has been successfully achieved. Water collection experiments and droplet shedding processes were documented for various samples, including superhydrophobic copper plates, hybrid wettability hemispherical bumps and hybrid wettability linear bump copper plates. As depicted in Figure 9.7b, when a droplet interacts with a superhydrophobic surface, minute water droplets initiate and expand into the fissures on the surface. Subsequently, the expanding droplets merge with neighboring droplets to form larger ones until they are swiftly dislodged by gravity, coalescing numerous tiny droplets along the way to rejuvenate nucleation sites on the superhydrophobic surface. Droplet shedding on superhydrophobic surfaces occurs more rapidly and is smaller in magnitude compared to other wettable surfaces.

Diverse fog collection behaviors are observed on mixed wettability surfaces featuring two distinct raised structures. Droplets initiate and expand more rapidly on hydrophilic projections compared to superhydrophobic surfaces. This is primarily attributed to two factors. First, the hydrophilic nature of the bumps facilitates the initiation and expansion of droplets on these surfaces. Second, the three-dimensional structure of the raised features captures more droplets than a flat surface, thereby accelerating the expansion of droplets on them [64]. The diffusion flux of droplets is higher in the raised area compared to the planar area of the same height. Moreover, the bump itself expedites the aggregation of droplets, as droplets are denser at the bump and the velocity vector at the apex of the bump is larger [55]. As illustrated in Figure 9.7c, the proximity to the bump correlates with an increased presence of minute droplets. The initiation of droplet formation in and around the bump occurs more densely, resulting in larger droplets. In this case, the surface of the linear bump structure demonstrates enhanced droplet capture and aggregation efficiency, enabling droplets to coalesce into larger droplets in a shorter duration. As illustrated in Figure 9.7g, the linear bump comprises a curved tip and two slopes with an inclination of approximately 24°. When the droplets expand to a specific size, this structure can propel the droplets along the slopes to the sides of the bumps. Simultaneously, the bumps are hydrophilic, and the hydrophilic surface has a high affinity for the droplets. Consequently, droplets at the tip of the Hybrid-LB sample migrate along the slope and accumulate at the edges of the bump (Figure 9.7g). As the droplets continuously accumulate on both sides of the bumps, the droplets will contact the superhydrophobic surface at intervals between the bumps. The low adhesion of the superhydrophobic surface, combined with the gravitational force acting on the droplets, will cause the large droplets to slide downward. The anisotropy of the linear bumps causes droplets to slide along a consistent trajectory and collect all droplets along the way.

In addition, according to Aizenberg, projections on the dorsal surface of the desert beetle optimize the rapid growth of localized droplets by aggregating the apical vapor diffusion flux. Furthermore, specific morphologies on smooth surfaces have been shown to effectively promote water condensation and enhance the vapor diffusion flux [65].

9.2.4 Patterned Surfaces

The above describes a series of amphiphilic/hydrophobic surfaces inspired by desert beetles that demonstrate a higher fog collection efficiency than a single wettability surface. The above cases primarily focus on the design of surface wettability, but the hydrophilic and hydrophobic regions are often disordered. By addressing the three key aspects of fog collection, it is possible to enhance efficiency by designing the hydrophilic and hydrophobic areas in an orderly manner. Specifically, creating a well-patterned wetting surface can improve the condensation, growth, and removal of fog droplets, representing a significant optimization strategy for the development of fog collectors. Consequently, in recent years, research on patterned surfaces for fog collection has become increasingly abundant.

Song et al. [66] obtained a superhydrophobic surface by spinning PDMS and graphene (PDMS/G) onto a superhydrophilic glass substrate, and then prepared a series of patterning surfaces between hydrophilic and hydrophobic by cutting the coating. In the fog collection test, it was found that the droplets on the superhydrophilic surface tended to form a continuous a water film, which prevented the subsequent collection of droplets.

On superhydrophobic surfaces, individual droplets condense into larger water droplets, resulting in an accumulation of water on the surface. However, on biological surfaces with wettable patterns, tiny water droplets are directed toward more hydrophilic areas. Square, conical and circular wettability patterns are used to observe the water collection process. Due to the presence of surface energy gradients and gravity, tiny droplets on the superhydrophobic regions to migrate toward the hydrophilic regions. The preparation of a conical pattern surface, due to the presence of a shape gradient, creates a Laplacian pressure on the droplet, thereby further improving the one-way transport of the droplet. Therefore, in the same time, the fog collection efficiency of the surface with the tip pattern is higher than that of the other two patterns. In addition, patterned surfaces collect more water than uniformly wettable surfaces due to surface energy gradients and gravity. In the same hydrophilic cone pattern, Sun and coworkers [67] deposited dopamine in the hydrophobic region using the immersion method to increase the hydrophobic hydrophilic water level point and strengthen the fog droplet trapping process.

Zhang et al. [68] prepared a biomimetic pattern grid with triangular superhydrophobic regions and hydrophilic channels by polymer-laser ablation for water collection from fog streams. The preparation process is shown in Figure 9.8a. Following laser sintering, different polymer films may exhibit varying degrees of wettability due to their unique elemental compositions and microstructures. Additionally, laser ablation enables the precise control over the pattern geometry, size, and position. When designed and constructed correctly, the stainless-steel

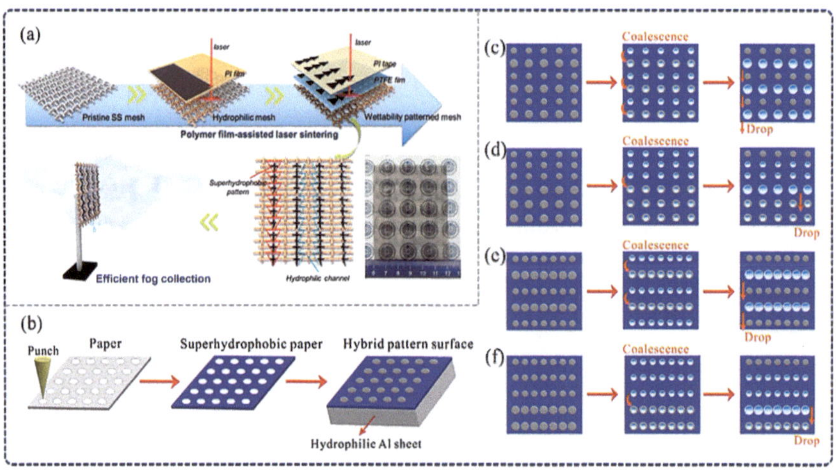

Figure 9.8 (a) The preparation diagram of wettability patterned mesh. Source: Reproduced with permission from [68] / American Chemical Society. With the assistance of a suitable polymer film, the surface wettability of the mesh could be selectively controlled from hydrophilicity to superhydrophobicity. The as-prepared mesh could be applied in efficient fog collection. The insert digital image shows a piece of wettability patterned mesh in a large size. (b–f) The fabrication process of the hybrid superhydrophobic–hydrophilic pattern surfaces (b) and dynamic fog-harvesting mechanism (c–f) of the 1.5-2a (c), 1.5-2b (d), 1.5-2c (e) and 1.5-2d (f) surfaces. Source: Reproduced from [69] / with permission of Elsevier.

mesh achieves an impressive collection rate of up to 177.65 mg cm^{-2} h^{-1}, representing an increase of nearly 16 times compared to the original stainless-steel mesh. Furthermore, the mesh exhibits high stability not only in long-term applications but also in multiple regeneration cycles.

Sun et al. [69] provide a simple and low-cost method for fabricating mixed superhydrophobic-hydrophilic patterned surfaces. The smooth aluminum alloy substrate serves as the hydrophilic region, and the superhydrophobic adhesive paper with adjustable pore sizes is utilized for the superhydrophobic region. The efficiency of fog trapping is assessed by adjusting and combining hole sizes on the surfaces of the superhydrophobic-hydrophilic hybrid patterns. To compare the hybrid superhydrophobic - hydrophilic pattern surfaces with mixed aperture and single aperture, six different hybrid surfaces with combined aperture were fabricated, denoted as 1.5-2a, 1.5-2b, 1.5-2c, 1.5-2d, 1.5-2.5a and 1.52.5b, respectively. The best fog efficiency of single aperture hybrid superhydrophobic-hydrophilic pattern surface is 192.5 mg cm^{-2} h^{-1}, which surpasses that of uniform superhydrophobic and hydrophilic surfaces. In addition, the fog trapping efficiency of the prepared combined aperture hybrid superhydrophobic-hydrophilic pattern surface is 88.3% higher than that of the single aperture surface. The dynamic fog collection mechanism shows that the optimal combined aperture mixing surface can maximize the number of merged droplets in a certain time to collect additional droplets, thus optimizing the fog collection efficiency. As shown in Figure 9.8c,d, the designed combination aperture arrangement has a great influence on the fog collection efficiency of 1.5–2 series surfaces. The fog collection efficiency of alternating (1.5-2a and 1.5-2c surfaces) surfaces with combined aperture superhydrophobic – hydrophilic patterns is better than that of sequential (1.5-2b and 1.5-2d surfaces). Among these, the surface with a 1.5–2 mm aperture has the best fog collection efficiency of 362.5 mg cm^{-2} h^{-1}.

Zhou et al. [70] successfully prepared mixed wettable polypropylene/graphene composite films (UPP/GNS) using extrusion compression molding and surface modification methods (Figure 9.9). The micro/nanostructure on the surface of the film consists of microcolumns and super-hydrophobic nanoparticles. Following the removal of the superhydrophobic nanoparticles from the top of the microcolumns, the hydrophilic top, superhydrophobic bottom, and side walls are formed. The contact angle of the surface of the UPP/GNS film with hybrid wettability is $160 \pm 3°$. The film exhibits excellent performance in accelerating droplet condensation and aggregation, which is conducive to droplet transfer in fog-collecting systems. The fog collection efficiency is 1251 mg cm^{-2} h^{-1}, due to the synergistic effect of the 3D mixing wettability and the abundance of tiny droplet trapping areas. Importantly, the excellent chemical resistance, passive anti-icing properties, active photothermal anti-icing, and sufficient durability allow the UPP/GNS film to maintain a stable fog collection efficiency under various external disturbances, such as salt, acid/alkali solutions, and high temperatures. The UPP/GNS fog collector proposed in this work offers a promising solution for the manufacture of efficient, economical, and mass-produced fog collectors and for alleviating global water scarcity.

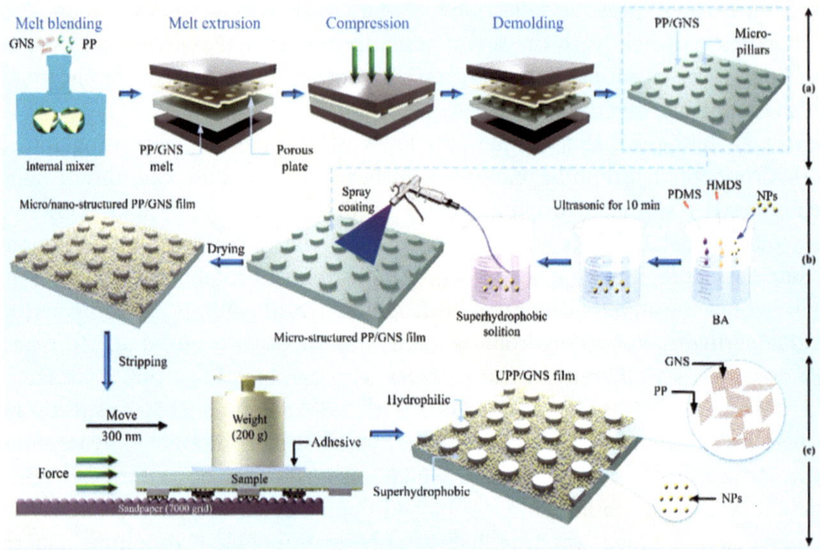

Figure 9.9 Schematic illustration of the preparation process of polypropylene/graphene nanosheets (UPP/GNS) films. Source: Reproduced from [70] / with permission of Elsevier.

Wettability gradient and patterned surface are two important strategies to control the wettability of solid surfaces and are common methods to achieve droplet directional transport [71]. Hou and coworkers [72] prepared a novel wettability gradient patterned surface (WPGS) using a copper plate as the substrate through the one-step anodizing (MGAO) method (Figure 9.10). During oxidation process, the mask surface cannot be oxidized except for the areas exposed to the electrolyte, thus obtaining a wettable mode. In addition, due to the current density and oxidation time gradient caused by electrolyte drain, a wettable gradient surface is formed on the surface from T to B, which not only improves droplet trapping performance but also maintains efficient droplet transport on hydrophobic substrates. What's more, the regenerated surface can further accelerate the wettable area by pushing the collected water droplets to more droplets, utilizing the synergies of multiple driving forces to achieve a continuous water collection cycle.

Bhushan and coworkers [73] used glass as the base material to prepare a hydrophilic triangle pattern surrounded by super hydrophobic edges, and studied the collection efficiency of fog on the surface. (Figure 9.11) When exposed to fog, droplets accumulate in the hydrophilic zone, grow, and condense into larger droplets. Eventually, they are large enough to contact the superhydrophobic boundary, triggering transport motion. Under Laplacian pressure, the droplet self-drives transport. The collected water moves more slowly over a triangular pattern with a larger top angle, and larger water droplets are fed into the collector.

Inspired by the fog collection mechanism of desert beetles and spider silk, Bai et al. [27] developed a coated surface with a star-shaped wettability pattern for fog collection (Figure 9.12). Compared with other homogeneous wettability

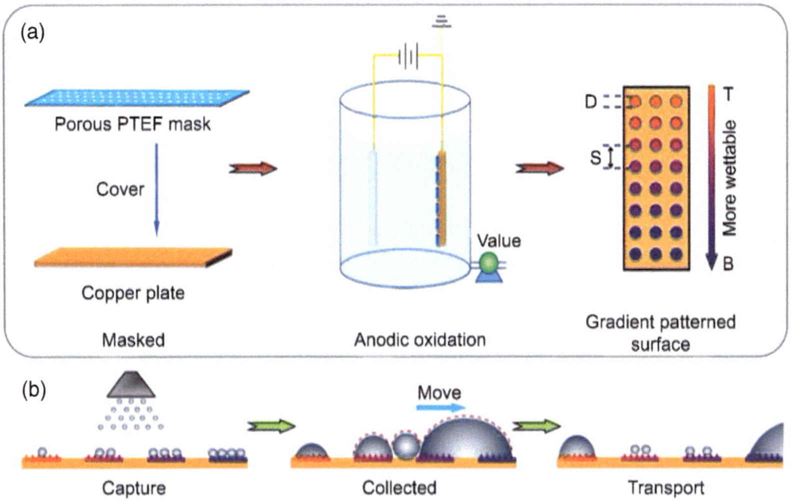

Figure 9.10 (a,b) Schematic illustration of the fabrication process of the surface with wettable patterns and gradient (WPGS) and the fog collection process on it: (a) Fabrication process of WPGS by a mask-based gradient anodic oxidation (MGAO). During the oxidation process, the masked surface cannot be oxidized except for the areas exposed to electrolytes. Wettable patterns are obtained. Besides, a wettable gradient is formed on the Surface from T to B due to current density and oxidation time gradient induced by emptying of the electrolyte; (b) Three main steps of the fog collection process, capture, collect, and transport. The droplets captured from the foggy wind coalesce into larger ones on the surface. Then the droplets are transported directionally toward more wettable regions taking advantage of the wettable gradient, which could accelerate the regeneration of the surface, leading to enhanced water collection efficiency (T: Top area and B: bottom area of the copper plate during the MGAO process, D: diameter of the circle-shaped hydrophilic pattern, S: separation distance between patterns center-to-center). Source: Reproduced from [72] / with permission of American Chemical Society.

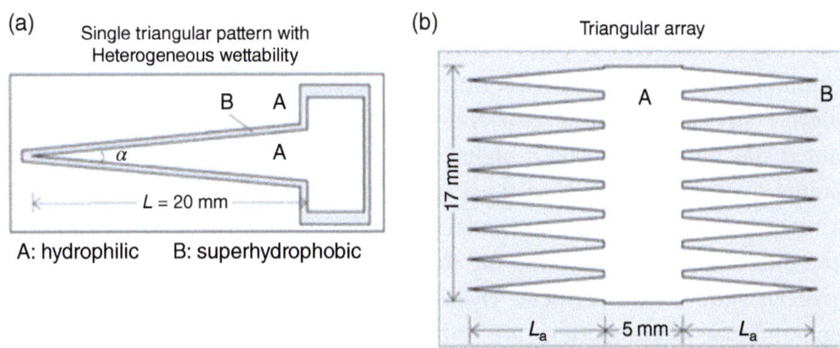

Figure 9.11 (a,b) Schematics of (a) the sample with a single triangular pattern and (b) a reservoir with an array of triangular patterns. Source: Reproduced from [73] / with permission of The Royal Society.

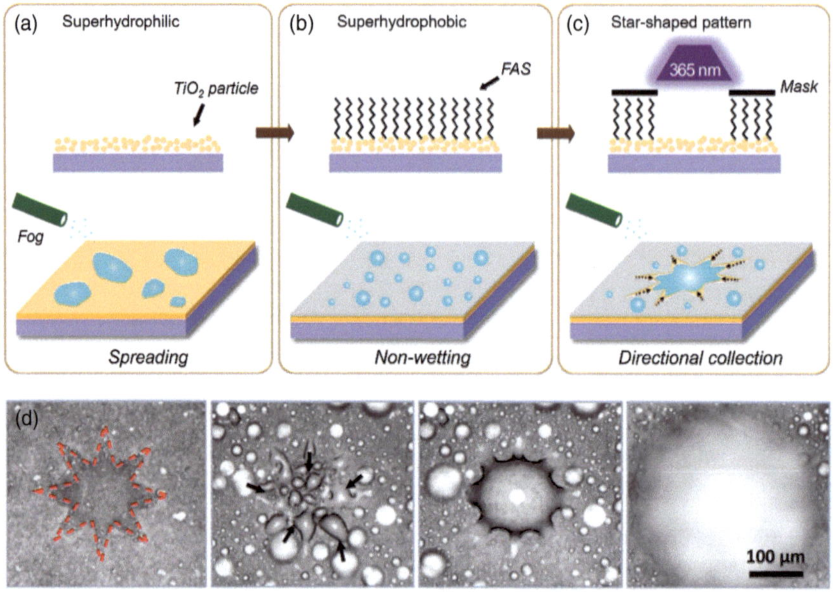

Figure 9.12 (a–d) Schematic illustration of the fabrication process for bioinspired surfaces with star-shaped wettability patterns and their performance in fog collection. (a) Superhydrophilic surface composed of TiO_2 nanoparticles, where fog droplets spread (the bottom). (b) Superhydrophobic surface modified with FAS showing the non-wetting property to fog droplets (the bottom). (c) Bioinspired gradient surface with a star-shaped wettability pattern. It is realized by illuminating the FAS-modified film under UV light with a photomask. Fog droplets are collected directionally towards the star-shaped region, which is more wettable (the bottom). (d) Different fog-collecting processes on surfaces with star-shaped wettability patterns. Source: Reproduced with permission from [27] / John Wiley & Sons.

superhydrophilic and superhydrophobic surfaces and circular patterns, this surface with a star pattern enables more efficient fog collection by integrating surface energy gradients and Laplacian pressure gradients. In Figure 9.12d, due to the "wedge" structure, the droplets move and spread toward the center of the star pattern. After the droplet is released from its original location, a new collection cycle begins immediately (see arrow). Therefore, in the same period, the surface with a star pattern collects water more efficiently than a circular pattern. Moreover, the five-pointed star pattern showed a higher collection rate than the star chart with other points. In addition, because the Laplace pressure gradient involved in the "collect" step is highly sensitive to the length scale, surfaces with smaller pattern sizes are more efficient at collecting fog than those with larger pattern sizes, even if the pattern shapes are identical. Overall, the experimental results show that the shape and size of the pattern are critical for enhancing the efficiency of fog collection.

Wang et al. [71a] used laser printing technology to achieve to achieve rapid, large-area, and highly flexible preparation of hydrophilic superhydrophobic patterned surfaces y (for example, metal and ceramic substrates). The design uses

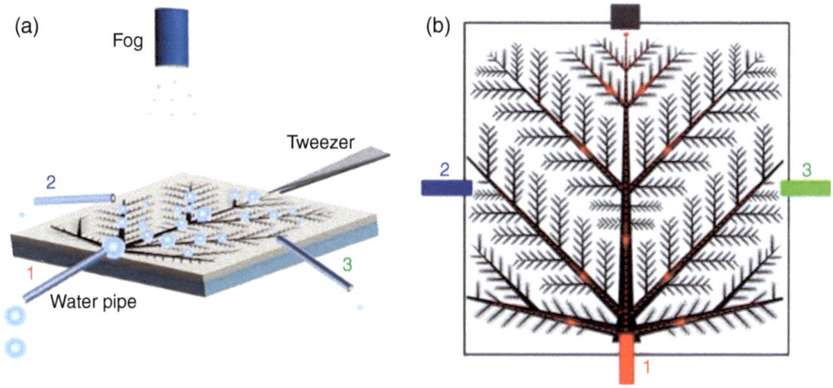

Figure 9.13 (a) Schematic illustration of the method used to quantitatively measure the fog collection efficiency of different surfaces. (b) Design of the tree pattern. Points 1, 2, and 3 represent the locations where the water pipes were placed, and the red arrow represents the driving direction of the water droplets. Source: Reproduced from [71a] / with permission of American Chemical Society.

laser writing directly to etch a hydrophilic tree pattern onto a superhydrophobic surface (Figure 9.13). By regulating the parameters of the instrument, the tree can exhibit both conical branches and hierarchies, with its dimensions controllable from microns to millimeters. This design also exploits surface energy gradients and Laplacian pressure gradients, enabling the tree-like layered conical pattern to propel minute droplets directionally and ultimately collect water at the intended location. Yin et al. [74] used femtosecond laser irradiation to prepare a hydrophilic/super-hydrophobic copper network with a high water collection rate. Previous literature has demonstrated the feasibility of hydrophobic/hydrophilic materials in fog collection processes. However, special devices, complex processes, and the limitations of substrate materials severely limit its application. Therefore, a simple and low-cost method is needed to prepare superhydrophobic-hydrophilic composite pattern surfaces on various substrates.

Similar to the above tree pattern, Guo et al. [75] inspired by desert beetles and leaf veins, successfully designed a vein-like patterned surface with mixed wettability for fog collection through a simple, low-cost, and environmentally friendly route. The design advantages of this pattern include two aspects: on the one hand, it can easily capture fog droplets through a wide range of hydrophilic points on the surface of the material, and expand the contact area between the surface and the fog flow. Secondly, the vein pattern facilitates the rapid transport and removal of water. This combination of multiple mechanisms significantly enhances the fog collection capability. Figure 9.14a shows the process of preparing a patterned surface on a zinc plate. First, the zinc plate is covered with PVC tape as a mask, and then the zinc plate base is sprayed with a super hydrophobic coating. After the mask is removed, hydrophilic copper nanoparticles are deposited. As shown in Figure 9.14b, by comparing six groups of materials with different wettability and patterning surfaces, it is found that materials with mixed wettability and veinlike

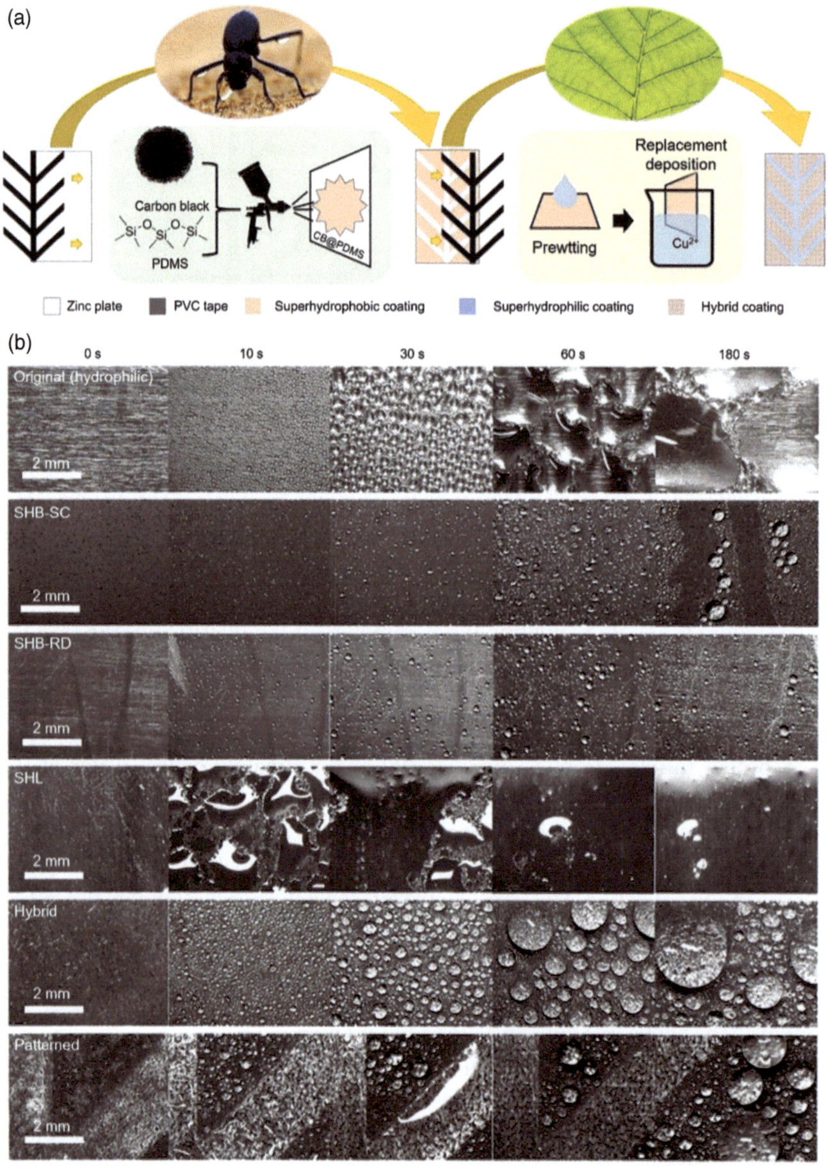

Figure 9.14 Design of a venation-like patterned surface with hybrid wettability and images of the harvesting progress. (a) Illustration of wettability-hybrid coating with a venation-like pattern preparation. (b) Images of the harvesting progress of the surfaces with different wettabilities. Source: Reproduced with permission from [75] / American Chemical Society.

patterning surfaces have obvious advantages in fog capture, growth, and removal, resulting in excellent fog collection ability. The maximum weight of water trapped per unit surface area is defined as the water-collecting capacity of the surface. When the samples were placed vertically, the water collection capacity of the wettable mixed surface was found to be 151.2% and 78.6% higher than that of SHL and SHB-SC, respectively. Weighing the water collected over 15 min for each sample, it was found that compared to the original, SHB-RD, SHB-SC, SHL and hybrid samples, the water collection rate of patterned wettability-hybrid sample increased by 83.9%, 225.6%, 116.8%, 103.4% and 15.5%, respectively. The results show that the wettability hybrid coating and vein pattern can improve fog collection efficiency.

Two superhydrophobic surfaces were obtained via spray-coating (SHB-SC) and replacement deposition (SHB-RD), respectively. The superhydrophilic surface (SHL) was fully covered with a water film in less than 60 s. In contrast, the superhydrophobic surface hardly captured any drops in the beginning due to its excellent water repellency. As for the wettability-hybrid surface, the fog drops were easily captured by the dense hydrophilic points from the beginning.

Inspired by the unique hydrophilic and hydrophobic patterns on the backs of Namib Desert beetles, Wang et al. [3b] designed a fine, veinlike pattern surface on an aluminum plate to facilitate the collection of water droplets, allowing them to coalesce and subsequently fall off. The entire preparation process is illustrated in Figure 9.15. Initially, a suspension is sprayed onto the surface of the substrate using a mask of different shapes, thus creating a hydrophobic pattern surface of varying shapes. Secondly, the drop coating method is employed to prepare the radiation cooling layer. In contrast to previously reported methods, this work also incorporates a radiation-cooled bionic coating into the catchment layer, derived from biomimetic Saharan silver ants that are capable of regulating their body temperature at extreme temperatures in the African desert. Using $MgHPO_4{\cdot}0.78\ H_2O$ as the functional particle of the radiation cooling layer, a cooling coating was prepared on the back of the water collection layer. It allows radiant cooling during the day to 3.5°C below ambient temperature and 5.4°C below primary aluminum plate temperature. The coating is applied to the back of the water-harvesting material, offering further possibilities for reducing the rate of water evaporation and increasing the efficiency of water harvesting. Figure 9.16a shows the temperature change of P(VDF-HFP)/$MgHPO_4{\cdot}0.78\ H_2O$ coating and aluminum sheet during the period from 9 : 30 to 16 : 30.

As shown in Figure 9.16b,c, the fog-collecting capacities of six groups of surfaces with different patterns are compared. Of the six patterns, the patterned surface with an 8 mm diameter hydrophobic point has the lowest water collection efficiency. This stands in stark contrast to the distribution of hydrophilic and hydrophobic areas on the backs of Namib desert beetles. Patterns IV and V contains hydrophilic points and hydrophobic sides of different diameters. Under the same area ratio (hydrophobic area: hydrophilic area = 1 : 1.5), the water collection efficiency of the hydrophilic point diameter of 4 mm is higher than that of the hydrophilic point

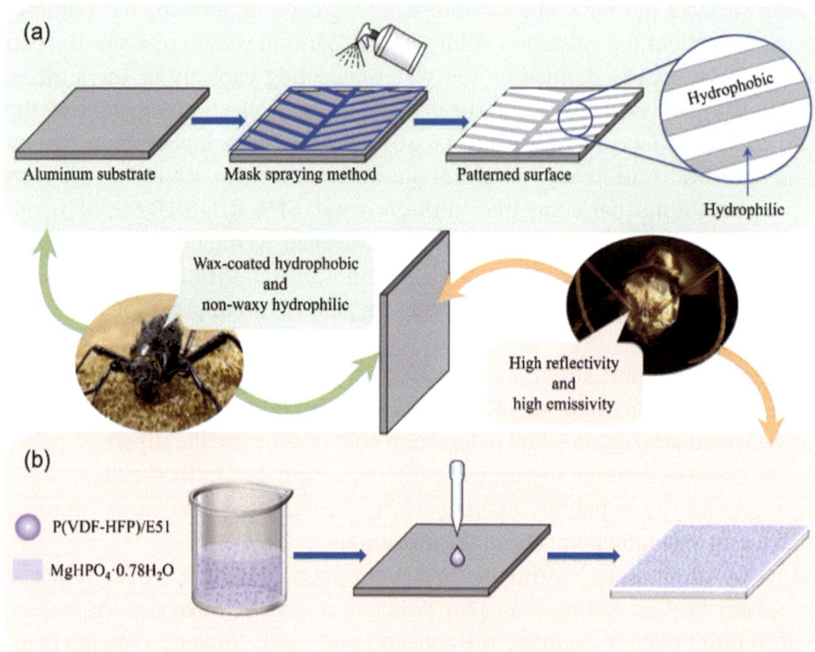

Figure 9.15 Schematic diagram of the preparation process of water collection device based on double-sided function. (a) A patterned hydrophilic-hydrophobic surface is obtained by spraying an Al_2O_3/TiO_2 hydrophobic suspension onto one side of the aluminum sheet. (b) Using $MgHPO_4{\cdot}0.78\ H_2O$ as a functional particle and P(VDF-HFP) as an adhesive to prepare the radiation cooling coating on the other side of the aluminum sheet. Source: Reproduced from [3b] / with permission of Elsevier; Reproduced with permission from [75] / American Chemical Society.

diameter of 6 mm. Of the six patterns, the type VI vein pattern has the highest water collection efficiency at 786.15 $mg\,cm^{-2}\,h^{-1}$, and also has the quickest initial drop, occurring in approximately 88 s. The second most efficient pattern is the type II vein-patterned surface (760.08 $mg\,cm^{-2}\,h^{-1}$). The two vein-like patterns, type VI and Type II, share a similarity in that they feature hydrophilic stripes as water channels. Figure 9.16d is an optical image of the growth process of water droplets on a vein pattern. At 10 s, water can be observed dripping from the mainstream in the middle. The change in the position of droplets 1 and 2 at 20 s and 30 s indicates that they are transported along the tributary main road to the main road. A similar situation occurred with drops 4, 5, and 9. From 50 to 60 s, droplets 6 and 7 converge and flow into tributaries. From 60 to 70 s, droplet 8 follows the mainstream to the bottom and away from the surface. Additionally, the self-cleaning properties of the hydrophobic surface allow for continuous nucleation and growth of new droplets, thus enhancing water collection. [76] At the same time, the experiment also proved that after 10 collections, the average water collection efficiency of the surface with the vein pattern can reach 1604.61 $mg\,cm^{-2}\,h^{-1}$.

In the most advanced fog collection materials, a single bionic surface is far from achieving optimal fog collection efficiency. In recent years, as researchers have

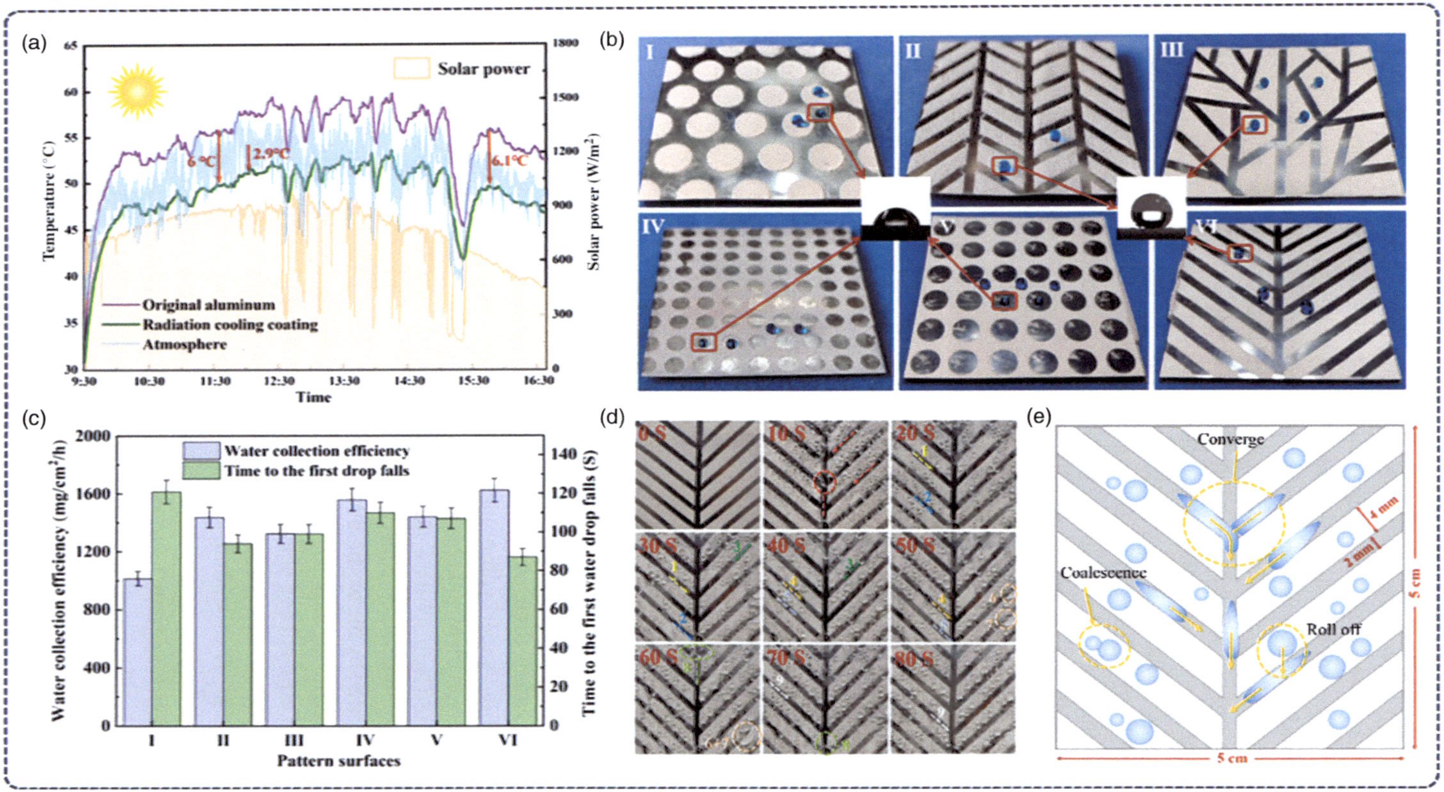

Figure 9.16 Radiative cooling layer boosting hydrophilic-hydrophobic patterned surface for efficient water harvesting. (a) Temperature change of P(VDF-HFP)/$MgHPO_4 \cdot 0.78\ H_2O$ coating and raw aluminum sheet from 9 : 30–16 : 30. The weather conditions on that day were sunny with a temperature range of 23–32°C and humidity of about 39%. (b) Optical images of water droplets on hydrophilic or hydrophobic sites. The water contact angle images only stand for their wettability and do not correspond to their optical image seriatim. (c) Water-harvesting efficiency of six different patterned surfaces and the time of first droplet drop. (d) Optical image of water collection on a surface with a Vein-like pattern and lines of the same color show the same water drop at different locations. (c) A diagram of water droplets growing on a Vein-like patterned surface. Source: Reproduced with permission from [3b] / Elsevier.

gained a deeper understanding of fog collection, many efficient fog collection materials have been achieved through a combination of like patterned surfaces, wettability gradient surfaces, lubricating oil injection surfaces, and Janus materials [77]. For example, Zhang et al. [78] designed a multi-bioinspired pattern fog collection system with hydrophilic nanofiber bulges and hydrophobic smooth substrates using copper mesh as the substrate through electrospinning technology and low surface energy modification. Among them, the protrusion is conducive to fog trapping and droplet growth and condensation. Nanofiber capillary channels are used to accelerate the directional transport of coalesced micro droplets. Hydrophobic substrates are used to facilitate droplet removal [1b, 79]. The resulting patterned nanofiber fog collector exhibits excellent water collection efficiency and long-term durability, enabling rapid directional delivery of tiny droplets and efficient water collection. In the following sections, we will introduce several cases of multi-bionic enhancement of fog collection, and more multi-bionic cases will be summarized and discussed in the last section of this chapter.

Zhong et al. [80] prepared Janus copper mesh with hydrophilic and hydrophobic hybrid wettability patterns and asymmetric wettability through simple liquid phase modification. The asymmetric wettability on both sides of the copper mesh can drive the directional transport of droplets and enhance the rapid removal of droplets, and the nanoscale hydrophilic and hydrophobic hybrid pattern enhances fog capture. The synergistic effect between the two enhances the fog capture, water supply and water removal capabilities of the fog collector. Compared to the corresponding superhydrophilic membrane, the fog collection effect of Janus membrane is improved by ~800%. The preparation process of the aerosol collection material is shown in the Figure 9.17a,b. First, the superhydrophilic copper mesh was prepared by ammonia alkali etching. Next, the etched copper mesh is placed horizontally at the bottom of the beaker and soaked in *n*-octadecyl mercaptan/ethanol solution. Four Janus copper networks with different surface wettability were prepared by controlling modification time, sample placement, and mixed solution concentration: superhydrophilic/superhydrophilic, (hydrophilic + hydrophobic)/hydrophilic, superhydrophobic/superhydrophobic, and hydrophilic/hydrophilic surfaces. In fog collection tests, the (hydrophilic + hydrophobic)/hydrophilic Janus material was found to have the highest water collection rate.

Fu et al. [81] inspired by the wettability and structural characteristics of lizard skin and leaves as well as the fog storage capacity of honeycomb, proposed a multi-bionic design strategy for efficient fog collection (Figure 9.18a). The physical and chemical forms of the melamine sponge were modified by the heat treatment process and spraying method. A collaborative wettability three-dimensional fog collector with high fog-collecting capacity was prepared. Besides, the fog collection strategy employed in this work features hydrophobic bumps and hydrophilic depressions, driven by chemical and structural gradients. The powerful water transport capacity of sponges is utilized to optimize the deposition, transfer, and transport processes of water droplets, and greatly improve the collection capacity of fog.

In the fog collection experiment, the behavior of droplet movement on the sample was further observed, encompassing the stages of droplet capture, growth, and

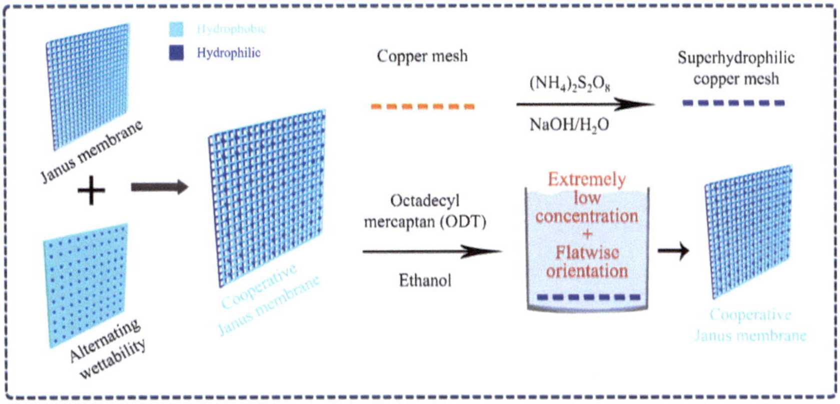

Figure 9.17 Schematic diagram for the preparation process of the hydrophilic-hydrophobic/hydrophilic Janus copper mesh. Green chains with red balls represent the molecules of octadecyl mercaptan (ODT). The position the copper mesh adopted has a great effect on the reaction between ODT and $Cu(OH)_2$. When the copper mesh is flat, the number of ODT molecules on the upsides (obverse) is more than the downside (reverse). Since the concentration of ODT is low, the molecules cannot penetrate the obverse timely after the reaction reaches equilibrium. Thus, the restricted ODT molecules preferentially react with the obverse side. When the copper mesh vertically stands on the liquid, there is no such difference. Source: Reproduced from [80] / with permission of The Royal Society of Chemistry.

removal, On the unpatterned and unmodified surface, the droplet movement is relatively simple, and it is directly absorbed by the hydrophilic surface, forming a water film on the surface (Figure 9.18f). This process slows the collection process of fog collection. In the fog-trapping strategy presented in this work, droplets accumulate in the superhydrophobic region, taking on a shape similar to a ball cap. Once the droplet reaches a certain size, gravity enables it to roll effortlessly along the linear hump of the superhydrophobic surface, ultimately reaching the hydrophilic valley. Once the edge of the hydrophilic valley is reached, the three-dimensional structure with excellent adsorption capacity rapidly adsorbs and transports droplets into the structure (Figure 9.18d). The entire collection cycle is completed in only 45 s (Figure 9.18e), which is an efficient fog collection strategy. This is due to two factors. First of all, the wettability of hydrophobic/hydrophilic regions on both sides differs significantly. Driven by a chemical gradient, the hydrophobic region has a large roughness under the action of nano silica, and the sponges in the hydrophilic region become more hydrophilic under the action of dopamine accumulation. There are obvious small spherical structures on the hydrophilic framework, which can enhance the capillary effect and promote the rapid transport of water inside the sponge. Secondly, driven by the structural gradient of the hump/valley structure, a reasonable linear bulge angle can create a terrain with droplet capture and increase the droplet capture area. As the velocity vector of fog droplets at the bulge is denser, and the velocity of fog droplets at the top of the bulge is higher [1c], the raised structure can capture more droplets at a faster rate [1b]. The hydrophilic zone facilitates the rapid movement of water droplets from the hydrophobic zone,

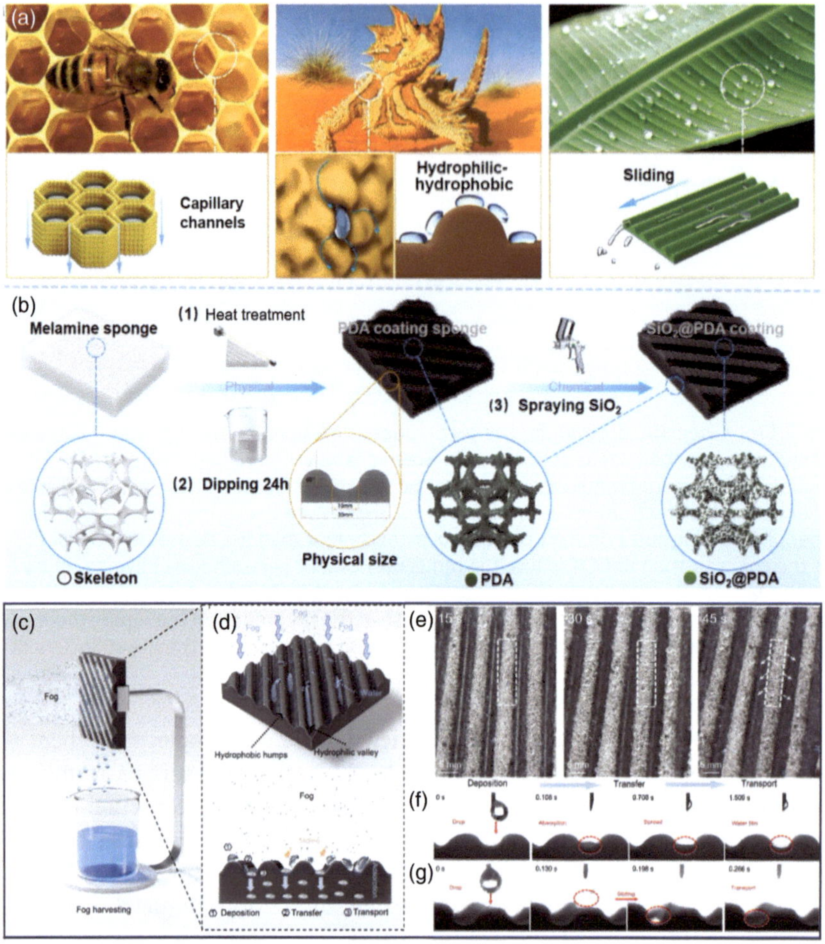

Figure 9.18 Design of multiple-biomimetic fog harvester and its corresponding fog collection performance. (a) Inspired by the natural world, the storage capacity of honeycombs, the wettability of desert lizards, and the water-collecting structures of banana leaves. (b) The biomimetic structure was prepared by using a melamine sponge as the substrate from a physical–chemical perspective. (c) Diagram of fog harvester module, (d) Sectional view of water collection module design. Fog-harvesting mechanism of collector, schematic diagram and optical photograph, schematic diagram. (e) Optical picture of surface water droplets. Study of droplet behavior on sponge surface, (f) original sponge, and (g) superhydrophobic/hydrophilic linear hump/valley sponge. Source: Reproduced with permission from [81] / Elsevier.

allowing them to spread quickly within the hydrophilic area and be transported to the interior of the sponge. On this linear peak-valley structure, the synergistic effect of superhydrophobic and hydrophilic wettability effectively accelerates the renewal rate on the surface of the collected material [77c], which is conducive to the continuation of droplet deposition in the future. In addition, when the water drops in the hydrophilic region, driven by the capillary force and wettability, the water droplets will be quickly transported to the sponge, which can effectively delay the formation of water film (Figure 9.18g). In addition, when the collection angle is slightly adjusted to 45°, two collection modes, capillary force-driven adsorption and gravity-driven drop, can be achieved through the combined action of gravity and capillary force, thus achieving efficient fog collection.

Zhou et al. [5] designed an integrated system comprising a conical spine and a Janus membrane with micro/nanostructures, inspired by cactus spine and trichome. The conical spine efficiently captures, aggregates, and transports tiny droplets, while the Janus membrane can further control the droplet transport process. The combination of the two designs fully utilizes the synergistic effect between the Laplacian pressure difference and the droplet aggregation-release surface energy, thus obtaining a higher fog collection rate. Inspired by lotus leaves and cacti, Liu and coworkers [82] also prepared a Janus functional film for fog collection using copper mesh as the base material, in which the conical micro-columns at the apex are superhydrophobic, while the remaining surfaces are hydrophobic. Water droplets are deposited, aggregated, and directed to a conical microcolumn at the bottom, where they pass through a water film on the copper surface. The asymmetric structure and wettability give the Janus membrane a fog collection efficiency of $7.05\,g\,cm^{-2}\,h^{-1}$. This research holds significant value for the design and development of fluid control equipment. In the next section, we will focus on the Janus fog collector, which uses wettability differences to enhance the fog collection process.

9.2.5 Janus Surfaces

Janus materials are widely used in various applications, including fog collection, due to the surface energy difference between the front and back surfaces and the asymmetric wetting characteristics on the porous surface, which can directionally transport water droplets from the hydrophobic surface to the hydrophilic surface [83]. Unlike most water droplets, which are transported by simple gravity, the asymmetric wettability on both sides of the Janus material creates a wettability gradient driving force that accelerates the directional penetration of trapped droplets, which greatly improves the efficiency of fog collection. For example, Feng and coworkers [84] prepared the Janus surface from an aluminum sheet as a substrate, inspired by biology. The fog collection is enhanced by 125% compared to conventional Janus membranes. The Janus film also exhibits excellent stability, which guarantees efficient fog collection over a long period. Wu and coworkers [85] prepared a hydrophobic/hydrophilic Janus aluminum sheet with a double gradient conical micropore array for efficient fog collection by femtosecond laser drilling and subsequent selective surface modification. Compared with the superhydrophilic membrane, the Janus membrane showed a 209% enhancement and a 75% reduction in the re-evaporation rate in fog collection. In this section, we will review the Janus surfaces reported so far and analyze the process of enhancing fog collection.

Tang et al. [86] designed a new type of Janus material to achieve rapid fog collection according to the droplet transport mode of capillary channels. Under the action of Laplace driving force, the droplets on the nanoneedles pass through the energy barrier of the hydrophobic layer and transport the nearest water layer in a directional manner, which plays a role in speeding up the droplet transport. In their work, micro-nanowires were first grown on the surface of copper foam by chemical etching. Then, a layer of PVDF particles was coated on the copper foam substrate with micro nanowires by spraying method. According to the spraying amount, the samples were named WCJM-1, WCJM-2, WCJM-3, WCJM-4, and WCJM-5, respectively. Eventually, Janus foams with micro-nanowire channels were prepared. Different from the traditional (super) hydrophilic surface and (super) hydrophobic surface, the Janus foam with micro-nanowire channels shows the following advantages: (i) Droplets nucleate on the superhydrophobic side, and the trapping process is carried out in the form of droplet condensation, which facilitates heat exchange and promotes the fog trapping process; (ii) The micro-nanowire channels enable the rapid transfer of trapped droplets to the superhydrophilic side for storage; (iii) Water droplets are transported unidirectionally from the superhydrophobic side to the superhydrophilic side through the action of capillary forces. On the contrary, the reverse transport of droplets from the superhydrophilic side to the superhydrophobic side needs to overcome a large breakthrough pressure. During the saturation stage of fog collection using a Janus membrane in the micro-nanowire channel, droplets flow into the collector under the influence of gravity, with saturation occurring on the superhydrophilic side. In the process of water fog collection, there are two condensing states: film condensation and drop condensation. As shown in Figure 9.19a, the surface of the sample exhibits membrane-like condensation, where trapped droplets accumulate and converge to form a liquid film. The formation of liquid film will hinder the heat exchange process of fog collection in the subsequent condensation stages, thereby affecting the condensation of fog capture in the later stage. In the late stage of film condensation, fog capture is mainly through small droplets colliding and adsorbing on the surface of liquid film for fog collection. As the amount of collected increases, it falls into the collector due to the force of gravity. This drip condensation mode is shown in Figure 9.19c. The superhydrophobic surface has fewer nucleation sites and exhibits weak surface adhesion. As a result, small droplets struggle to adhere to wettable surfaces, which will also affect subsequent processes of continuous condensation and droplet growth. In the end, only a small amount of fog is captured,

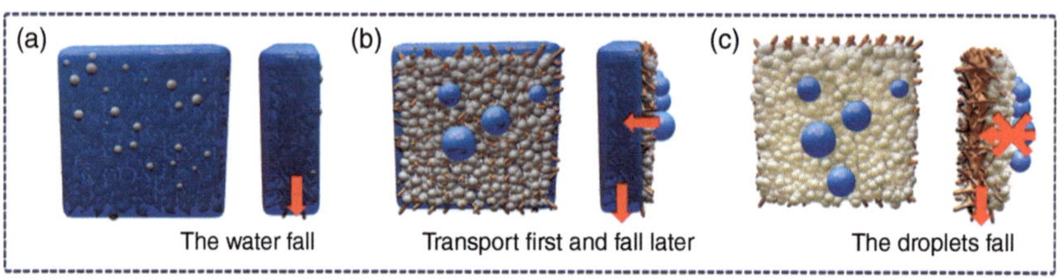

Figure 9.19 (a) Schematic diagram of fog collection of WCJM-1 and WCJM-2. The droplets coat the sample to form a liquid film. (b) Schematic diagram of fog collection of WCJM-3. The fog is captured on the superhydrophobic side (SHB) and transported to the superhydrophilic side (SHH) for collection. (c) Schematic diagram of fog collection of WCJM-4 and WCJM-5. The droplets cannot penetrate the superhydrophobic membrane. Source: Reproduced from [86] / with permission of American Chemical Society.

resulting in low fog collection efficiency. As shown in Figure 9.19b, this surface (WCJM-3) exhibits an efficient fog collection mode. The water fog is collected by drip condensation on the superhydrophobic side. When the droplets come into contact with the micro-nanowire structure, the droplets are quickly transported to the superhydrophilic side driven by capillary forces and provide a new condensation site for subsequent droplet condensation. At the same time, as the fog collection time increases, the micro-nanowire structure can be infiltrated by droplets, and the transport resistance of subsequent droplets can be reduced. As a result, the WCJM-3 surface achieves the highest fog collection efficiency of 1.1 $g\,cm^{-2}\,h^{-1}$.

As mentioned in the previous section, the preparation of cutting-edge Janus materials, is often a combination of multiple bionics. Specifically, by maintaining opposing wettability on the surface, the geometric structure characteristics and wettability characteristics of the bionic biological surface are used to further accelerate the capture of droplets and promote the condensation of droplets on the surface to achieve efficient fog collection. For example, Zhou et al. [4] also attempted to develop novel Janus materials. The mixed wettability of the beetle was combined with the contrasting wettability of the upper and lower surfaces of lotus leaves. ZnO nanorods were uniformly grown on $Cu(OH)_2$ nanorods by one-step hydrothermal method. Subsequently, a grafted with low concentration PDMS on the surface of the ZnO nanorods by photocatalytic reaction. The hybrid hydrophobic-hydrophilic surface and Janus copper foam (HB − HL + JCF) have been successfully prepared. Compared to the original copper foam, superhydrophobic copper foam, and super-hydrophilic copper foam, Janus copper foam showed excellent fog-trapping ability. Compared to the original copper foam, Janus copper foam has improved the efficiency of fog collection by about 209%. The whole preparation process is shown in Figure 9.20a. First, superhydrophilic copper foam covered with $Cu(OH)_2$ nanowires was prepared by alkali ammonia etching. Secondly, the superhydrophilic copper foam surface with CuO–ZnO hierarchical micro-nanostructure was prepared by hydrothermal method. Finally, Janus copper foam with hydrophilic and hydrophobic interphase surface was prepared by spraying method.

In the fog collection experiment, the original copper foam, superhydrophobic copper foam, superhydrophilic copper foam and Janus copper foam was selected as the control group. As shown in Figure 9.21b, when droplets begin to deposit on the hydrophobic surface, not all droplets can be captured and collected. Some droplets will bounce directly off the hydrophobic surface, and some droplets will evaporate. Only a fraction of the droplets is captured, accumulating and growing larger as they are transported directionally to the hydrophilic side, where they ultimately fall under the influence of gravity. As mentioned above, the entire collection process of fog can be divided into three parts, which are the capture of fog, condensation and transportation. It is worth noting that this Janus copper foam is designed to efficiently implement all three processes. Figure 9.21b–f is an optical photo of different wettability surfaces collected by fog over 0–30 min. In the first stage of fog collection, the capture of fog mainly depends on the collision between the droplets and the surface, and can only be effectively captured by the surface when the kinetic energy of the droplets is completely consumed. Interestingly, the boundary effect is effectively

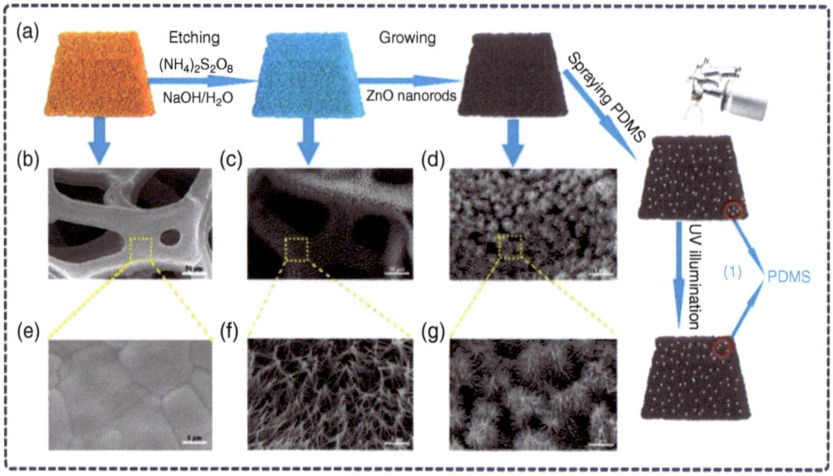

Figure 9.20 (a) Schematic of the hybrid hydrophobic-hydrophilic surface and Janus copper foam (HB − HL + JCF) preparation. The ZnO nanorods grew on $Cu(OH)_2$ nanowires by hydrothermal method and the PDMS brush grafted on ZnO nanorods by light-catalyzed reaction. (b) The SEM image of the original copper foam. (c) The SEM image of $Cu(OH)_2$ nanowires on copper foam after immersing in ammonia solution. (d) The SEM image of ZnO nanorods grew on $Cu(OH)_2$ nanowires. (e–g) The magnification SEM images of (b–d). Source: Reproduced with permission from [4] / Elsevier.

reduced by the water drain point formed by tiny droplets on the Janus copper foam drain water surface. In the second stage, trapped droplets are already present on the surface. In the following collection process, when the droplets approach the surface at high speed, regardless of their kinetic energy, these external fogs can be effectively captured once the droplets collide with the droplets formed on the surface. In addition, under asymmetric wettability, the droplets can be quickly removed, enabling a sustainable fog collection process.

As shown in Figure 9.22a–c, as the collection time increases, the number of droplets captured on the surface of superhydrophobic copper foam gradually increases, while water droplets on the surface of the original copper foam gradually deposit and form a liquid film. In contrast, water droplets on the hydrophobic surface of Janus copper foam partially disappeared, and the existence of hydrophilic points accelerated the capture and transport of droplets, further demonstrating the continuous cycle of the fog collection process. The Janus copper foam combines the two main characteristics needed to capture fog: the adsorption and removal of droplets. Hydrophilic sites are required for the adsorption of small droplets, while hydrophobic sites are conducive to the removal of larger droplets. At the same time, the asymmetric wettability distribution further promotes the transport and removal of droplets and has a good water collection effect. In contrast, the water collection in homogeneous wettable copper foam is mainly removed by gravity-induced separation, which further increases the probability of water re-evaporation. The quality of fog collected by different wettability copper foams in the laboratory for 1 h was compared. Janus copper foam had the highest total fog collection (Figure 9.22d), about 209% higher than the original copper foam collection.

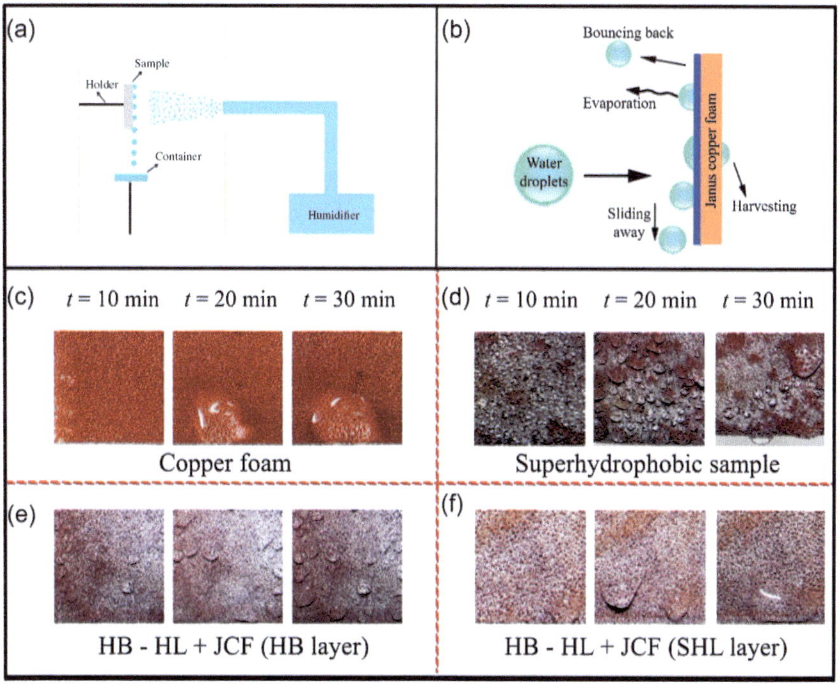

Figure 9.21 The fog-collecting performances of the HB – HL + JCF. (a) The schematic illustration of the fog-harvesting system. (b) The schematic illustration of water transport routes during water harvesting. (c–f) Images of the harvesting progress of the copper foam, SHB copper foam, and the HB – HL + JCF (both HB layer and SHL layer) in 10, 20, and 30 min. Source: Reproduced with permission from [4] / Elsevier.

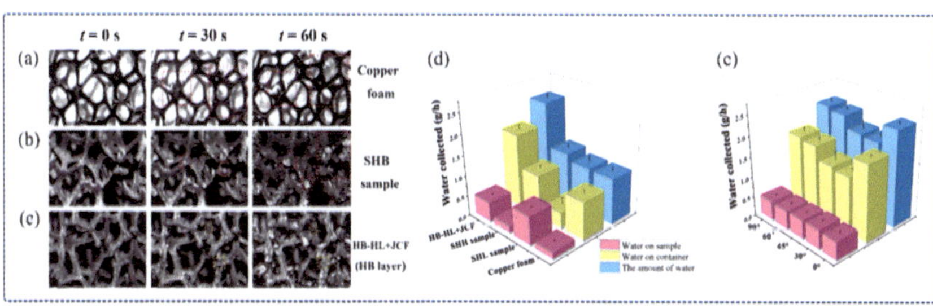

Figure 9.22 The fog-collecting performances of the raw copper foam, superhydrophobic, and HB – HL + JCF. (a–c) A comparison of the fog capture between the surfaces of copper foam (a), SHB copper foam (b), and HB – HL + JCF (c) in 0, 30, and 60 s, marked with red marks and the yellow marks representing the appearance and disappearance of water droplets, respectively. (d) The total water and water collected on the sample, container in 1 h on the fog collection tests of all samples (copper foam, SHL copper foam, SHB copper foam, and HB – HL + JCF). (e) Water was collected using the as-prepared HB – HL + JCF at different inclined angles. Source: Reproduced with permission from [4] / Elsevier.

In addition, the size of the fog-collecting surface and the angle of placement also greatly affect the efficiency of fog collection. When the sample is positioned vertically at 90°, it faces the fog outlet during peak fog volume, maximizing both the contact area with the sample and the fog collection (Figure 9.22e). The horizontal placement of the specimen maximizes the gravitational action, and the driving force generated by the asymmetric wettability further promotes unidirectional penetration of the droplets, reducing the possibility of re-evaporation. As a result, the fog collection of Janus copper foam is relatively large when placed at 0° and 90°. The size of the sample also affects the effectiveness of fog collection. In the fog environment, the fog deviates from the flow path line at the one-dimensional material, but the deviation is much smaller than the two-dimensional plane. When a two-dimensional surface is placed within the fog stream, the fog experiences a substantial deviation, preventing most of the fog from reaching the surface. Therefore, as the size of the Janus copper foam decreases, the deviation of the fog flow is further minimized. Consequently, more droplets reach the surface of the Janus copper foam per unit area. The reduced volume of the Janus copper foam significantly diminishes the fog flow deviation, thereby facilitating an efficient water supply. As shown in Figure 9.23, Wang and coworkers [87] developed a three-layer Janus fog collector by sandwiching a hydrophilic polyester net between two superhydrophobic polyester nets. In the traditional double-layer Janus fog collector, when the hydrophobic surface is exposed to the fog flow, the intercepted droplets can be transported from the hydrophobic side to the hydrophilic side. When the hydrophilic surface is exposed to a fog stream, the Janus fog collector acts only as an ordinary hydrophilic net, losing its ability to direct droplet transport. For the three-layer Janus atomizer, no matter which direction the fog comes from, the three-layer mesh can always be used as a Janus collector to effectively collect fog droplets.

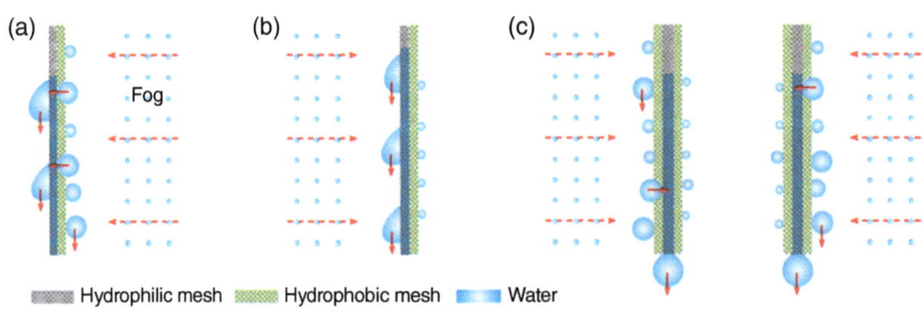

Figure 9.23 (a–c) Schematics of the Janus and sandwiched fog collectors. (a) Schematic of the dual-layer Janus fog collector when the hydrophobic face is exposed to the fog stream. The intercepted droplets can be directionally transported from the hydrophobic to the hydrophilic side. (b) Schematic of the dual-layer Janus fog collector when the hydrophilic face is exposed to the fog stream. In this case, the Janus fog collector just serves as a common hydrophilic mesh, losing the ability of directional droplet delivery. (c) Schematics of the three-layer sandwiched fog collector. No matter which direction the fog is from, the three-layer meshes can always serve as a Janus collector to efficiently harvest the fog droplets by taking advantage of directional droplet transport from the hydrophobic to hydrophilic layer. Source: Reproduced from [87] / with permission of Elsevier.

Fu et al. [88] concentrated on the synergistic effect of needle-like surface wettability to boost the collection effectiveness of fog while constructing Janus water collection materials. Taking advantage of wetting properties such as cactus needles, desert beetle backs, and banana leaf stripes, they proposed a fog-dew collection system consisting of conical needles, conical holes, and Janus sponges (Figure 9.24c,d). The acicular structure of the cactus is conical, and the diameter of the spines gradually increases from tip to end. In addition to gravity drive, this structure also has Laplacian additional pressure driving force [89]. This makes it easier for the droplets to move in the direction of the spines [13, 90].

The backs of desert shell beetles are hybrid hydrophilic and hydrophobic wettability [91]. In addition, banana-leaf-like stripes are hydrophobic and have low slip angle characteristics, which can quickly transport droplets and achieve rapid removal of droplets on the surface [92]. Inspired by this, Fu et al. designed hydrophilic bumps and delivery grooves on the tapered needles throughout the system and divided them using super-hydrophobic regions (Figure 9.24b). This structure provides additional trapping sites for the fog during circulation, and the hydrophobic separation also effectively slows down the formation of a water film. Importantly, on the mixed pattern surface, the circular pattern can construct a 3D topography that captures water, thereby increasing the droplet-trapping area. The stripe pattern can intercept falling droplets, reduce the impact between falling and immature droplets through strategically planned drainage, and then significantly enhance the fog collection capability of the surface (Figure 9.24e). According to the Kelvin equation, concave forms have a lower vapor pressure than convex forms and are more conducive to agglomeration. However, the transport of droplets in a trench channel still leaves behind a water film, which often results in water filling the channel without efficient capture. As a result, the concave and convex surfaces can capture droplets through the bumps and transport them in the fluted channels. Under the concave and convex structure, a larger groove is indirectly formed between the two bumps, allowing the fog flow to remain on the surface more effectively and be captured continuously and efficiently (Figure 9.24f). In addition, combined with the Janus sponge with a hydrophobic surface (Figure 9.24c), both sides of the Janus sponge showed superhydrophilicity approaching 0° and superhydrophobicity approaching 151°, respectively. In addition, the Janus sponge uses conical pores as connection points to increase the driving force of the liquid, separating the droplets and transporting them to the sponge for storage. The proper assembly angle and wettability coordinate to accelerate the speed of droplets entering the Janus sponge. With the synergistic effect of all levels of structure, droplets are rapidly captured, transported, and stored as they pass through the entire system (Figure 9.24d). The collection system also demonstrated an excellent dew collection capability of $1.18\,g\,cm^{-2}\,h^{-1}$ in ultra-cold environments, which many fog collection systems cannot achieve. The whole collection system collects fog dew from four steps: capture, transport, separation, and storage, which can realize portable freshwater collection in remote coastal or inland areas, this design can provide ideas for large-scale production of fog dew collection systems in the future.

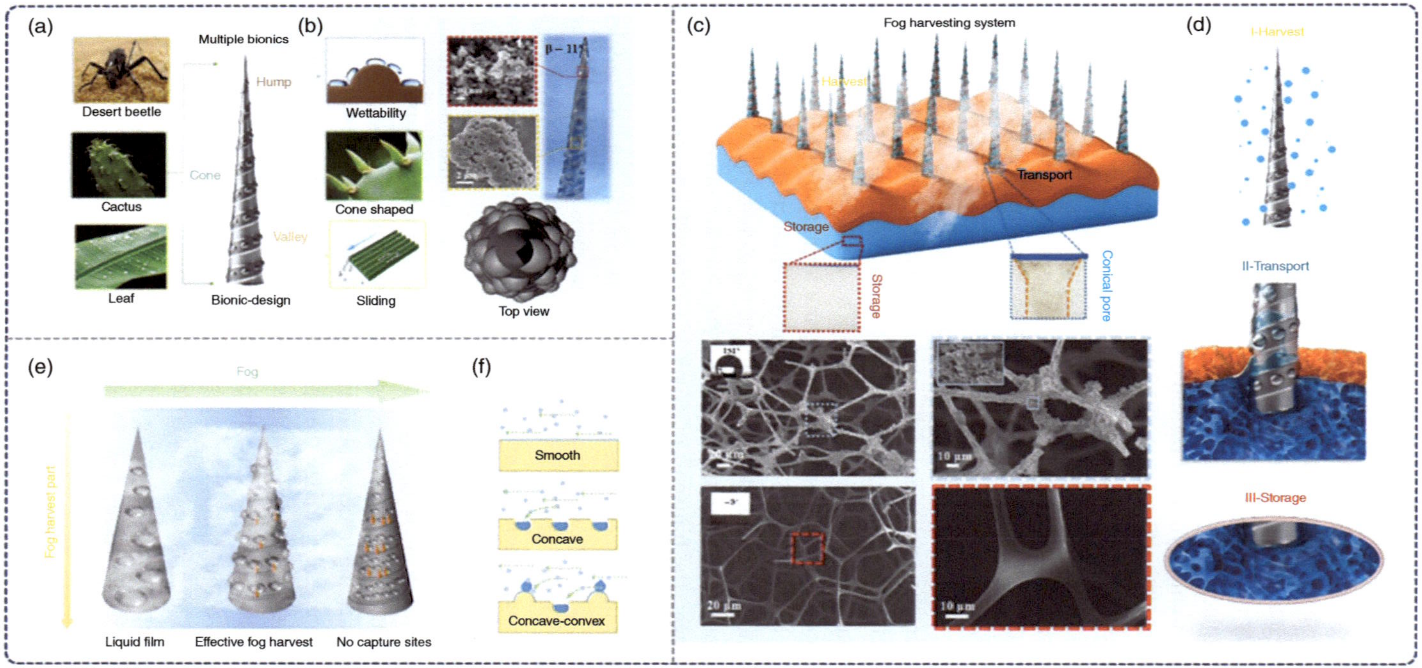

Figure 9.24 (a) Schematic diagram of an efficient fog-trapping conical needle inspired by a variety of fog-collecting creatures. (b) The hump-groove surface morphology and hybrid wettability of the conical needle and the top view. (c) The concept of the overall fog capture system and the SEM images of the transportation and storage parts of the Janus sponge. (d) The process of fog collection includes (I) droplet capture, and transport, (II) droplet separation, and transport, and (III) storage. (e) Surface phenomena during fog collection on planar, hump, and hump-groove structures. (f) The hump-groove structure facilitates fog collection. Source: Reproduced with permission from [88] / Elsevier.

Zhang et al. [93] reported a switchable slippery-superhydrophilic copper sheet with a spine structure (SLP-SHL@SP-Cu). The upper surface of this structure is smooth, while the lower surface is super-hydrophilic. The wetting properties of various plant and animal surfaces, such as cactus, bluegrass, lotus leaves, and pine cones, were integrated into the SLP-SHL@SP-Cu structure by surface modification and laser etching. The synergistic optimization of slip wettability, Laplace pressure gradient, and Janus wettability of the structure enables rapid droplet transport. The resulting SLP-SHL@SP-Cu has a water collection rate (WHR) of 7421 $mg\,h^{-1}\,cm^{-2}$ and long-term durability. Importantly, SLP-SHL@SP-Cu can respond intelligently to humidity due to the use of sensitive hydrogel fibers. When droplets flow onto SLP-SHL@SP-Cu, they intelligently switch from a two-dimensional flat structure to a three-dimensional spiral structure for fog collection. Once the fog flow near the collector is reduced, it can switch back to a two-dimensional flat structure to help store water. In addition, SLP-SHL@SP-Cu's intelligent design enables a high water storage rate (82%) at 60°C for more than 25 h.

Wang and coworkers [94] made metal-mesh Janus membranes (MM-JMs) by lithography and electroplating instead of laser ablation (Figure 9.25). The team comprehensively analyzed the effects of metal Janus membrane structural parameters (spacing, thickness, conical shape, and surface wettability) on fog collection efficiency, and found that the thinner the membrane thickness, the more hydrophobic surface, can accelerate water transport and reduce water re-evaporation, helping to improve fog collection efficiency. The work not only elucidates microscopically how individual structural parameters of Janus membranes affect fog collection performance but also provides guidance for designing efficient Janus fog collectors to sustainably collect water from the air, promising a breakthrough in clean water production.

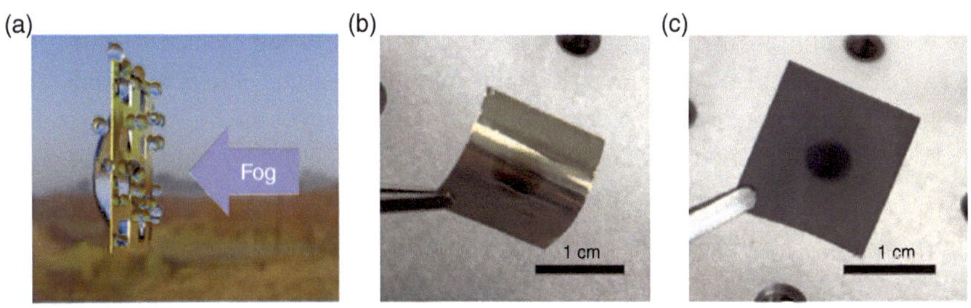

Figure 9.25 Fabrication of ultrathin planar metal-mesh Janus membranes (MM-JMs). (a) Schematic diagram of fog harvesting through an MM-JM in arid areas. Photos of (b) the hydrophobic Au side and (c) the superhydrophilic CuO side of the MM-JM. The membrane is so thin that it is semi-transparent and flexible. Source: Reproduced with permission from [94] / Elsevier.

9.3 Design of Surface Geometry

9.3.1 Linear Structure

Inspired by the ability of Sarracenia trichoid microstructure to trap fog and transport water, Wan et al. [95] used H62 brass as a substrate to prepare layered groove structures on copper wires by wire discharge machining. The experimental results show that, compared with a smooth surface, the surface with a layered groove structure has better water transport performance and higher fog collection efficiency. The water transport and fog collection ability are associated with the low rib number of the multistage micro-groove surface.

Based on the fog-water collection behavior of dragon fruit leaves, Zhong et al. [14] focused on the efficient fog-water supply process dominated by its one-dimensional linear structure (Figure 9.26). The leaves of dragon fruit covered by one-dimensional micron spines showed high efficiency in fog collection, which could capture tiny droplets in fog through spines on the leaf surface. Then, the captured tiny droplets grow rapidly on the hydrophilic surface at the tips of the spines (Figure 9.26b). Notably, compared to cactus spines and spider silk, this plant did not show a special drive for water to move droplets during water collection, but it did show an efficient fog-water supply mechanism, further illustrating the importance of the fog-water supply phase. Due to the one-dimensional linear microstructure and surface wettability of copper wire similar to foliar thorns, Zhong et al. performed biomimetic preparation based on copper wire and studied its fog collection process.

The wettability of copper wires with diameters of 50, 400, and 1000 μm and lengths of 2 cm was changed to a hydrophobic/hydrophilic phase after chemical modification. The results of the fog collection experiment show that the copper wire on the surface of hydrophobic/hydrophilic interphase wettability (SHB-SHH) does not show an obvious advantage in fog collection. It has been generally proved that the hydrophilic/hydrophobic phase mode enhances fog-collecting ability in the two-dimensional plane, indicating that one-dimensional linear materials and two-dimensional planar materials have obvious differences in the process of fog collection. The difference between the amount of fog collected by copper wires with different surface wettability is less than 30 mg. Relatively speaking, the amount of super-hydrophilic copper wire collected is small; The original copper wire collects the most water at a diameter of 400, 1000 μm. It is worth noting that the original copper wire has a similar wettability to the thorns on the leaves of the dragon fruit plant. It can be seen that the surface wettability has a certain effect on the fog collection behavior of copper wire, but the influence of the diameter of copper wire on the collected water is far greater than the surface wettability. For the original copper wire, the difference between the amount of water collected by different diameter copper wires can be as much as 60% (50, 400, and 1000 μm diameter copper wires collect ~120, ~300, and ~400 mg, respectively, within 30 min). In addition, a

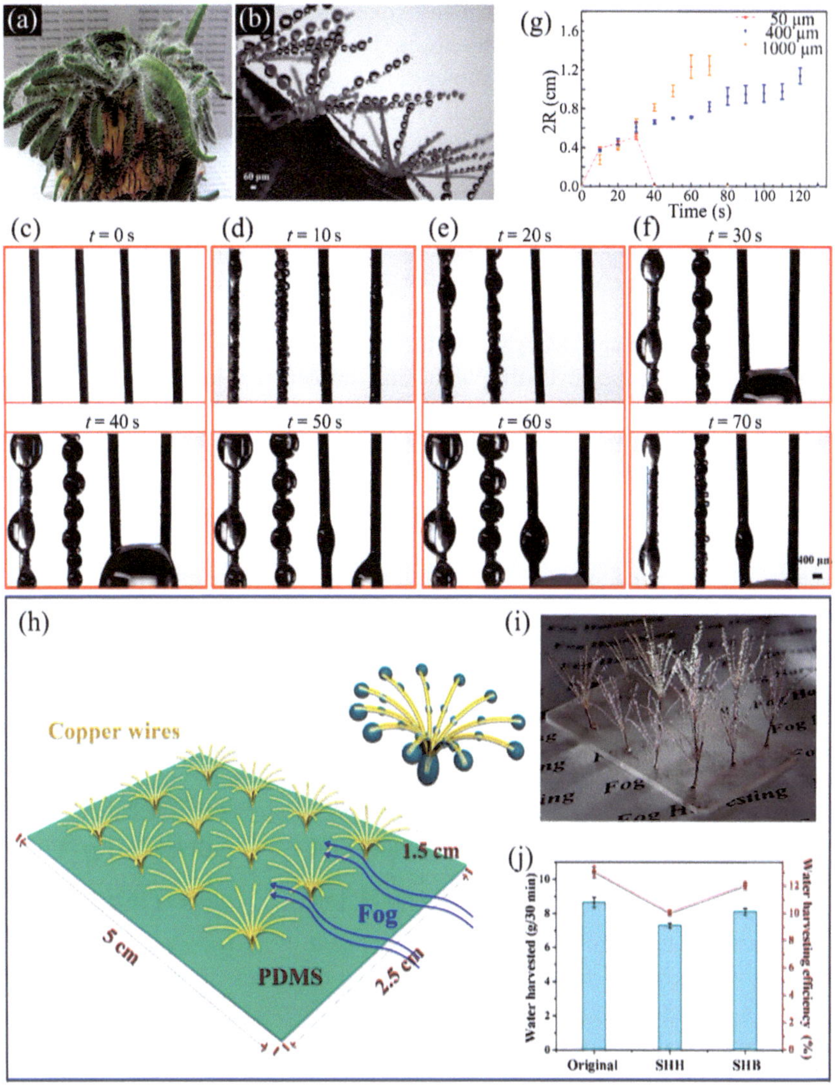

Figure 9.26 Inspired by pitaya, the bionic fog collector is used for fog collection. (a) Optical image of pitaya. (b) The pitaya covered with microthorns showed excellent fog harvesting in 20 s with a fog velocity of ~60 cm s^{-1}. (c–f) Time-lapse images of the process of water harvesting on original hydrophilic, SHB, SHB-SHH, and SHH copper wires from left to right, respectively (scale bar: 400 μm). (g) Droplet growth on superhydrophobic copper wires. (h) Schematic diagram of the 3D fog collector. (i) Fog collector made up of thorns (original copper wire) and supporter (PDMS) during fog harvesting. (j) Amount of water and the fog-harvesting efficiency of fog collector made up of original copper wire, SHH copper wire, and SHB copper wire. Source: Reproduced with permission from [14] / American Chemical Society.

single copper wire with a length of 2 cm and a diameter of 400 μm can harvest about 0.3 g of water every 30 min; In the same case, a single copper wire with a diameter of 1000 μm collects 0.4 g of water. This shows that the one-dimensional linear micron structure has an excellent ability to collect fog compared to the two-dimensional plane. This excellent fog collection capability is closely related to its efficient fog supply mechanism.

Through the analysis of the enhancement effect of fog supply, it can be found that the micron one-dimensional linear structure has a very obvious advantage in fog capture and fog supply. As shown in Figure 9.26c–f, firstly, the surface of copper wire with different wettability has an obvious difference in water capture. Copper wires with hydrophilic and hydrophobic surfaces can effectively trap fog droplets, and their surfaces can be covered by tight tiny droplets within 10 s. The droplets with higher velocities are more likely to be trapped by the superhydrophobic copper surface due to energy considerations. This is further enhanced by the excellent fog supply mechanism of the one-dimensional linear geometry, which enables a greater amount of fog to reach the copper wire surface. When the amount of fog reaching the surface of the copper wire increases, the amount of water finally collected will also be greatly increased. In addition, water droplets grow more rapidly on the surface of the linear copper wire, which of course also benefits from the advantages of one-dimensional material for water supply. The growth curve of water droplets on the superhydrophobic copper wire with a diameter of 50, 400, and 1000 μm is shown in Figure 9.26g. The diameter of the surface water droplets grows to ~0.5 mm within 30 s, and all of them are collected from the copper wire within 2 min. This growth pattern is rapid. This is because the droplets carried by the flowing air can effectively merge with the droplets (or liquid film) captured on the surface of the copper wire, and the rapid merging process accelerates the growth rate of the droplets. Therefore, the copper wire with one-dimensional linear micron-level geometry has an efficient fog supply mechanism in the fog fluid, and its efficiency can reach 100 times that of the corresponding two-dimensional plane.

Based on the excellent fog collection performance of one-dimensional micron copper wire reinforced by an efficient fog supply, Zhong et al. further designed a three-dimensional fog collector (Figure 9.26h). The role of various parts of the dragon fruit plant structure in fog collection: micron-scale spines with a certain arrangement capture fog in the fluid, and then their leaves drain the harvested water out of the system. Based on this structure, more spines are incorporated in the form of "grass" above the base of the PDMS (Sylgard 184). The spines were prepared from raw copper wire, superhydrophilic copper wire, or superhydrophobic copper wire (Figure 9.26i,j). The amount of water collected by the collector can reach 8 g in 30 min, and the collection efficiency of the fog is greater than 10%, which makes it impossible to reach the common two-dimensional planar fog device. Moreover, such high efficiency can meet practical applications. Based on an efficient fog supply with a one-dimensional linear microstructure, plastic fibers with high flexibility can build fog collectors with ultra-high efficiency and mechanical durability in a denser arrangement, which is essential for ensuring access to fresh water in arid regions.

9.3.2 Conical Structure

The curvature gradients of spider silk and cacti cause the surface droplets to be transported directionally under Laplace pressure. Inspired by this, many fog collectors with curvature gradients have been prepared. The cone [96] exhibits a typical curvature gradient, and several research groups have also combined geometric curvature gradients with wettability gradients to enhance droplet transport and removal rates. For example, Zheng and coworkers [8] fabricated a conical copper wire with a roughness gradient using a controlled electrochemical corrosion method with periodic current gradients.

As shown in Figure 9.27a, the average roughness (Ra) of positions 1, 2, and 3 on the copper wire is 80.4, 154.0, and 63.1 nm, respectively. The copper wire can realize the directional transport of the droplets, to complete the timely removal of the droplets and continuous collection of fog. This is due to the dual driving force provided by the conical gradient and the wettability gradient for droplet transport. Additionally, the experiment demonstrates that the ability to collect fog is also influenced by the tilt angle of the copper wire.

In addition to designing geometric gradients and wettability gradients to speed up droplet transport and removal, the research team also added a fog collection base to further enhance droplet removal efficiency and minimize droplet evaporation, considering the entire process of fog collection. Cao et al. [48a] combined a hydrophobic conical array with a hydrophilic cotton substrate to prepare a cactus-shaped spherical fog collection device, which greatly improved the process of water drop capture and removal. The selection of an appropriate base material is critical for the water drop removal process. Considering the excellent water collection effects of the superhydrophobic-hydrophilic copper needle and the hydrophilic slip surface, combining the two can simultaneously achieve the capture, condensation, transport, and removal of water droplets while maximizing water collection efficiency. Therefore, Zhou et al. [48b] inserted a conical superhydrophobic hydrophilic copper needle vertically into the hydrophilic glide surface to explore the overall fog collection effect of this combination (Figure 9.27b–e). During the first collection cycle, the droplets are captured simultaneously by the superhydrophobic-hydrophilic copper needle and the hydrophilic sliding surface (Figure 9.27b). After the superhydrophobic surface at the tip of the copper needle quickly captures the droplets, the droplets rapidly move back and merge to become larger under the action of the Laplacian pressure difference. When the droplet is transported to the junction of hydrophilic and hydrophobic surfaces, the droplet movement slows down until a water film is formed in the hydrophilic region at the tail of the copper needle, and the droplet moves faster driven by the wettability gradient. Subsequently, water droplets moving to the tail of the copper needle contact the water film formed on the hydrophilic sliding surface creating a water bridge between them. The oil layer on the base surface reduces the sliding friction of droplets on the surface and accelerates the removal of droplets. The first period for collecting water lasts approximately 30 s. Figure 9.27e shows a schematic

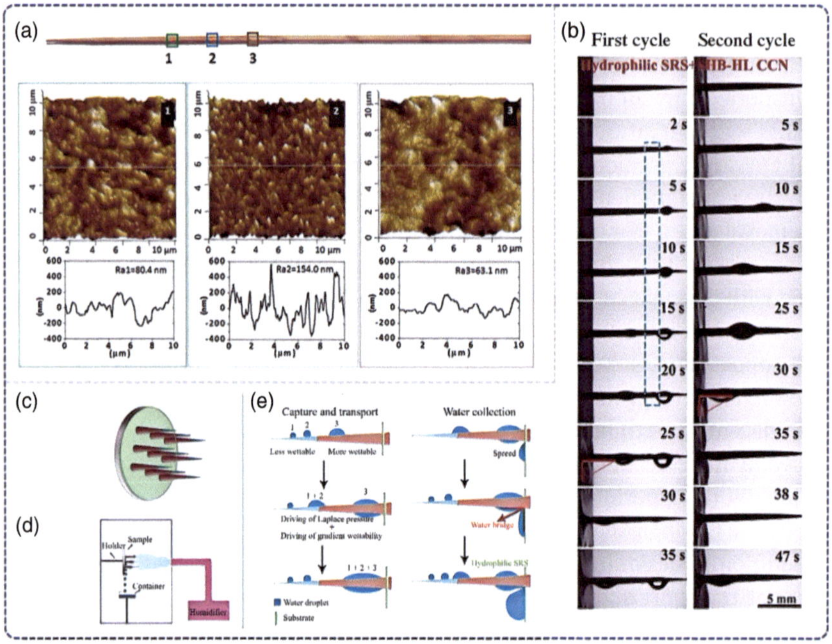

Figure 9.27 (a) Surface properties of periodic roughness-gradient conical copper wire (PCCW), including optical image (Scale bar, 1 mm) and AFM images of different sites within one roughness period of PCCW. Source: Reproduced with permission from [8] / American Chemical Society. The average roughness (Ra) of 1, 2, and 3 sites is 80.4, 154.0, and 63.1 nm, respectively. (b–e) The fog collector consists of superhydrophobic-hydrophilic conical copper needles (SHB-HL CCN) and a hydrophilic slippery rough surface (SRS). Source: Reproduced with permission from [48b] / American Chemical Society. (b) In situ observation of droplet collection on hydrophilic SRS + SHB – HL CCN (the first cycle). For the second cycle, the needle was moved back by 0.2 cm. As the length of the needle increases, the droplet transport time during water collection also increases (scale bar: 5 mm, the range of the blue box is around the hydrophilic-hydrophobic junction). The red triangle marks a water bridge, accelerating the rapid removal of water droplets. (c) The schematic demonstrates the newly designed fog collector: a new system integrated by conical copper needles with gradient wettability and hydrophilic slippery rough surfaces. (d) The schematic shows the fog-harvesting system. Fog collection process: (e) The whole fog-harvesting process of the hydrophilic SRS + SHB – HL CCN (water capture, transport, coalesce and remove). Note that: in the collecting stage, when a water droplet contacts the hydrophilic slippery rough surfaces, a water bridge forms, facilitating rapid droplet removal.

diagram of water droplets being transported on a superhydrophobic-hydrophilic copper needle and a hydrophilic sliding surface. In the second fog collection cycle, at 25 s, the copper needle reappeared with the water bridge on the base surface. The whole fog collection process is completely continuous. Compared to traditional one-dimensional conical or two-dimensional planar fog collector, this fog collector can quickly transport and store collected water droplets while constantly releasing new nucleation sites for the next round of fog collection.

9.3.3 Spindle Structure

The periodic spindle knot of the spider mesh is a key factor in its successful water collection. In recent years, researchers have successfully prepared a series of bionic artificial spider silk by imitating spider silk and conducted in-depth research on its mechanism [97]. The surface energy gradient provides the driving force of surface wetting, making the water droplets more inclined to move towards the more hydrophilic region. Additionally, the geometric structure gradient can generate Laplace additional pressure on the water droplets to drive the droplet to move in the direction of larger diameter. The hydrophilicity of spider silk and the nanoscale fiber structure on its surface further enhance the ability of this droplet to be transported directionally. This excellent directional droplet transport enables the spider mesh exhibit efficient fog-collecting behavior. Additionally, a large number of biomimetic artificial fibers have been prepared by using various methods, including electrodynamics [14, 98], dip coating [99], microfluidic [65b, 100], and flow coating [50, 101].

Tian et al. [102] used the microfluidic method to manufacture cavity microfibers with spindle nodules on a large scale. The structure of the spindle junction provides both geometric driving force and wettability driving force for the directional transportation of water droplets, resulting in excellent fog-collecting ability of the biomimetic fog collector. The artificial spider silk consists of periodically expanded silk and two main axis fibers. The expanded silk is composed of hydrophilic nanofibers, which facilitate the condensation of tiny water droplets. As the water continues to condense, the expanded filaments contract into periodic spindle nodules. Curvature gradients and roughness gradients are generated along the junction to the spindle junction, both of which contribute to water transport and merging. At the same time, Tian et al. [59] cleverly immersed nylon fibers in a polymer solution and then horizontally pulled them out. Due to Rayleigh instability, a string of polymer droplets fell on the fibers and became spindle nodules after drying (Figure 9.28a).

Bai et al. [104] customized the surface energy of the spindle junction by selecting different polymers, such as polyvinyl acetate (PVAc), polymethyl methacrylate (PMMA), polystyrene (PS), and polyvinylidene fluoride (PVDF). The intrinsic water contact angles of these polymers were 56.7°, 68.4°, 93.3°, and 92.7°, respectively. The results demonstrate that by optimizing several factors of the curvature, chemical gradient, and roughness gradient of the fiber surface, it is possible to precisely control tiny water droplets (tens of picoliters) in a controlled direction ("towards" or "away" from the junction). At the same time, the surface roughness (porous nanostructure) of the spindle junction can be designed by phase separation. This work considers both the surface energy gradient, controlled by the water contact Angle between the spindle junction and the fiber, and the Laplace pressure, influenced by the size and curvature of the spindle junction. It demonstrates shows that water droplets can move from the hydrophilic to the hydrophobic region due to a combined effect. In addition, to study the effect of spindle junction geometry

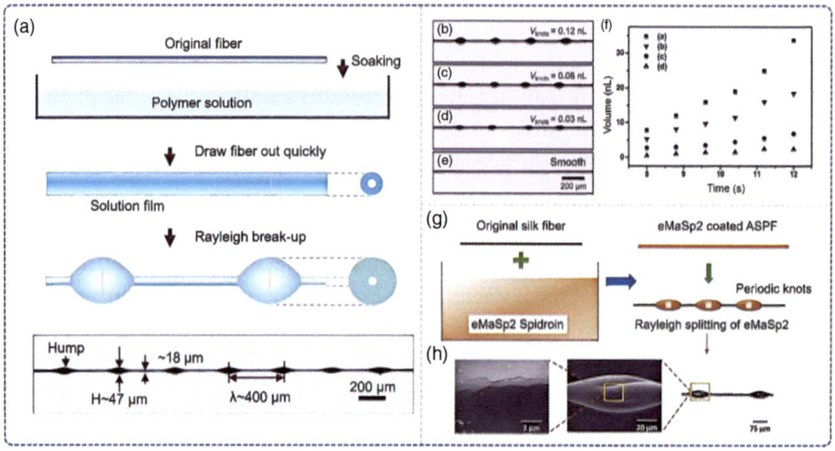

Figure 9.28 (a) Scheme of fabricating bioinspired fiber with hump structure. Source: Reproduced with permission from [59] / John Wiley & Sons. (b–f) The water collection performance of bioinspired artificial spider silks (BASs). Source: Reproduced with permission from [26] / John Wiley & Sons. (b–e) Optical images of BASs used to investigate their water collection ability. Each BAS contains four spindle knots with an average size of (b) 0.12 nL, (c) 0.08 nL, (d) 0.03 nL, respectively, and are compared with (e) smooth nylon fiber. (f) BASs in (b–e) are compared with their water collection ability under a fog flow with a velocity of ~75 cm s^{-1}. The total volume of all the water droplets along the BAS is calculated at different time. The scatters show that on each BAS, the water volume increases with time. Within the same time interval, the BAS with larger spindle knots collects more than that with smaller spindle knots. Source: (g, h) Reproduced with permission from [103] / John Wiley & Sons.

on water-collecting capacity, the team also examined how different viscosity and surface tension solutions, as well as fiber extraction rates, affected the size of bionic spider silk. [26] The results showed that all the bionic spider silk had the same length and included four spindle knots that were similar distances from each other. Furthermore, it was found that bionic fibers with large spindle knots collected more water than bionic fibers with small spindle knots (Figure 9.28b–f).

The influence of spindle node periodicity on droplet suspension performance was further studied by Hou et al. [105]. They discovered that biomimetic fibers have a higher water-collecting capacity compared to ordinary fibers, enabling the rapid formation of larger droplets at specific locations. Additionally, due to the longer three-phase contact line, fibers with two spindle junctions exhibit greater suspension capacity than those with a single spindle junction. The roughness and curvature characteristics of the spindle structure on biomimetic spider silk were studied by Huang et al. [106], who proposed a new capillary adhesion model. This model provided a fresh perspective for studying the adhesion mechanism of larger pearl water droplets. In addition, many studies have made efforts to enhance the potential of fog collection materials in achieving efficient water collection performance through different approaches such as polymer coatings, fiber thickness, spindle junction diameter, and distance between spindle junctions [107]. It has been found that increasing the fiber diameter, the size of the spindle junction, and the periodicity all lengthen the length of the three-phase contact line, resulting in a higher volume threshold to promote larger droplet aggregation.

In a recent study, Hu and coworkers [103] used whole-filament protein fibers with periodic junctions as the main material for biomimetic spider silk. The material is lighter than synthetic polymers, and it exhibits the highest level of the corresponding water-collecting properties among previously reported biomimetic fibers. Degummed silk fibers have high absorbency. After the surface modification, the formed spindle junction enables directional water collection. During preparation, fibers were horizontally immersed in a 10% concentration solution of eMaSp2 snails then steadily pulled out to form a film (Figure 9.28g,h). Due to Rayleigh's instability, it was treated with methanol steam in a controlled environment to obtain faux fiber with eMaSp2 bionic spindle junction. A significant surface energy gradient is generated between the eMaSp2 junction and the porous surface and fiber, which helps improve the stability of the three-phase contact line. In the water collection test, the maximum volume threshold reached with the degummed silk fiber at 250 s is 2.2 μL. In comparison, ASPF reaches 6.6 μL at 690 s, which is 3 times that of degumped silk fibers. The water-harvesting efficiency is expressed in terms of the volume mass index (VTMI) of the fiber mass, and the bionic fiber is reported to be 100 times more efficient than the highest reported water-harvesting efficiency. The PVDF-coated nylon fiber weighs 252 times more than ASPF, but it can only collect twice the volume of water (13.1 μL). [98] It is worth noting that while the bionic fiber has a relatively high water collection efficiency in its initial stage, it cannot achieve continuous spontaneous water collection. Once it reaches maximum capacity, water droplets will hang on the spindle junction and can only be slowly removed by gravity, which affects subsequent water collection.

Chen et al. [108] utilized the spraying method and etching process to create a specific curvature gradient and surface energy gradient on nylon mesh, aiming to capture and collect fog (Figure 9.29a) Figure 9.29b,c show SEM images of the surface morphology of the intersections, where the curvature gradually increases from the spindle junction to the junction. SEM images on the spindle junction show dense mesh and synapses forming regions with high surface free energy, where the water contact angle (WCA) is about 70°. SEM images of the contacts reveal show sparse networks and synapses, which form regions of low surface free energy (WCA = 97°). WCAs on PMMA smooth surfaces and untreated nylon surfaces are 84° and 99°, respectively. In the middle between the spindle junction and the junction shown in Figure 9.29d,e, the density of both the network and synapses is slightly lower than that of the spindle junction but higher than that of the other junction. Therefore, at the intersections of nylon mesh fibers, the short distance from the contact to the spindle junction creates a wettability gradient, which extends across the entire nylon mesh and further enhances the macro wettability effect.

Figure 9.29f–m shows a typical wetting process in which water condenses, aggregates, and transfers at the spindle junction. In the initial stages of water collection (Figures 9.29f,g), water droplets randomly condense on the surface, with some areas covered by small water droplets less than 30 μm in diameter. As water collection proceeds, smaller droplets merge into larger droplets at 30 s (Figure 9.29h).

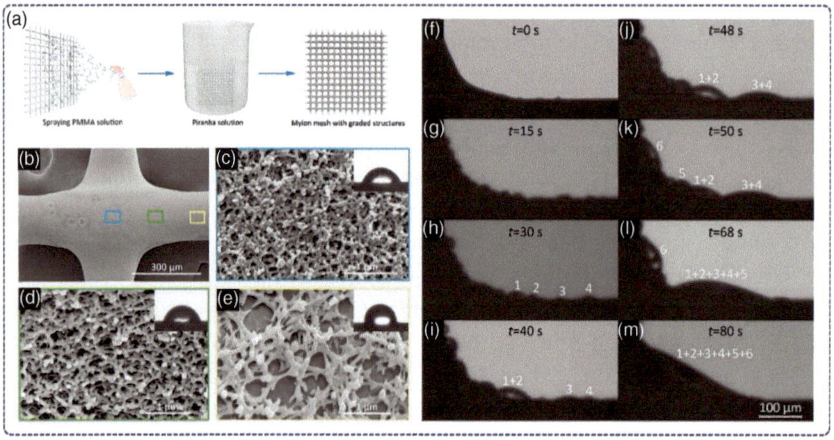

Figure 9.29 Design and characterization of NMGS and optical photos of water collection. Source: Reproduced with permission from [108] / Elsevier. (a) Schematic diagram of NMGS fabrication. (b) SEM image of the intersection of NMGS. The locations of the knot, joint, and middle site are marked with blue, yellow, and green squares respectively. (c) SEM image of the knot. The contact angle at this location is 70°. (d) SEM image of the middle site. The contact angle at this location is 82°. (e) SEM image of the joint. The contact angle at this location is 97°. (f–m) Optical photos of water collection on the knot at a fog flow rate of 30 g s^{-1} m^{-2}. The water drops in the photos are marked by numbers 1–6.

Large droplets 3 and 4 at the junction begin to transport towards the spindle junction at 50 s (Figure 9.29k). Water droplets located on either side of the spindle node gradually move closer together, and finally, these large water droplets merge and wet the spindle node (Figures 9.29l,m). The forces of surface energy gradient and curvature gradient do not play a full role in the growth of small droplets until large droplets grow. When droplets nucleate on the surface of PMMA, there is only a slight difference in wetting and curvature on both sides due to the strong pinning effect, thus micro-droplets are not influenced by the driving force of the surface gradient. Until the tiny droplets increase and combine, when the size exceeds a certain value, the wettability of the two sides of the droplet will be significantly affected by the driving force of the curvature gradient and roughness gradient. However, when the water drops fall completely on the contact, the surface curvature change is almost zero, which is mainly affected by the driving force of the surface free energy gradient, such as droplets 3 and 4 (Figure 9.29j). When the droplet is located between the contact and the spindle junction, it can be driven by the surface free energy gradient and the Laplace pressure difference, such as droplets 1 and 2 (Figure 9.29i), making it more easily driven compared to a droplet subjected to only one driving force. The experimental results show that the nylon mesh exhibits excellent performance in water collection, mainly due to its spindle structure with a large trapping area, surface wetting gradient, and curvature gradient characteristics. During the initial stage of water collection, both the nucleation density of the hydrophilic nylon net and that of the nylon net with a single curvature gradient were higher than those of the original nylon net. At the fog flow rate of 110 g s^{-1} m^{-2}, the stable water collection rate of the nylon net in water collection was higher than that of the original nylon net of 1390 mg cm^{-2} h^{-1} and remained at 1688 mg cm^{-2} h^{-1}.

9.4 Intelligent Design for Bionic Surface Fog-Harvesting Material

The previous sections have elucidated the development of bioinspired hierarchical structures and the effect of wettability on fog collection efficiency. The synergistic relationship between structure and chefogry determines superior water fog collection efficiency. Consequently, manipulation liquid droplets over a gradient surface texture are essential for advancing high-performance water fog-collecting materials. The reversible conversion of wettability and the rate of droplet motion can be autonomously governed by external stimuli. Specifically, the microtopography, together with the surface chemical composition, fluctuates under environmental conditions, enabling the material's surface wettability to transition reversibly from an ultralow contact angle (superhydrophobic) state to a super-lubricious state and subsequently managing the velocity of droplet motion. Intelligent water fog materials are capable of adjusting to their surroundings, making them more responsive for fog applications. Common external stimuli include stress, magnetic field, temperature, light, etc. The following explanations are based on different responses to external stimuli.

9.4.1 Stress Responsive

Typical responsive materials include Polydimethylsiloxane (PDMS) and Polyurethane (PU). As an efficient technique, tensile strategies are particularly advantageous for manipulating flexible materials, possessing vast potential applications including adjustable transparency, adaptive adhesive control, and droplet manipulation. [109] The elasticity polydimethylsiloxane films were used as the foundational material by Zhu and coworkers [110] to fabricate a smart, reconfigurable Janus film with controllable fog collection efficiency through ultrafast laser drilling. The distance between each pair of adjacent micropores on the Janus film's surface was 200 μm. On Laser irradiation PDMS surface (front face), the water contact angle value is 154°, conversely, the other side (backside) was modified with a highly surface energy superhydrophilic reagent and integrated with laser induced bonding to transform it into hydrophilic, which exhibited a contact angle of 3°. The preparation process is illustrated in Figure 9.30a. As the internal structure malfunctions due to laser exposure, varying strain under different tensile stress conditions can be easily facilitated by adjusting and applying stress. With the activation of wetting driving force, the Janus film acquires an exclusive interplay amongst multiple complementary polar structures, allowing droplets to automatically penetrate from hydrophobic surfaces to hydrophilic ones. As shown in Figure 9.30b for different strains ($\varepsilon = 0\%$ and 200%), the fog collection behaviors of Janus films vary. Due to synergistic drive from the wetting driving force (F_{wet}) and Laplace pressure

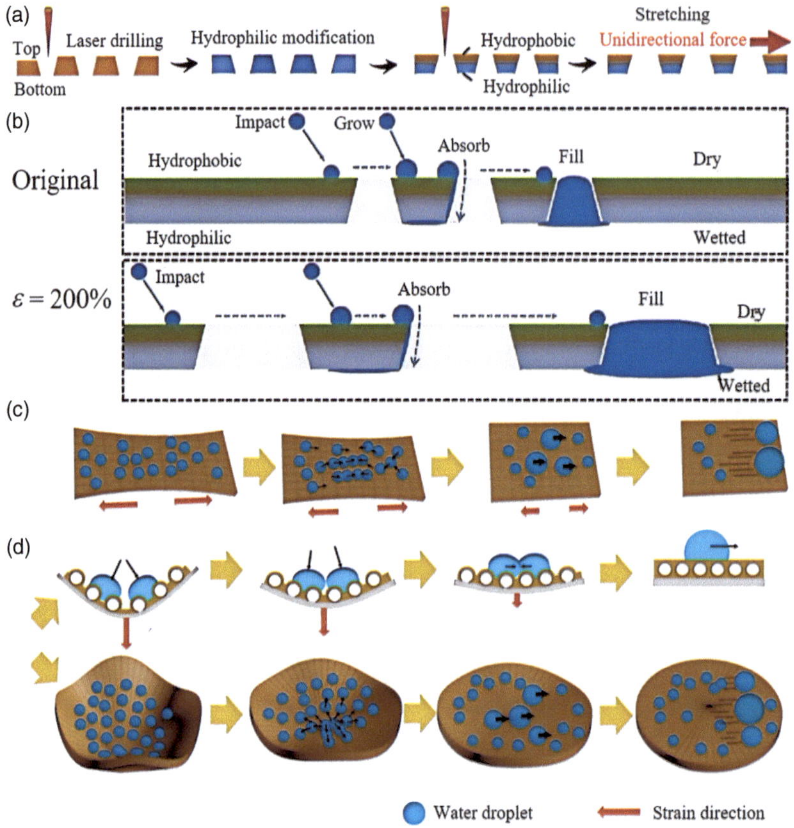

Figure 9.30 (a) Laser fabrication process of Janus membrane. (b) Fog collection behavior at different stresses ($\varepsilon = 0\%$, 200%). Source: Reproduced from [110] / with permission of John Wiley & Sons. (c) Schematic diagram of enhanced droplet polymerization. (d) Agglomeration of two adjacent droplets around the drag point. Source: Reproduced from [109a] / with permission of American Chemical Society.

($\Delta F_{\text{Laplace}}$), fluid droplets are simultaneously absorbed and penetrated once they interact with the hydrophilic zones on the less hydrophilic surface. This rapid penetration process regenerates the fresh dry surface prone to droplet adherence and condensation, thereby sustaining the continuous fog collection process. Water tends infiltrate at the edges rather than the center of concavities. Analysis reveals that the fog collection efficiency of samples subjected to 200% stretching increases by approximately 67% compared to their original state. Similarly, the fog collection rate for samples with 200% strain ascends proportionately with uniaxial tension, reaching as high as about 79%. These intelligent Janus films have potentials beyond fog collection; they may also operate in specific scenarios, including dynamic fog flux adjusters.

Furthermore, the super-hydrophilic nature of flexible substrate surfaces can be reversibly adjusted by altering the geometric structure of the solid surface without changing its chemical composition. The corresponding fluid motion behavior of the super-slippery surface changes simply by stretching or flexing it. The transition in wetting properties regarding the super-slippery surface under mechanical strain is continuous and predictable. The relaxed state of the super-slippery surface facilitates its mechanical extension due to low adhesion between fluids and lubricants. Although neither the interfacial tension nor contact angle of the super-slippery surface fluctuates during the expansion process, the roughness is reduced. Concurrently, under stress, the lubricant gets deposited into the cavities of the material. As a result, the adhesion strength between the driving droplet and solid/lubricant increases considerably, allowing for stationary droplets on the surface. Under mechanical tensile action, the wetting condition of the material's surface transitions from super-slippery to unsmooth. Super-slippery porous surfaces with isotropy exhibit adjustable anisotropic wetting properties. Jiang and coworkers [111] synthetically produced anisotropic microgroove structured super-slippery surfaces through mechanical straining. By asymmetrically stretching the microgrooves, unidirectional droplet repulsion can be achieved due to scalloped recesses forming at different ends. Inspired by this unidirectional droplet sliding, Xin and coworkers [109a] developed an efficient water collection system using high-elasticity films transformed under strong wind operation. Chemical affinity was established between TPU and PFPE through the introduction of a fluorinated polymer coating. In addition to regulated morphology, the prepared membranes exhibited a higher tolerance for stretch strains, with a fracture elongation rate of approximately 200%. Based on this, it is possible to enhance droplet coalescence through strain relaxation. Specifically, when stretched films gradually return to their original shape, water droplets that are inconsistently positioned on the surface overlap due to their close proximity. Once the overlapping water beads reach a threshold for slip, they may fall (Figure 9.30c). This phenomenon enhances droplet coalescence and sliding. By applying external tensile stress acting upon them (wind), real-time manipulation of droplet sliding and coalescence can be achievable on the prepared films (Figure 9.30d). Additionally, these films possess a retention capability which prevents water from being blown away by winds while ensuring effective water collection in highly windy environments.

The mechanical tensile stress response provides a direct and efficient approach to regulate the surface wettability of materials. Mechanical tensile stress is confined to elastic materials; therefore, factors like process cost, mechanical performance, and durability need to be considered when engineering stress-responsive materials.

9.4.2 Magnetic Responsive

Magnetic response materials incorporate magnetic particles such as iron and cobalt. Various studies have suggested that a slippery surface inspired by pitcher plants could exhibit superior potential for fluid transportation [112]. Owing to the rapid responsiveness, recoverability, and remote operationality of external stimulus responsive materials, anisotropic wetting can be regulated using surfaces that respond magnetically, electrically, thermally, or photothermally.

The Guo group [113] successfully fabricated a magnetically deformable PDMS@Fe_3O_4 surface via the magnetic particle assisted molding (MPAM) approach. This surface is composed of an array of microfibers infused with lubricating liquid, and the orientation of these microfibers can be controlled through the movement of a magnet. This enables the transition between superhydrophobic states and slippery states on the wetted surfaces, as illustrated in Figure 9.31a,b. Additionally, this switchable wetting property based on the magnetic response of PDMS@Fe_3O_4 fabric can also be programmed for fog collection research. As shown in Figure 9.31c–f, due to its high surface area, the superhydrophobic model sheet can effectively trap water droplets from air swiftly. Furthermore, it was observed that numerous small droplets were completely immobilized and adhered to the fabric surface (Figure 9.31d). Upon gradually increasing the droplet volume and reaching critical dimensions, excess condensate water slid off the fabric surface under gravity (Figure 9.31f), indicating a transition from superhydrophobicity to slipperiness. An external magnetic field can reestablish the superhydrophobic state of the surface, thus enhancing fog collection efficiency. Moreover, high durability is an advantageous feature of biomimetic moistening surfaces. Herein, the resistance against mechanical fatigue of samples can further be characterized through switches in the motion of micropillars (Figure 9.31g–i). In this study, we manipulated the direction of micropillar movement from horizontal to vertical configuration at a frequency of 40 Hz. The study found that when Fe_3O_4 comprises up to 75% within PDMS@Fe_3O_4 fabric, it can achieve reversible transitions more than 40,000 times between the slipperiness and the superhydrophobicity in fog collection applications. Slight variations in contact angle and sliding angles of the sample surface were noted (Figure 9.31h), attributed to the excellent adhesion strength of PDMS. PDMS@Fe_3O_4 fabrics are capable of repetitive transitions between superhydrophobicity and slipperiness, exhibiting superior durability, that significantly contributes to their anti-slip performance. This work provides valuable insights for both fundamental research and practical application.

The efficacy of fog collection is influenced significantly by the size and properties of the fog droplets as well as the wind velocity. Biological and biomimetic collectors have demonstrated their effectiveness under naturally turbulent foggy winds or moderate artificial fog flow rates. However, in certain regions with persistent fog, low-velocity fog tends to accumulate on surfaces and traverses air spaces, which inherently undermine the capacity of fog collectors for collecting fog. Consequently, recapturing fog from near static atmosphere remains a formidable challenge even for existing strategies for fog collection. Liu and coworkers [13] fabricated a novel material capable of collecting fog regardless of wind conditions by integrating cactus spine structure and a magnetically responsive conical array. Here, they utilized techniques such as mechanical drilling and template replication before introducing cobalt magnetic particles (CoMP). Approximately 200 mg of CoMP, with an average diameter of approximately two micrometers, were dispersed into each mechanical hole of the array. Subsequently, a liquid PDMS prepolymer containing a curing agent was poured onto the filled template augmented with CoMP using a vacuum pump.

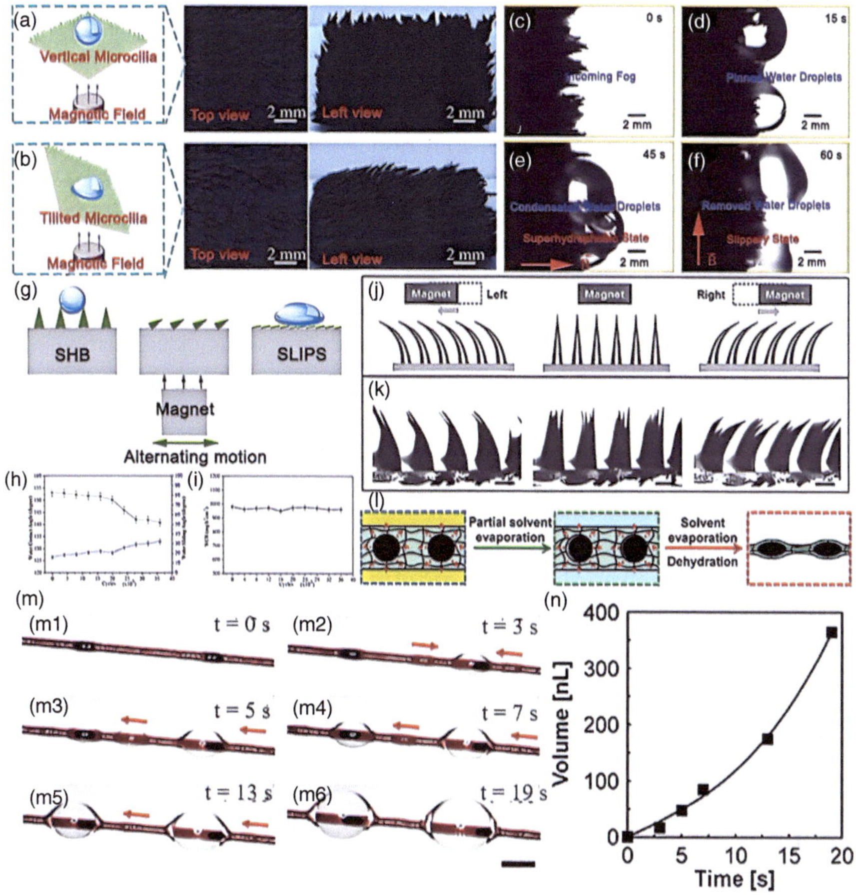

Figure 9.31 (a, b) Optical microscope images of column arrays in different magnetic field directions, where the microcolumn shape can range from an upright direction (superhydrophobic state) to a flat direction (slippery state). (c) The direction of entering the fog perpendicular to the sample surface. (d–f) The fog collection of the microcolumn in the superhydrophobic state. (g) The removal of condensate from the surface in the smooth state. (h–i) Schematic diagram of fatigue tests performed to evaluate the robustness of PDMS@Fe_3O_4 fabrics. (h) WCA on superhydrophobic surfaces and WSA on smooth surfaces after fatigue tests. (i) Stability of water collection after fatigue tests 40,000 cycles. Source: Reproduced with permission from [113] / The Royal Society of Chemistry. (j) Schematic diagram of the response of a magnetically induced cone array to a magnetic field. (k) The response of a cone array to a magnetic field is recorded by a high-speed CCD camera. Source: Reproduced from [13] / with permission of John Wiley & Sons. (i) Schematic diagram of the preparation process of spindle magnetic microfibers. (m) Optical images showing the water collection process on dehydrated microfibers and the magnetic spindle junction; and (n) the relationship between the water collection time and the water collection time of a single spindle junction. Source: Reproduced with permission from [65b] / American Chemical Society.

Concurrently, ferromagnets measuring 60 mm in diameter and 50 mm in thickness were prepared. During degassing and solidification stages, the sample was placed on top of a permanent magnet with a surface field strength of almost 0.9 T, ensuring vertical alignment of the CoMP within the holes. The schematic illustration (Figure 9.31j) depicts how the conical array responds to the magnetic field. Using a high-speed CCD (charge coupled device) camera, we captured the motion of the array responding to the magnetic field (Figure 9.31k). When the magnet was directly applied to the specimen, the spine-like cone of the cactus became saturated perpendicularly. Thus, in the presence of a moving magnetic field, the array could achieve accomplish spontaneous and continual fog collection. However, without a magnetic field, almost no droplets were deposited on the array. Thus, a symbiosis relationship between the spine-style architecture of the cactus and the cactus and the magnetic responsiveness of the flexible conical array enables magnetic induced fog collection even in windless conditions. Furthermore, more extensive ferromagnetic fog collectors can be configured to meet pressing demands for basic water usage promptly. This system has significant potential for application, especially in areas with both stagnant weather conditions and abundant fog.

Motivated by the remarkable properties of spider silk, various advanced functional microfibers have been engineered from biomaterials such as polymethyl methacrylate (PMMA), azobenzene, poly (isopropyl acrylamide), alginate and proteins. [57b, 61, 114] The incorporation of polymer spindles within these microfibers mimics the structural configuration of a spider's web-like webs, offering significant potential for water collection, intelligent catalysis, filtration, and sensing. Nonetheless, whilst each type of individual ultrafine fiber exhibits exceptional functionality across numerous areas, achieving sophisticated three-dimensional integration with controlled assembly remains challenging. Therefore, it is crucial to develop spider silk-like microfibers that exhibit controllable assembly behaviors for various enhanced applications such as water collection and cell culture. For instance, Chu et al. [65b] devised an ingenious strategy to fabricate magnetically controllable spider silk-like microfibers incorporating magnetic spindle knots, illustrated in Figure 9.31l. These microfibers are composed of biocompatible calcium alginate capable of responding to external 2D magnetic fields by aligning their orientation. The spinning behavior originates from the interaction between the 2D field and superparamagnetic Fe_3O_4 nanoparticles embedded in the magnetic knot. The demonstrates that ultrafine fibers exhibit outstanding magnetic responsiveness. By optimizing the ultrafine fibers with oleic acid modified magnetic nanoparticles, promising properties were demonstrated during local stirring and mixing in micron/microfluidic mixers. The transformation of magnetic substances within the material offers provides guidance for pattern identification and assembly capacity. Consequently, manipulating various 2D and 3D magnetic fields enables patterning these microfibers. The exceptional magnetic guidance assembly performance allows the microfibers to form spider-web-like structures for fog collection. Studies examining single dehydrated microfiber fog collection demonstrate minor expansion when tiny drops come into contact with the fiber, followed by condensation occurring at the spindle knot or other positions. As illustrated in Figure 9.31m, even when the fiber is inclined, drops positioned lower can defy gravity and move towards nearby spindles, a phenomenon that also occurs on horizontal microfibers. In comparison, no comparable fog-collecting outcome was observed over non-dehydrated fine fibers due to the absence of spindle structure. Since the size of the spindle knot can be adjusted by altering the microfluidic flow rate, it is possible to enhance the water-collecting ability by employing larger spindle knots. The relationship between the volume collected via a single spindle knot and the collecting duration shows in Figure 9.31n, indicating that artificial cobwebs fabricated using microfibers offer robust fog collection capabilities. Therefore, it becomes evident that microfibers can be utilized for versatile and adaptable structures in complex assemblies for water collection. Additionally, since the uniform spindle knots along the microfibers possess equal water-collecting abilities, increasing the number of included spindle knots in the pattern could enhance water collection efficiency.

Despite the advantageous effects of magnetic response in controlling fluid droplets, there is still significant room for improvement in achieving transformations in wettability and fine-tuning adhesion.

9.4.3 Temperature Responsive

Thermally responsive materials, such as paraffin wax, 12-hydroxystearic acid (12-HSA), 4-cyano-4′-pentylbiphenyl (5CB), and silicone-based oil, etc. Xin and coworkers [115] prepared a versatile and straightforward sponge-like cotton fabric capable of independently absorbing and discharging water from humid environments. This capability is governed by temperature fluctuations over a routine diurnal range in deserts. The hydrophilic surface of the cotton fabrics was modified using poly(*N*-isopropylacrylamide) (PNIPAAm) and brushes through a "grafting" approach. At a specific temperature (+32°C), temperature-sensitive PNIPAAm undergoes a coil-to-pellet transition (from hydrophilic to hydrophobic, respectively) in aqueous solution, which is denoted as a lower critical solution temperature (LCST). The combination of temperature-responsive polymers and the fabric's highly coarse surface enables reversible switching between superhydrophobicity and superhydrophilicity, which can be repeated many times (Figure 9.32a). The PNIPAAm molecule is predominantly hydrophilic below the LCST or in the presence of water, mainly because amide groups tend to form intermolecular hydrogen bonds with surrounding water molecules (Figure 9.32b). Conversely, above the LCST, hydrophobic isopropyl methyl tends to establish polymer-polymer interactions, which are more energetically favorable at elevated temperatures (Figure 9.32c). This transition is accompanied by a change in configuration from the coil to the pellet. By adjusting the temperature below the LCST, water can be captured from fog or humid atmosphere. This material and concept provide a simple and versatile solution for the collection, unidirectional flow, and purifying water harvested from the atmosphere.

In addition, Zheng and coworkers [114a] had alternatively employed *N*-isopropylacrylamide polymers to fabricate a strand having similar magnitude and curvature characteristics to wet spider silk. By combining the surface texture of the strand,

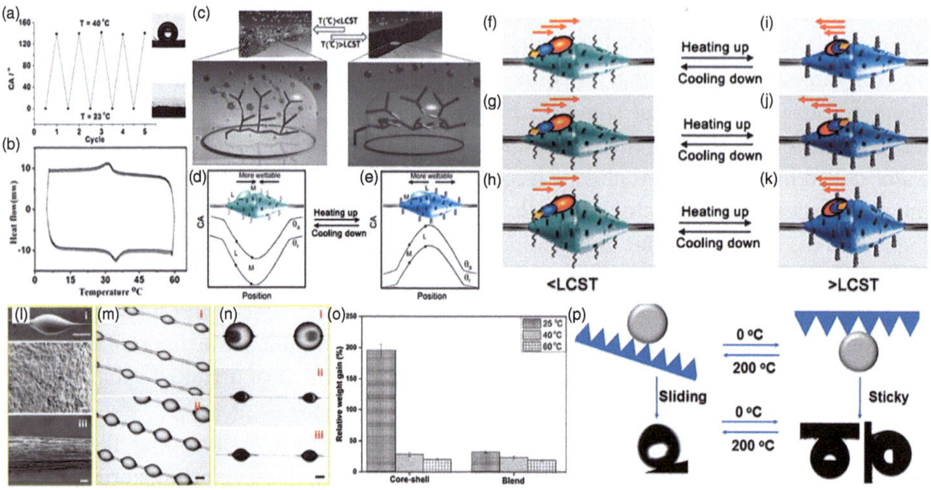

Figure 9.32 (a–c) Reversible response behavior of PNIPAAm cotton fabrics: (a) PNIPAAm cotton fabrics at 23 and 40°C. (b) Differential scanning calorimetry temperature maps showing LCST phase transitions in continuous cycles. (c) Water harvesting ("superhydrophilic state") through hydrogen bonding between PNIPAAm and PNIPAAm molecular bonding through water release ("superhydrophobic state"). Source: Reproduced from [115] / with permission of John Wiley & Sons; Reproduced with permission from [1c] / Elsevier. (d) at < LCST, FW and FL drive the droplet in the same direction. (e) At > LCST, FW and FL drive the droplet in opposite directions. (f, g, and h) at < LCST, the droplet moves completely to the spindle. (i, j, and k) At > LCST, the apex angle of the spindle junction is small, and at low relative humidity, the droplet is partially away from the spindle junction. Source: Reproduced from [114a] / with permission of The Royal Society of Chemistry. (l) Scanning electron microscope images of the microfibers were displayed separately. (m) The water capture performance of the microfibers before and after placement in a fog environment. (n) The thermally triggered water convergence behavior of the microfibers and the NIPAM spindle. Source: Reproduced with permission from [116] / John Wiley & Sons. (o) Comparison of moisture absorption efficiency of fibers at different temperatures. Source: Reproduced from [117] / with permission of Elsevier. (p) Schematic diagram of switchable adhesion due to thermal change. Source: Reproduced with permission from [66] / The Royal Society of Chemistry.

its curvature dimension, and its thermoresponsive wetting behavior, micron-sized water droplets can be directed along the strand. The temperature response is derived from poly(*N*-isopropylacrylamide)-*b*-polymethyl methacrylate (PMMA-*b*-PNIPAM) strands, while the change in wetting behavior is caused by the conformational shift of the PNIPAM molecular chain at various temperatures. Below the lower critical solution temperature, the spindle node exhibits hydrophilicity. The wetting layer instigated by the texture layer can propel the droplet towards the hydrophilic spindle junction. Above 40°C, the spindle junction displays hydrophobicity, causing the droplet to migrate towards it. Below the LCST, both wetting layer force and Laplace force direct the droplet in the same direction. The droplet undergoes violent deformation, and the M edge is strongly propelled (Figure 9.32d). Above the LCST, the wetting layer force and Laplace force steer the droplet in the opposite direction. The droplet undergoes minimal deformation, and the movement of the

M edge is impeded. The M and L edges correspond to the "more wet" and "less wet" sides, respectively (Figure 9.32e). Below the LCST, the droplet migrates entirely to the spindle junction (Figure 9.32f–k). Above the LCST, there is a decrease in apex angle for the spindle junction, and at low relative humidity, partial detachment of the droplet from the spindle junction (Figure 9.32i). At high relative humidity, the droplet is completely detached from the spindle junction (Figure 9.32j). For the higher apex angle of the spindle junction, the droplet is invariably partially detached from the spindle junction (Figure 9.32k). This research has significant implications for advancing intelligent materials to regulate fluid delivery or achieve efficient liquid collection.

Specific temperature-responsive poly(*N*-isopropyl acrylamide) (PNIPAM) polymers can also be utilized to create heterostructured microfluidic spindles, which are suitable for dynamic heat-induced water collection and release. Zhao and coworkers [116] successfully fabricated a multifunctional spindle-knotted microfiber. The fibrous spindle nodes were examined by electron microscopy. It is evident from Figure 9.32l-ii that the spindle junction surface exhibits a rough nanostructure, while the fibrous surface is relatively smooth (Figure 9.32l-iii). This leads to the formation of a wetting layer from the spindle node to the fiber joint, and Laplacian pressure difference due to the unique structure of the spindle junction. The synergy of these two forces expedites the capture of water fog by spindle pairs. It is apparent from Figure 9.32m that the spindle junction serves as an efficient water fog collection system. Spindle knots made of temperature-responsive material PNIPAM in a humid environment adhere to and accumulate water droplets. As the temperature rises, the maximum suspension capacity decreases due to the contraction of the spindle junction, so the droplets on the spindle junction detach. When the temperature returns to room temperature, these spindle knots expand again for the subsequent water collection cycle (Figure 9.32n). Therefore, intelligent water collection systems constructed with temperature-responsive microfibers may provide innovative strategies for water collection in regions with significant temperature variations between day and night.

The Avinash team [117] fabricated core-shell textures and incorporated nanofibers for temperature-responsive cellulose acetate-poly(*N*-isopropyl acrylamide) (CA-PNIPAM). At room temperature, the water contact angles of the core-shell fibers and the composite fibers were $19.8 \pm 1.2°$ and $7 \pm 0.9°$, respectively. The difference in wettability in this instance was attributed to the divergence in surface construction morphology. As the temperature ascended above the LCST of PNIPAM (45°C), the contact angles of both fibers surpassed 120°, indicating that the thermal response attributes of both samples could be harnessed for atmospheric water collection with temperature as an external stimulus. Compared with composite fibers, core-shell fiber absorption diminished more markedly with temperature (Figure 9.32o). This is due to the exposed PNIPAM layer in the core-shell fiber-shell component, which exhibits a propensity to modify its wettability. This study illustrates that surface morphology plays a critical role in the wettability and water collection capacity of electrospun fibers.

Liu and coworkers [66] developed a water-wicking surface by utilizing a biomimetic tip pattern. The pattern composition comprises polydimethylsiloxane (PDMS) and graphene. Owing to alterations in pore size and chemical composition, laser-etched porous graphene surfaces can generate substantial wettability variations between 200 and 0°C, as illustrated in Figure 9.32p. The congruence of geometric structure and adjustable wettability plays a pivotal role in water collection. Surfaces with tip-shaped wettability patterns can swiftly propel minute water droplets into more water-permeable areas, significantly enhancing water collection efficiency. Furthermore, due to the efficient congruence between the superhydrophobic and hydrophilic regions of the unique tip pattern, unidirectional water delivery on the PDMS/G surface heated at 200°C can be accomplished. This study offers novel insights into designing temperature-adjustable materials with interphase wettability, which can enhance the fog collection efficiency of engineering liquid collection apparatus and achieve unidirectional liquid delivery, with potential applications in microfluidics, medical devices, and condenser design.

Although many temperature-responsive water fog collection materials have been studied, most are made from a small number of temperature-sensitive materials, and more sensitive materials should be introduced.

9.4.4 Light Responsive

Light responsive materials encompass ZnO nanorods, P3HT/PCBM, and binary systems, alongside poly(3-hexylthiophene-2,5-diyl) (P3HT). Zheng's research team [27] capitalized on the photo responsiveness of titanium dioxide to fabricate water-harvesting materials with distinctive wetting morphologies. Initially, they fabricated a material for a superhydrophobic titanium dioxide substrate, subsequently illuminating this surface with an ultraviolet lamp, while positioning masks of varying shapes between the lamp and the surface. Given that ultraviolet irradiation can alter the surface wettability of titanium dioxide (from superhydrophobic to superhydrophilic transitions), it is feasible to generate circular, quadrangular, pentagonal, hexagonal and octagonal superhydrophilic morphologies on the superhydrophobic substrate (Figure 9.33a–c). Compared to uniform superhydrophobic and superhydrophilic surfaces, these surfaces with superhydrophilic patterns exhibit superior water collection efficiencies. Among all hydrophilic patterned surfaces, the pentagram-shaped surface collects the maximum amount of water by mass and achieves the optimal collection rate due to the synergistic effect of the surface energy layer and Las Plas layer (Figure 9.33d–h).

Furthermore, Zheng's research team [114b] has observed a similar phenomenon on photo responsive azo dye-doped poly(methyl methacrylate) P(MMA-azo) spindles. The wettability layer can guide droplets to the wettable portion. After UV irradiation, the spindle surface becomes more hydrophilic. During fog collection, tiny droplets initially gather randomly on the spindle. As the droplets grow in size and merge, one side of the droplets migrates away from the spindle. Conversely, following visible light irradiation, the polymer assumes a hydrophobic state and undergoes surface roughness amplification. Under the synergistic influence of roughness, curvature, and light response, wettability can modify the direction of droplet motion, which can also be corroborated as illustrated in Figure 9.33i. In one specific instance, $D_{ultraviolet} > D_{visible}$, indicating that more water accumulates on the bionic fibers post UV irradiation. This research offers insight into novel layered surface designs that facilitate the propulsion of minute droplets in the desired direction, potentially benefiting applications such as micromotors, heat transfer, and water-harvesting apparatuses.

9.4.5 Humidity Response

In addition to examining light response, Zheng's research team [53] developed beaded heterostructure fibers (BSHFs) with alternating hydrophilic and hydrophobic surfaces, and demonstrated reversible responses to environmental alterations. As relative humidity fluctuated, the fibers exhibited segmented expansion and contraction behaviors in the "bead" portion, while the "string" portion remained unaltered. When the relative humidity escalated from 55% to 100%, the diameter of the spindle knot could be expanded from 7.2 to 9.5 μm, but the joint diameter remained unaltered (Figure 9.33j,k). The expansion behavior of the spindle knot is analogous to that of spider silk in a humid environment. Conversely, the swollen spindle knot contracts as relative humidity varies from 90% to 35%. This one-dimensional BSHF has significant implications for integrated functional devices and holds potential for numerous important properties and applications.

9.4.6 Environmental Response

Tang et al. [118] fabricated a Janus polyvinyl alcohol sponge (JPVA) incorporating shape memory alloy springs into an ecologically sensitive water fog collector. Heating prompts the memory alloy to undergo a phase transition at the phase transition point. Upon heating the bi-directional memory alloy spring to 60°C, it undergoes rapid transformation and compression, instigating the expulsion of water from the JPVA rapidly. The escalated water droplet quantity confers on the JPVA, making it relatively dry, facilitating the collection of water from moist air. Moreover, it also possesses light responsiveness. The genuine morphology of Janus PVA sponge with bi-directional memory alloy springs is illustrated in Figure 9.33l, depicting (right) and devoid of (left) CNT. The former apparatus is identified as JPVA + LTR due to its light and thermal response functionality of a fusion of CNT and a bi-directional memory alloy spring. When the aforementioned two apparatuses are exposed to an infrared lamp, as depicted in Figure 9.33m,n, the bi-directional memory alloy spring compression with decoration can be discerned (right, Figure 9.33o). Owing to the exceptional photothermal conversion capacity of carbon nanotubes, it is simpler for the bi-directional memory alloy spring to attain the deformation of the critical temperature (50°C, Figure 9.33n). Simultaneously, the spring temperature and ambient temperature of the spring utilizing carbon nanotubes are appreciably higher than those without carbon nanotubes, demonstrating the remarkable effect of reaching the critical temperature of the bi-directional memory alloy spring (50°C) on a

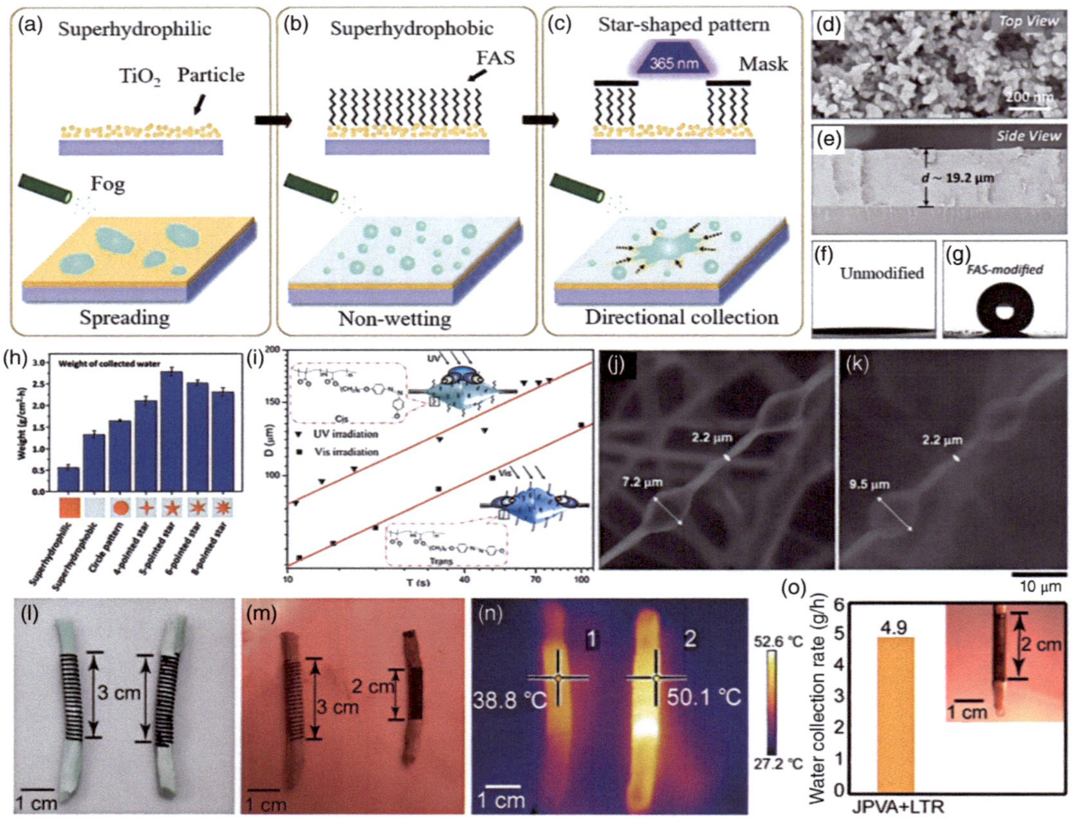
(a) Superhydrophilic
TiO2 Particle
Fog
Spreading
(b) Superhydrophobic
FAS
Non-wetting
(c) Star-shaped pattern
365 nm
Mask
Directional collection
(d) Top View
200 nm
(e) Side View
d ~ 19.2 µm
(f) Unmodified
(g) FAS-modified
(h) Weight of collected water
Weight (g/cm²·h)
Superhydrophilic
Superhydrophobic
Circle pattern
4-pointed star
5-pointed star
6-pointed star
8-pointed star
(i) D (µm)
T (s)
UV irradiation
Vis irradiation
Cis
Trans
UV
Vis
(j) 7.2 µm
2.2 µm
(k) 9.5 µm
2.2 µm
10 µm
(l) 3 cm
1 cm
(m) 3 cm
2 cm
1 cm
(n) 1
2
38.8 °C
50.1 °C
1 cm
52.6 °C
27.2 °C
(o) Water collection rate (g/h)
4.9
JPVA+LTR
2 cm
1 cm

Figure 9.33 (a–c) Schematic diagram of the fabrication process of bionic surfaces with star-shaped wettability patterns: (a) Superhydrophilic surface nanoparticles consisting of TiO_2; (b) Superhydrophobic surface modified with FAS and (c) bionic gradient surface with star-shaped wettability patterns. (d) Top view SEM image of the surface structures on the TiO_2 film, where clusters of TiO_2 nanoparticles can be observed. (e) Side view SEM image of the spin-coated TiO_2 film with a thickness of ~19.2 μm. (f) Optical image showing the superhydrophilic property on an unmodified TiO_2 film with CA < 5°. (g) Optical image showing the superhydrophobic property on an FAS-modified TiO_2 film with CA > 150°. (h) Fog-collecting efficiency of different wettability surfaces. Bioinspired surfaces with wettability patterns are more efficient than surfaces with uniform wettability. Star-shaped pattern is more efficient than that with circle-shaped pattern. Source: Reproduced with permission from [27] / John Wiley & Sons. (i) Linear relationship between condensation droplet diameter (*D*) and condensation time (*T*). Source: Reproduced from [114b] / with permission of The Royal Society of Chemistry. (j, k) Segmental swelling behavior of BSHF observed within ESEM. Source: Reproduced with permission from [53] / John Wiley & Sons. (l) Collection device with a memory alloy spring of pristine shape (left) and a spring surface decorated with carbon nanotubes (right). (m) Photograph the device in infrared light. (n) Thermal imaging of the device under infrared light; and (o) JPVA + LTR water collection measurements under infrared light. Source: Reproduced with permission from [118] / American Chemical Society.

sunny day with a lower ambient temperature (27.2°C), thereby ensuring the value of its application. Since the Janus PVA sponge is compressed by the bi directional memory spring, the corresponding WCR measurement is $4.9\,g\,h^{-1}$, as illustrated in Figure 9.33o. The aforementioned information substantiates that the water collection device possesses a light response capability, rendering it more suitable for practical applications.

9.5 Summary and Outlook

9.5.1 Aerodynamic-Based Fog Collection Efficiency

The geometric topography of non-phase change bionic fog- and water-harvesting systems positively influences the aerodynamics of fog flow interaction between material surfaces. Fog is a two-phase flow composed of 10 μm microdroplets and streaming air, where the streaming air can create resistance to the motion of fog droplets and variations in the gas phase can influence the trajectory of individual fog microdroplets. Isotropic and directional droplet motions on surfaces or cylindrical objects may be developed by unbalanced surface tension or oppositely directed capillary forces, which may be generated by gradient surface wettability or anisotropic structures. On gradient wettability surfaces, wettable surfaces exhibit smaller forward contact angles and higher surface tension, while non-wettable surfaces exhibit larger backward contact angles and lower surface tension, and unbalanced forces are generated at the extremities of the droplets to propel the droplets in motion [119].

Zhong et al. [1c] have carried out some research on diffusive processes. Evaluations were conducted on the fog collection behavior on non-uniform and

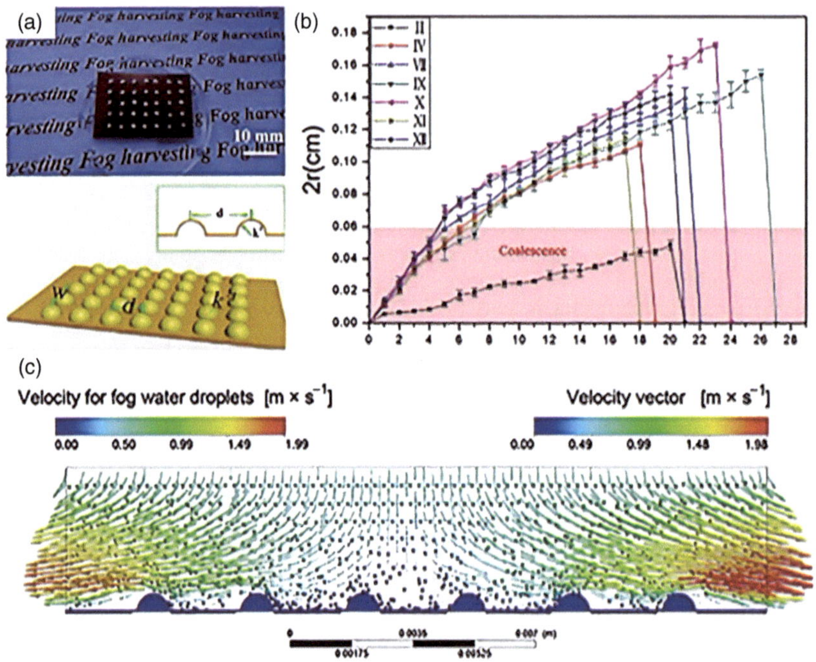

Figure 9.34 (a) Schematic design of the bionic surface of the desert beetle. (b) Growth rate profiles of water droplets on superhydrophobic horizontal plane samples II, superhydrophobic buffer-like samples IV, VII, X, XI, and XII, and on hydrophilic buffer-hydrophobic perimeter samples IX; and (c) Transient velocity vectors and velocities of these microscopic droplets around bumps and planes, with numerical magnitudes differentiated by different colors. Source: Reproduced with permission from [1c] / Elsevier.

planar copper sheet surfaces whilst retaining identical surface wetting properties (Figure 9.34a,b). By embellishing the planar surface with millimeter-scale projections, a 2.83-fold enhancement in fog collection efficacy was attained. Furthermore, as the count of projections augmented, the volume of collected water augmented to a maximum and then decreased. This signifies that the arrangement of projections also impacts fluid dynamics (Figure 9.34c). The projections impede the flow of air and the trajectory of microdroplets. Air interference provides projections with more opportunities to trap droplets, which results in accelerated droplet growth on projections compared to planar surfaces. This implies that optimizing surface geometry alone can also improve fog collection effectiveness. Similar outcomes were demonstrated for millimeter-scale elevated material surfaces.

The Guo group team [120] derived fog collectors with distinct Kirigami configurations and wettability gradients (Figure 9.35a–i), and realized 1.57, 1.63, and 1.82-fold amplifications in the total volume of collected fog (Figure 9.35j,k). One of the most notable differences between the three trials was the unique W-cone configuration in Sample 3. This structure resulted in more bumps on Sample 3's surface than other samples, resulting in a larger solid–liquid interface area, facilitating nucleation and growth of smaller liquids on the surface. In addition, because Sample 3 is subjected

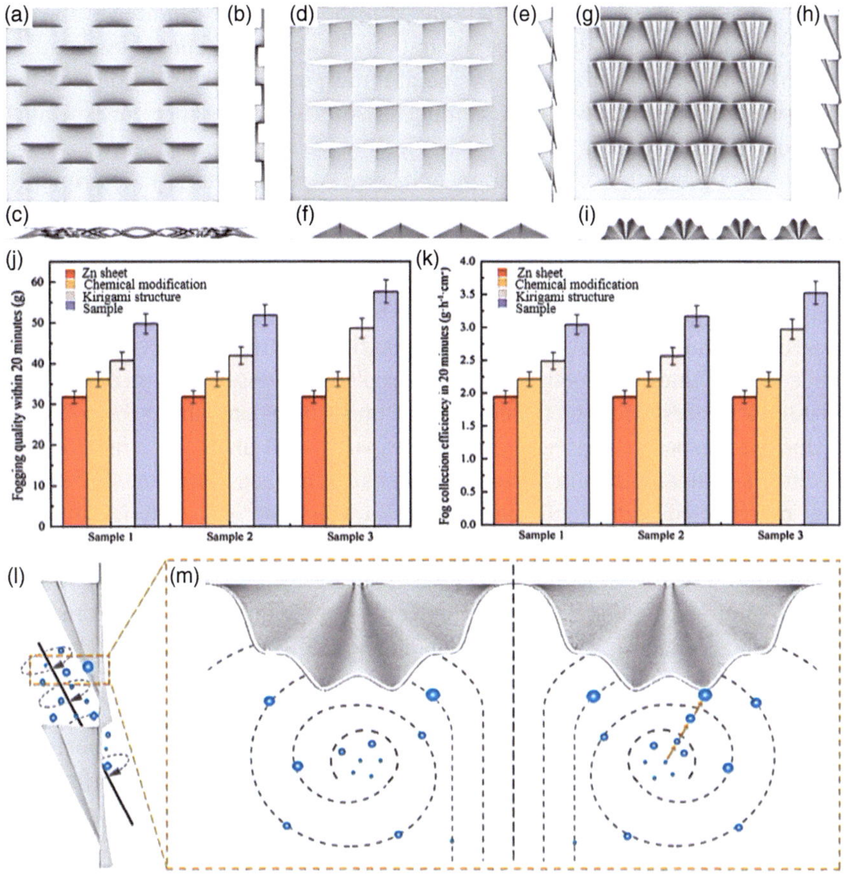

Figure 9.35 (a), (d), and (g) are front views of Sample 1, Sample 2, and Sample 3, respectively. (b), (e), and (h) are right views of Sample 1, Sample 2, and Sample 3, respectively; and (c), (f), and (i) are top views of Sample 1, Sample 2, and Sample 3, respectively. (j) Fog collection mass for three samples over 20 min. (k) Fog collection efficiency for three samples over 20 min. (l) Side view of the vortex; (m) Top view of the vortex. Source: Reproduced from [120] / with permission of American Chemical Society.

to higher forces during the fabrication process, stresses are generated due to substrate bending and deformation, resulting in different potential energies between convex and concave portions. Therefore, this particular property makes the raised portion more susceptible to nucleation [121]. Besides the aforementioned factors, there are numerous other factors that positively contribute to enhancing the fog collection capacity of the sample. Among all the factors contributing to the increased fog collection capacity, the authors contend that the most critical factor is the formation of a vortex in the hollow portion of the sample (Figure 9.35l,m). This vortex effectively blocks droplets and pushes them further towards the superhydrophilic surface of the test. This process helps to increase the effectiveness of capturing small droplets on the superhydrophilic surface, allowing more small droplets to nucleate and grow on the surface, and ultimately allowing more fog to be captured by the fog collector. In addition, the W-cone configuration allows the directional transport of droplets. Under gravity, droplets grow and are transported effortlessly to reach the superhydrophobic surface while overcoming the surface pinning effect for expedited collection.

The surface geometries of commonly employed capture materials are primarily categorized into one-dimensional linear (filaments, ridges) and two-dimensional planar surfaces. The aerodynamics displayed when fog flow impacts one-dimensional linear and two-dimensional substrates are divergent. The two-dimensional surfaces induce a greater deviation of the microdroplets from the one-dimensional linear configuration. Consequently, the quantity of droplets captured by the surface is less than the total volume of droplets flowing to the surface. This deviation decreases when the diameter of the one-dimensional linear fog collector is reduced to a specific threshold. Droplets expand ~100 times more rapidly in a one-dimensional wire than in a two-dimensional plane. Fog collection efficiencies (mass of collected water/total mass of fog flowing over the material surface) can be further enhanced up to 13% for one-dimensional wire fog collectors by utilizing three-dimensional structures comprising one-dimensional wires. Inclined columnar arrays generate faster and longer directional liquid transport on two-dimensional surfaces. To further augment the velocity and distance of directional liquid diffusion, more complex structures with both forward diffusion forces and backward resistance are necessary. Microgrooves with wedge-shaped arcs at the base, resembling the circumference of a hogwash, emerge as a potential solution, with gradient wedge angles that amplify liquid diffusion in the forward direction and sharp edges that inhibit liquid diffusion backward [9, 122].

Wang et al. [123] conducted some work on the effect of angle, modification, and structure on fog collection performance to evaluate the fog collection behavior of fog collectors with different bifurcation structures (Figure 9.36a–l). Due to variations in the number of bifurcations among samples, there were differences in fog collection efficiencies, which improved by 9.49, 10.72, and 12.17 times, respectively. The more bifurcations, the more hydrophilic nucleation sites in the initial fog concentration, which is favorable for the nucleation growth of microdroplets. Increasing bifurcation reduces the angle at each tip and increases the Laplace pressure difference on the surface, facilitating merging growth and rapid droplet transport while overcoming surface pinning effects faster.

9.5.2 From Mono-Bionic to Multi-Bionic Fog and Water Collection Devices

9.5.2.1 Single Bionic-Inspired Fog Collection Device

A fog collection apparatus, inspired by the hydrophilic-hydrophobic microarchitecture found on the back of a desert beetle's nanoscale fabric, has been developed for a wide range of applications. Various water-attracting surfaces with hydrophilic-hydrophobic zones integrated with hydrophilic-hydrophobic zones were fabricated by spraying, chemical alteration, laser direct inscription, and

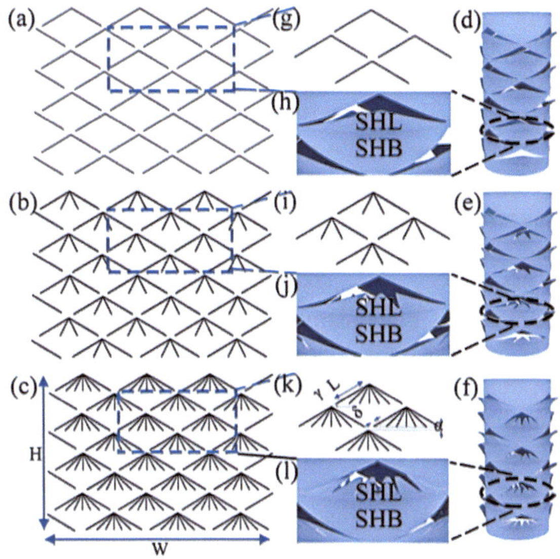

Figure 9.36 Sample structure; (a), (b), and (c) are CAD sketches of samples 1, 2, and 3, respectively; (d), (e), and (f) are drawings of the physical models of samples 1, 2, and 3, respectively; (g), (i), and (k) are partial enlargements of the CAD of the samples, respectively; (h), (j), and (l) are partial enlargements of the models, in which the tip position has superhydrophilicity at the upper portion, and superhydrophobicity at the lower portion, respectively. Superhydrophilic in the upper part and superhydrophobic in the lower part of the tip position. Source: Reproduced from [123] / with permission of Elsevier.

weaving. Lai's team [126] utilized a weaving technique to fabricate economical polyester fabrics with a Cu-superhydrophobic–superhydrophilic pattern for use as a fog collector. The optimal fog collection effect ($1432.7\,\text{mg}\,\text{h}^{-1}\,\text{cm}^{-2}$) was demonstrated when the superhydrophilic and superhydrophobic zones coexisted in an appropriate proportion. Efficient fog collection requires high thermal conductivity [127], and materials such as Cu, Al, and Fe have been previously used [128]. Here, we employ a one-step copper mirror reaction to selectively deposit copper onto a patterned fabric (Cu SHB-SHL). The accomplishment was attributed to its bionic three-dimensional structure, the augmented condensation properties of the copper coating, and the logical distribution of wetting units. The device structure utilizes a wide spectrum of available textile materials and is optimally suited for large-scale industrial production through validated manufacturing techniques. This technology represents a straightforward, cost-effective and efficient method for harvesting water from fog. The spraying method, although simplistic and cost-effective, forms random patterns with variations that result in substantial deviations in fog collection efficiency. In contrast, the photolithographic mask method rectifies pattern differentiation defects and demonstrates superior potential for controlled patterning, bringing us one step closer to our objective of delivering fresh water at minimal cost and on a grand scale to regions of the world that require it [129].

Due to the collective influence of Laplace pressure disparity, conical spines, and wetting gradient instigated by hydrophilic trichomes, the cactus displays superb fog collection capability. Numerous cactus-like synthetic fog collectors can be created through electrochemical etching, paper cutting, and 3D printing [130]. Cao [124] streamlined the intricate 3D complex structure of the cactus into a 2D hollow plate structure based on paper cutting, while fully preserving its ability to directionally and continuously transport droplets (Figure 9.37a). Anisotropically shaped wax bubble kirigami can replicate the function of cactus thorns, i.e. efficiently capturing droplets through droplet-directed self-propulsion and promptly refreshing the collection interface. Fluid simulations demonstrate that thinner spines and smaller vertex angles yield higher forward flow velocities and thus superior fog capture. The orientation of the triangular spine does not influence the directional transport of droplets from the apex to the base. Conversely, the arrangement of the spines significantly impacts the transport velocity. Compared to the symmetric arrangement, an alternate arrangement results in a shorter droplet transportation time and a smaller critical size of rolling droplets. This can be attributed to differences in contour lengths along the three-phase contact line. The resistance experienced by droplets is directly proportional to the length of their contact line, and spines arranged alternately exhibit shorter contour lengths (Figure 9.37b,c). The arrangement of the spines affects fog collection efficiency in terms of both droplet size and transport velocity (Figure 9.37d). At a fog flow rate of $\sim$220 cm s^{-1}, the cactus can accumulate water at a rate of $\sim$4000 mg cm^{-2} h^{-1}, which is 1.6 and 11 times higher than that of the harp collector and plate collector. Moreover, the cost of this paper-based substrate and construction process is substantially reduced to nearly \$0.5 m^{-2}. This work presents a logical design for advanced fog collectors and unlocks more opportunities for developing functional materials from 3D to 2D for microfluidics, condensation, and liquid collection. Most of the aforementioned fog collectors are implemented on external surfaces and are susceptible to environmental factors. Concurrently, in order to optimize the water delivery pathway of gravity-driven fog collectors, Jiang and coworkers [131] proposed a synergistic multistage hydrophilic (3H) fog collection surface with self-pumping droplet suction function. Direct water delivery depends entirely on the surface energy release of suspended spherical droplets. The 3H fog collection surface consists of upright steel needles, melamine resin hydrophilic foam, and hydrophobic silica strips, resulting in an augmented fog collection capacity, which is four times higher than the original hydrophilic foam surface and two times higher than the hydrophilic/hydrophobic surface without hierarchical structure [124]. The fabricated fog collector captures droplets on one side and transports droplets within the device on the other side. More crucially, since the entire water transport process transpires inside the device, it does not impede the water transport process even under severe natural conditions. Enhanced pathways for water retention overcome the drawbacks of conventional systems. The protruding structure effectively traps fog, which is then

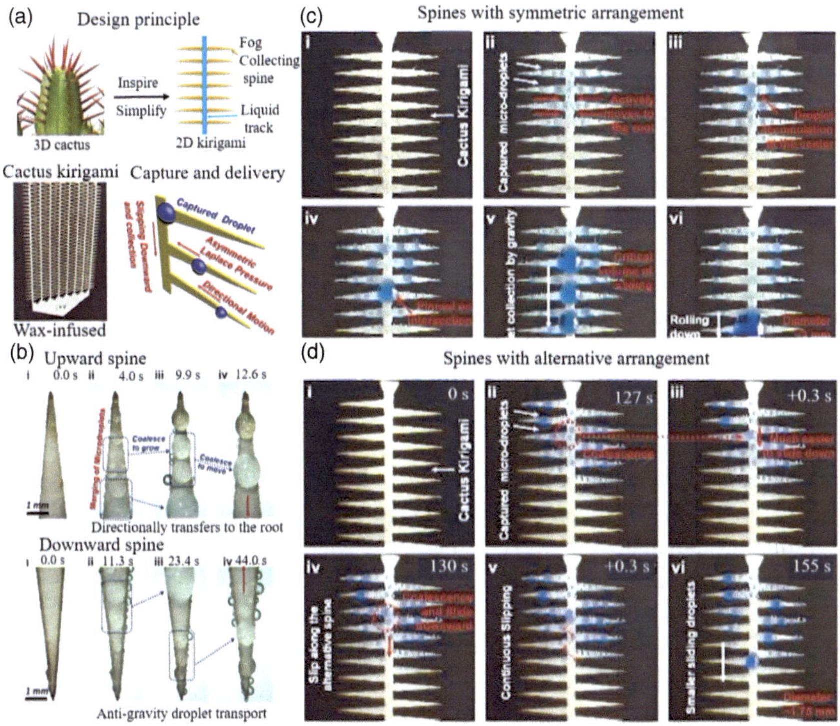

Figure 9.37 (a) 3D cactus turned into 2D paper cutout material and (b) directional movement behavior of water droplets on triangular spines. (c,d) movement behavior of water droplets on symmetrically aligned spines and alternately aligned spines. Source: Reproduced from [124] / with permission of The Royal Society of Chemistry; Reproduced with permission from [26] / John Wiley & Sons.

absorbed by the hydrophilic foam propelled by a wettability gradient. The further incorporation of a striped water barrier fosters unidirectional water transport, even in the presence of gravity. Driven by surface energy, this 3H fog collector enables a gravity-independent process for efficient fog capture, directional water delivery and rapid water storage. Additionally, the device demonstrates exceptionally high fog collection efficiency by accumulating 7 mL of water (21 000 mg h^{-1} cm^{-2}) in 20 min.

The construction of fog collectors based on spider silk is hindered by the need to devise a "spindle junction/joint" structure in order to achieve effective fog collection properties. The surface topography and dimensions of spindle junctions and joints have a significant impact on the efficiency of fog collection. Zheng [125] employed a flexible microfluidic parallel nozzle strategy to fabricate rugged spindle junctions' microfibers based on $CaCl_2$ and chitosan. This methodology facilitated the large-scale fabrication of biomimetic spindle junctions featuring a roughness gradient between spindle junctions which featured a roughness gradient between the spindle junctions and joints, as well as the surface topography of the spindle junctions. By adjusting the $CaCl_2$ content and flow rate, it was possible to manipulate the dimensions of both the spindle junctions and joints, thereby altering the efficiency of water fog collection. The spindle junction microfibers fabricated with chitosan and calcium alginate as primary raw materials exhibit robust mechanical properties and corrosion resistance in acidic (pH = 5) and alkaline (pH = 9) environments. During the water collection process, minute moisture droplets can be adsorbed on the surface of the spindle junction microfibers, and subsequently the droplets can be amalgamated and transported from the splices to the spindle junction segments. The findings indicate that the surface topography of the spindle junction directly influences the water collection efficacy, with greater roughness gradient resulting in superior water collection efficiency. Specifically, fog collection efficiency diminished from 4570 to 4230 $mg\,h^{-1}$ when the distance between two spindle junctions was augmented from 1.43 to 1.67 mm. This parallel nozzle microfluidic technique offers a cost-effective and adaptable method for the preparation of bio-inspired rugged spindle junction microfibers with high biocompatibility, which holds potential applications in large-scale water harvesting, sustained drug delivery, and targeted water harvesting. Similarly, transporting droplets within the structure was accomplished by a fusion of electrostatic spinning and fluid coating materials, and biomimetic nanofibrous bump fibers (BNFs) with a homogenous outer shell were synthesized by Zheng and coworkers (Figure 9.38a,b) [50]. The spindle junction composed of nanofibers possesses a higher specific surface area, which can enhance the efficiency of fog trapping. Moreover, the spindle junction can also transport droplets within the BNF due to the formation of capillary channels between the nanofibers. The combination of internal and external transport further augments the velocity at which captured fog is transported. Owing to the audacious innovation of utilizing spindle junction fibers as electrostatically spun collectors, the BNF exhibits a distinctive structure with randomly distributed nanofibers on humps and aligned on joints. The nanofibers are stacked at the collector to form a randomly distributed structure, and the nanofibers at the joints are attracted and oriented towards the collector (Figure 9.38c). There was a substantial difference in the roughness of the joints comprising oriented nanofibers and the knots consisting of randomly arranged nanofibers (Figure 9.38d). The gradient of surface energy between spider junctions and joints propels droplets towards the rougher spider junctions. The fog trapping process of the three different lengths of BNF webs is illustrated in Figure 9.38e, with the longer BNF webs exhibiting a higher water collection efficiency. Figure 9.38f displays digital photographs of droplet formation (left) and transport (right) in three modes, including two crossed fibers, three crossed fibers, and mesh fibers. It can be observed that the droplet separation times are 21.50, 31.20, and 0.16 s, respectively. This difference in separation time can be attributed to the ability of the mesh fibers to concentrate surrounding droplets towards the center and accelerate droplet growth.

Compared to superhydrophobic–superhydrophilic composite surfaces, the biomimetic superslip surfaces prepared using piggyback bit excellent fog trapping and fog collection efficiency due to higher thermal conductivity and lower contact

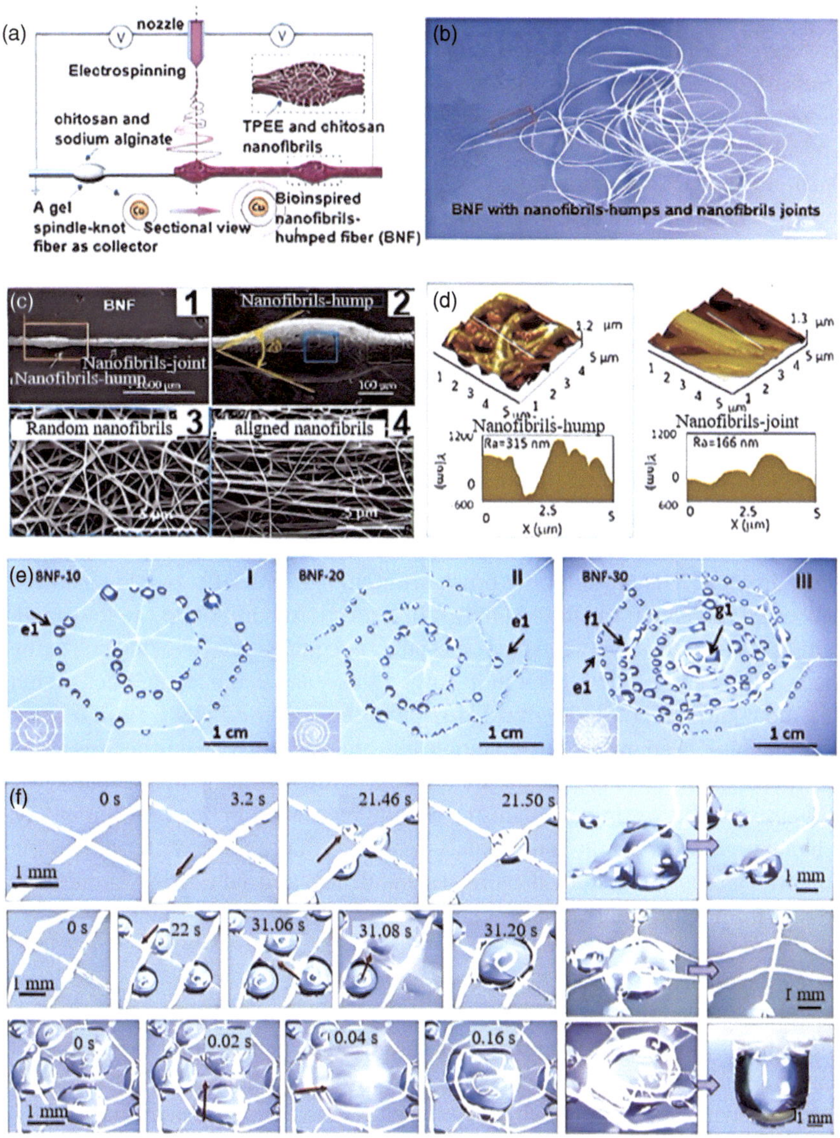

Figure 9.38 (a) BNF preparation process; (b–d) digital photographs, SEM images, and AFM images of BNF. (e) Water trapping behavior of BNF with different helical gaps; and (f) fog trapping (left) and separation (right) process of two cross-fibers, three cross-fibers, and reticulated fibers. Source: Reproduced with permission from [125] / John Wiley & Sons.

angle hysteresis. However, the fog collection efficiency of the bionic superslip surface decreases with the loss of lubricant. Our team has conducted extensive research based on bionic superslip surfaces. Most of the ideas used hydrothermal, chemical etching, and spraying methods to construct a rough substrate and then infuse the lubricant to prepare a stable bionic superslip surface. The above-mentioned

Fan et al. constructed hollow ZnO nanotubular rough structures on the surface of zinc sheets [133] and used WO_3 nanoparticles and an inorganic binder to spray the substrate and then modified it with a low surface energy substance to prepare a biomimetic superslip surface [134]. The prepared surfaces exhibit excellent chemical and physical stability, and demonstrate outstanding performance in droplet nucleation and removal even after extreme conditions, such as knife scraping, sandpaper abrasion, and strong acid/base immersion treatments. The fog collection efficiency of the bionic superslip surfaces is also high under extreme conditions. Additionally, when modifying silicone oil on surfaces with specific geometrical structures, bionic superslip surfaces showed better fog collection efficiency compared to bionic superslip planar surfaces. Xiao and coworkers [132] utilized cutting, water bath, and spraying methods to combine macroscopic patterns and microstructures, and prepared smooth "fog trapping brushes" with the surface of alumina microneedles (Figure 9.39a–f). The smooth surface facilitates fog capture and transport while minimizing droplet escape from the surface, thus promoting droplet merging. Curved aluminum panels exhibit greater fog trapping capability than rectangular aluminum panels due to reduced displacement of surface droplets. The fog-trapping brush with curved pins 4.0 mm wide, showed excellent fog collection capability (445 $mg\,h^{-1}\,cm^{-2}$). In addition to the four main organisms mentioned above, some special but equally effective fog-collecting devices have been reported, inspired by organisms such as the ferruginous orchid [135], the mammon [136], the lizard [137], and the shorebird [138]. Despite great efforts that have been made, current bionic fog- and water-harvesting devices based on the above-mentioned organisms are not able to fully replicate their capabilities, and their water-harvesting mechanisms need to be studied in depth. Additionally, the biology of other plants and animals in nature that possess water-harvesting capabilities needs to be studied more systematically, and fabricating biomimetic materials from interesting organisms still faces great challenges.

9.5.2.2 Binary Bionic-Inspired Fog Collection Device

Individual bio-inspired wettable surfaces exhibit limited fog collection capacity due to their unique bio fog collection characteristics [139]. Fog collection processes based on nano desert beetle, cactus, and spider silk surfaces occur on the outer surface and slide downward when the volume of captured droplets is greater than a critical size. Captured droplets invariably evaporate during transport. Bionic super-slip surfaces are proficient in rapid droplet transport, yet possess subpar droplet capture capabilities. By emphasizing the advantages of multiple organisms within a single device, it can effectively counterbalance the aforementioned limitations of a single organism-inspired fog collection device. However, striking a balance between optimizing biological properties and streamlining the device preparation process remains a challenge that warrants further exploration [140]. With the advancement of research, the remarkable fog collection efficiency of the biomimetic fog collection device amalgamating two organisms through an ingenious design has been widely reported [141].

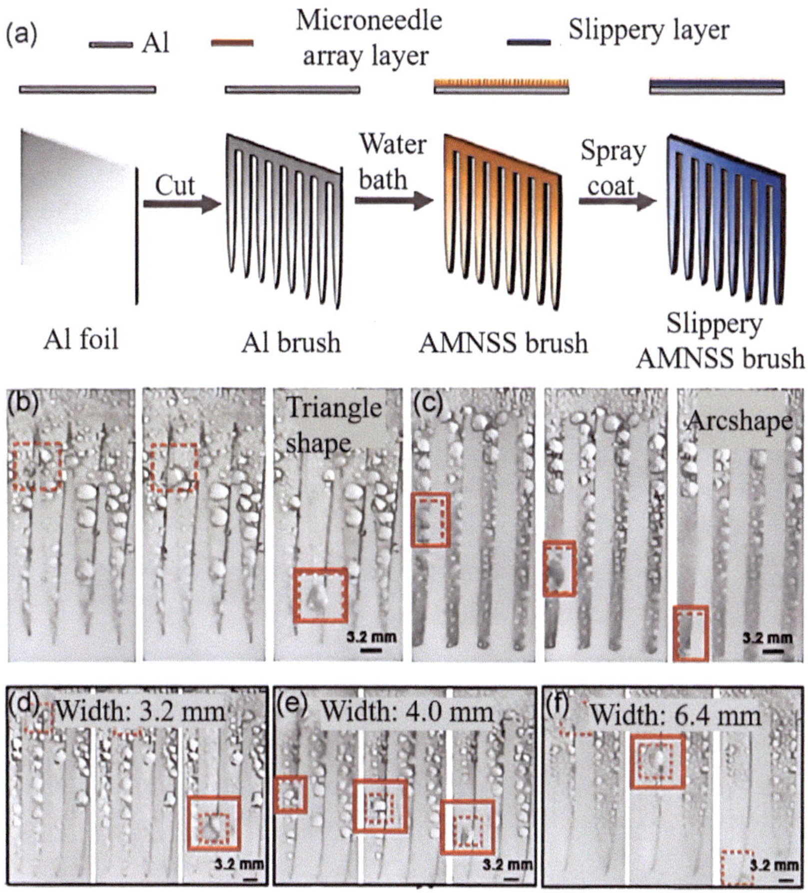

Figure 9.39 (a) Preparation of smooth alumina microneedle structured surfaces. (b,c) Dynamic fog collection process on surfaces of different shapes (d–f) with different unit widths. Source: Reproduced with permission from [132] / Elsevier.

Zhao's team [142] inspired by cactus conical spines and desert beetles prepares hybrid membranes with asymmetric structure and anisotropic wettability via electrostatically spun layers of poly(vinylidene fluoride)-hexafluoropropylene copolymer nanofibers (PNFs), and subsequently grew a layer of $Cu(OH)_2$ nanoneedles (CNNs) on the surface of the copper mesh through electrochemical anodic oxidation. This resulted in hybrid membranes that feature an interleaved structure composed of CNNs and PNFs (Figure 9.40a). The SEM images of the samples after each step are illustrated in Figure 9.40b–d. The surface of the original Cu mesh was smooth with an average pore size of approximately 75 μm, which was subsequently coated with a layer of uniformly distributed PNFs via electrostatic spraying, followed by the anodic growth of a dense CNN layer on the Cu mesh. Hydrophilic CNNs could capture fog droplets, and hydrophobic PNFs facilitated the transport of captured droplets. Under the influence of the wetting gradient, small droplets in the CNN/PNF staggered region are transported directly from the

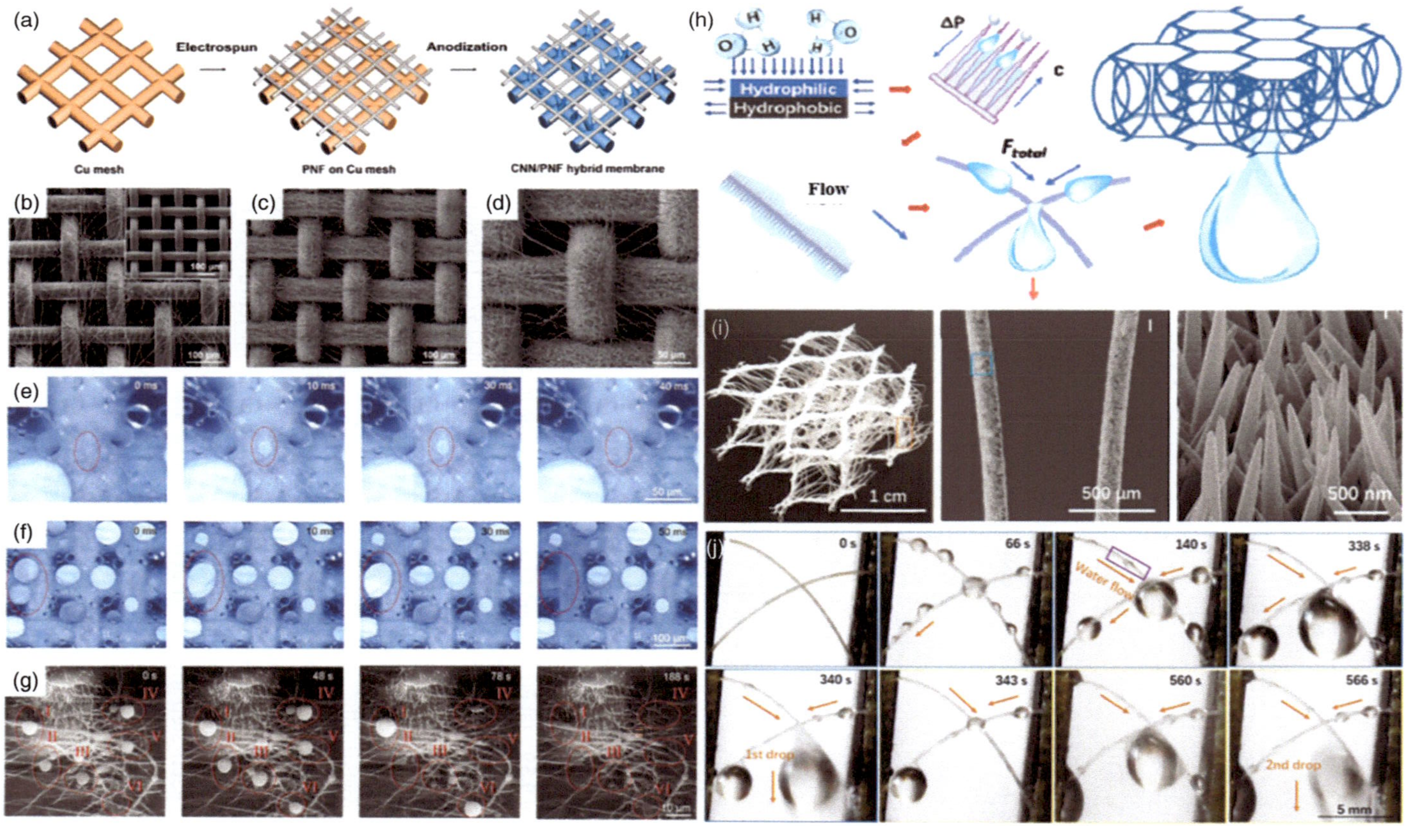

(a)
Electrospun
Anodization
Cu mesh
PNF on Cu mesh
CNN/PNF hybrid membrane
(b)
(c)
(d)
(e)
0 ms
10 ms
30 ms
40 ms
50 μm
(f)
100 μm
(g)
0 s
48 s
78 s
188 s
I
II
III
IV
V
VI
10 μm
(h)
Hydrophilic
Hydrophobic
ΔP
c
Flow
F total
(i)
1 cm
500 μm
500 nm
(j)
0 s
66 s
140 s
338 s
Water flow
340 s
343 s
560 s
566 s
1st drop
2nd drop
5 mm

Figure 9.40 (a) Hybrid membrane design scheme with wettability and topological anisotropy. (b–d) SEM images of PNF overlaying a pristine Cu grid, inset shows the pristine Cu grid. (e–f) Dynamic fog trapping behavior of CNN/PNF hybrid membranes. (g) ESEM images of droplet condensation and transport in a saturated vapor atmosphere. Source: Reproduced with permission from [142] / The Royal Society of Chemistry. (h–i) Schematic diagrams of a fog collection device with a 3D fiber network inspired by spider webs and cactus spines. (j) Fog droplets on a 3D fiber network with coalescence and separation behavior. Source: Reproduced with permission from [143] / American Chemical Society.

PNF side to the CNN side (Figure 9.40e–f). Droplets suspended on the PNF adhere to other small droplets until they contact CNN and are absorbed (Figure 9.40g), and the hybrid membrane was reported to exhibit an exceptional fog collection efficiency of 69 600 mg h^{-1} cm^{-2}. Zheng and coworkers [143] were inspired by cactus conical thorns and spider webs, and decorated 3D fiber networks with hydrophilic zinc oxide (ZnO) nano cones (Figure 9.40h–i). The high specific surface area multilayer structure exposed to the fibers, coupled with the hydrophilic decoration, renders the fibers more captivating to water molecules [144]. Due to the difference in Laplace pressure, these nanocones can trap fog droplets and generate agglomerated droplets, thereby enhancing droplet transport efficiency. Importantly, a novel mechanism reveals that nanocone-decorated fibers are wetted, i.e. they form a water film and promptly disintegrate into droplets due to forces associated with Rayleigh instability. The ensuing low retention surface enables the formation of a swift and continuous water flow instead of the conventional intermittent process. Consequently, this structure effectively addresses the issue that captured droplets are not readily transported. The fibers are inclined during the water collection process, and under gravity, the thickness of the water film is different at both ends such that the water film can be transformed into continuous water droplets. The dynamic fog collection process is depicted in Figure 9.40j. This continuous fog transport is due to the advantages of cross-structured and hydrophilic ZnO nanocones integrated onto a decorated 3D fiber network. Minute water is more likely to form a continuous stream on N3D instead of remaining static due to Laplace force, increased droplet weight, and Rayleigh instability. Furthermore, due to its conical structure with a large specific surface area that is exposed to moisture and exhibits hydrophilic properties, it enables rapid capture of water molecules on the N3D surface while expediting both fog trapping and water aggregation processes. According to statistics, this fabricated device has an impressive fog collection capacity reaching up to 3600 mg h^{-1} cm^{-2}. The mass of water collected in 2 h could exceed 240 times the weight of the original 3D mesh devoid of nanocones. This bio-inspired ZnO nanocone-decorated 3D fiber network has potential applications in the collection of fog in production or life, e.g. in cooling water towers and agricultural irrigation systems, or even in water-scarce countries for water recondensation. Jiang and coworkers [56] emulated the architecture of cactus spines and bottle grass trichomes to establish hierarchical structures on copper plates. An origami methodology was also utilized to transform the 2D spider web-like fog collector into a 3D cactus-like fog collector (Figure 9.41a).

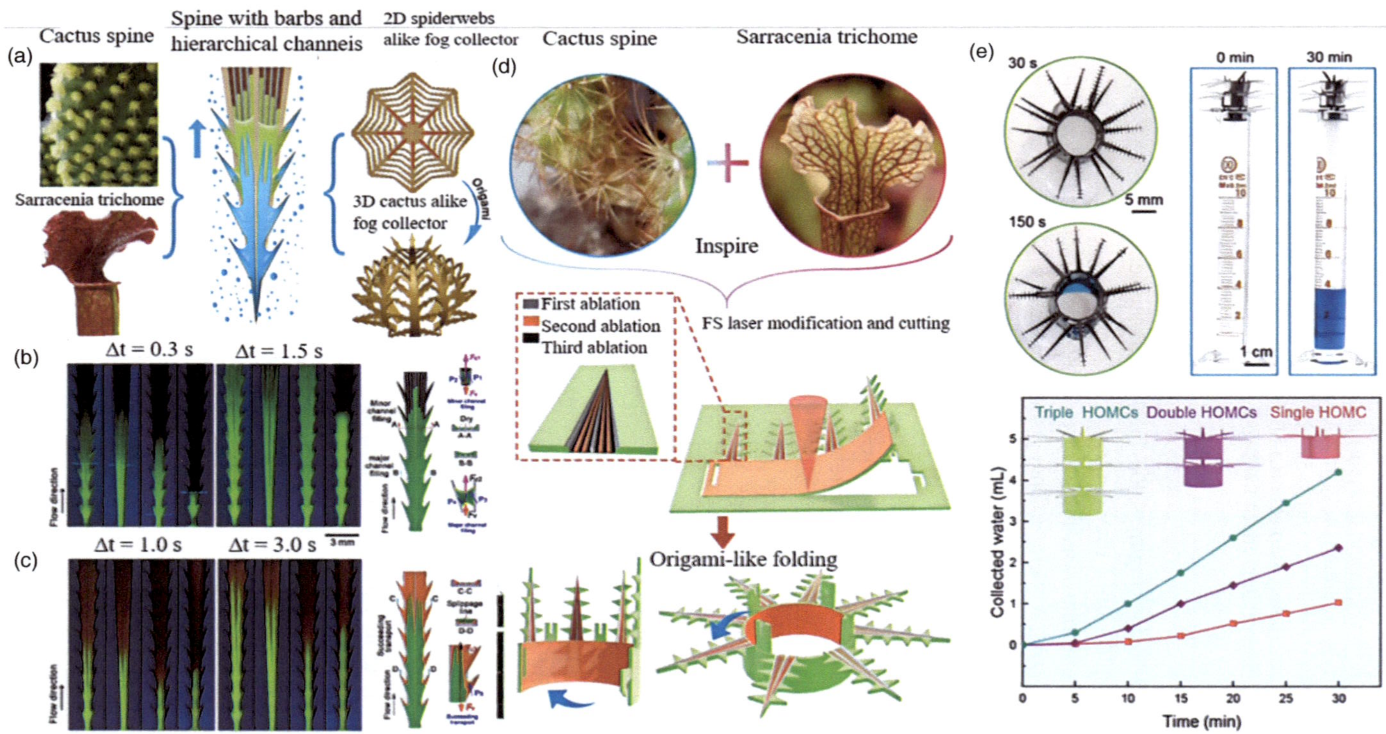

Figure 9.41 (a) Schematic diagram and fog collection capacity of a 3D cactus-like bionic fog collection device. (b,c) Droplet transport process on different structured surfaces in dry and wet conditions. Source: Reproduced with permission from [56] / American Chemical Society. (d) scheme and procedures for the fabrication of layered origami fog and water collector. (e) Fog collection and water storage performance of the fog and water collector. Source: Reproduced with permission from [145] / The Royal Society of Chemistry.

By evaluating transport processes along spines with barbs and layered channels, spines with layered channels, spines with barbs and grooves, as well as spines with barbs under dry and wet conditions (Figure 9.41b,c), it was found that spines with barbs and layered channels demonstrated the most rapid droplet transport rates. Collectively, bionic fog collection apparatuses inspired by cactus and bottle grass trichomes have achieved fog collection efficiencies as high as $560\,mg\,h^{-1}\,cm^{-2}$. While majority of fog collection systems focus on resolving droplet capture and transport issues, limited research has considered water storage efficiency. Li and coworkers [145] fabricated a hierarchical origami fog-water collector by fusion of three-step femtosecond lasers and origami folding (Figure 9.41d). It comprises three components, a circular spine array with spines and a gradient channel for water collection, an annular thin wall with internal superhydrophilic grooves for water storage, and a bracket for securing and attaching to the other parts. The performance of the apparatus in terms of fog collection and water storage is illustrated in Figure 9.41e. The triple fog and water collection device could accumulate 4.2 mL of fog droplets in 30 min (fog flow rate $\approx 2.33\,mg\,cm^{-2}\,s^{-1}$). Its water storage capacity was examined by mounting it on a rose stem, and some water was still retained in the fog collection device after 120 min, while no water was stored in the original rose stem after 90 min. The prepared fog collection device not only displayed high efficiency in collecting fog but also exhibited outstanding capacity for storing water.

9.5.2.3 Ternary Bionic-Inspired Fog Collection Device

The integration of the unparalleled attributes of three organisms into a single bionic fog collector can effectively counterbalance the limitations of bionic fog collectors inspired by a singular organism [146]. However, the methodology for fabricating tertiary bionic fog-collecting devices frequently necessitates intricate preparation due to the multi-scale integration of physical/chemical parameters. Three-dimensional printing is frequently employed as a viable alternative for constructing tertiary bionic fog-harvesting structures due to its submicron fabrication precision and rapid production speed.

Wang and coworkers [147] devised novel asymmetric fibers employing droplet microfluidics and 3D printing. The periodic bumps on the foreface of the fibers were derived from the leaves of the aquatic plant Nelumbo nucifera, while the absorbent fiber body mirrored the hairs of mountain coriander leaves. Additionally, and the superhydrophobic–superhydrophilic phase pattern on the exterior of the fiber was derived from the nanofabricated desert beetle (Figure 9.42a). The findings indicate that the concavity-valley structure facilitates continuous fog collection. To further investigate the impact of structure on fog collection, we compared the dynamic fog collection process with smooth fiber mesh, symmetric fiber mesh, and asymmetric fiber mesh. In the case of smooth and symmetric fiber grids, captured droplets remained on the fiber surface rather than sliding downwards due to higher contact angle hysteresis and no Laplace pressure difference, whereas substantial droplet transport processes transpired on asymmetric surfaces. Fiber grids, with fog collection efficiencies of up to $195\,L\,m^{-2}\,day^{-1}$ ($8250\,mg\,h^{-1}\,cm^{-2}$), could satisfy the potable water requirements of 26 individuals. Chen and coworkers [148] fabricated a

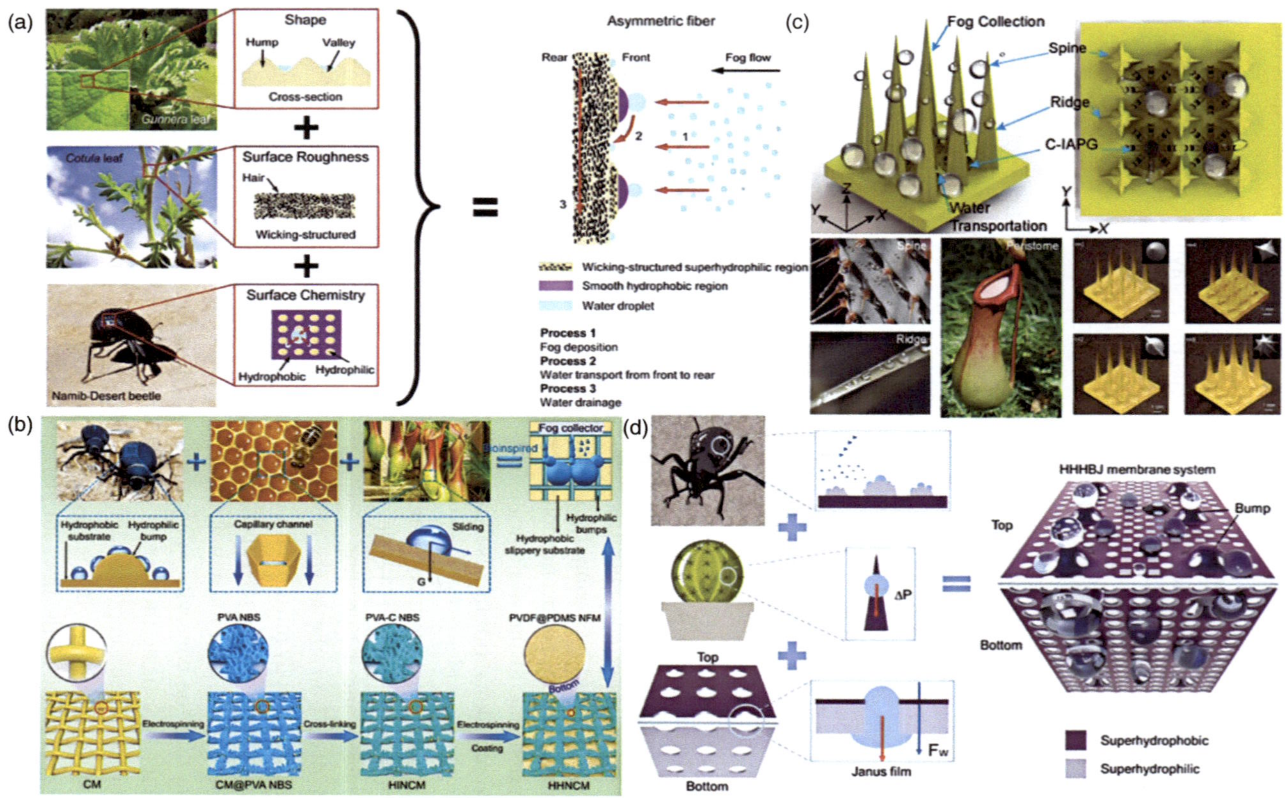
(a)
Gunnera leaf
Shape
Hump
Valley
Cross-section
Cotula leaf
Surface Roughness
Hair
Wicking-structured
Namib-Desert beetle
Surface Chemistry
Hydrophobic
Hydrophilic
Asymmetric fiber
Rear
Front
Fog flow
Wicking-structured superhydrophilic region
Smooth hydrophobic region
Water droplet
Process 1
Fog deposition
Process 2
Water transport from front to rear
Process 3
Water drainage
(c)
Fog Collection
Spine
Ridge
C-IAPG
Water Transportation
Spine
Peristome
Ridge
(b)
Hydrophobic substrate
Hydrophilic bump
Capillary channel
Sliding
Fog collector
Bioinspired
Hydrophilic bumps
Hydrophobic slippery substrate
PVA NBS
PVA-C NBS
PVDF@PDMS NFM
Electrospinning
Cross-linking
Electrospinning
Coating
CM
CM@PVA NBS
HINCM
HHNCM
(d)
ΔP
Top
Bottom
Janus film
HHHBJ membrane system
Bump
Top
Bottom
Superhydrophobic
Superhydrophilic

Figure 9.42 (a) A graphic illustration of asymmetrical fibers derived from the hairs of Fenella grass leaves, Coriander leaves, and Namib Desert beetles. Source: Reproduced with permission from [147] / Elsevier. (b) An artificial fog-water gathering apparatus inspired by the basal grooves of desert grass leaves, cacti spines, and Nepenthes. Source: Reproduced with permission from [148] / American Chemical Society. (c) An artificial fog-water gathering apparatus derived from Namib Desert beetles, honeycomb mesh structures, and Nepenthes hyperslip surfaces. Source: Reproduced with permission from [78, 148] / American Chemical Society. (d) An artificial fog-water gathering apparatus informed by Namib Desert beetles, cacti spines, and Janus structures, HHHBJ films. Source: Reproduced with permission from [77c] / American Chemical Society.

biomimetic fog water collector inspired by diverse organisms. It comprises a surface with adjustable longitudinal ridges (inspired by desert grass blades), a cactus-like spine structure, and a hogweed-like bottom groove embellished with curved diagonal curved pits via 3D technology (Figure 9.42b–d). Initially, droplets are captured at the tips of the spine and the spine and flow to the base under Laplace force. The accumulated droplets are subsequently efficiently conveyed to the base reservoir through the curved and inclined curved pit grooves. The apex of ridges with greater curvature exhibits superior fog capture capacity [48b, 149]. The optimal fog collection of the sample was achieved when there were 4 longitudinal ridges. However, as the number of ridges increased from 4 to 8, the efficiency of fog collection showed a reverse trend due to the reduced angular distance between neighboring ridges, which impeded the fog collection effect of water capture and water transport propelled by the surface energy gradient. The curved and inclined curved pit notch can amalgamate Laplace pressure, capillary force, and gravity to exhibit the swiftest droplet transport effect (Figure 9.42b). By judiciously optimizing the stinger and bottom grooves, 0.218 ± 0.049 g of water (1090 mg h^{-1} cm^{-2}) can be harvested in 7200 s. Ding and coworkers [78] reported a hydrophilic-hydrophobic nanofiber-copper mesh (HHNCM) (Figure 9.42c) by a three-stage process of electrostatic spinning, which integrates the dimpled structure of the surface of the nanofabricated desert beetles to enhance the droplet deposition efficiency, a honeycomb grid of nanofiber capillaries to regulate the direction of fog droplet transport, and an ultra-slip surface of hogwash to accelerate the droplet transport speed are three attributes. According to the study, the entire process of fog capture and transportation can be accomplished within 5 min. Moreover, the fog capture capacity of the sample was sustained at a high level of approximately 1110 mg h^{-1} cm^{-2}. The fog collection efficiency of the HHNCM was maintained at a high level even after 20 cycles. Inspired by the aforementioned prepared bionic fog collection device, the authors propose an inventive application concept of utilizing HHNCM as a fog collection window. Open all windows at night to capture water droplets in the water stream and then accumulating them spontaneously into a water container, while keeping windows closed during the day can effectively circumvent volatilization of the captured water. This application idea not only saves water for agriculture, but also effectively converts natural resources into usable resources, thus realizing sustainable development.

Consequently, scientists have devoted substantial effort to address the issue of liquid droplet evaporation on material surfaces. Drawing inspiration from the Janus structure of lotus leaves, the conical configuration of cactus thorns, and the concave-convex surface of desert beetles, Zhang et al. [77c] devised a layered hydrophilic/hydrophobic/concave-convex Janus (HHHBJ) ultrathin film on aluminum foil through selective femtosecond laser ablation and surface modification (Figure 9.42d). Droplets entrapped in the hydrophobic zone converge directionally towards the hydrophilic zone to generate a water film. This water film is then conveyed to the underside surface for storage under the joint influence of gravity and Laplace force engendered by the tapered holes when the water film reaches a sufficient size and contacts the tapered holes. A total of 7.6 g of fog water can be harvested in 40 min at vertical fog flow ($\approx$11 400 mg h^{-1}).

9.5.2.4 Quaternary Bionic-Inspired Fog Collection Device

An effective combination of different water-harvesting mechanisms is a practical approach to enhance efficiency. Multi-bionic is not just a simple combination of different material properties, the difficulty of manufacturing multi-bionic fog collection devices increases exponentially with the increase of the combination of properties, and in most cases, the fog collection efficiency of multi-bionic fog collection devices cannot match the production cost, which leads to a few reports on multi-bionic fog collection devices.

Li et al. [150] ingeniously fabricated Janus membranes possessing heterogeneous wettability and interpenetrating architectures, drawing inspiration from the asymmetric hierarchical architecture of cactus thorns, the interlocked hydrophobic-hydrophilic composition on the reverse of the nanofabricated desert beetles, the porous super-slippery surface of hogweed, and the directional pumping capabilities of aquatic plants (Figure 9.43a). The polybio-inspired Janus membrane was fabricated utilizing a fusion of spraying and etching fabrication methodologies, exhibiting efficient fog collection through its fluid unidirectional operating behavior. Hybrid Janus membranes comprising of $Cu(OH)_2$ nanofibers (CNFs) and microtetrapods ZnO (ZnO-tetrapods) were directly ablated to form arrays of regular asymmetric raised micropores and lubrication layers on the substrate of Janus CNFs/ZnO-tetrapod membrane. The inhomogeneous hydrophobic/hydrophilic structure engenders an energy gradient and Laplace pressure difference, which synergistically contributes to enhancing the performance of unidirectional liquid transport by amplifying the high hydrophobic-hydrophilic pressure difference through asymmetric wettability micropores. Furthermore, alternating arrays of CNFs and ZnO tetrapods with multiple superwettability can be fabricated by altering surface free energy and surface roughness. Under the suitable concentration of low-energy perfluoropolyether (PFPE) lubricant, the local distribution state of PFPE on the surface of polydimethylsiloxane (PDMS) films transitioned from a continuous oil layer to a dispersed distribution. This transition is attributed to the migration of PFPE from the base to the apex of the nanofibers, resulting in alternating arrays of the tip with hydrophilic properties, intermediate region has low-adhesion hydrophobic properties, and the base have super-slippery properties.

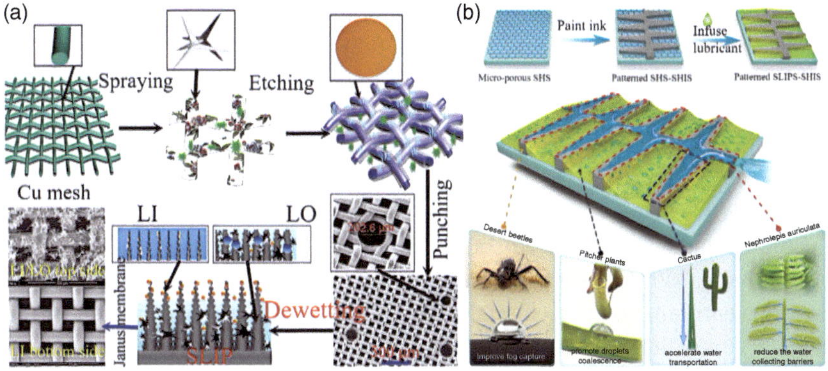

Figure 9.43 (a) Preparation process of Janus membrane with heterogeneous wettability and interpenetrating structure. Source: Reproduced from [150] / with permission of Elsevier. (b) Quaternary bionic fog collection device inspired by nanofabricated desert beetles, self-pigmented super-slip surfaces, and self-pigmented kidney ferns for cactus spines. Source: Reproduced with permission from [151] / Elsevier.

Consequently, small droplets are captured and attracted from the fog by asymmetric microstructures akin to cactus spines and alternating hydrophobic/hydrophilic projections similar to those found on the nanofiber desert beetle in reverse. Droplets of varying sizes can be pumped downward directionally through hollow channels of projections from diverse directions and distances by capillary drive akin to aquatic plants and ultimately collected. It is noteworthy that the droplet driving force originates solely from capillary drive rather than external energy input and droplet coagulation or agglomeration, resulting in the desired efficiency. It has also been demonstrated that smooth surfaces exhibit significant versatility and recoverability in droplet capture, pumping and collection during high-efficiency fog collection. It can be deduced that such multi-bio-inspired hybrid Janus membrane surfaces will offer novel insights for the large-scale production of functional membranes with multiple super-wettability, which hold promising applications in fluid manipulation that do not necessitate additional energy, such as mitigating water scarcity and analyzing biochemical organic fluids.

Wang and coworkers [151] employed a two-step approach to construct a quaternary bionic fog-water collection apparatus. Initially, they produced a microporous interface was produced through the technique of water droplet templating and superhydrophobic enhancement. Subsequently, they fabricated microporous superhydrophobic films from predetermined patterns. As illustrated in the illustration, the superhydrophobic–superhydrophilic pattern was derived from nano fabric desert beetles for effective fog collection, the superslip surface was inspired by hogweed for rapid droplet transport, the wedge-shaped superhydrophilic array was influenced by cactus spines for directional droplet routing, and the alternate alignment array inspired by kidney ferns decreased collection hurdles (Figure 9.43b). Comparatively, the wedge pattern exhibited the highest water collection efficacy of $2166 \pm 71\ \mathrm{mg\,h^{-1}\,cm^{-2}}$ at a flow rate of $45–50\ \mathrm{cm\,s^{-1}}$. Zhou et al. [153] proposed a novel design methodology for the fabrication of multi-biologically-inspired superslip surfaces originating from nanobucket desert beetles, cacti, aquatic plants, and hogweed, which synergistically integrates fog trapping, condensation, and transportation with superior performance compared to other synthetic surfaces. Without necessitating additional energy input or droplet aggregation, this smooth surface inspired by various organisms exhibits optimal droplet pumping and water-harvesting efficiency. The water-harvesting efficiency of this unique water-harvesting apparatus was enhanced by approximately 257.9% compared to a single pristine zinc sheet of approximately 2086.67 $mg\,h^{-1}\ cm^{-2}$. The prepared substrate displayed exceptional stability and weathering resistance in diverse environments. Therefore, this logical design strategy and corresponding theoretical simulations can be extensively utilized in the field of water harvesting, etc. Li et al. [152] successfully fabricated a quaternary biomimetic Janus membrane with both asymmetric microstructures and opposing wettability by emulating the asymmetric structure of cactus thorns, hydrophilic/hydrophobic surfaces on the back of desert beetles, fusiform nodes of spider silk, and asymmetric wettability similar to lotus leaves through chemical etching and electrostatic spinning techniques. The material is chemically etched onto a metal substrate. The material is chemically etched on the surface of metallic copper mesh to generate micron-sized "needle" structure, and subsequently electrostatically spun a bead-like superhydrophobic electrostatic spinning membrane. This allows the micron-sized needle structure to penetrate the electrostatic spinning membrane, constructing a nanofiber membrane/microneedle copper mesh composite water-collecting material with asymmetric microstructure and hydrophilic and hydrophobic attributes (NF/MN Cu mesh), achieving efficient fog collection. The specific preparation procedure is depicted in Figure 9.44a. In order to change the wettability of the original copper mesh and increase its specific surface area, we constructed $Cu(OH)_2$ micro-needles on the copper mesh surface by alkali etching method (Figure 9.44b–g). The preparation process amalgamates the advantages of low cost and straightforward process, which offers a novel concept to alleviate the scarcity of freshwater resources. The results indicated that the fog collection efficiency of NF/MN Cu mesh was significantly higher than that of other samples, and the quantity of water collected was as high as 65.081 $g\,min^{-1}\ cm^{-2}$. After repeating the test for each sample 10 times, the water collection data remained stable, allowing for a continuous fog collection process. The material not only incorporates both wettability gradient and shape gradient but also impedes the transfer of water from the hydrophilic side to the hydrophobic side in Janus membrane material, reducing water evaporation and effectively improving water retention rate. The synergistic effect of the two factors enables efficient fog collection and provides a novel concept to alleviate freshwater resource crisis.

9.5.3 Observations on the Evolution of Non-physical Phase Alteration Bionic Fog Collection

Resource scarcity has emerged as a significant global issue, drawing considerable attention from researchers to organisms capable of extracting water droplets directly

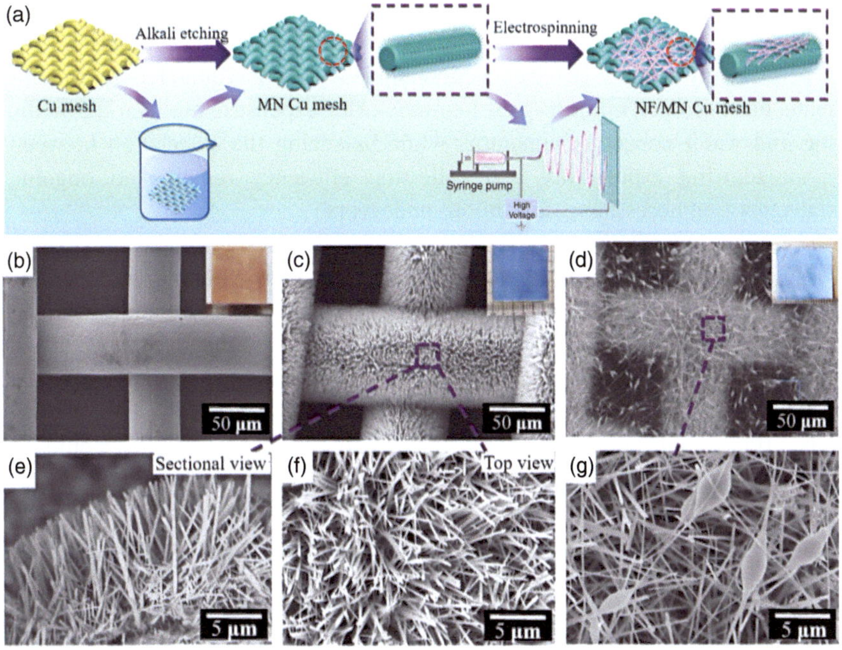

Figure 9.44 (a) Flowchart of preparation of NF/MN Cu mesh. (b–d) SEM images of Cu mesh, MN Cu mesh, and NF/MN Cu mesh at 500×. (e) SEM cross-sectional image of MN Cu mesh. (f,g) SEM images at 5000× corresponding to c–d, respectively. Source: Reproduced with permission from [151, 152] / Elsevier.

from fog streams. Efficient and multifunctional bionic fog-water-harvesting devices have demonstrated considerable potential in addressing water scarcity due to their substantial theoretical fog extraction capacity, affordable cost and versatile application scenarios. Despite considerable advancements, there is still a deficiency in systematic research on bionic fog-harvesting devices for extensive commercial implementation. The manufacturing process of these devices continues to face challenges such as labor-intensive procedures, intricate structures, and reduced efficiency. Nevertheless, these challenges can be reasonably mitigated through strategic approaches such as fabrication methodologies, surface/interface engineering, multifunctional combinations, etc., such as fabrication, liquid phase modification, magnetorheological stretch lithography, and commercialization methods such as 3D printing. To facilitate the large-scale commercialization of bionic fog-water-harvesting devices, careful consideration is still required in the following aspects:

1) Currently, the conceptual designs of bionic fog and water collection apparatuses primarily rely on previously studied organisms, while certain organisms with superior fog collection efficiency remain unexplored. Further endeavors are necessary to unravel the survival strategies of additional flora and fauna that exhibit proficient water collection abilities.

2) The combination of diverse water collection mechanisms is not merely a juxtaposed amalgamation; the optimization of diffusion dynamics/aerodynamics synergistically to enhance fog collection efficiency is also a quandary to contemplate. Furthermore, mitigating the manufacturing expense of multivariate bionic fog and water collection apparatus while balancing the correlation between manufacturing expenditure and collection efficiency remains an ongoing trajectory that necessitates continuous endeavors.
3) Theories and mechanisms underlying fog collection must be further elucidated. Despite an extensive array of fog-harvesting apparatuses being documented, scant attention has been accorded to the configurational correlation between sample architecture and fog-harvesting efficacy, e.g. the optimal ratio of hydrophilic and hydrophobic regions in the fog-harvesting apparatus, inspired by the Namib Desert beetle, remains inconclusive.
4) Operational application procedures are typically conducted in complex outdoor environments, such as deserts and water-scarce areas. Harsh conditions including strong winds, sandstorms, and inclement weather can potentially cause damage to fog-water-harvesting apparatuses. Therefore, it is crucial to place increased emphasis on enhancing the durability and stability of fog collection devices.
5) The majority of fog collection systems currently tested in the laboratory operate under high humidity conditions (70–99%) dissimilar to their real-world application (extremely low humidity in arid regions). Future studies ought to devote more attention to low-humidity scenarios.
6) Formulate a uniform assessment criterion for fog collection efficiency. External variables such as the velocity of the fog stream, temperature, and humidity also influence fog collection efficiency. Experimental conditions and parameters documented in the literature vary, making it challenging to compare the overall performance of these fog collection devices, which is less persuasive for fog collection efficiency. Most experiments typically establish a control group, and the superiority of the surface of the prepared material is gauged by comparing the difference in fog collection efficiency between the optimal group and the control group.
7) Enhancement of material surface properties. During prolonged fog collection, moist surfaces are prone to bacterial colonization. Additionally, the functional coating may detach from water droplets, potentially contaminating the captured fog and rendering it unsuitable for potable water collection

The continuity of the nucleation-growth-transportation process of droplets is challenging to achieve on conventional material surfaces used for fog-water collection, thereby hindering the optimization of fog-water collection efficiency. Hence, the focus of future research lies in developing biomimetic multifunctional superwetting materials with diverse biological characteristics. Each step involved in fog collection is intricate and meticulous, necessitating an understanding of both static wetting behavior and dynamic transportation processes on material surfaces as a prerequisite for achieving efficient fog collection. Furthermore, while bionic fog-harvesting aims to procure potable/lifestyle water, current apparatuses primarily emphasize fog-harvesting rate (the quantity of water harvested per unit area per unit time), neglecting comprehensive studies on water-harvesting volume/total quantity corresponding to the collected fog. Future research should strive towards enhancing the design of fog collection apparatuses and bridging the gap between fundamental research and applied technology. Achieving efficient and cost-effective preparation of fog-harvesting equipment through further innovations remains crucial for sustainable human development strategies.

References

1 (a) White, B., Sarkar, A., and Kietzig, A.-M. (2013). *Applied Surface Science* 284: 826. (b) Park, K.-C., Kim, P., Grinthal, A. et al. (2016). *Nature* 531: 78. (c) Zhong, L., Zhu, H., Wu, Y., and Guo, Z. (2018). *Journal of Colloid and Interface Science* 525: 234.

2 Wang, B., Zhang, Y., Liang, W. et al. (2014). *Journal of Materials Chemistry A* 2: 7845.

3 (a) Peng, Z. and Guo, Z. (2022). *Biomaterials Science* 10: 5831. (b) Wang, Q., Yang, F., Wu, D., and Guo, Z. (2023). *Colloids and Surfaces A: Physicochemical and Engineering Aspects* 658: 130584.

4 Zhou, H., Jing, X., and Guo, Z. (2020). *Journal of Colloid and Interface Science* 561: 730.

5 Zhou, H., Zhang, M., Li, C. et al. (2018). *Small* 14: 1801335.

6 Zheng, Y., Bai, H., Huang, Z. et al. (2010). *Nature* 463: 640.

7 Ju, J., Bai, H., Zheng, Y. et al. (2012). *Nature Communications* 3: 1247.

8 Xu, T., Lin, Y., Zhang, M. et al. (2016). *ACS Nano* 10: 10681.

9 Chen, H., Zhang, P., Zhang, L. et al. (2016). *Nature* 532: 85.

10 Bohn, H.F. and Federle, W. (2004). *Proceedings of the National Academy of Sciences* 101: 14138.

11 (a) Adera, S., Alvarenga, J., Shneidman, A.V. et al. (2020). *ACS Nano* 14: 8024. (b) Jing, X. and Guo, Z. (2019). *ACS Applied Materials & Interfaces* 11: 35949.

12 Anand, S., Paxson, A.T., Dhiman, R. et al. (2012). *ACS Nano* 6: 10122.

13 Peng, Y., He, Y., Yang, S. et al. (2015). *Advanced Functional Materials* 25: 5967.

14 Zhong, L., Zhang, R., Li, J. et al. (2018). *Langmuir* 34: 15259.

15 Ju, J., Yao, X., Yang, S. et al. (2014). *Advanced Functional Materials* 24: 6933.

16 Nioras, D., Ellinas, K., Constantoudis, V., and Gogolides, E. (2021). *ACS Applied Materials & Interfaces* 13: 48322.

17 (a) Lei, J. and Guo, Z. (2020). *Nanoscale* 12: 6921. (b) Parker, A.R. and Lawrence, C.R. (2001). *Nature* 414: 33.

18 Xu, W., Lan, Z., Peng, B. et al. (2015). *RSC Advances* 5: 812.

19 Dai, X., Sun, N., Nielsen, S.O. et al. (2018). *Science Advances* 4: eaaq0919.

20 Zou, J., Wang, P.F., Zhang, T.R. et al. (2011). *Physics of Fluids* 23: 044101.

21 Azad, M.A.K., Ellerbrok, D., Barthlott, W., and Koch, K. (2015). *Bioinspiration & Biomimetics* 10: 016004.

22 Ju, J., Xiao, K., Yao, X. et al. (2013). *Advanced Materials* 25: 5937.

23 Gou, X. and Guo, Z. (2020). *Langmuir* 36: 8983.

24 Park, K.-C., Chhatre, S.S., Srinivasan, S. et al. (2013). *Langmuir* 29: 13269.

25 Xia, D., Johnson, L.M., and López, G.P. (2012). *Advanced Materials* 24: 1287.

26 Bai, H., Ju, J., Sun, R. et al. (2011). *Advanced Materials* 23: 3708.

27 Bai, H., Wang, L., Ju, J. et al. (2014). *Advanced Materials* 26: 5025.

28 Wen, R., Li, Q., Wu, J. et al. (2017). *Nano Energy* 33: 177.

29 Feng, J., Zhong, L., and Guo, Z. (2020). *Chemical Engineering Journal* 388: 124283.

30 (a) Wong, T.-S., Kang, S.H., Tang, S.K.Y. et al. (2011). *Nature* 477: 443. (b) Boreyko, J.B., Polizos, G., Datskos, P.G. et al. (2014). *Proceedings of the National Academy of Sciences of the United States of America* 111: 7588.

31 (a) Wang, S., Liu, K., Yao, X., and Jiang, L. (2015). *Chemical Reviews* 115: 8230. (b) Ge, P., Wang, S., Zhang, J., and Yang, B. (2020). *Materials Horizons* 7: 2566. (c) Nishimoto, S., Mano, T., Kameshima, Y., and Miyake, M. (2010). *Chemical Physics Letters* 500: 86.

32 Tran, M.Q., Yamaguchi, Y., Yamatoya, K. et al. (2018). *Materials Letters* 213: 51.

33 Liu, H., Feng, L., Zhai, J. et al. (2004). *Langmuir* 20: 5659.

34 Kontos, A.I., Kontos, A.G., Tsoukleris, D.S. et al. (2007). *Thin Solid Films* 515: 7370.

35 Vernardou, D., Kalogerakis, G., Stratakis, E. et al. (2009). *Solid State Sciences* 11: 1499.

36 Lai, Y., Lin, C., Wang, H. et al. (2008). *Electrochefogry Communications* 10: 387.

37 Ye, J., Yin, Q., and Zhou, Y. (2009). *Thin Solid Films* 517: 6012.

38 Nishimoto, S., Kubo, A., Nohara, K. et al. (2009). *Applied Surface Science* 255: 6221.

39 Permpoon, S., Houmard, M., Riassetto, D. et al. (2008). *Thin Solid Films* 516: 957.

40 Choi, W.-K. (2007). *Surface and Coatings Technology* 201: 8099.

41 Kim, H., Jung, S.J., Han, Y.H. et al. (2008). *Thin Solid Films* 516: 3530.

42 Chan, C.M., Ko, T.M., and Hiraoka, H. (1996). *Surface Science Reports* 24: 1.

43 (a) Yanagisawa, K., Sakai, M., Isobe, T. et al. (2014). *Applied Surface Science* 315: 212. (b) Li, D., Qian, C., Gao, S. et al. (2017). *International Journal of Refrigeration* 79: 25. (c) Wang, F.-C., Yang, F., and Zhao, Y.-P. (2011). *Applied Physics Letters* 98: 053112.

44 Wang, X., Zeng, J., Yu, X. et al. (2019). *RSC Advances* 10: 282.

45 Timmons, C.O. and Zisman, W.A. (1966). *Journal of Colloid and Interface Science* 22: 165.

46 (a) Liu, T.L. and Kim, C.-J.C. (2014). *Science* 346: 1096. (b) Tuteja, A., Choi, W., Ma, M. et al. (2007). *Science* 318: 1618. (c) Boreyko, J.B. and Chen, C.H. (2009). *Physical Review Letters* 103: 184501.

47 (a) Lafuma, A. and Quéré, D. (2003). *Nature Materials* 2: 457. (b) Miljkovic, N., Enright, R., Nam, Y. et al. (2013). *Nano Letters* 13: 179. (c) Wenzel, R.N. (1936). *Industrial & Engineering Chefogry* 28: 988.

48 (a) Cao, M., Ju, J., Li, K. et al. (2014). *Advanced Functional Materials* 24: 3235. (b) Zhou, H., Jing, X., and Guo, Z. (2020). *Langmuir* 36: 6801. (c) Xing, H.T., Cheng, J., Zhou, C.L. et al. (2018). *Micro & Nano Letters* 13: 1068.

49 Thakur, N., Ranganath, A.S., Agarwal, K., and Baji, A. (2017). *Macromolecular Materials and Engineering* 302.

50 Liu, Y.F., Yang, N., Gao, C.L. et al. (2020). *ACS Applied Materials & Interfaces* 12: 28876.

51 Díaz, J.E., Barrero, A., Marquez, M., and Loscertales, I.G. (2006). *Advanced Functional Materials* 16: 2110.

52 Jin, Y., Yang, D.Y., Kang, D.Y., and Jiang, X.Y. (2010). *Langmuir* 26: 1186.

53 Tian, X.L., Bai, H., Zheng, Y.M., and Jiang, L. (2011). *Advanced Functional Materials* 21: 1398.

54 (a) Dong, H., Wang, N., Wang, L. et al. (2012). *ChemPhysChem* 13: 1153. (b) Song, C., Zhao, L., Zhou, W.B. et al. (2014). *Journal of Materials Chefogry A* 2: 9465.

55 Joanna, K.K., Ura, D.P., Gajek, M. et al. (2020). *ACS Applied Materials & Interfaces* 12: 1665.

56 Wang, J., Yi, S., Yang, Z. et al. (2020). *ACS Applied Materials & Interfaces* 12: 21080.

57 (a) Chen, Y., He, J., Wang, L. et al. (2014). *Journal of Materials Chefogry A* 2: 1230. (b) Bai, H., Tian, X., Zheng, Y. et al. (2010). *L. Jiang* 22: 5521. (c) Feng, S.L., Hou, Y.P., Chen, Y. et al. (2013). *Journal of Materials Chefogry A* 1: 8363.

58 Gao, Y., Wang, J., Xia, W., and Mou, X. (2018). Z, Cai. *ACS Sustainable Chemistry & Engineering* 6 (6): 7216.

59 Tian, X.L., Chen, Y., Zheng, Y.M. et al. (2011). *Advanced Materials* 23: 5486.

60 Lin, Y.C., Hu, Z.Y., Zhang, M.X. et al. (2018). *Advanced Functional Materials* 28.

61 Kang, E., Jeong, G.S., Choi, Y.Y. et al. (2011). *Nature Materials* 10: 877.

62 Zhan, Y., Yu, S., Amirfazli, A. et al. (2021). *Colloids and Surfaces A: Physicochemical and Engineering Aspects* 629: 127441.

63 Zhang, M., Zheng, Z.Y., Zhu, Y.Q. et al. (2022). *Chemical Engineering Journal* 433.

64 Cui, L.-Y., Liu, H.-P., Zhang, W.-L. et al. (2017). *Journal of Materials Science & Technology* 33: 1263.

65 (a) Zhao, J., Chen, L.-j., Yu, K. et al. (2015). *Transactions of Nonferrous Metals Society of China* 25: 824. (b) He, X.-H., Wang, W., Liu, Y.-M. et al. (2015). *ACS Applied Materials & Interfaces* 7: 17471.

66 Song, Y.Y., Liu, Y., Jiang, H.B. et al. (2018). *Nanoscale* 10: 3813.

67 Fan, J., Yang, Z., Sun, W. et al. (2022). *Colloids and Surfaces A: Physicochemical and Engineering Aspects* 650.

68 Zhang, L., Sun, J., Han, P. et al. (2022). *ACS Applied Polymer Materials* https://doi.org/10.1021/acsapm.2c01686.

69 Sun, R., Zhao, J., Liu, C. et al. (2022). *Progress in Organic Coatings* 171: 107016.

70 Zhou, W.l., Wu, T., Du, Y. et al. (2023). *Chemical Engineering Journal* 453.

71 (a) Wang, M., Liu, Q., Zhang, H. et al. (2017). *ACS Applied Materials & Interfaces* 9: 29248. (b) Wang, S., Wang, C., Peng, Z., and Chen, S. (2019). *The Journal of Physical Chefogry C* 123: 1798. (c) Wang, S., Peng, Z., Li, J. et al. (2020). *Computational Materials Science* 175: 109616.

72 Xing, Y., Shang, W., Wang, Q. et al. (2019). *ACS Applied Materials & Interfaces* 11: 10951.

73 Song, D. and Bhushan, B. (2019). *Philosophical Transactions: Mathematical, Physical and Engineering Sciences* 377: 20190128.

74 Yin, K., Du, H., Dong, X. et al. (2017). *Nanoscale* 9: 14620.

75 Guo, J., Huang, W., Guo, Z., and Liu, W. (2022). *Nano Letters* 22: 3104.

76 Chen, X., Wu, J., Ma, R. et al. (2011). *Advanced Functional Materials* 21: 4617.

77 (a) Zhang, L., Wu, J., Hedhili, M.N. et al. (2015). *Journal of Materials Chemistry A* 3: 2844. (b) Zhou, H., Jing, X., and Guo, Z. (2020). *Nanoscale Advances* 2: 1473. (c) Su, Y., Chen, L., Jiao, Y. et al. (2021). *ACS Applied Materials & Interfaces* 13: 26542. (d) Zhu, H., Cai, S., Zhou, J. et al. (2021). *Water Research* 206: 117759. (e) Liu, H., Chen, C., Wen, H. et al. (2018). *Journal of Materials Chefogry A* 6: 18839.

78 Zhang, Y., Meng, N., Babar, A.A. et al. (2021). *Nano Letters* 21: 7806.

79 (a) Zhang, X., Sun, L., Wang, Y. et al. (2019). *Proceedings of the National Academy of Sciences of the United States of America* 116: 20863. (b) Zhang, Y., Meng, N., Babar, A.A. et al. (2021). *ACS Applied Materials & Interfaces* 13: 36587. (c) Zhao, J., Zhu, W., Wang, X. et al. (2020). *ACS Nano* 14: 1045.

80 Zhong, L., Feng, J., and Guo, Z. (2019). *Journal of Materials Chefogry A* 7: 8405.

81 Fu, Y., Ai, S., Guo, Z., and Liu, W. (2023). *Journal of Colloid and Interface Science* 649: 646.

82 Song, Y.Y., Yu, Z.P., Dong, L.M. et al. (2021). *Langmuir* 37: 13703.

83 (a) Knapczyk-Korczak, J., Zhu, J., Ura, D.P. et al. (2021). *ACS Sustainable Chefogry & Engineering* 9: 180. (b) Mohd, G., Majid, K., and Lone, S. (2022). *ACS Applied Materials & Interfaces* 14: 4690. (c) Yang, H.-C., Hou, J., Chen, V., and Xu, Z.-K. (2016). *Angewandte Chemie International Edition* 55: 13398.

84 Liu, C., Lu, C., Zhan, H. et al. (2021). *Advanced Materials Interfaces* 8.

85 Ren, F., Li, G., Zhang, Z. et al. (2017). *Journal of Materials Chefogry A* 5: 18403.

86 Tang, X., Zhang, F., Huang, J., and Guo, Z. (2021). *Langmuir* 37: 13778.

87 Li, J., Li, W., Han, X., and Wang, L. (2021). *Journal of Colloid and Interface Science* 581: 545.

88 Fu, Y., Wu, L., Ai, S. et al. (2023). *Nano Today* 52: 101979.

89 Heng, X., Xiang, M., Lu, Z., and Luo, C. (2014). *ACS Applied Materials & Interfaces* 6: 8032.

90 Yi, S., Wang, J., Chen, Z. et al. (2019). *Advanced Materials Technologies* 4: 1900727.

91 Wang, Y., Zhang, L., Wu, J. et al. (2015). *Journal of Materials Chefogry A* 3: 18963.

92 Zhang, Y., Wu, L., Zhao, X. et al. (2018). *Advanced Energy Materials* 8: 1801139.

93 Yu, Z., Zhang, J., Li, S. et al. (2022). *Advanced Functional Materials* https://doi.org/10.1002/adfm.202210730.

94 Chen, L., Li, W., Gan, Z. et al. (2022). *Journal of Cleaner Production* 363.

95 Wan, Y., Xu, J., Lian, Z., and Xu, J. (2021). *Colloids and Surfaces A: Physicochemical and Engineering Aspects* 628.

96 Gurera, D. and Bhushan, B. (2019). *Journal of Colloid and Interface Science* 551: 26.

97 (a) Thakur, N., Ranganath, A.S., Agarwal, K., and Baji, A. (2017). *Macromolecular Materials and Engineering* 302: 1700124.

98 Chen, Y., Li, D., Wang, T., and Zheng, Y. (2016). *Scientific Reports* 6: 19978.

99 (b) Bai, H., Tian, X., Zheng, Y. et al. (2010). *Advanced Materials* 22: 5521.

100 (a) Lin, Y., Hu, Z., Zhang, M. et al. (2018). *Advanced Functional Materials* 28: 1800163. (b) Ji, X., Guo, S., Zeng, C. et al. (2015). *RSC Advances* 5: 2517. (c) Shang, L., Wang, Y., Yu, Y. et al. (2017). *Journal of Materials Chefogry A* 5: 15026. (d) Peng, Q., Shao, H., Hu, X., and Zhang, Y. (2017). *Macromolecular Materials and Engineering* 302: 1700102. (e) Tian, Y., Wang, J., and Wang, L. (2018). *ACS Applied Materials & Interfaces* 10: 29219.

101 Xue, Y., Chen, Y., Wang, T. et al. (2014). *Journal of Materials Chefogry A* 2: 7156.

102 Tian, Y., Zhu, P., Tang, X. et al. (2017). *Nature Communications* 8: 1080.

103 Venkatesan, H., Chen, J., Liu, H. et al. (2020). *Advanced Functional Materials* 30: 2002437.

104 Bai, H., Ju, J., and Zheng, Y. (2012). *L. Jiang.* 24 (20): 2786.

105 Hou, Y., Chen, Y., Xue, Y. et al. (2012). *Langmuir* 28: 4737.

106 Huang, Z., Chen, Y., Zheng, Y., and Jiang, L. (2011). *Soft Matter* 7: 9468.

107 (a) Wang, S., Feng, S., Hou, Y., and Zheng, Y. (2015). *Macromolecular Rapid Communications* 36: 459. (b) Chen, Y., Wang, L., Xue, Y. et al. (2013). *Scientific Reports* 3: 2927.

108 Chen, W., Li, S., and Guo, Z. (2021). *Materials Letters* 291: 129546.

109 (a) Wang, Y., Qian, B., Lai, C. et al. (2017). *ACS Applied Materials & Interfaces* 9: 24428. (b) Liu, M., Liu, F., Xu, X. et al. (2019). *ACS Applied Materials & Interfaces* 11: 16914. (c) J.-N. Wang, Y.-Q. Liu, Y.-L. Zhang, J. Feng, H. Wang, Y.-H. Yu, H.-B. Sun, 2018, 28, 1800625; (d) Lin, G., Chandrasekaran, P., Lv, C. et al. (2017). *ACS Applied Materials & Interfaces* 9: 26510.

110 Su, Y., Cai, S., Wu, T. et al. (2019). *Advanced Materials Interfaces* 6: 1901465.

111 Zhang, P., Liu, H., Meng, J. et al. (2014). *Advanced Materials Interfaces* 26: 3131.

112 (a) Leslie, D.C., Waterhouse, A., Berthet, J.B. et al. (2014). *Nature Biotechnology* 32: 1134. (b) Zhang, P., Chen, H., Zhang, L. et al. (2016). *Journal of Materials Chefogry A* 4: 12212.

113 Li, D., Huang, J., Han, G., and Guo, Z. (2018). *Journal of Materials Chefogry A* 6: 22741.

114 (a) Hou, Y., Gao, L., Feng, S. et al. (2013). *Chemical Communications* 49: 5253. (b) Feng, S., Hou, Y., Xue, Y. et al. (2013). *Soft Matter* 9: 9294. (c) Rising, A. and Johansson, J. (2015). *Nature Chemical Biology* 11: 309.

115 Yang H., Zhu H., Hendrix M.M.R.M. et al. (2013). *Advanced Materials*, 25: 1150.

116 Shang, L., Fu, F., Cheng, Y. et al. (2017). *Small* 13: 1600286.

117 Thakur, N., Baji, A., and Ranganath, A.S. (2018). *Applied Surface Science* 433: 1018.

118 Tang, J., Peng, L., Chen, D. et al. (2022). *ACS Applied Materials & Interfaces* 14: 2202.

119 (a) Chaudhury, M.K. (1992). *G. M. Whitesides* 256: 1539. (b) S. Daniel, M. K. Chaudhury, J. C. Chen, Science, 2001, 291, 633; (c) Lorenceau, E. and Quéré, D. (2004). *Journal of Fluid Mechanics* 510: 29.

120 Wang, X., Guo, Z., and Liu, W. (2023). *Langmuir* 39: 8926.

121 (a) Xu, L., Shyu, T.C., and Kotov, N.A. (2017). *ACS Nano* 11: 7587. (b) Zhao, R., Lin, S., Yuk, H., and Zhao, X. (2018). *Soft Matter* 14: 2515. (c) Cui, J. and Poblete, F.R. (2018). *Y. Zhu* 28: 1802768. (d) Park, J.J., Won, P., and Ko, S.H. (2019). *International Journal of Precision Engineering and Manufacturing-Green Technology* 6: 147.

122 (a) Higuera, F.J., Medina, A., and Liñán, A. (2008). *Physics of Fluids* 20. (b) Ponomarenko, A., QuÉRÉ, D., and Clanet, C. (2011). *Journal of Fluid Mechanics* 666: 146.

123 Wang, X., Zhang, W., Li, S. et al. (2023). *Chemical Engineering Journal* 472: 145042.

124 Bai, H., Zhao, T., Wang, X. et al. (2020). *Journal of Materials Chefogry A* 8: 13452.

125 Liu, Y., Yang, N., Li, X. et al. (2020). *Small* 16: 1901819.

126 Yu, Z., Zhang, H., Huang, J. et al. (2021). *Journal of Materials Science & Technology* 61: 85.

127 (a) Wang, X., Zeng, J., Yu, X. et al. (2019). *Applied Surface Science* 465: 986. (b) Fan, Y., Fan, J., and Wang, C. (2019). *Journal of Materials Science & Technology* 35: 2163. (c) Lai, Y., Huang, J., Cui, Z. et al. (2016). *Small* 12: 2248.

128 (a) Sun, J., Chen, C., Song, J. et al. (2019). *Journal of Colloid and Interface Science* 535: 100. (b) Zhu, T., Li, S., Huang, J. et al. (2017). *Materials & Design* 134: 342. (c) Wu, Z., Xu, Q., Wang, J., and Ma, J. (2010). *Journal of Materials Science & Technology* 26: 20.

129 Zhu, R., Liu, M., Hou, Y. et al. (2020). *ACS Applied Materials & Interfaces* 12: 50113.

130 (a) Bai, F., Wu, J., Gong, G., and Guo, L. (2015). *Advanced Science* 2: 1500047.

131 Bai, H., Zhang, C., Long, Z. et al. (2018). *Journal of Materials Chefogry A* 6: 20966.

132 Cheng, Y., Zhang, S., Liu, S. et al. (2021). *Journal of Cleaner Production* 315: 127862.

133 Fan, H. and Guo, Z. (2020). *New Journal of Chefogry* 44: 15483.

134 Fan, H. and Guo, Z. (2021). *Journal of Colloid and Interface Science* 591: 418.

135 Raux, P.S., Gravelle, S., and Dumais, J. (2020). *Nature Communications* 11: 396.

136 Elbaum, R., Zaltzman, L., Burgert, I., and Fratzl, P. (2007). *Science* 316: 884.

137 Comanns, P., Withers, P.C., Esser, F.J., and Baumgartner, W. (2016). *Journal of Experimental Biology* 219: 3473.

138 Prakash, M., Quéré, D., and Bush, J.W.M. (2008). *Science* 320: 931.

139 Dai, Z., Chen, G., Ding, S. et al. (2021). Magnetically responsive film decorated with microcilia for robust and controllable manipulation of droplets. *ACS Applied Materials & Interfaces* 13 (1): 1754–1765.

140 Yu, Z., Li, S., Liu, M. et al. (2022). *Journal of Materials Chefogry A* 10: 304.

141 (a) Wang, X., Zeng, J., Li, J. et al. (2021). *Journal of Materials Chefogry A* 9: 1507. (b) Maji, K., Das, A., Dhar, M., and Manna, U. (2020). *Journal of Materials Chefogry A* 8: 25040. (c) Liu, M., Peng, Z., Yao, Y. et al. (2020). *ACS Applied Materials & Interfaces* 12: 12256.

142 Hu, R., Wang, N., Hou, L. et al. (2019). *Journal of Materials Chefogry A* 7: 124.

143 Li, C., Liu, Y., Gao, C. et al. (2019). *ACS Applied Materials & Interfaces* 11: 4507.

144 (a) Dhanalakshmi, A., Palanimurugan, A., and Natarajan, B. (2018). *Materials Science and Engineering: C* 90: 95. (b) Xu, L., Hu, Y.-L., Pelligra, C. et al. (2009). *Chefogry of Materials* 21: 2875.

145 Tang, X., Liu, H., Xiao, L. et al. (2021). *Journal of Materials Chefogry A* 9: 5630.

146 (a) Zhang, W., Gao, J., Deng, Y. et al. (2021). *Advanced Functional Materials* 31: 2101068. (b) Qin, H., Zhang, Y., Jiang, J. et al. (2021). *Advanced Functional Materials* 31: 2106269.

147 Zhu, P., Chen, R., Zhou, C. et al. (2021). *Chemical Engineering Journal* 415: 128944.

148 Liu, L., Liu, S., Schelp, M., and Chen, X. (2021). *ACS Applied Materials & Interfaces* 13: 29122.

149 Fisher, L.R., Gamble, R.A., and Middlehurst, J. (1981). *Nature* 290: 575.

150 Li, D., Fan, Y., Han, G., and Guo, Z. (2021). *Chemical Engineering Journal* 404: 126515.

151 Feng, R., Song, F., Xu, C. et al. (2021). *Chemical Engineering Journal* 422: 130119.

152 Li, H.Y., Zhang, Z.B., Ren, Z.T. et al. (2023). *Chemical Engineering Journal* 455.

153 Zhou, H., Jing, X., Li, S., and Guo, Z. (2021). *Chemical Engineering Journal* 417: 129215.

10

Adsorption-Based Atmospheric Water Harvesting

In the previous chapters, this book has discussed the bionic prototypes in detail, the basic principles of fog collection, the factors affecting the fog collection process, the mechanical properties of fog collection materials, and the application of different surface treatment methods in fog collection. These contents mainly focus on the latest progress in the field of non-phase change fog collection inspired by natural organisms. However, in nature, fog is not a resource that is continuously present. Therefore, atmospheric water collection technology has emerged as an important parallel or complementary technology based on liquid water resources for the production of fresh water. In this chapter, the book will focus on the commonly used substrates, structural design, and surface hydrophilicity adaptation in non-phase change fog collection systems, in order to elaborate on the current development status of phase change water collection technology.

10.1 Introduction

The emergence of industrialized societies has intensified the scarcity of freshwater resources, presenting a significant challenge for humanity globally [1]. Expert forecasts indicate that by 2030, global water consumption will attain 6.9 trillion cubic meters. By 2050, per capita water availability is expected to decline to two-thirds of current levels, leaving approximately 2 billion to 7 billion individuals unable to meet their daily water requirements [2]. Consequently, water scarcity is about to become one of the most critical global crises [3]. In response to this crisis, countries have implemented various measures, such as urging citizens to conserve water, enhancing water distribution systems for inter-basin transfers, minimizing water pollution, and promoting reforestation. Nonetheless, these actions fail to tackle the underlying issues or improve the overall freshwater reserves. Consequently, the development of innovative technologies for sourcing more freshwater has emerged as a prominent area of research [4].

Water on Earth occurs in the atmosphere, seas, rivers, lakes, marshes, soil, glaciers, permafrost, deep below the Earth's crust, and among flora and fauna.

Bioinspired Materials for Fog Harvesting: Principles, Design, Preparation, and Applications, First Edition. Zhiguang Guo and Weimin Liu.

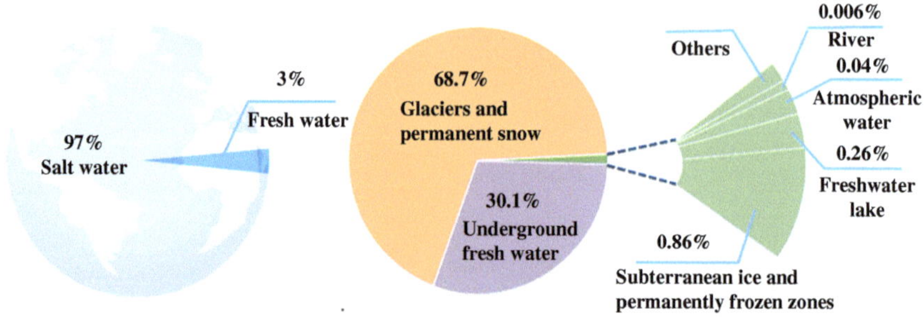

Figure 10.1 Distribution of the world's saltwater resources and freshwater resources.

They metamorphose into one another, collectively forming a hydrosphere that envelops the Earth, including a total volume of 1.4 billion cubic kilometres of water. Of the total, around 97% constitutes saltwater, while just 3% represents freshwater, as seen in Figure 10.1. Within this limited freshwater supply, over 68% is comprised of glacial and permanent snow, making it challenging to use. 30% consists of fresh groundwater. The excessive extraction of this resource has resulted in several ecological and environmental issues, including the persistent decline of groundwater levels, the depletion of aquifers, ground subsidence, and the degradation of water quality. These two provide over 99% of freshwater content. Rivers and freshwater lakes remain the primary source of water for humans. The atmosphere contains around 12,900 cubic kilometres of viable freshwater resources or 0.04% of the total freshwater capacity.

Atmospheric water vapor mostly comes from the seas, with a secondary contribution from evaporation from continental surfaces. Solar radiation incident on the ocean surface is absorbed as thermal energy, resulting in the evaporation of water. Water vapor ascends from the sea surface, is carried by the wind, and is conveyed across the land. Upon cooling to the dew point, water vapor condenses into liquid, resulting in the precipitation of freshwater as rainfall. The atmospheric water vapor cycle consists of the processes of evaporation, condensation, precipitation, and further evaporation, as seen in Figure 10.2. All current manufactured distillation devices are diminutive reproductions of this natural process.

Water exists in three states: solid, liquid, and gas, often referred to as the three phases of water. Figure 10.3a shows the three-phase transformation (liquid, solid and gas) diagram of water. In phase change under certain conditions of temperature and pressure, the state of existence of water can change. Figure 10.3b represents the relationship between the three phases of water with temperature and pressure [5].

Currently, mainstream water generation methods such as desalination by water evaporation, atmospheric water harvesting (AWH) involve three-phase processes of water. Desalination techniques include both phase change and non-phase change procedures, including Multi-stage Flash (MSF), Multiple Effect Boiling (MEB), Vapor Compression (VC), and freezing, which are categorized as phase change processes. Reverse osmosis (RO), electrodialysis (ED), and other membrane processes are non-phase transition methods. The conventional multi-stage flash

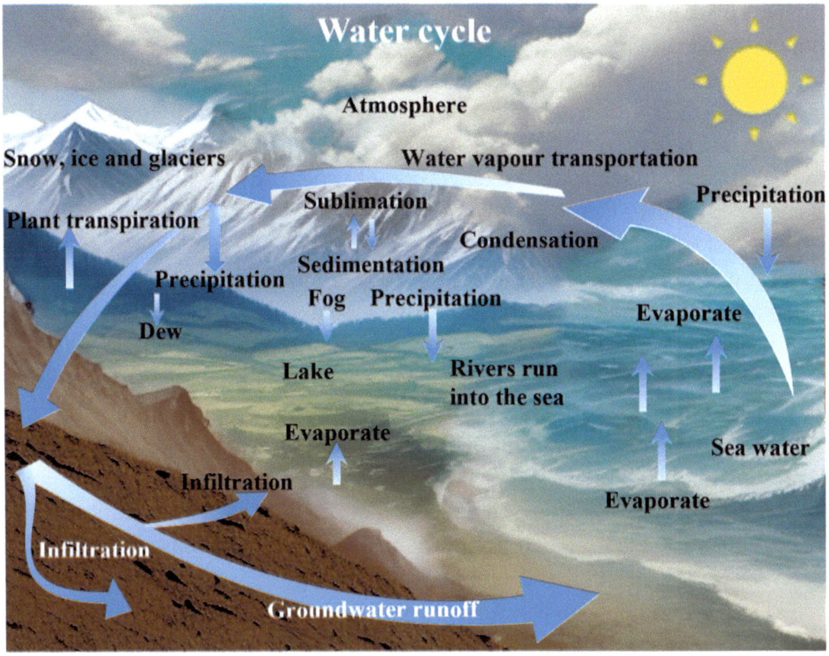

Figure 10.2 The natural water cycle diagram mainly involves the evaporation of ocean water. Then the water vapor is transferred to inland rivers and soil through rain, fog and dew, and eventually returns to the sea.

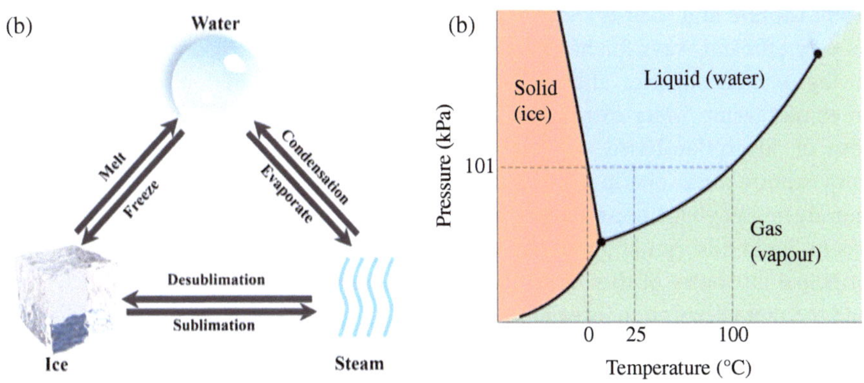

Figure 10.3 (a) Three-phase transformation (liquid, solid and gas) diagram of water. (b) Relationship between the three phases of water with temperature and pressure. Source: Reproduced from [5] / with permission of Springer Nature.

and multiple effect boiling techniques primarily serve to convert saltwater from liquid to gas by pressure decrease and temperature elevation, constituting a phase transition from liquid to gas. Freezing is a separation process associated with the liquid-solid phase transition phenomena. Ice crystals of pure water occur in the salt solution when the brine temperature decreases to the freezing point due to changes in salinity. The ice crystals may be mechanically extracted from the concentrated solution, purified, and re-melted to get pure water.

In membrane processes such as RO and ED, which do not involve a phase change, RO requires electricity or mechanical power to operate a pump that elevates the pressure of the saline solution to the necessary level, the same time, ED also demands electricity to facilitate the ionization of water. Desalination methods are often energy-intensive. Consequently, desalination facilities are often limited in scale because of substantial energy requirements and capital expenditures. Furthermore, desalination may adversely affect the ecosystem. Significantly, water-scarce countries are often landlocked and lack direct access to saltwater desalination, hence limiting its extensive use. Atmospheric water vapor, as a significant but underutilized water resource, has garnered considerable study interest owing to its substantial freshwater content and widespread availability, alongside the application of atmospheric water-harvesting methods to extract this water. AWH approaches primarily include cooling condensation collection, fog water collection, and adsorption-based water capture. Fog collection is often limited to certain regions and necessitates elevated relative humidity, while cooling condensation depends on power for water collection and exhibits a high energy consumption ratio, hence restricting its broader applicability. Adsorption-based atmospheric water harvesting (SAWH) utilizes intermolecular interactions to induce airborne water molecules to aggregate into liquid water clusters on the material's surface, which are then collected. Nevertheless, the adsorbed water is often not immediately useful and often needs an extra re-evaporation and condensation step. Solar energy is a renewable energy source that needs no further energy input. Consequently, adsorption-based water capture and solar-driven water desorption AWH technologies have emerged as very efficient ways for harvesting atmospheric moisture in dry environments, owing to their superior efficiency, minimal energy consumption, and capability to extract water under extremely low humidity circumstances. The fundamental idea of adsorption-based solar-driven atmospheric water-harvesting technology is to capture atmospheric water vapor at night and release the water by evaporation during daylight hours. The whole process experiences a phase change from gas-to-liquid adsorption and liquid-to-gas-to-liquid desorption. Consequently, the principal attributes of this AWH technology are the effective adsorption of water and the rapid desorption by evaporation.

10.2 Evaporation Phenomenon and Materials for Solar Water Evaporation

10.2.1 The Evaporation Phenomenon

The molecular thermal movement explains the process of water evaporation at the microscopic level. At any temperature, certain molecules of water possess sufficient kinetic energy to disrupt the hydrogen bonds between them, overcome the intermolecular forces, and change from the liquid phase to vapor; conversely, water vapor in the atmosphere can contact the liquid surface, establish new hydrogen bonds, and change to liquid water. The two processes occur simultaneously, but typically, there is a lower concentration of water vapor in the atmosphere, resulting in a greater number of water molecules transitioning to the gaseous state rather than remaining liquid, which is macroscopically seen as a decrease in liquid water, known as evaporation.

Consider a sealed container filled with water at a constant temperature, with a vacuum present above the water surface illustrated in Figure 10.4b. Consequently, as a result of the thermal motion of water molecules, some molecules at the liquid's surface continuously detach from the liquid and transition into the atmosphere as water vapor (evaporation), thus exerting pressure in the air owing to the presence of water vapor. This pressure will compel a segment of the water vapor molecules next to the liquid surface to revert to the liquid state (liquefaction), hence reducing the overall rate of water molecule transition from liquid to gas. When the partial pressure of water vapor reaches a sufficient magnitude, the rate of water molecules transitioning from the liquid to the air equals the rate of their return from the air to the liquid. At this juncture, the air becomes saturated with water vapor, referred to as saturated vapor at this temperature, and the associated partial pressure of water vapor is referred to as the saturation vapor pressure. The variation in saturated vapor pressure escalates with temperature shown in Figure 10.4c. Consequently, when the actual vapor pressure is inferior to the saturation vapor pressure, the system exhibits net evaporation; conversely, when the actual vapor pressure exceeds the saturation vapor pressure, the system demonstrates net liquefaction; both are equivalent when the system attains dynamic equilibrium. In Figure 10.4a, an open system in everyday circumstances, the concentration of water vapor in the atmosphere is below 1%, resulting in the real partial pressure of water vapor being less than 1% of the atmospheric pressure. The standard weather forecasts indicate that relative humidity is typically in the tens of per cent range. Relative humidity refers to the partial pressure of water vapor compared to the saturation vapor pressure of water at a given temperature. The partial pressure of water vapor consistently remains

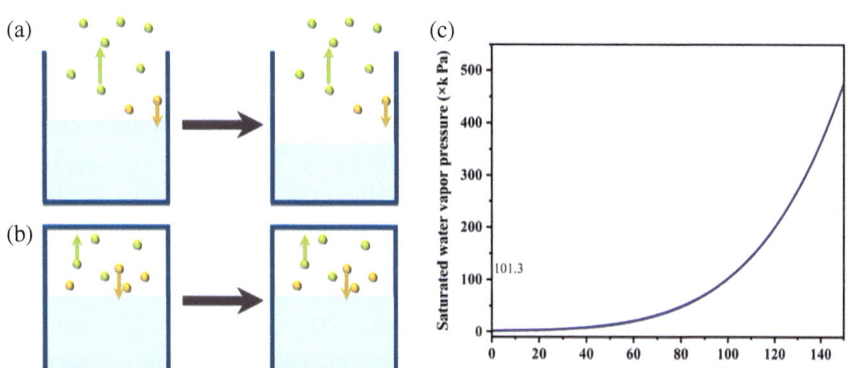

Figure 10.4 (a) Evaporation process in an open system. (b) Evaporation process in a closed system. (c) Variation of the saturated vapor pressure with temperature.

below the saturation vapor pressure, preventing liquefaction due to cold, which is a macroscopic display of evaporation. A comparable process can transpire in solid water, as the water molecules, despite being constrained within the crystal lattice, retain sufficient kinetic energy to transition into water vapor; however, the molecular kinetic energy is significantly diminished, resulting in a considerably slower rate of phase transition. The process of water evaporation is influenced only by two factors: the concentration of water molecules in the air (partial pressure of water vapor) and the temperature of the liquid water (average kinetic energy of water molecules). Temperature modulation is an efficient method for controlling the evaporation properties of water. As temperature increases, water molecules more readily disrupt their organized structure, enabling them to acquire more energy to transition from the liquid surface into the atmosphere as amorphous water vapor.

10.2.2 Solar Absorbent

The essential aspect of solar-powered SAWH desorption is the transformation of solar energy into heat, which may efficiently boost the surface temperature and facilitate water evaporation. Incident light must be absorbed, while the material's reflectance and emissivity should be reduced. Recent advancements in photothermal materials have revealed several interesting candidates for solar light absorption. This includes semiconductor particles, polymers (such as polypyrrole and polyaniline), carbonaceous materials (including graphene, carbon black, and carbon nanotubes), and metallic substances. Significant absorption of the whole solar spectrum, cost-effectiveness, and enduring stability are critical factors in the selection of an appropriate solar absorber. The light absorption capacity and evaporation rates of the reported solar absorbers are shown in Table 10.1.

Semiconductor particles have remarkable light absorption properties and thermal stability, enabling them to effectively capture solar energy for the conversion of sunlight into thermal energy, thereby promoting water vapor condensation [7]. These materials have found extensive applications in atmospheric water harvesting, desalination, and various other fields. For instance, drawing inspiration from the distinct hydrophobic-hydrophilic pattern observed on the dorsal surface of a nano cloth desert beetle, Guo and [8] engineered layered CuO@TiO_2-coated surfaces featuring alternating hydrophobic and hydrophilic chemical motifs for atmospheric water collection through photocatalytic ultraviolet activation of titanium dioxide, showcasing remarkable water-harvesting efficiency. This innovative approach offers novel perspectives on developing hybrid hydrophilic surfaces conducive to efficient water-harvesting processes. Furthermore, Guo et al. [9] introduced a light-absorbing spongy hydrogel embedded with nanoscale solar energy absorbers (LASH), synthesized through the in-situ gelation of poly (vinyl alcohol) (PVA) integrated with uniformly dispersed titanium dioxide (Ti_2O_3) nanoparticles, exhibiting superior photothermal conversion efficiency. The LASH structure achieved an unprecedented water vapor generation rate of 3.6 $kg\,m^{-2}\,h^{-1}$, coupled with the removal of over 99.9% of salt ions from seawater through solar desalination during the evaporation process, showcasing enduring stability. This groundbreaking study

Table 10.1 A summary of the evaporation rates for some reported representative solar absorbers.

Solar absorber	Optical absorption	Evaporation rates (kg m^{-2} h^{-1})
CB nanoparticles	~95	4.32
rGO/MXene	~93.5%	2.09
rGO-supported Fe_3O_4	~90%	1.12
CNTs@SiO_2	~98%	1.50
MWCNT	~97%	4.3
Plant-derived carbon Nanospheres	~92%	1.53
Graphitic carbon spheres	~90%	1.29
rGO-Ag	~91%	2.02
PPy-PDA	~97%	1.77
PDA	~86%	1.36
Graphene	N.A.	0.77
Ti_2O_3 nanoparticles	~90%	3.6
Polyaniline	~95%	1.6
Ag/PPy	~88.7%	1.37
MoCX nanocrystals	~83.6%	1.59

Source: Reproduced from [6] / with permission of Elsevier.

paves the way for advanced solar thermal applications, heralding new possibilities in the realm of sustainable water treatment and energy conversion technologies.

Conjugated polymer materials [13] consist primarily of a π-conjugated backbone of carbon atoms hybridized through SP^2, a structural characteristic that imparts these materials with split energy levels and adjustable bandgaps akin to those found in inorganic semiconductors [14]. These materials demonstrate significant potential for robust light absorption, effectively harnessing solar energy and converting it into heat, thereby significantly enhancing the rate of water evaporation [15]. Due to their exceptional light absorption properties and efficiency in thermal energy conversion, conjugated polymer materials find diverse applications in critical domains such as atmospheric water harvesting and desalination. For example, Maity et al. [10] developed an AWH device featuring an inner layer composed of alginate (Alg) and polyaniline (PANI) networks alongside an outer layer of thermo-responsive poly(N-isopropylacrylamide) (PNIPAM) for water adsorption at low humidity. This AWH system, as illustrated in Figure 10.5a–c is capable of harvesting 6.5 L of water per kilogram in a single daily absorption/desorption cycle under sunlight and functions efficiently at relative humidity levels as low as 17% without requiring any additional external energy input.

Zhou et al. [16] engineered a solar-powered polymeric moisture-absorbent hydrogel, utilizing a preparation technique that integrates solution-exchange and freeze-drying methods, primarily aimed at atmospheric water harvesting.

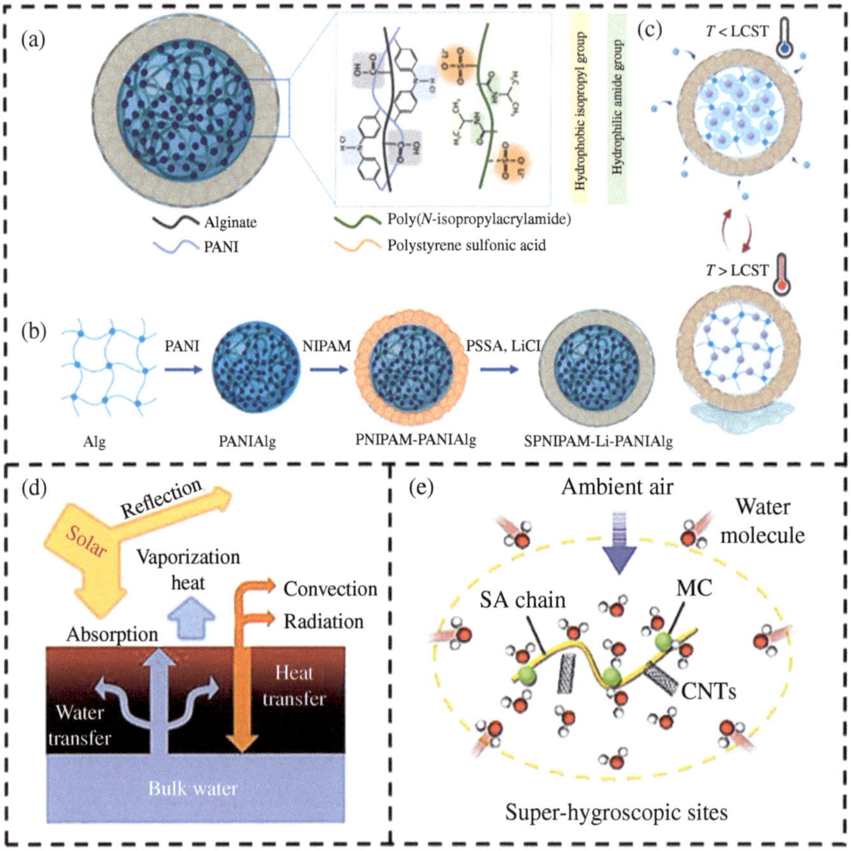

Figure 10.5 (a–c) Conceptual drawings of the design of the AWH system with a schematic cross-section of the spray-lithium titanium hydride hydrogel showing the core-shell structure; schematic diagram of the synthesis process of the spray and schematic diagram of reversible water vapor adsorption below the LCST and desorption above the LCST of the developed AWH hydrogel, respectively. Source: Reproduced from [10] / with permission of John Wiley & Sons. (d) Schematic diagram of energy balance and mass transfer in a solar desalination system. Source: Reproduced from [11] / with permission of John Wiley & Sons. (e) Working mechanism diagram of superabsorbent aerogel. Source: Reproduced from [12] / with permission of John Wiley & Sons.

This hydrogel demonstrates high performance with a high vapor capture capacity (up to $5.86\,g\,g^{-1}$ at 90% RH and 25 °C) and rapid photothermal conversion efficiency (up to 37.4 °C within 10 min under $0.5\,km^{-2}$ light at 25 °C). In real outdoor conditions (0.11–$1.01\,kW\,m^{-2}$), the water-harvesting capacity meets the minimum daily water intake requirement for an adult, reaching $1.125\,L\,kg^{-1}$, thus offering an effective solution to water scarcity. Li et al. [11] fabricated a composite hydrogel with coordinately self-assembled nanofibers, where the self-assembled network facilitates continuous microchannels for efficient mass transfer, thereby preventing salt accumulation. The uniform loading of polypyrrole (PPy) within this network achieves effective photothermal conversion. As depicted in Figure 10.5d, a computational model of energy balance based on the solar evaporation process was established to elucidate the complex interplay between the specific physical processes. Under solar irradiation, the material maintained a stable evaporation rate of 2.85 $kg\,m^{-2}\,h^{-1}$ during continuous desalination of 20% brine, providing insights for the development of advanced solar desalination systems.

Carbon-based materials [17] exhibit broadband absorption and excellent optical properties [18] due to their unique electronic structure and continuous energy levels generated by hybridized bonds [19]. With a low reflectivity (about 0.85) [20], they can effectively absorb sunlight and convert it into thermal energy, resulting in significant solar-thermal energy conversion potential [21], which is promising for applications in atmospheric water harvesting and other fields. For instance, Brassard and colleagues [22] enhanced the synthesis process of nanoporous sponges (NPS) based on carbon adsorbents by integrating pyrolysis and carbon dioxide into a single step. This advancement led to the creation of an improved NPS that consistently achieved a water absorption capacity of 0.47 $g\,g^{-1}$ at 95% relative pressure, demonstrating exceptional durability. The scalability of this material is anticipated to address pressing issues like water scarcity. Anjali et al. [23] introduced a solid adsorbent comprising a graphene oxide/poly(vinyl alcohol) (GO/PVA) gel infused with a desiccant for atmospheric water harvesting. The novel gel exhibited water collection capabilities of 0.289 and 1.2 $g\,g^{-1}$ at relative humidity levels of 30% and 75%, indicating the potential of GO-based gel for practical applications. In a groundbreaking approach, Zhang et al. [12] converted magnesium chloride into a highly hygroscopic magnesium complex (MC) and combined it with sodium alginate and carbon nanotubes (SA/CNTs) to fabricate a composite hydrogel with a photothermal effect. The mechanism, depicted in Figure 10.5e, involves the super-hygroscopic MC capturing water vapor from the atmosphere on the hydrophilic SA chain, storing the absorbed water in an expanded state. At the same time, the carbon nanotubes serve as the photothermal element, enabling the aerogel to release water under solar irradiation. The composite hydrogel achieved impressive water absorption capacities of 5.43 and 0.27 $kg\,kg^{-1}$ at 95% and 20% relative humidity, respectively, across a broad humidity range. This innovative design offers a promising avenue for realizing autonomous, high-efficiency solar-assisted water harvesting. Song et al. [24] developed a nanoporous carbon material derived from metal–organic framework (MOF) carbon, serving as a carbonaceous adsorbent with efficient solar thermal heating and high thermal conductivity, facilitating rapid desorption. The optimized structure of this adsorbent featured 40% adsorption sites and a pore size of −1.0 nm to minimize diffusion resistance, resulting in superior adsorption kinetics. As a result, a fast-cycling collector based on MOF-derived nanoporous carbon achieved a water production rate of 0.18 $L\,kg^{-1}\,h^{-1}$ at 30% relative humidity following a single exposure to sunlight. This strategy is poised to steer the development of high-yield solar-driven AWH systems.

The photothermal effect observed in metals primarily manifests on their surfaces are due to their inherent opacity. Metals such as gold, silver, and platinum possess a high concentration of free electrons capable of absorbing light energy. Plasmonic metal nanoparticles exhibit efficient heat generation under sunlight exposure, attributed to their surface plasmon resonance (SPR) phenomenon, which induces resonance in free electrons upon interacting with incident light [25]. Upon excitation by light, these electrons transition from lower to higher energy levels, resulting in the generation of hot electrons [26]. Subsequently, these hot electrons dissipate their energy into the surroundings through an electron-phonon relaxation process. Consequently, metal nanoparticles featuring localized surface plasmon resonance hold promise for solar thermal energy conversion.

For instance, Liu et al. [27] have effectively developed a novel solar-driven evaporator by incorporating reduced graphene oxide (rGO) and silver (Ag) nanowires as a photothermal conversion agent within a sodium alginate (SA) hydrogel matrix. When subjected to typical solar irradiation, the evaporator exhibited outstanding evaporation capabilities, achieving an impressive evaporation rate of $2.02\,kg\,m^{-2}\,h^{-1}$. This research stands to provide valuable insights for advancing the real-world implementation of solar evaporation technology. Nonetheless, the narrow-band absorption characteristics of plasmonic metals constrain their utility in broad-spectrum light absorption [28]. This limitation can be effectively addressed through the application of template-assisted self-assembly techniques, which have been extensively investigated [5]. For instance, He et al. [29] successfully self-assembled MOF-801 into a durable hierarchical porous monolith utilizing combined sol-gel phase separation and acid-thermal post-processing methods, achieving optimized porosity levels of up to 59.7%, specific surface areas of $977.8\,m^2$, and microporous volumes of $0.234\,cm^3\,g^{-1}$. The monolithic MOF-801 demonstrated exceptional hygroscopic properties at 30% RH, offering a viable approach for large-scale industrial AWH through monolithic MOF molding. Hu et al. [30] introduced a straightforward and effective polyvinylpyrrolidone (PVP) -assisted self-assembly approach for producing core-shell MOF-801@MIL-101(Cr) (M-8010) superparticles. Leveraging the synergistic effects between the shell MOF-801preconcentration function and the core MIL-101 high water storage capacity, the resulting superparticles exhibited 225% and 390% higher water absorption capacities compared to pure MOF-801 and MIL-101, respectively, at 8% relative humidity. This study presents innovative concepts for designing MOF ultra-particles optimized for high-performance water collection in arid regions. Moreover, Luo et al. [31] functionalized glass fiber paper with duplexed metal–organic frameworks continuous glass fiber (CGF)-mixed-MOFs (Al) and tailored a series of duplexed sinks comprising MIL-160(Al) and aluminum fumarate on the CGF surface using an in-situ synthetic self-assembly technique. The water uptake of CGF-mixed-MOFs $(Al)_5$ at 30% RH, 25% RH, and 20% RH reached 0.37, 0.31, and $0.25\,g\,g^{-1}$, respectively, demonstrating its potential to capture increased freshwater from the atmosphere under arid environmental conditions.

10.3 Progress in Sorbents for SAWH

The adsorption of water is classified as physical adsorption or chemisorption, depending on the nature of the interaction between the adsorbent and the water molecules. Chemisorption is accompanied by the formation and breaking of chemical bonds, which require considerable activation energy and can, therefore, only be driven by low-level heat. In this regard, only a few chemisorbents have been selected for the SAWH system, such as LiCl and $CaCl_2$. Physical adsorption is caused by van der Waals forces present between the adsorbent and adsorbate molecules without any breaking or formation of chemical bonds. Physical adsorbents have the advantages of low heat of adsorption, low activation energy, high adsorption/desorption rate, and good reversibility. Therefore, they are widely used in dehumidification and purification processes. Physical adsorbents support multilayer adsorption, and their adsorption capacity is closely related to the specific surface area and surface properties. To increase the specific surface area, physical adsorbents are always prepared with special morphology and structure, such as one-dimensional vesicular structure, two-dimensional planar porous structure, and three-dimensional porous structure. Common physical adsorbents include traditional silica gel and zeolite, new metal–organic frameworks (MOFs), polymer gels, and other porous materials. Moreover, due to the high water absorption of lithium chloride and calcium chloride, they are usually also prepared together with physical adsorbents to form composite adsorbents. They are synthesized by various methods, including sol-gel, solvothermal synthesis, microemulsion, spray drying, and molecular self-assembly.

Adsorption-based atmospheric environmental water-harvesting materials are capable of harvesting water at lower ambient humidity by adsorption and/or absorption. The release process consists of energy input, e.g. heating, and the harvested moisture can be efficiently collected, benefiting from adequate moisture accumulation. The adsorption requires a strong force between the material and the moisture. The stronger the force, the greater the adsorption, but at the same time, the higher the energy consumption required for desorption [32]. Therefore, the selection of suitable adsorbent materials is crucial and commonly used adsorbents are classified into the following categories (Figure 10.6).

10.3.1 Conventional Absorbent Materials

(i) **Silica gel.** A conventional adsorbent in traditional materials is silica gel, a partly dehydrated polymeric silicate characterized by silanol and siloxane groups on its surface. The adsorption capacity test system is illustrated in Figure 10.7. It is extensively used as a desiccant owing to its high adsorption rate and low regeneration temperature of 150 °C [2, 34]. Silica gel with a higher density of silica-oxygen hydrogen bonding has better moisture adsorption when silanol groups are used as the main adsorption sites. The results show that the first layer of water molecules is adsorbed on the silica gel surface due to hydrogen bonding between water molecules and silanol groups; when multiple layers are adsorbed, hydrogen bonding clusters are formed on the surface. In other words, water molecules in the gas phase and those immobilized on the silica gel surface are also hydrogen-bonded. The heat of adsorption of silica gel is about 10.5 kcal/mol, which is equivalent to the heat of condensation of water vapor [35]. According to

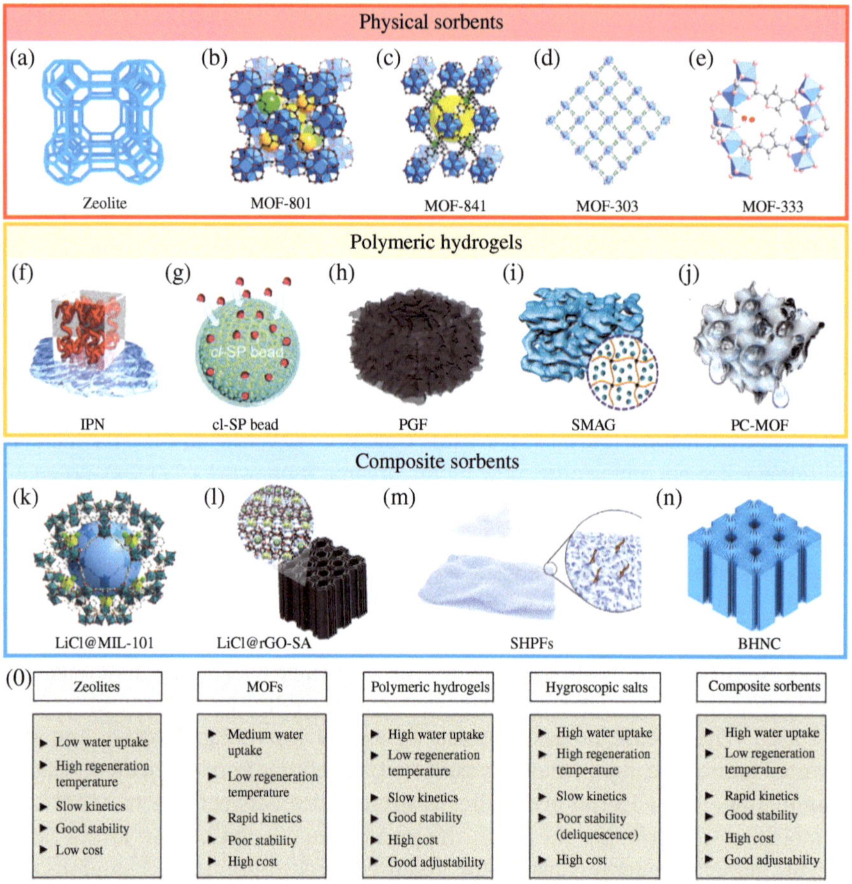

Figure 10.6 Progress in adsorbent materials, including (a–e) physical adsorbents, (f–j) polymeric hydrogels, and (k–n) composite adsorbents and (o) their properties. Source: Reproduced from [33] / with permission of Elsevier.

the adsorption properties of silica gel, its water adsorption capacity depends on the density of α-(OH). For conventional amorphous silica gel, the density of silanol groups falls within a narrow range of 4.3–5.8 OH nm^{-2}, and when the temperature exceeds 200 °C, the silanol groups will fall off the silicon surface; therefore, care should be taken to control the drying temperature. In addition, as the pore size increases, the moisture absorption gradually increases straight to the upper limit ($0.4\,g\,g^{-1}$). Silica gel itself cannot play an important role in water vapor adsorption due to the limitation of some inherent defects. Silica gel has a very low adsorption capacity of up to 40% and poor thermal stability compared to chemisorbents. In addition, microporous silica gel has a low adsorption rate at low ambient humidity. It is difficult to desorb when microporous silica gel has a high desorption rate and limited water adsorption [36], and constantly repeated adsorption and desorption will lead to a significant decrease in the adsorption capacity of silica gel for water [37]. The higher the

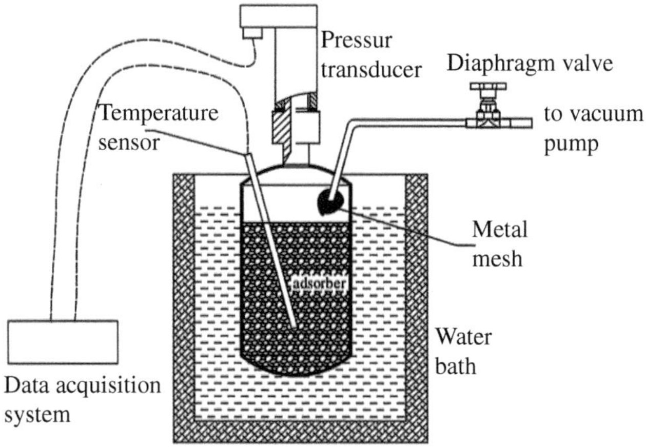

Figure 10.7 Illustrate the Silica gel adsorption capacity test system. Source: Reproduced from [37] / with permission of Elsevier.

switching frequency between adsorption and desorption, and the higher the cycling frequency, the greater the decrease in adsorption capacity. The decrease in adsorption capacity slows down as the cycling frequency increases. Overall, the adsorption capacity of silica gel is affected by many factors, such as the number and type of silanol groups, specific surface area, and microporous particle size distribution. The adsorption degradation of silica gel is the result of a combination of these factors. That is, if an adsorption characteristic parameter deteriorates, it does not mean that the adsorption capacity decreases. For example, the adsorption capacity does not always decrease with a decrease in pore volume or specific surface area because the number of free silanol groups may increase. Similarly, the adsorption capacity of silica gel decreases when the micropores are clogged with solid particles, even though the free silanol groups do not change.

(ii) **Zeolite.** Zeolite is a natural or manufactured aluminum-bearing silicate mineral with a homogeneous and adjustable pore structure that adsorbs and stores water molecules or other polar molecules. Zeolites have many polyhedral units, including cubes, hexagonal, square prisms, and octahedra. These structural units consist of silicon tetrahedra and aluminum oxide tetrahedra. Each of the two silicon tetrahedra is connected by a dot, meaning that they share a bridging oxygen atom. Alumina tetrahedra, however, cannot be connected by a bridging oxygen. For silicon tetrahedra, silicon atoms can be replaced by aluminum atoms to form aluminum oxide tetrahedra. The aluminum atom provides only three electrons, resulting in an alumina tetrahedron with no negative charge. Alkali metals and alkaline earths can neutralize the electrons in alumina tetrahedra. By substituting aluminum/silicon atoms, several types of molecular sieves can be obtained.

(iii) **Commercial zeolite molecular sieves can be classified into three types.** Type A, type X, and type Y. When zeolite comes into contact with moist air, water molecules are adsorbed onto the pore surface or inside the pores of the zeolite due to intermolecular interactions such as van der Waals forces, hydrogen bonding, and electrostatic forces. The diameter of the zeolite pores is usually comparable to the size of the water molecules, and this size match allows the water molecules to easily enter the pores and be adsorbed. The moisture adsorption capacity of a zeolite depends on the size and shape of its pores, the number of pore channels, and the nature of the functional groups in the pore walls. When the relative humidity is below 30%, the adsorption capacity and rate of adsorption of zeolite are significantly higher than those of silica gel and activated alumina. Zeolite is also the only available high-temperature adsorbent, with an adsorption capacity of 15% even at 100 °C and a relative humidity of 1.3%, which is more than 20 times higher than that of silica gel. The adsorption curves of zeolite molecular sieves were studied by thermogravimetric analysis and mass spectrometry. Its equilibrium adsorption capacity was greatly affected by the ambient temperature and humidity. The physical adsorption degree of zeolite molecular sieves decreased sharply when the temperature increased.

In contrast, the saturated vapor pressure increased rapidly with the increase of ambient temperature, and thus, the water absorption rate increased significantly. In general, the equilibrium adsorption capacity increases and then decreases as the ambient temperature increases. Unfortunately, the regeneration temperature of zeolite molecular sieves reaches a very high temperature of about 300 °C. Low thermal conductivity is another obvious drawback of zeolite molecular sieves. The equivalent thermal conductivity of the zeolite particle-filled layer is only 0.07–0.16 W/m. Improvements are currently being made by adding high thermal conductivity materials and compressed zeolite molecular sieves. The addition of power-expanded graphite can greatly increase the thermal conductivity, but the synthesis technique is very complicated [38].

(iv) **Hygroscopic Salt**. Hygroscopic salt is a low-cost but highly effective desiccant, and the mechanism of hygroscopicity of hygroscopic salt is mainly based on its ability to interact with water molecules by its chemical properties. Hygroscopic salts usually contain ions capable of attracting water molecules, such as sodium, potassium, calcium, magnesium, and other cations, as well as chlorine, bromine, iodine, and other anions. Thus, common hygroscopic salts such as calcium chloride ($CaCl_2$), lithium chloride (LiCl), and magnesium chloride ($MgCl_2$) are capable of absorbing 1–3 times their weight in water molecules at moderate humidity. Although hygroscopic salts have a high water absorption rate, a fatal drawback is that they eventually dissolve in the collected water, converting from a solid to a salt solution liquid. This results in reduced absorption kinetics and leakage of the salt solution may corrode the system and make the recovery of the adsorption difficult [39]. To solve this problem, the most popular strategy is to incorporate hygroscopic salt into a designed porous matrix where the loaded salt plays a major role in trapping water molecules, and the porous matrix acts as a container for the final liquefied salt. The water absorption properties of salt matrix composites depend on the choice of salt, the loading content, and the design of the matrix.

Hygroscopic salts such as lithium chloride, magnesium chloride, and calcium chloride can access water through the hydration reaction. Still, there are problems with particle surface agglomeration and passivation layer formation during the hydration process, which reduces the permeability of water vapor and decreases the adsorption capacity. In addition, while higher salt loadings tend to result in higher water uptake, excessively high salt levels can lead to dramatic salt leakage during water collection, especially under conditions of high ambient humidity.

10.3.2 Novel Adsorbent Materials

10.3.2.1 Metal–Organic Frame Materials

MOFs are newcomers in polymer adsorbents. It is an organic–inorganic hybrid material with intramolecular pores formed by the self-assembly of organic ligands and metal ions or clusters through coordination bonds. The reticulated porous structure in MOFs facilitates the improvement of the heat transfer efficiency, which is well suited for solving the water scarcity crisis [40]. Moreover, their exceptional porosity and the diversity of their chemical compositions allow tuning their water-absorbing properties in a variety of ways [41]. Initially, MOFs were not used for atmospheric water harvesting due to their poor hydrolytic stability. As research has progressed, this issue has been addressed, and many MOFs have been shown to have high hydrostability. MOF structures can be designed to have the right balance between hydrophobic and hydrophilic pore environments to allow for the capture of water from desert air and the release of water with minimal energy requirements [42]. Specifically, the feasibility of adsorption is determined by the interaction of the ligand with the vapor. At the same time, the coordination of the metal center with the water molecules influences the stability of adsorption. Meanwhile, the adsorption capacity is mainly influenced by the internal surface area at low humidity and the pore volume at high humidity [43]. In the process of atmospheric water adsorption by MOF, the stability of the MOF structure allows the material to keep its performance stable in hundreds of experiments. Through structural design and adjustment, some MOFs with high efficiency of atmospheric water collection were prepared, and some representative MOFs are introduced one by one in Figure 10.8.

UIO-66 is a metal–organic framework material with a unique pore structure and excellent chemical stability. UiO-66 is a 200 nm^{-1} white powder prepared by using zirconium as the central atom and terephthalic acid for coordination, with a Brunauer-Emmett-Teller (BET) specific surface area of about 1000 $m^2\ g^{-1}$ [45]. UiO-66 is one of the most time-stable MOFs, which can be stored for a long period. UiO-66 is one of the most time-stable MOFs, which can be stored stably for a long period and is also very stable in aqueous and acidic conditions. In addition, UiO-66 has high thermal stability (thermal decomposition temperature > 400 °C).

The majority of UiO-66 needs to be prepared by solvothermal methods, with most processes requiring synthesis at temperatures in the range of 80–120 °C, typically lasting 24 h or more. While these methods are useful and highly reproducible

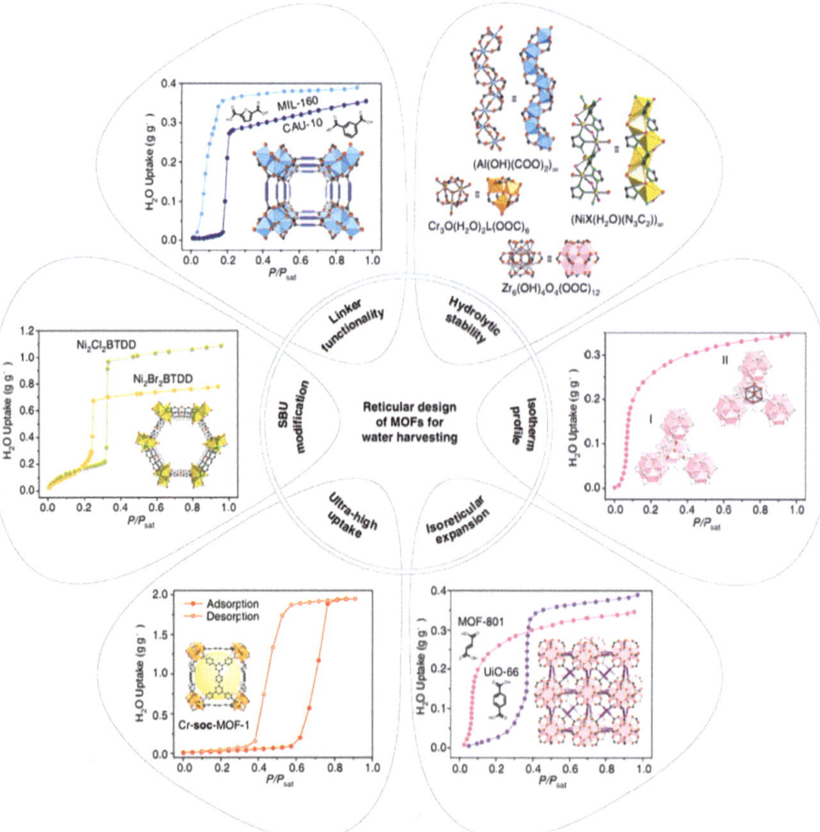

Figure 10.8 Various MOFs adsorbents for atmospheric water harvesting. Source: Reproduced from [44] / with permission of Springer Nature.

for routine UiO-type synthesis, they are not ideal for incorporating thermosensitive materials into the framework - for example, the encapsulation of metal nanoparticles, which lose much of their catalytic activity upon thermally induced aggregation. Although some attempts to synthesize UiO-66 at lower temperatures via solution synthesis are currently reported, room temperature synthesis has resulted in materials that have not yet been achieved with the same crystallinity and porosity as those synthesized at higher temperatures [46]. UiO-66 is an example of a material that contains structural defects, such as the lack of organic linkers. However, the defective framework may exhibit more desirable behavior. It was shown that the number of defective sites can be easily adjusted by varying the p*K*a of the modulator and the amount of modulator present. Temperature may also be a suitable tool to control the number of defects, with the number of missing linker defects decreasing as the synthesis temperature was increased from 100 to 220 °C. It is hypothesized that due to the increased instability of the Zr(IV)—O bond, higher temperatures could correct bonding errors in the periodic structure, allowing the formation of "more perfect" crystals. Still, their thermal stability decreases as the number of defects increases [47]. Since the defective regions have larger pore sizes and more connection sites introducing the number of defects significantly affects the water adsorption properties, leading to a sharp increase in water adsorption, many researchers have focused on the modulated synthesis of UiO-66 to control the defects. It has been reported to demonstrate that the combined use of trifluoroacetic acid and hydrochloric acid during synthesis leads to more defects in UiO-66, and the subsequent activation gives the material a more open framework with higher catalytic activity. It has also been experimentally demonstrated that hierarchical porous UiO-66 was happily fabricated by linker pyrolysis, using long-chain monocarboxylic acids as modifiers to give defect-containing MOFs [48]. Although much progress has been made, it is still very challenging to develop simple and efficient methods for the preparation of layered multivacancy UiO-66.

Conventional UIO-66 does not have high water absorption and is more often studied for gas storage and separation. To enhance the hygroscopic properties of UiO-66, researchers have modified UiO-66 by adding hydrophilic side groups to the framework to make it more suitable for water adsorption in dry environments, of which NH_2-UiO-66 is one of the best in terms of hygroscopic properties. It is a light-yellow powder with a particle size of 100–500 nm. The modification does not alter the stability properties of the material. Still, an obvious limitation of this approach is the reduction of the free pore volume, and the BET specific surface area of NH2-UiO-66 decreases to approximately 650–800 $m^2\ g^{-1}$ [49]. Modified MOFs are synthesized by a completely green method that avoids the use of organic solvents. The researchers employed a mild synthesis method using an aqueous solution of acetic acid as a modifier and 2-aminoterephthalic acid as a linker. When the reactants were mixed in an aqueous solution at room temperature, a yellow MOF precipitate was formed immediately and became crystalline within 24 h. The nanocrystalline powders obtained by optimizing the reaction conditions based on the pretreatment of the cationic precursor, the moderator-diluent ratio, and the reaction length have a high specific surface area comparable to the products obtained by conventional solvent thermal synthesis [50].

The first MOF to be applied in a natural sunlight-driven SAWH was MOF-801 (Zr) in Figure 10.9, whose viability under real desert conditions was confirmed in 2018. MOF-801, which, like UiO-66, uses zirconium atoms as the central atoms but chooses fumaric acid as the metal–ligand, is a white powder with a particle size of 200 nm to 1 μm. MOF-801 is stable in aqueous solution and acidic conditions, can be stable in air for several weeks, and has a thermal decomposition temperature greater than 400 °C [49b]. In the structure of MOF-801, there are two different types of micropores of tetrahedral and octahedral shapes. When a small amount of water is injected into the crystal, the formation of tetrahedrally and cubically arranged aqueous aggregates in these pores can be observed, respectively, each bonded together by hydrogen bonds [51]. Not only that, but MOF-801 is one of the best hygroscopic metal–organic frameworks available, with a moderate water uptake of 0.31 $g\,g^{-1}$ at 30% RH. Although the desorption process of MOF-801 can be directly driven by solar heating, the enthalpy of adsorption of MOF-801 is as high as 60 $kJ\,mol^{-1}$. The researchers synthesized MOF-801 powder based on experimental

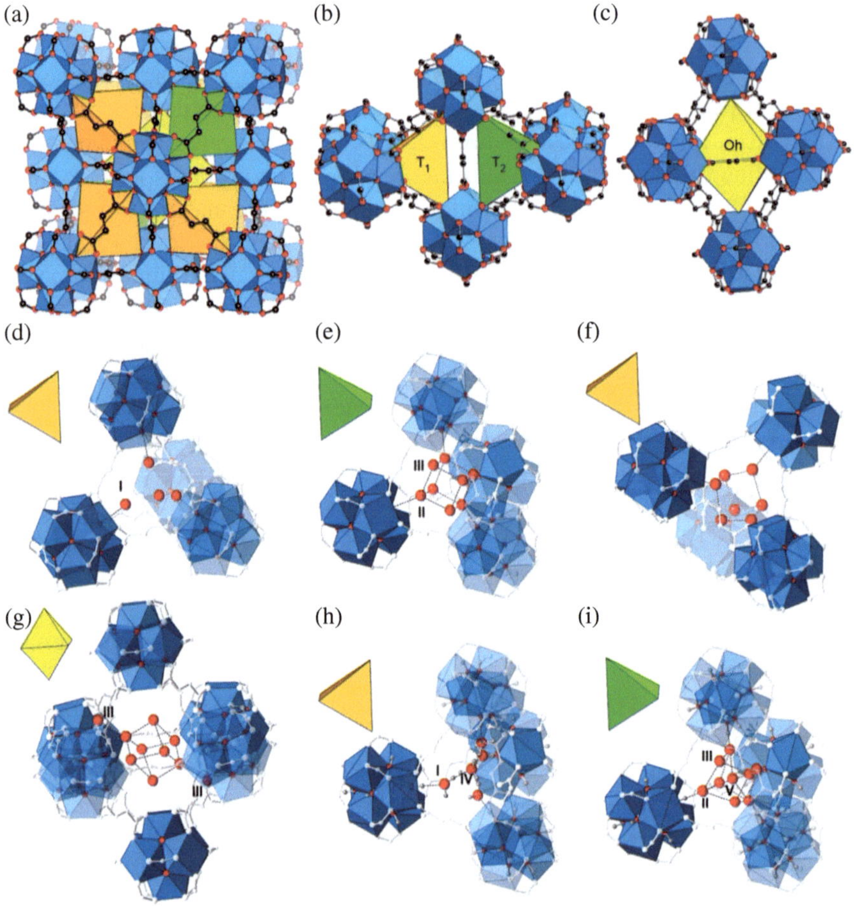

Figure 10.9 The structure of MOF-801 consists of three symmetry-independent cavities (a), two of tetrahedral shape, T1 and T2 (b), and the other of octahedral shape (c); single-crystal diffraction studies on moisture-exposed crystals (d–e) show that water molecules are initially adsorbed in the tetrahedral cavities through the formation of hydrogen bonding with the SBUs (sites I and II) and among them (site III). The same arrangement was found in crystals exposed to 100% RH, as well as incomplete cubic clusters (f); the octahedral cavities were also filled, and several adsorption sites were identified, forming hydrogen bonds with site III (g). In the microcrystalline MOF-801-P, neutron diffraction studies (h–i) revealed similar adsorption sites I, II, and III, as well as other sites IV and V located in the center of the tetrahedral cavity. Source: Reproduced from [49b] / with permission of American Chemical Society.

reports and then activated it (removal of solvent from the pores) by heating it under a vacuum at 150 °C for 24 h. The powder was then heated to a thickness of 0.5 mm, and the solvent was removed from the pores. The powder penetrated a porous copper foam with a porosity of about 0.95 at a thickness of 0.41 cm. Activation of MOF-801 filled about 85% of the copper foam pores, and the copper foam substrate enhanced structural rigidity and heat transfer. The device captures water from the atmosphere under ambient conditions using low heat from natural sunlight below one sun. It is also capable of capturing 2.8 L of water per kilogram of MOF per day at relative humidity levels as low as 20% relative humidity and requires no additional energy input [51a].

In 2019, a cheaper aluminum-based MOF-303 was developed and showed good water absorption at low humidity. MOF-303 [Al(OH)(PZDC)] 1-H-pyrazole-3,5-dicarboxylate (PZDC) is a 500 nm to 1 μm lamellar particle obtained by using aluminum as the central atom and 3,5-pyrazolylenedicarboxylic acid (H_2PDC) as the ligand. The specific surface area of BET can be up to 900–1100 m^2/g, with good water stability and high thermal stability (thermal decomposition temperature > 400 °C) [52]. Not only that, the MOF-303 atmospheric water-harvesting system developed to be able to perform continuous and multiple water-harvesting cycles per day, even at low ambient humidity can still adsorb atmospheric water molecules and release water under mild heating. Experimental tests have shown that it has a daily ambient water collection mass of 1.3 L/Kg at 32% RH and 27 °C [42b]. In addition, the highly open structure of MOF-303 makes it more suitable for both the adsorption and desorption phases. The water adsorption of MOF-303 shows type IV isotherm with a low inflexion point at $\alpha = 0.13$, a plateau at $P/P_0 = 0.3$, minimal hysteresis, and excellent durability to withstand 150 repeated adsorption cycles [42b].

Considering the great potential of MOF-303 as a water-harvesting material and various other water-related applications, researchers have tried to develop a series of green and reliable synthesis methods using different equipment to prepare this useful MOF [53]. Specifically, four synthesis methods involving different equipment requirements (solvothermal, reflux, vessel, and microwave) were proposed, and all of the synthesis schemes are water-based, making them safe and friendly to both researchers and the environment (Figure 10.10). Typically, the solvothermal method requires about 24 h to synthesize MOF-303, while the reflux and vessel methods can reduce the time to 4–8 h [52]. Using the microwave-assisted method, the reaction time can be further reduced to only 5 min. The yield of MOF-303 can be increased from 38% to 94% by slight modification of the reported aqueous reflux synthesis process for aluminum–molybdenum. With an optimized 3 : 1 M ratio between sodium hydroxide and H2PZDC connector, the synthesized MOF-303 has a high degree of crystallinity. In addition, this approach allows for a substantial increase in reaction yield for large-scale production of MOF-303, which can be scaled up for synthesis to a 91% yield of 3.5 kg per batch at a uniform particle size, and the mass-synthesized MOF-303 has a highly similar crystallinity and water-absorption capacity to the corresponding material produced in the laboratory [43a, 54]. The obtained MOFs have greater tunability in the operating relative humidity range (16%), regeneration temperature (14 °C), and desorption enthalpy (5 kJ mol^{-1}) than previously reported. In addition, the synthesis was highly productive (≥90%) and scalable (~3.5 kg) in water with excellent temporal and spatial yields without compromising the crystallinity, porosity, and water-harvesting properties of the skeleton. This study demonstrates a poly-MOF system based on readily available starting materials. The results show that the framework has a greater tunable range in terms of atmospheric hygroscopic humidity range, regeneration temperature, and adsorption enthalpy. Overall, these advances led to a

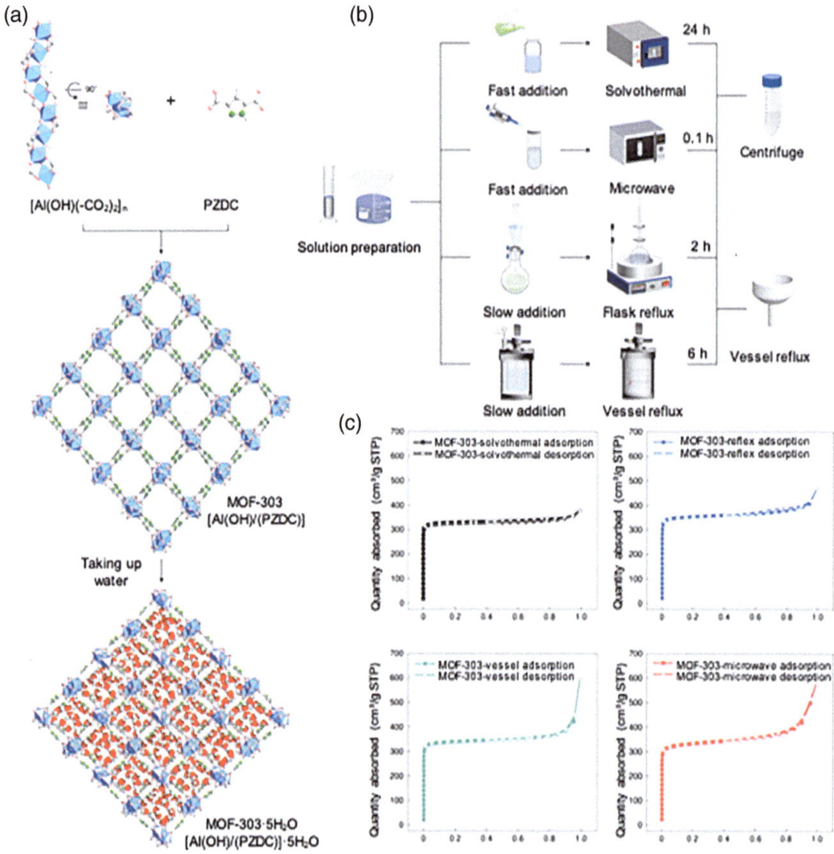

Figure 10.10 (a) Crystal structure of MOF-303 constructed from infinite $[Al(OH)(-CO_2)_2]_n$ rod SBUs stitched together by 1-H-pyrazole-3,5-dicarboxylate (PZDC) linking units. (b) Various methods for the synthesis of MOF-303, along with the reaction time of each. (c) Water vapor sorption isotherms of MOF-303 samples measured at 25 °C. Source: Reproduced from [52] / with permission of Springer Nature.

more energy-efficient and versatile water-harvesting system under arid conditions. Importantly, a procedure was developed for the synthesis of these MOFs that allows them to be easily produced on a kilogram scale using water as a solvent, with high temporal and spatial yields and without significant impact on their water-harvesting properties.

MILs materials are another more specialized type of MOFs materials, which can be made using trivalent transition metal ions (e.g. Fe, Al, and Cr) coordinated with carboxylic acid-based ligands (terephthalic acid, homotrimellitic acid). The structure of this type of MOFs can be transformed between macroporous and narrowly porous forms in response to external factors. This action is called "respiration". When water molecules are adsorbed, the water and the skeleton are connected by hydrogen bonding and a strong electrostatic force is generated, which increases the total amount of water adsorbed. Two representative MOFs of the MIL class are MIL-100 (Fe) and MIL-101 (Cr). MIL-100(Fe) has two different mesoporous cages with free diameters of 2.5 and 2.9 nm, accessed through microporous hexagonal and pentagonal windows with diameters of 0.86 and 0.56 nm, respectively [55]. Therefore, MIL-100(Fe) MOFs can be used as effective adsorbent materials due to their unique structure. The BET specific surface area of MIL-100(Fe) is much larger, up to 1900 $m^2\ g^{-1}$. It is stable in air, aqueous solution, and acidic conditions, even if the crystalline structure of MOF particles remains unchanged after 3 years of age in the air. Under this condition, it is surface area significantly Its surface area decreased significantly (46%), and its thermal decomposition temperature was greater than 280 °C [56]. Therefore, the high hydrothermal stability and excellent adsorption properties of MIL-100 make MIL-100(Fe) a candidate for many potential applications. It should be noted that MIL-100(Fe) with zeolite-MTN topology is an advanced functional material because its central metal ion is iron, which is inexpensive, biocompatible, and environmentally friendly compared to other metals (e.g. Co, Cr, etc.). This has led to the popularity of MIL-100(Fe) as one of the most promising MOFs for practical applications [57].

However, the conventional method of synthesis of MIL-100(Fe) is cumbersome, and generally, hydrothermal synthesis of MIL-100(Fe) is carried out at temperatures above 150 °C [57, 58]. In this method, iron precursors (iron salts or iron powders), homotrimer acid, and auxiliary reagents are mixed and transferred to an autoclave in the presence of vapor in the autoclave where a self-assembly reaction occurs between the iron trimer and the 1,3,5-benzenetricarboxylic acid (BTC), where hydrofluoric acid (a corrosive toxicant) is used as a mineralizing agent to increase the degree of crystallinity of MIL-100(Fe) [59]. However, the method still has some drawbacks, such as long synthesis time, stringent high temperature, high pressure and toxicity of auxiliary reagents, the difficulty of precisely controlling the conditions of the hydrothermal synthesis method, and the risk of the synthesis process limiting its large-scale application. Therefore, it is important to develop a method for the synthesis of MIL-100(Fe) without the addition of hydrofluoric acid under mild conditions. Based on the fact that microwave irradiation can achieve molecular-level heating and cause a rapid and uniform thermal reaction, the microwave-assisted hydrothermal method is an alternative method to shorten the synthesis time [56]. The first microwave-assisted method was used to synthesize MIL-100(Fe) in 2012, which shortened the reaction interval from several days to 6 min. In addition, like microwave-assisted, ultrasound-assisted synthesis methods generate high energies that provide relatively unusual reaction conditions, such as short time intervals and very high temperatures and pressures in aqueous solutions. Thus, ultrasonic irradiation is considered a potential method to aid in the synthesis of porous materials. To achieve the synthesis of highly crystalline MOFs materials using the least amount of energy, the room temperature synthesis method becomes an ideal synthetic route. It has been shown that the synthesis of MOFs with trivalent metal ions as ligand centers can be achieved at room temperature using reduced metals as precursors by the oxidative radical-promoted facile synthesis method

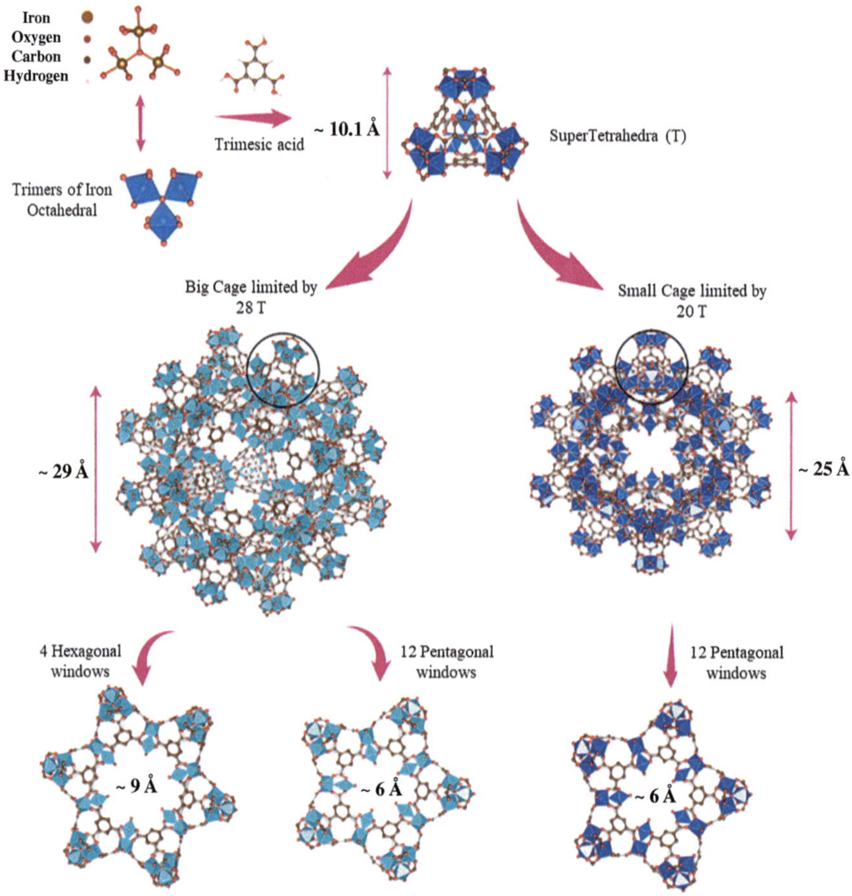

Figure 10.11 Schematic structure of MIL-100(Fe)MOF. Source: Reproduced from [60b] / with permission of Elsevier.

employing p-benzoquinone for the generation of HOO-radicals, which accelerates the crystallization of MIL-100(Fe) under ambient conditions. The samples obtained by this method have structural properties similar to those of the conventional hydrothermal method, and the yields can reach 88%. Structural properties and the yield can reach 88%. Moreover, this method of synthesizing MIL-100(Fe) at room temperature is simple and environmentally friendly, which greatly reduces the harmful effects of conventional synthesis techniques (Figure 10.11) [60].

MIL-101(Cr) is a light green powder with a diameter of 500–800 nm, whose structural units are metal clusters coordinated with the organic ligand terephthalic acid, forming a three-dimensional spatial network structure. Metal–organic frameworks consisting of supertetrahedral elements with large specific surface area and pore volume, high-temperature resistance and air/water stability, stable in air for several months, resistant to strong acids and weak bases in solution, and with a slightly higher thermal decomposition temperature (>300 °C) [61].

MIL-101(Cr) has attracted much attention due to its large surface area, large pore volume, and hydrophilicity. Different synthesis and activation conditions have significant effects on the morphology, specific surface area, yield, structural stability, and crystallinity of MIL-101(Cr) materials [62]. Therefore, the synthesis method is an important factor affecting the properties of MIL-101(Cr). The main synthesis methods currently available are hydrothermal synthesis, solvothermal, microwave-assisted, and template methods [63]. Porous nanomaterials with high crystallinity and excellent properties can usually be obtained using hydrothermal synthesis, which is the most traditional method in the synthesis of MIL-101(Cr). It consists of transferring a mixed solution of $Cr(NO_3)_3$-$9H_2O$, terephthalic acid (H_2BDC), deionized water, and hydrofluoric acid into a stainless-steel reaction vessel lined with polytetrafluoroethylene and heating the reaction at 220 °C for 8 h. The final product was then purified sequentially with ammonium fluoride and ethanol and dried in a vacuum oven. The extensive use of hydrofluoric acid for the synthesis of MIL-101(Cr) requires additional protective equipment and safety precautions due to its high toxicity and volatility, which undoubtedly increases the cost of the synthesis [64]. Many scientists used other acidic additives (acetic acid, hydrochloric acid, etc.) instead of hydrofluoric acid in the reaction and successfully synthesized MIL-101(Cr) with excellent properties, high specific surface area, and porosity. The solvothermal method consists of reacting metal salts, organic ligands, and solvents (non-aqueous or organic solutions) in a certain ratio to produce MOF crystals. This method is usually carried out at higher temperatures or under vapor conditions so that the reaction occurs through intermolecular contact, and the presence of high temperature and pressure allows for higher solubility of the metal salt in the solvent and faster reaction rates. The microwave-assisted synthesis method involves the synthesis of MIL-101(Cr) by rapid microwave heating under microwave conditions in a hydrothermal environment [65]. This method improves the synthesis efficiency and reduces the synthesis time of MIL-101(Cr), which is mainly due to microwave heating of the solvent, which increases the rate of nucleation. This method improves the reproducibility of the preparation [66]. In general, the template method allows the synthesis of hierarchical porous MIL-101(Cr), which improves the properties of MIL-101(Cr) materials and extends their applications. In addition, using templates to occupy space in the crystalline framework of MIL-101(Cr), removing the templates produces additional mesoporous or microporous structures, which helps to enhance its adsorption properties. N_2 adsorption–desorption isotherm of MIL-101(Cr) is a type IV isotherm, with a two-step occurrence of the relative pressure $P/P_0 = 0.2$, which is caused by two different sizes of mesoporous in MIL-101(Cr) channels caused by the two different sizes of mesopores in MIL-101(Cr) [67]. Compared with microporous MOFs, mesoporous MOFs usually absorb large amounts of moisture but are difficult to apply under arid conditions because the step point of the adsorption curve is backward. Based on the combination of air temperature and humidity regulation and adsorption properties, the researchers proposed to construct a cold source on the adsorption side to change the humidity environment in and around the pore channel so that the material can achieve moisture trapping even under arid and low-humidity conditions. The backward step point can be migrated to a more forward position. The hydrothermally synthesized MIL-101 material is coated onto a highly thermally conductive metal substrate to form a dense coating based on a bonded coating process, which is simple and easy to perform and reduces the contact thermal resistance between the irregular material and the metal. To validate the research idea, a semiconductor chip was used to construct a cold source, and the temperature of the cold source was precisely controlled by adjusting the voltage. As indicated by the adsorption curves, the moisture absorption of the material was stabilized at a low level at 25 °C and 30% RH. After the introduction of an appropriate cold source, the moisture adsorption increases significantly, reaching $1.05\,g\,g^{-1}$ at 14 °C. Since the moisture adsorption equilibrium point of the material crosses the step range, the further reduction of the cold source temperature does not significantly increase the moisture adsorption. The lower the temperature of the cold source, the higher the rate of moisture uptake due to the more pronounced increase in the difference in moisture concentration between the local humidity and the pores, despite the weakening of molecular diffusion. Since this strategy does not change the proper properties of the material itself, when the cold source disappears, the material can spontaneously tend to the equilibrium site in the low humidity state and achieve self-regeneration. The desorption rate can be accelerated by 10–30 times at a small temperature difference (5–20 °C) with the introduction of an appropriate heat source. Based on the adsorption characteristics, the adsorption curves corresponding to different cold source temperatures were projected onto the ambient humidity axis, and the adsorption curves were realized to migrate under cold source construction. By changing the humidity environment in and around the adsorbent pores, the cold-source construction strategy, targeting S-curve adsorbents, significantly improves the moisture uptake and, at the same time, effectively overcomes the detrimental effects of adsorption of hot stripes compared to the conventional adsorption process. This strategy allows the material to absorb significantly more moisture in arid environments than single adsorbents reported in the existing literature focusing on airborne water extraction in arid regions [68].

Gordeeva et al. synthesized MIL-160 by hydrothermal method using 2,5-furandicarboxylic acid, $AlCl_3$-$6H_2O$, and sodium hydroxide as raw materials. The structure of MIL-160 is formed by inorganic AlO_6 octahedra, which are linked by carboxylic acid groups of ligands (2,5-furandicarboxylic acid) to form cis-angle shared chains. It possesses both strong and weak adsorption sites. Ferreira et al. evaluated the potential of MIL-160(Al) as a water absorbent. They studied the adsorption equilibrium of water vapor at three different temperatures: 303, 323, and 343 K. The adsorption equilibrium isotherms were V-shaped according to the International Union of Pure and Applied Chemistry (IUPAC) classification, confirming the hydrophilicity of the adsorbent (Figure 10.12) [69]. The hydrophilicity of MIL-160(Al) is due to the interaction of water molecules and hydroxyl groups present in the pores, as pointed out in the study by Cadiau et al. [70].

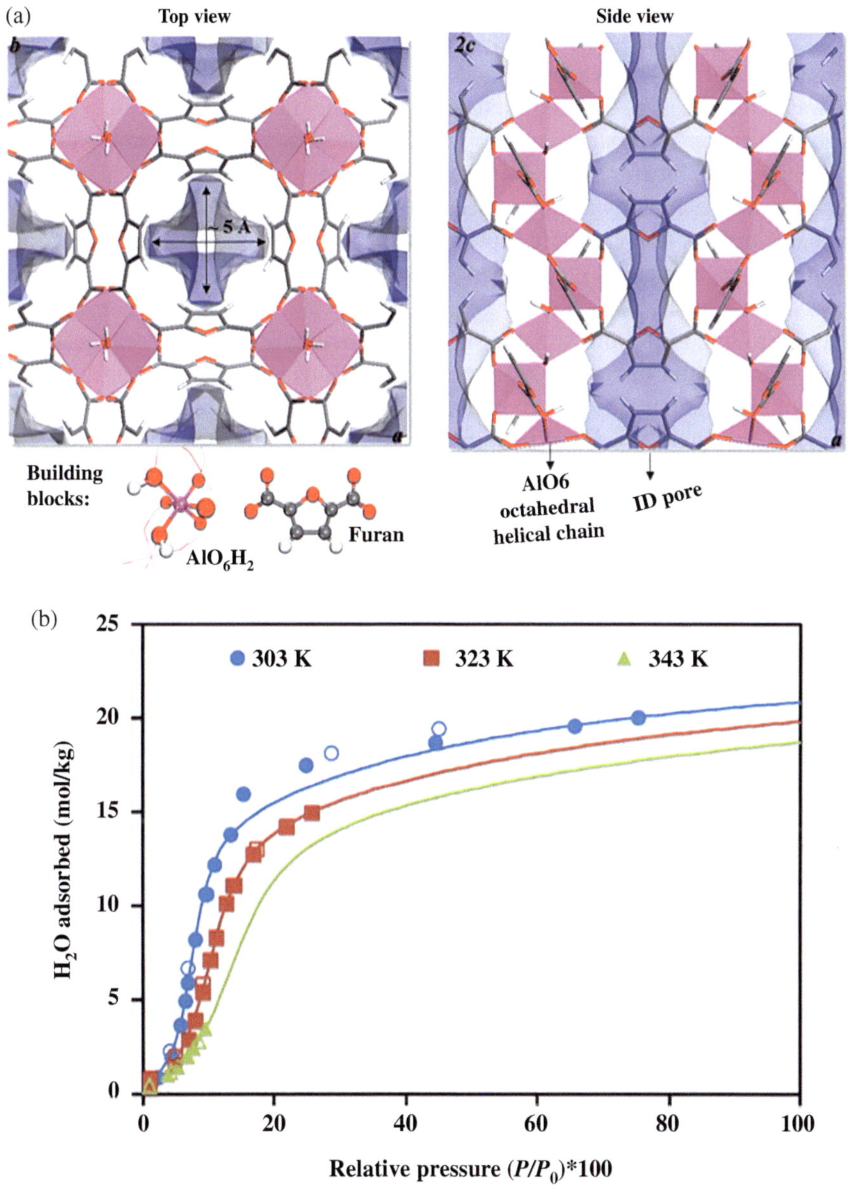

Figure 10.12 (a) Top and side views of MIL-160(Al) (1 × 1 × 2) structure. The Al, O, C, and H atoms are represented by pink, red, grey, and white colors, respectively. The isosurface in blue color shows the accessible pore volume that consists of rectangular-shape sinusoidal channels. (b) Water vapor adsorption–desorption equilibrium isotherm on MIL-160(Al) at 303, 323, and 343 K as a function of relative pressure. Source: Reproduced from [69] / with permission of Springer Nature.

10.3.2.2 Polymer Gel-Based Adsorbents

Polymer gels are widely used in delivery systems and tissue engineering due to their excellent swelling, encapsulation, and delivery capabilities. In addition, their use in water resources has gained increasing attention in recent years due to their high adsorption, reversible swelling, antimicrobial properties, biocompatibility, and biodegradability [6, 71]. Hydrogels, a bioactive material with a high affinity and adsorption capacity for water, have been used to design and develop efficient water vapor adsorption that can store captured water in cross-linked polymer networks. Hydrogels are composed of polymers and water, presenting a three-dimensional crosslinked network, which is formed through covalent bonding or non-covalent interaction build-up. Non-covalent interactions mainly include physical entanglement, electrostatic interactions, supramolecular interactions, hydrogen bonding, hydrophobic interactions and coordination interactions. Hydrogel-based composites have been able to achieve the modulation of special functions to obtain more functional characteristics such as maximum equilibrium swelling, high absorption, uniform softness, high durability, and light stability. Hydrogels swell during the water trapping process, so their water absorption is not limited by their original pore volume, resulting in a high absorption rate. In addition, highly tunable molecular structures, such as hydrophilic functional groups and ionic sites, allow hydrogels to further enhance water absorption. These unique properties make hydrogel materials promising for atmospheric water-harvesting applications [72]. Hydrogels are classified into natural and synthetic hydrogels according to their synthetic raw materials. Natural polymer hydrogels (e.g. chitosan, sodium alginate, and cellulose) and synthetic polymer hydrogels (e.g. polyvinyl alcohol and polyacrylamide) are used to synthesize hydrogels with different functional properties. Hydrogels, as well as aerogels formed by freeze-drying hydrogels, are used extensively in atmospheric water harvesting.

Chitosan is a biologically active substance abundant in nature and is commonly used in the preparation of low-cost, stable, antimicrobially active, mechanically efficient, biodegradable, and environmentally friendly hydrogels. Chitin, the precursor of chitosan, is the most widely occurring biopolymer in nature after cellulose and is found in a range of eukaryotic species such as crustaceans, insects, and fungi. Chitin is a polymer of *N*-acetyl-D-glucosamine, which is known as chitosan when it undergoes deacetylation, i.e. β-1,4-D-glucosamine [73]. The molar fraction of n-acetylated repeating units is defined as the degree of acetylation (DA) , while the percentage of β-1,4-D-glucose repeating units in polysaccharides is defined as the degree of deacetylation (DD). DD is a key parameter for chitosan, as previous studies have reported that chitosan with higher DD exhibits stronger biological effects and higher water solubility. This is because a higher DD indicates a higher concentration of amino groups in the molecule, and protonation of amino functional groups is essential to show the biological effects and water solubility of chitosan. Therefore, chitosan can be used as a raw material for the preparation of stable, inexpensive, water-retentive, and water-absorbent hydrogels [74]. Chitosan gels and their composites are characterized by high porosity, large specific surface area, biodegradability, biocompatibility, and richness of properties, which provide good adsorption potential at a lower cost compared to traditional adsorbents such as activated carbon. Increasing the availability of chitosan functional groups can further improve the physicochemical properties [75]. Chitosan gels can be obtained by establishing physical cross-linking, which produces a physical hydrogel by establishing weak and reversible interactions, i.e. electrostatic forces and hydrogen bonding between the polymer parts. The advantages of using chitosan as a precursor to prepare the final gel are as follows: (i) chitosan is the second renewable biopolymer after cellulose and can be extracted from a wide range of organisms; (ii) chitosan chains are rich in amine and hydroxyl groups, and the presence of amine groups has a pH-responsive property, which makes chitosan gel preparation very simple by adjusting the pH of the solution to prepare chitosan physical gels using hydrogen bonding; (iii) the chitosan The amine groups on the backbone not only have cross-linking reactivity, but also can be modified simply by derivatization reaction, so various (macro) molecular cross-linking agents can be used to covalently cross-link or easily modified to introduce chemical cross-linking groups, and various chemical modifications such as grafting reaction can improve the physicochemical and functional properties of chitosan, which can be used to improve the performance of atmospheric water collection [76].

Alginate consists of 30–60% of brown algae. Due to their chemical and physical properties, they have been extensively studied due to their ability to form gels and particles. Sodium alginate is one of the best-known members of the hydrogel moiety, whose molecules consist of β-D-mannuronic acid (β-D-mannuronic, M) and α-L-guluronic acid (α-L-guluronic, G) linked by a $(1 \rightarrow 4)$ bond. Crosslinking polymer chains can design alginate hydrogels [77]. The physical and chemical properties of alginate hydrogels depend on the crosslink density, type of crosslink, and molecular weight. One suitable method for forming alginate hydrogels is intermolecular cross-linking, in which only the lauric acid group of the alginate is involved, along with a divalent cation such as calcium. Many divalent cations show a unique affinity for alginate. Still, calcium is more commonly used for alginate gelation, and in the presence of divalent calcium ions, sodium alginate can be used to build gels without any heating.

Cellulose is biocompatible, biodegradable, resourceful and renewable, mechanically strong, environmentally friendly, and the safest material on earth. In general, cellulose is a polysaccharide consisting of linear chains of β $(1 \rightarrow 4)$ linked d–d-glucose units. It is odorless and insoluble in water and most organic solvents. Cellulose chains are rich in hydroxyl groups, making it a hydrophilic material, and high inter- and intramolecular hydrogen bonding and van der Waals forces provide the basis for the preparation of cellulose hydrogels, which can then be obtained by removing the solvent by oven drying, freeze drying or supercritical drying [78]. Intrinsically hydrophilic cellulose-based gel materials also show considerable development and application in water harvesting and water evaporation [79].

Synthetic polymer hydrogels, including polyacrylamide (PAM) and poly N-isopropyl acrylamide (PNIPAM) hydrogels, have attracted much attention due to their simplicity of preparation and lower cost and are the most widely studied long-chain polymer hydrogels, with a large range of applications in bioengineering, flexible electronics, and energy devices [80], and have also been extensively studied in atmospheric water-harvesting technologies [81]. Among them, since the discovery of the thermal phase transition behavior of poly(PNIPAM) by Scarpa in 1967, PNIPAM has become a representative member of the growing family of thermoresponsive polymer-based smart materials. PNIPAMs are produced by the polymerization of the monomer N-isopropylacrylamide (NIPAM), characterized by the amide and isopropyl portions of the monomer structure, and they have a low critical solution temperature (LCST, – 32 °C) in aqueous media, and exhibit a drastic and reversible bulk phase transition by swelling/shrinking near LCST [82]. In addition to the size change during the bulk phase transition, this process is accompanied by changes in many other properties, such as hydrophilicity, transparency, and apparent electrostatic dielectric constant. At temperatures below the LCST of PNIPAM, hydrogen bonding between the hydrophilic amide groups and water molecules diminishes the hydrophobic interactions between the isopropyl groups, leading to the collapse of the PNIPAM chain. As the temperature increases, even above the LCST, the number of hydrogen bonds decreases, leading to enhanced hydrophobic interactions between the isopropyl groups, entropy-driven phase separation occurs, and the PNIPAM chains are dehydrated and aggregated into tightly packed globular conformations. The LCST of PNIPAM can be manipulated over a wide range of temperatures by co-polymerization incorporating either hydrophilic components or hydrophobic monomers, thereby effectively altering enthalpic contributions to the thermal response [83]. There is a creature in the ocean that can adapt to danger, the pufferfish, a fish in the family Tetraodontidae, which can inflate itself by inhaling water when threatened and releasing water to regaining its original shape when the threat of danger is gone. Thermally responsive PNIPAM is very similar to this biological adaptive behavior in nature. This thermo-responsive PNIPAM hydrogel has a unique potential for atmospheric water harvesting by switching adsorption and release of water by hydrophilic-phobic switching at lower LCST, which can efficiently regulate the transport and release of water molecules in the pores of the three-bit network of the gel [84], as shown in Figure 10.13. Zhang et al. [85] developed a photo-responsive core-shell structured hydrogel with both highly efficient hygroscopic performance and fast water release performance by using an open-pore sodium polyacrylate hydrogel (PAAS) as the shell layer, a temperature-sensitive, large pore size PNIPAM hydrogel as the core layer, and polydopamine nanoparticles (PDAs) as the photo-thermal conversion material. The atmospheric water-harvesting performance of the hydrogel was effectively improved, exhibiting excellent hygroscopicity over a wide range of humidity ($2.76\,g\,g^{-1}$, RH = 90%; $0.68\,g\,g^{-1}$, RH = 30%). The photothermal synergy between polydopamine nanoparticles and temperature-sensitive PNIPAM can rapidly increase the surface temperature of the core-layer PNIPAM hydrogel above LCST, which in turn effectively improves its water-release performance, exhibiting excellent water-release performance ($1.42\,kg\,m^{-2}\,h^{-1}$) at 1 solar intensity, as shown in Figure 10.14.

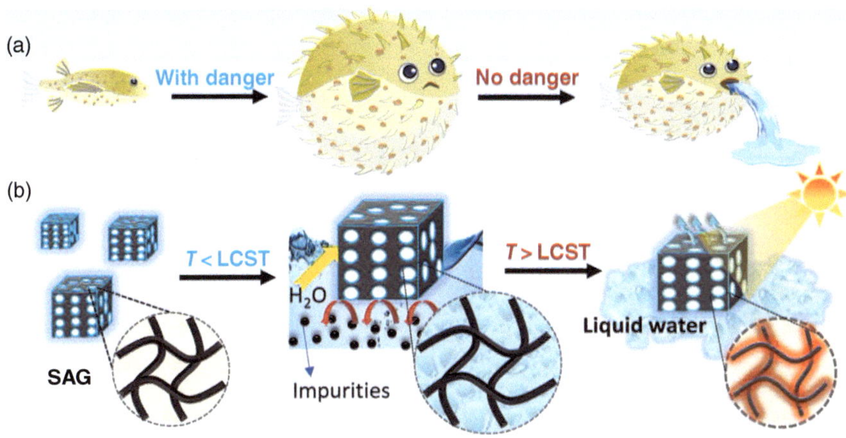

Figure 10.13 (a) Pufferfish change shape by absorbing and releasing water when threatened; (b) PNIPAM hydrogel adsorbs and releases water by switching between hydrophilic and hydrophobic at lower LCST. Source: Reproduced from [84] / with permission of John Wiley & Sons.

10.3.2.3 Composite Adsorbents

In practice, a single composite adsorbent always has various deficiencies, and it is usually necessary to combine two or more of the above adsorbents to form a multivariate composite adsorbent. For example, although hygroscopic salt has a high water absorption rate, it is easy to turn from a solid to a salt solution liquid under high humidity, resulting in the leakage of the salt solution, which is difficult to recover. To solve this problem, a common approach incorporates the hygroscopic salt into a designed porous matrix where the loaded salt plays a major role in trapping water molecules, and the porous matrix serves as a container to hold the final liquefied salt. Porous matrices include MOFs, aerogels, and hydrogels. Modification of hygroscopic salts into MOF can compensate for the lack of water absorption of MOF materials at high humidity. Similarly, hydrogel and aerogel materials are often used as substrates loaded with hygroscopic salts to improve overall water absorption. Various composite adsorbents in actual SAWH are described in detail below.

10.4 Composite Adsorbents in SAWH

10.4.1 MOF-Based Composite Adsorbent

MOFs are usually insufficiently hygroscopic at high humidity, so in recent years, researchers have developed many combinations of hygroscopic salts with metal-organic frameworks. An et al. [86] used $CaCl_2$ to modify MOF-808 by homogeneously dispersing it in 10 mL of aqueous solution, stirring for half an h, and then adding a certain amount of $CaCl_2$ and stirring for 12 h so that calcium ions and chloride ions penetrate the pores of MOF-808. Then, the composite adsorbent was centrifuged at 11,000 rpm for 3 min to remove the excess salt solution.

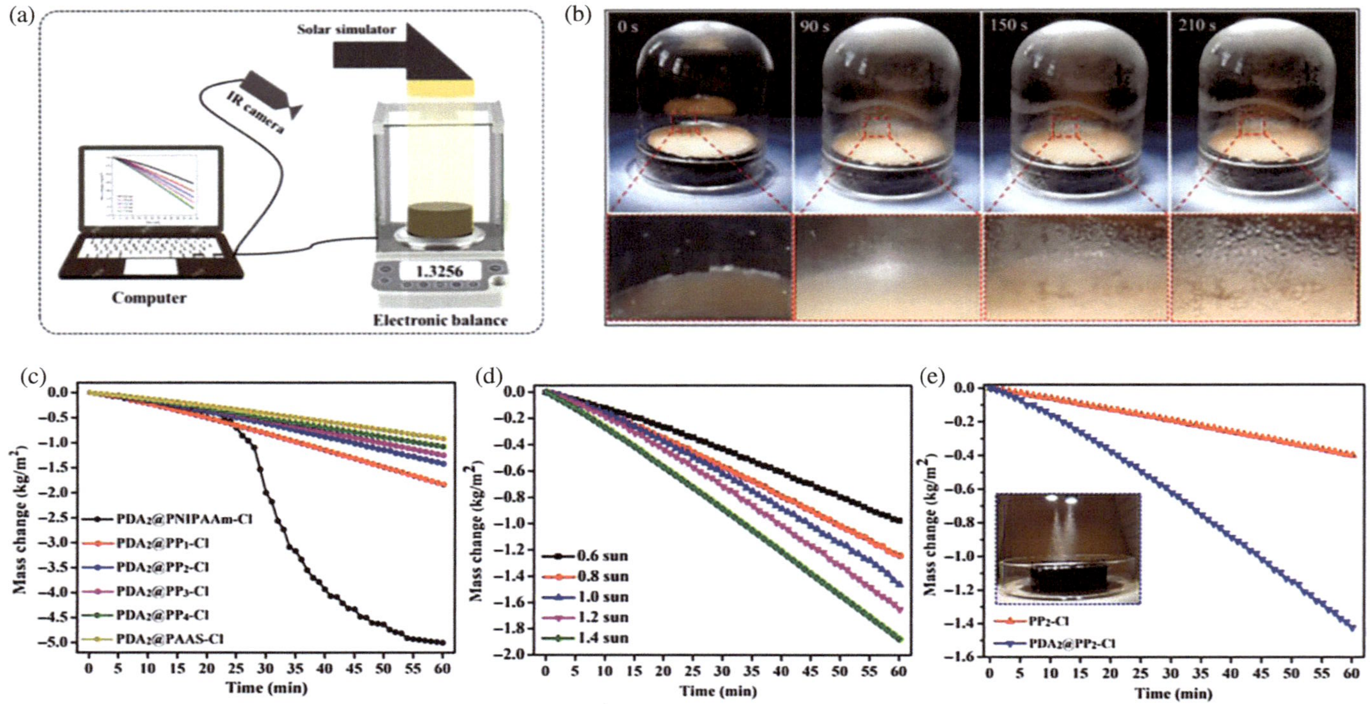

Figure 10.14 (a) Schematic diagram of the water evaporation test; (b) Photographs of water droplet formation on the wall of the device; (c) Mass change of PDA_2@PAAS-Cl, PDA_2@PNIPAM-Cl, and PDA_2@PP-Cl hydrogels under 1 sun; (d) Mass change of PDA_2@PP_2-Cl hydrogels under different light intensities; (e) Mass change of PP_2-Cl and PDA_2@PP_2-Cl hydrogels under 1 sun. Source: Reproduced from [85] / with permission of American Chemical Society.

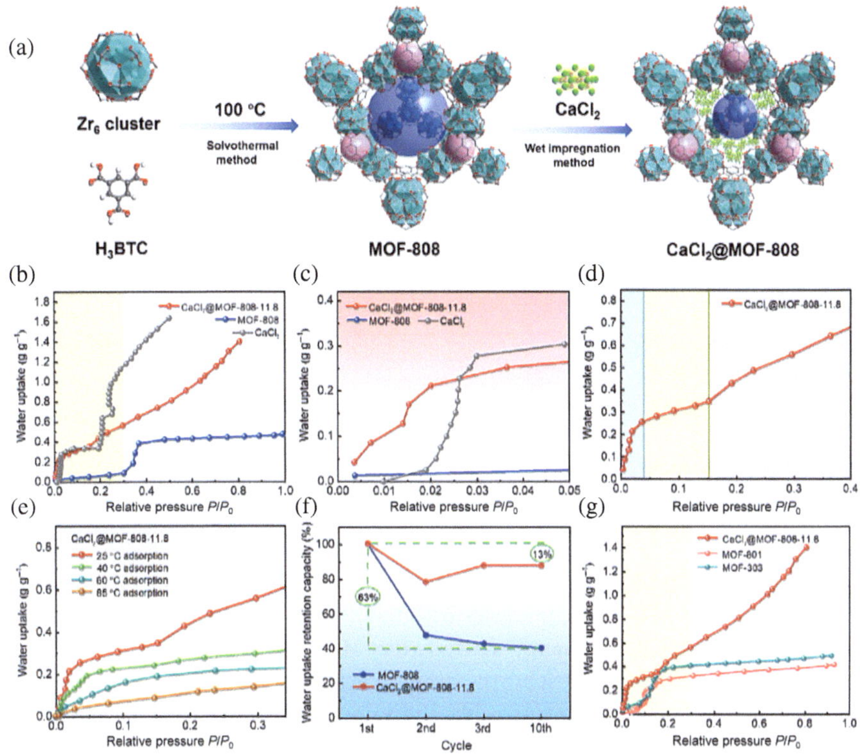

Figure 10.15 (a) Synthesis and morphological characteristics of the $CaCl_2$@MOF-808 composite adsorbent. (b–d) Water adsorption performance of $CaCl_2$@MOF-808-11.8, $CaCl_2$, and pristine MOF-808 at 25 °C. (e) Water adsorption isotherms of $CaCl_2$@MOF-808-11.8 at different temperatures. (f) Retained water uptake capacity of $CaCl_2$@MOF-808-11.8 after different numbers of adsorption–desorption cycles at 25 °C. (g) Comparison between the water adsorption isotherms of $CaCl_2$@MOF-808-11.8, MOF-801, and MOF-303 at 25 °C. Source: Reproduced from [86] / with permission of Elsevier.

Then, the composite adsorbent was quickly washed with distilled water to remove the remaining $CaCl_2$ solution adhering to the outer surface of MOF. After drying, the $CaCl_2$@MOF-808 composite adsorbent was obtained (Figure 10.15). The larger pores of MOF-808 acted as a porous matrix to confine the $CaCl_2$. In contrast, the guest molecules in the pores of $CaCl_2$ enhanced the hydrophilicity of the composite material and improved its water stability. The water absorption of the described composite adsorbent (water absorption: 0.56 g g^{-1} at 25 °C) was significantly higher than that of MOF-808 (water absorption: 0.08 g g^{-1} at 25 °C) at 30% relative humidity. $CaCl_2$@MOF-808-11.8 can capture water within 2 h and release it within 40 min, and the water production can reach 1.8 kg kg^{-1} per day. In the large-scale AWH experiments, the actual water-harvesting performance of the composite adsorbent is better than that of the standard water-harvesting materials, MOF-303 and MOF-801.

Single MOF-801 is limited by its limited workability, powder structure, and stability. To overcome these issues, MOF-801 was crystallized through the surface of macroporous poly(N-isopropylacrylamide-methacrylate) spheres, termed

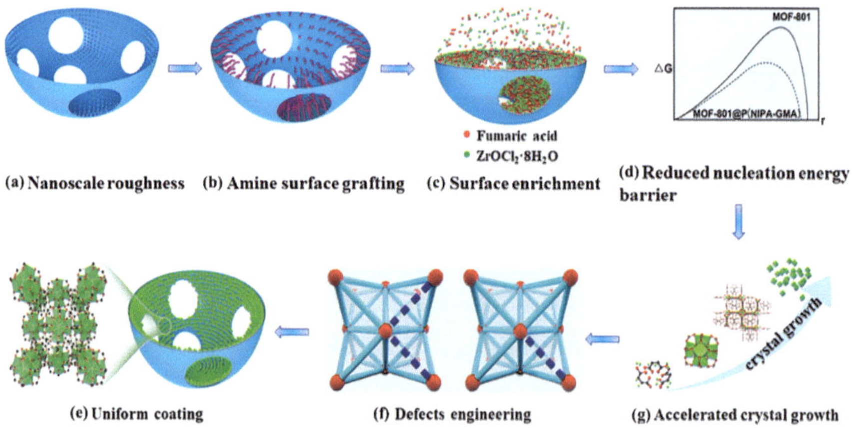

Figure 10.16 Schematic diagram of the in situ confined growth mechanism of MOF-801 crystals on the macroporous surface of a P(NIPAM-GMA) sphere. Source: Reproduced from [87] / with permission of John Wiley & Sons.

P(NIPAM-GMA), through an in-situ growth limitation strategy to form spherical MOF-801@P(NIPAM-GMA) complexes with temperature-responsive functionality (Figure 10.16) [87]. The energy barrier for MOF-801 nucleation is significantly reduced due to the enrichment on the surface of nanoscale macroporous P(NIPAM-GMA) substrates. As a result, the number of critical nuclei is greatly increased. As the number of available precursors is limited by pore confinement, the size of the grown crystals decreases to the nanoscale. Uniformly distributed nanocrystals with large surface-to-volume ratios and an abundance of structural defects serve as active sites for water adsorption. The temperature-responsive P(NIPAM-GMA) substrate becomes hydrophobic at temperatures above 62 °C, thereby promoting vacuum-supported water sorption. The composite provides unprecedented high water collection efficiency. Produced in kilogram quantities, the composite can capture 1.60 kg H_2O kg^{-1} of composite per day at 20% relative humidity between 25 and 85 °C.

In addition, it was reported that a water-stabilized MOF material with high pore volume, MIL-101(Cr), was selected as the porous matrix, and lithium chloride was filled into the MOF to prepare the composite adsorbent material [88]. The hydrophobicity of the MIL-101(Cr) material makes it difficult for its solid powder to be fully dispersed in the salt solution, and treatments such as ultrasonication are prone to destroying the structure of the MOF itself. The authors used an iontophoresis method, in which a quantitative amount of lithium chloride was added to the MOF suspension after hydrothermal synthesis and sufficient water washing, and a large amount of salt immersion and loading was realized after sufficient stirring. The residual salt solution was removed by centrifugation, followed by a quick wash using an alcohol-based reagent to maintain the internal salt solution while cleaning the salt solution remaining on the surface. Figure 10.17a–c show the synthesized pure MOF materials with better microforms and microporous pores, and Figure 10.17d–f shows the composite adsorbent materials uniformly filled with a large amount of lithium chloride inside. Figure 10.17g shows the nitrogen

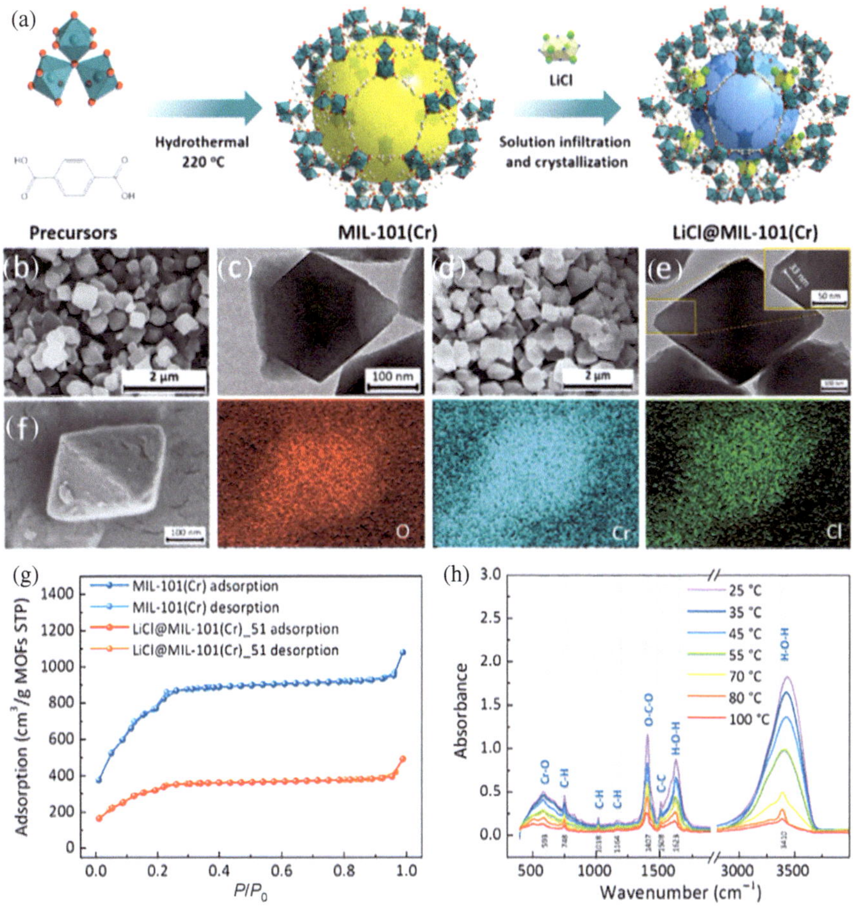

Figure 10.17 Synthesis and morphological characterization of LiCl@MIL-101(Cr) composite adsorbent. (a) Schematic of the fabrication process of LiCl@MIL-101(Cr). The porous MOFs of MIL-101(Cr) were first prepared by hydrothermal synthesis. Then, lithium chloride was encapsulated in MIL-101(Cr) by solution permeation and crystallization. The porous matrix of MIL-101(Cr) not only provides a large pore space for the storage of lithium chloride and water but also improves the mass transfer capacity of water vapor. Dark cyan polyhedra, Cr, red, O, grey, C, blue, Li; green, Cl; (b) SEM image of MIL-101(Cr); (c) TEM image of MIL-101(Cr); (d) SEM image of LiCl@MIL-101(Cr) showing that the MOF powder maintains a loose and porous structure; (e) LiCl@ TEM image of 1MIL-101(Cr) shows that encapsulation of LiCl retains some of the ordered nanopores of the MOF matrix; (f) energy dispersive X-ray spectroscopy (EDX) elemental mapping shows the uniform distribution of LiCl in the MIL-101(Cr) matrix; (g) at 77 K, the MIL-101(Cr) and LiCl@MIL-101(Cr)_51 N_2 gas adsorption–desorption isotherms indicate that despite the encapsulation of up to 51% LiCl in MIL-101(Cr), there is still enough pore space for vapor transfer and storage of liquefied water during AWH; (h) variable temperature FTIR patterns of LiCl@MIL-101(Cr)_51 at different temperatures of 2.3 kPa, indicating that there is no chemical interaction force. The strong hydrogen bonding bands indicate the high amount of water in the composite adsorbent. Source: Reproduced from [88] / with permission of John Wiley & Sons.

adsorption equilibrium curve. The pore volume of the pure MOF material is as high as 1.71 $cm^3\ g^{-1}$, and the BET specific surface area is as high as 3311 $m^2\ g^{-1}$. The pore volume of the composite material is still as high as 0.69 $cm^3\ g^{-1}$, and the BET specific surface area is as high as 1179 $m^2\ g^{-1}$ after 51% of the lithium chloride was filled in the composite material. Figure 10.17h shows a variable-temperature Fourier-transform infrared (FTIR) pattern under atmospheric pressure water vapor atmosphere, and all the peaks in the pattern except the hydrogen-bonding peaks are the same as the hydrogen-bonding peaks. Hydrogen bonding peaks correspond to the peaks of pure MOF material, indicating that there is no chemical interaction between lithium chloride and MOF matrix in the composite material; the higher hydrogen bonding indicates that the material is a potential air-water extraction material with significant differences in water vapor adsorption at different temperatures.

Ge et al. [89] prepared the binary nanocomposite MIL-101/mobil composition of matter (MCM-41) by mixing MIL-101 and ordered mesoporous silica MCM-41 to further enhance the performance (Figure 10.18). The highly ordered mesoporous channels of MCM-41 provide abundant nucleation sites, which are favorable for the

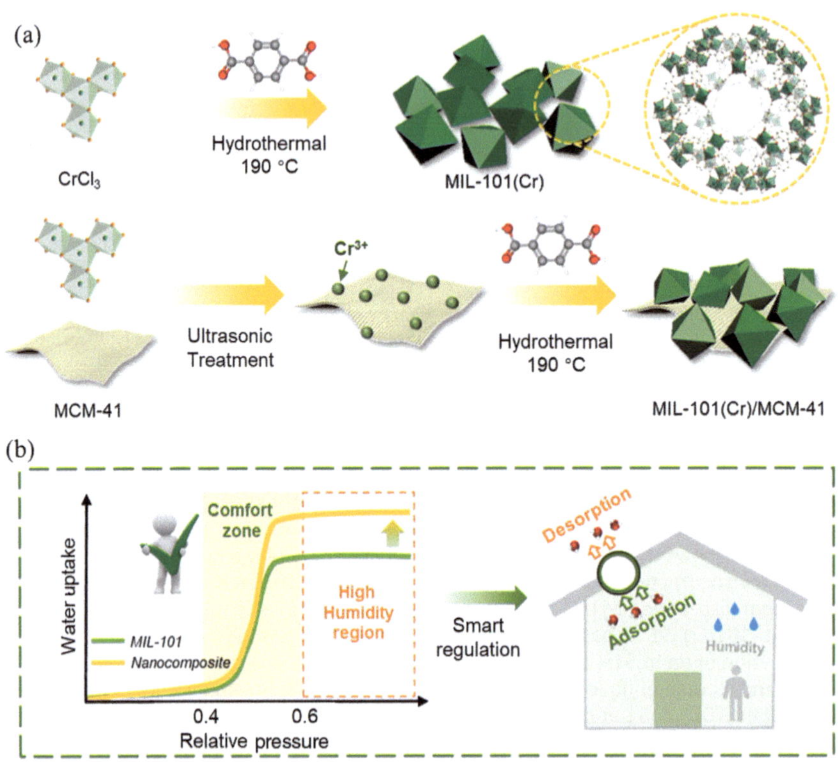

Figure 10.18 Design of the nanocomposite MIL-101/MCM-41 and its applications. (a) Schematic diagram of the template-guided tuning strategy. (b) Demonstration of the smart indoor humidity regulation. Source: [89] / John Wiley & Sons / CC BY 4.0.

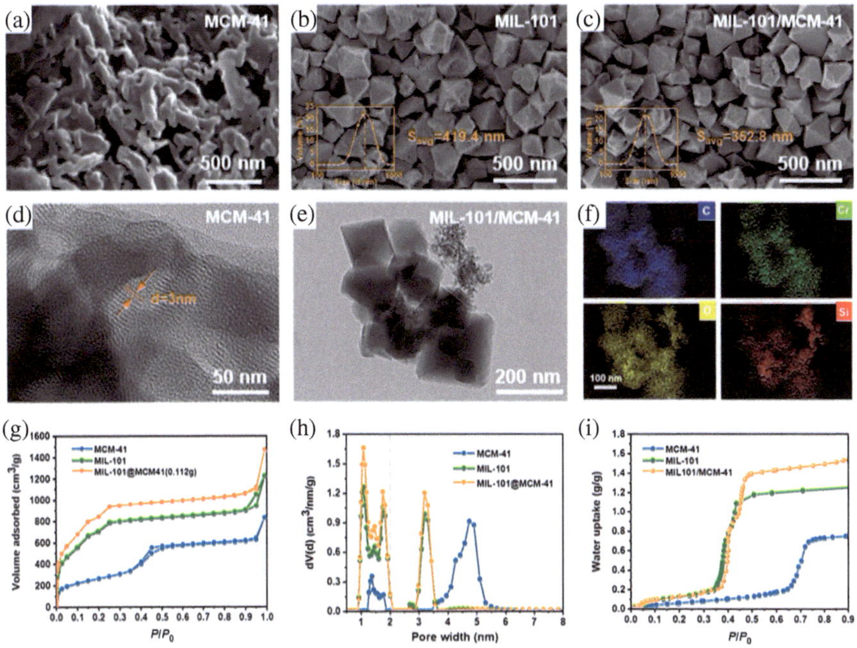

Figure 10.19 Characterizations of MIL-101(Cr)/MCM-41. (a–c) Scanning electron microscope (SEM) images of MCM-41, MIL-101, and MIL-101/MCM-41. (d–e) Transmission electron microscope (TEM) images of MCM-41 and MIL-101/MCM-41. (f) energy dispersive spectrometer (EDS) mapping showing the distribution of C, Cr, O, and Si in MIL-101/MCM-41. (g) X-ray diffraction (XRD) patterns, (h) Fourier-transform infrared (FTIR) spectra, and (i) Roman spectra of MCM-41, MIL-101, and MIL-101/MCM-41. Source: [89] / John Wiley & Sons / CC BY 4.0.

growth of MOF crystals. The finer particles and higher specific surface area endowed the nanocomposites with excellent water absorption performance and stability compared with the parent MOF, showing a specific surface area of 3188.14 $m^2\ g^{-1}$ and a water absorption rate of 1.53 $g\,g^{-1}$ at 90% RH (Figure 10.19).

Song et al. [24] reported MOF-derived nanoporous carbon with optimized hydrophilic sites and pore sizes, adsorbents with fast adsorption kinetics, and excellent photothermal properties for high-yield AWH. Figure 10.20a present synthesis scheme of MOF-derived nanoporous carbon. The optimized MOF-derived nanoporous carbon with 40% adsorption sites and ~1.0 nm pore size was calculated by density functional theory (DFT) theoretical modelling to have a reduced diffusive resistance of water while having superior adsorption kinetics. In addition, the carbonaceous adsorbent exhibited rapid desorption kinetics through effective solar heating and high thermal conductivity. This MOF-derived nanoporous carbon absorbed 0.18 $L\,kg^{-1}$ of water within 45 min and released all captured water within 10 min under single solar illumination, as shown in Figure 10.20b,c. For nanoporous and microporous adsorbents, previous studies have shown that the rate of water uptake is controlled by external osmosis and internal diffusion. In contrast, intrinsic and surface diffusion are complementary phenomena. The diffusion rate of water in

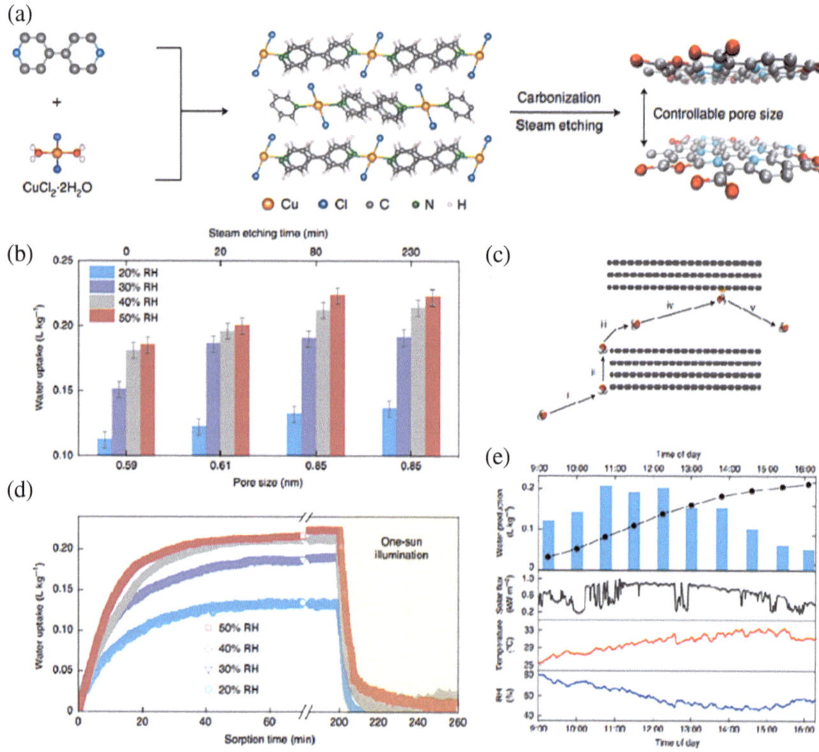

Figure 10.20 (a) Synthesis scheme of MOF-derived nanoporous carbon; (b) Comparison of water uptake of different nanoporous carbon samples at relative humidity (RH) of 20–50%; (c) Water adsorption–desorption test at 25 °C and RH = 20–50%. At 200 min, the samples were exposed to solar irradiation to release water; (d) AWH through the microporous adsorbent comprising a series of nanoscale water molecule transport phenomena: (i) bulk diffusion, (ii) surface diffusion, (iii) external penetration, (iv) internal diffusion and (v) water release. Grey spheres represent carbon; red spheres represent oxygen; white spheres represent hydrogen; yellow spheres represent nitrogen; (e) Schematic representation of water-harvesting productivity, solar flux, ambient relative humidity, and temperature for each cycle of MOF-derived nanoporous carbon. Source: Reproduced from [24] / with permission of Springer Nature.

nanoporous adsorbents was investigated by analyzing the diffusion energy barriers. As shown in Figure 10.20d, it is demonstrated that the adsorption–desorption process of adsorbed AWH for water requires five steps: (i) body diffusion: water adheres to the outer surface of the adsorbent particles; (ii) surface diffusion: water diffuses on the outer surface of the particles; (iii) external penetration: water enters into the nanopores of the adsorbent; (iv) internal diffusion: water diffuses through the nanopores and is adsorbed by the binding sites; (v) water release: water desorbs from the binding sites and leaves the adsorbent.

10.4.2 Polymer Gel-Loaded Hygroscopic Salt Composite Adsorbent

Research on molecular modification of polymer gels for atmospheric water harvesting has widely focused on the introduction of hygroscopic inorganic salts. The introduction of hygroscopic agents significantly improves the water molecule adsorption and desorption behavior of polymer gels, which in turn modulates the adsorption–desorption kinetics. The composite adsorbents of polymer gels with salts are used in the adsorption of atmospheric water collection with surface trapping and subsequent permeation. In the adsorption stage, water vapor undergoes the process of "diffusion capture - adsorption by salt (or formation of hydrates) - saturated solution - infiltration into the hydrogel - liquid water storage"; in the desorption stage, the water stored in the gel adsorbent undergoes the process of "desorption of the salt solution - evaporation - overall diffusion - release of water vapor - condensation - fresh water collection." Hygroscopic salt composite adsorbents are currently the most promising materials for engineering applications, and these adsorbents generally exhibit higher adsorption capacity over a wide range of relative humidity compared to MOFs and hydrogels. What's more, hygroscopic salt composite adsorbents have a high potential for application because of their high stability, low production cost, and simple immersion-drying preparation process, which has a large potential for large-scale production.

Lu et al. [71a] used chemically crosslinked polyacrylamide (PAM) hydrogels doped with LiCl to prepare PAM-LiCl hydrogels to investigate the effect of inorganic salts on adsorption and desorption during atmospheric water collection, as shown in Figure 10.21a–f. Specifically, the inorganic LiCl phase is the active hygroscopic center of PAM-LiCl during atmospheric water molecule adsorption. When exposed to moisture, the LiCl phase in PAM-LiCl will spontaneously capture and liquefy water molecules in the air. At the same time, the hydrophilic discoidal polymer chains will transform to a more swollen conformation when interacting with liquefied water, i.e. swelling PAM-LiCl to store water. The adsorption kinetics of PAM-LiCl is faster than that of pure hygroscopic agent LiCl, which may be attributed to the larger active hygroscopic sites, i.e. the microscale hydrogel network matrix helps to achieve the homogeneous distribution of LiCl in PAM-LiCl and avoid the LiCl crystal agglomeration, thus increasing the effective hygroscopic area and realizing rapid water absorption.

The interaction of the polymer chains of polymer gels with water plays a crucial role in the water release process. Depending on the strength of the interaction between water molecules and polymer chains, water molecules in hydrogels can exist in three states: bound water, intermediate water, and free water. Bound water is water that forms hydrogen bonds with the polymer chains and cannot be frozen because the binding structure remains below 0 °C. Free water is a water molecule with a configuration similar to that of the main body of water; intermediate water is located between bound and free water and exists in the form of weakly bound water.

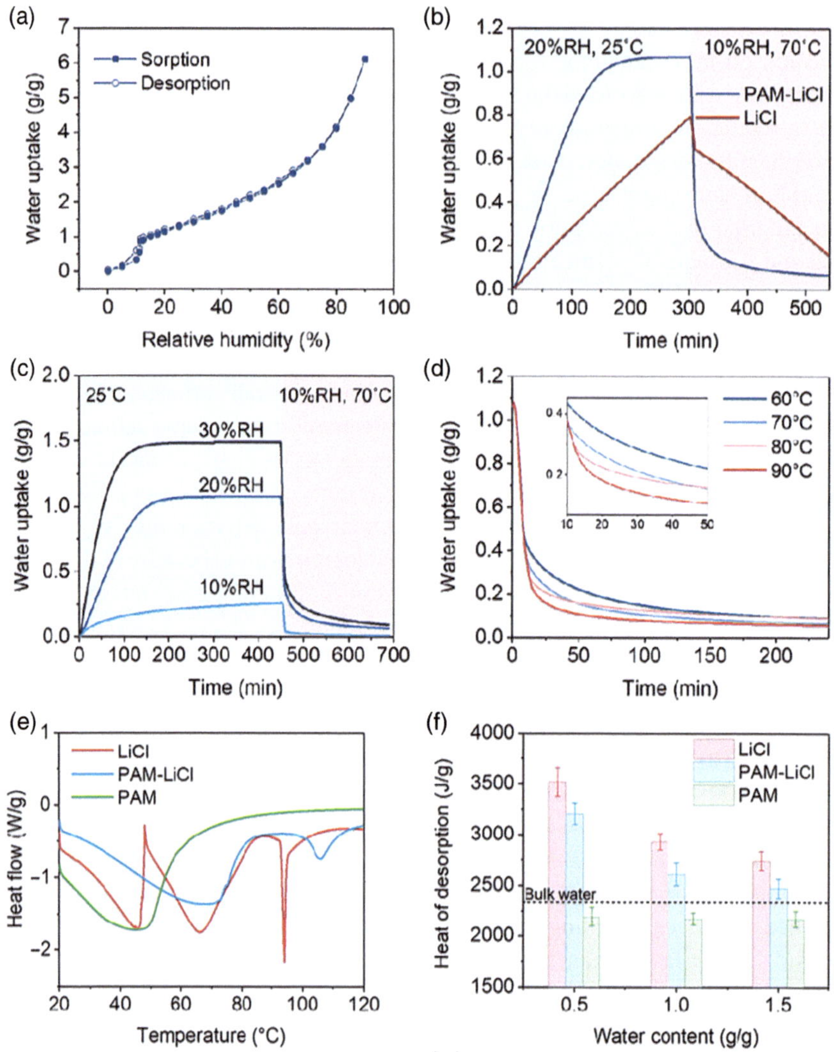

Figure 10.21 Adsorption/desorption properties of PAM-LiCl. (a) adsorption and desorption isotherms at 25 °C; (b) adsorption and desorption kinetics of PAM-LiCl and LiCl. Adsorption conditions: RH = 20%, 25 °C; Desorption conditions: RH = 10%, 70 °C; Desorption water vapor pressure is equal to the saturated water vapor pressure at 25 °C to simulate the actual working conditions; (c) Adsorption–desorption performances of PAM-LiCl at 25 °C, RH = 10%, 20%, and 30%. (d) Hydrolysis adsorption curves of PAM-LiCl at 60, 70, 80, and 90 °C. The inset shows the change in water absorption from the first 10 to 50 min. At all desorption temperatures, the PAM-LiCl samples were dried at the same water vapor pressure of 3.17 kPa, which corresponds to saturated vapor pressure at 25 °C. The RH of the PAM-LiCl samples at all desorption temperatures was 3.17 kPa. The RH at each temperature was 16% at 60 °C, 10% at 70 °C, 7% at 80 °C, and 4% at 90 °C; (e) heat flow curves of LiCl, PAM-LiCl, and PAM during desorption of water as determined by differential scanning calorimetry (DSC); (f) heat of desorption of LiCl, PAM-LiCl, and PAM. Source: Reproduced from [71a] / with permission of John Wiley & Sons.

Both free and intermediate water are freezable [90]. It has been shown that the heat of desorption of PAM gels is lower than that of integral water, suggesting that the evaporation of water molecules from within the polymer network is a lower energy process. Compared to the heat of desorption of LiCl at 0.5, 1.0 and 1.5 $g\,g^{-1}$ which were 3511, 2934 and 2734 $J\,g^{-1}$ respectively, the heat of desorption of PAM-LiCl was much lower at 3200 $J\,g^{-1}$ (0.5 $g\,g^{-1}$), 2608 $J\,g^{-1}$ (1.0 $g\,g^{-1}$) and 2468 $J\,g^{-1}$ (1.5 $g\,g^{-1}$). This data suggests that the presence of the PAM hydrogel network reduces the desorption energy of PAM-LiCl by about 300 $J\,g^{-1}$ on average; in other words, the interaction between PAM and water plays a crucial role in the moisture release process. Second, the dissolution of hygroscopic salts such as LiCl also affects the hydrogen bonding in water molecules. Li^+ has a high heat capacity of hydration of −514 $kJ\cdot mol^{-1}$, which disrupts the tetrahedral hydrogen-bonding structure when dissolved in water, and Cl^-, a moderate hydrogen-bonding acceptor, tends to replace OH—O with OH— and Cl^-, which renders water molecules involved in the hydration of Li or Cl ions non-freezable. As LiCl gradually absorbs water, the water molecules in the vicinity of the metal ion behave more like water with an ordered structure (freezable water). The interaction of freezable water with the adsorbent is weaker than that of non-freezable water, which makes adsorbents with a higher content of freezable water require less energy to desorb at the same moisture content.

Also, Zhao et al. [83a] proposed a super-absorbent gel (SMAG) synergistically combining a hygroscopic polymer with a water-retaining hydrophilic gel, which integrates a lithium chloride-modified polypyrrole (PPy-Cl) with hygroscopic properties and poly(isopropylacrylamide) (poly-NIPAM) as shown in the Figure 10.22a–g. The obtained SMAG consists of a poly-NIPAM framework, which enlarges the gel/air interfacial area and acts as a water channel during water release. The hydrophilic PPy-Cl facilitates its thorough mixing with NIPAM monomers to form an interpenetrating network after gelation. The microstructure of SMAG was characterized by scanning electron microscopy (SEM) . Uniformly distributed pores with widths ranging from a few meters to tens of micrometers can be observed as micrometer channels for water release. In addition, the wall surface contains embedded PPy-Cl nanoparticles instead of separate spherical pure PPy-Cl. Indicating that PPy-Cl nanoparticles were successfully integrated into the poly-NIPAM framework. The saturated water contents of the corresponding dried SMAGs were 6.7, 3.4, and 0.7 $g\,g^{-1}$ when the RH was 90%, 60%, and 30%, respectively. A two-step water transfer achieved the water uptake of the SMAGs: (i) the PPy-Cl trapped the water molecules in the air, and (ii) the condensed water was transferred to the poly-NIPAM network for storage.

Polymer hydrogels, when compounded with hygroscopic salts, can achieve the desired amount of water absorption at lower humidity levels. During the hygroscopic process, water vapor is first captured and liquefied by the hygroscopic component, and liquid water is subsequently transferred to the interior of the polymer hydrogel matrix for storage. Ideally, the high swelling tendency of the hydrogel matrix should facilitate this storage. However, the salting out effect leads to unwanted aggregation of polymer chains, which inhibits the swelling of

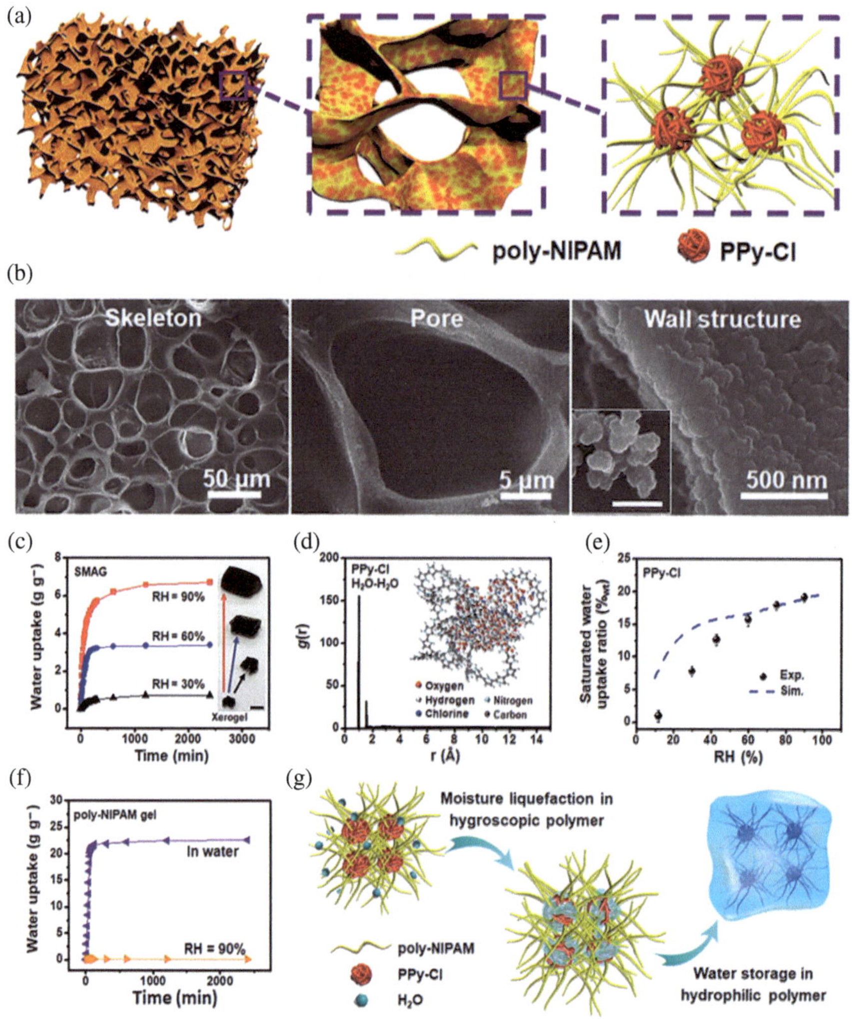

Figure 10.22 (a) Schematic of the skeleton, porous structure, and an interpenetrating network of poly-NIPAM and PPy-Cl clusters; (b) SEM images of a dried SMAG in different magnifications, showing the porous structure, which enhances the interaction between moisture and SMAG during the adsorbing process. (c) The water capturing the behavior of SMAG at RH of 30%, 60%, and 90%; Insets from bottom to top: corresponding dried and hydrated SMAG with equilibrated water uptake (scale bar: 1 cm). (d) The pair-correlation function (PCF) of water molecule pairs from molecular dynamic simulation; Inset: computing model of hydrated PPy-Cl shows the water aggregations in the central part of PPy-Cl clusters. (e) The experimental (black points) and simulated (blue dash) data of moisture absorbing based on PPy-Cl at various RH. (f) The liquid water (violet curve) and moisture (orange curve) absorption of the poly-NIPAM gel reveals a superior liquid water absorption and poor moisture capturing of the poly-NIPAM gel. (g) Schematic illustration of the moisture absorption enabled by SMAGs. Source: Reproduced from [83a] / with permission of John Wiley & Sons.

these hydrogels, limiting their water transfer and storage capacity and, ultimately, their water vapor adsorption capacity. Polyamphiphilic hydrogels not only achieve high water adsorption performance at low relative humidity due to the presence of water-absorbing ionic sites but the salting-out effect that amphiphilic hydrogels possess promotes the adsorption of water vapor by hygroscopic salts under the same conditions. In addition, because of ampholytic antipolymer electrolytes, the added salt can trap water vapor while improving the solubility and swelling properties of the hydrogels. Aleid et al. [91] synthesized PDMAPS-LiCl hygroscopic gels by synthesizing poly[2-(methacryloyloxy) ethyl]dimethyl-(3-sulfopropyl)ammonium hydroxide (PDMAPS) amphoteric hydrogels and embedding LiCl hygroscopic salts into PDMAPS as depicted in Figure 10.23a–i. LiCl not only gave the adsorbent a high water vapor adsorption capacity but also promoted the self-association dissociation

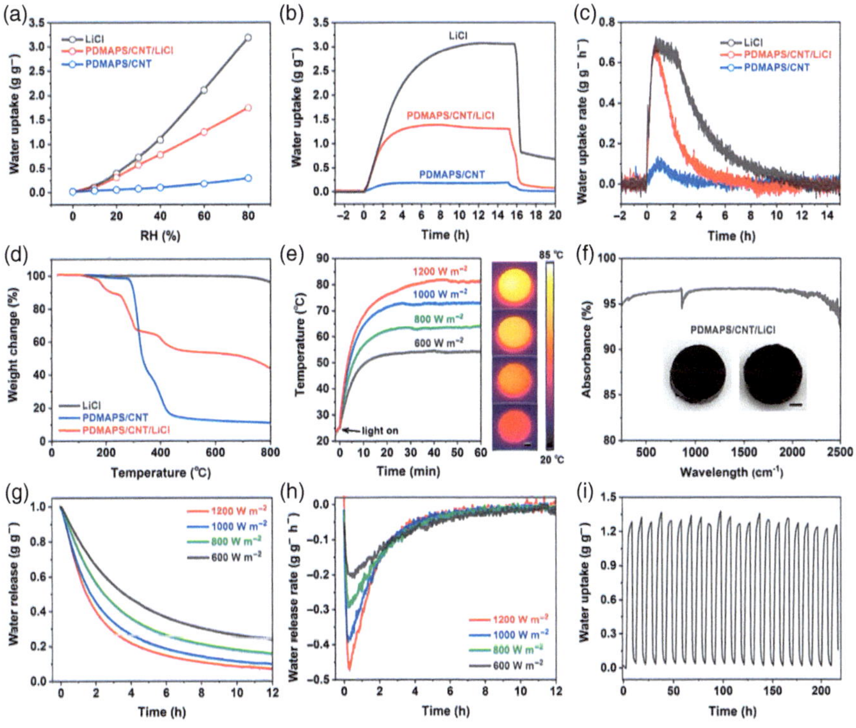

Figure 10.23 (a) Absorption isotherm at 25 °C, (b) static absorption–desorption curves, (c) static water absorption, (d) Thermal Gravimetric Analyzer (TGA) curves of pristine LiCl, PDMAPS/CNT and PDMAPS/CNT/LiCl. (e) Infrared thermograms of the surface temperature of PDMAPS/CNT/LiCl hydrogels with time under different light intensities. (f) UV–Vis–NIR (near-infrared) absorption spectra of PDMAPS/CNT/LiCl. The insets are photographs of both sides of the PDMAPS/CNT/LiCl hydrogel. Scale bars in (e) and (f) are 0.5 cm. (g) Desorption curves and (h) desorption rates of PDMAPS/CNT/LiCl hydrogels at different light intensities. (i) Twenty-two water vapor adsorption–desorption tests were performed on PDMAPS/CNT/LiCl hydrogels. The adsorption was carried out at 25 °C, 60% RH for 6 h, and the desorption was carried out at 80 °C, ~0% RH for 4 h. Source: [91] / American Chemical Society / CC BY 4.0.

of the cationic and anionic groups of PDMAPS. In addition, the PDMAPS-CNT-LiCl hydrogel maintained a high water adsorption–desorption capacity (~1.26 g g^{-1}) after 22 cycles, indicating a high cycling stability. The salting-out effect was evaluated and verified by DFT calculations, indicating that the salting-out effect gives the amphiphilic hydrogel matrix an enhanced swelling capacity, which results in a higher atmospheric water adsorption performance of the adsorbent [92]. It was also noted that the water desorption capacity and kinetics were improved by reducing the thickness of the PDMAPS-CNT-LiCl hydrogel.

Biomass aerogels as a porous material have also been shown to be effective in loading adsorbents such as hygroscopic salts. Zhang et al. [12] innovatively converted magnesium chloride ($MgCl_2$) into super hygroscopic magnesium complexes (MC), which can effectively solve the problems of salt deliquescence and agglomeration as shown in the Figure 10.24a–h. The MC was integrated with a photothermal aerogel consisting of sodium alginate and carbon nanotubes (SA/CNTs) to form a composite aerogel. SA contains abundant oxygen-containing groups (—OH,—COO, and=O), which make it a hydrophilic framework, enabling the prepared SA/CNTs/MC aerogel to store water for high-capacity water absorption. The synergistic combination of these components enables SA/CNTs/MC to absorb water vapor from the air, store the absorbed water in an extended form, and achieve solar-driven rapid water release. In addition, the layered porous structure of SA/CNTs/MC provided abundant water vapor channels, with a fast absorption rate of 0.84 kg kg^{-1} h^{-1} in the 1 h at 70% RH. SA/CNTs/MC showed high water absorption rates over a wide RH range, reaching 5.43 and 0.27 kg kg^{-1} RH 95% and 20%, respectively. The multistage porous structure enabled the prepared SA/CNTs/MC to be cycled 12 times per day at 70% RH, which corresponded to a water yield of 10.0 L kg^{-1}·day^{-1}.

The pore structure of the aerogel also has an important influence on the kinetics of water adsorption and desorption. Xu et al. [93] developed a vertically aligned nanocomposite adsorbent LiCl@rGO-SA by confining LiCl in a reduced rGO and sodium alginate (SA) matrix, as shown in Figure 10.25a. The LiCl@rGO-SA adsorbent integrates the chemisorption of LiCl, deliquescence of LiCl-H_2O, and adsorption of LiCl aqueous solution into one, and the water absorption rate is up to three times its mass. Ultra-high water adsorption capacities of 1.01, 1.5_2, and 2.76 g g^{-1} were obtained at RH of 15%, 30%, and 60% as shown in Figure 10.25b. Figure 10.25c curves of the dynamic water uptake process of pure LiCl under the same test conditions; Figure 10.25d the dynamic water uptake process of LiCl@rGO-SA at different temperatures (60, 70, 80, and 90 °C) curves of the dynamic desorption process of water from LiCl@rGO-SA; In addition, the vertically aligned and hierarchical pores of the rGO-SA matrix acted as water vapor transfer channels, providing the significant advantage of lower water transfer resistance, which enabled LiCl@rGO-SA to exhibit rapid adsorption–desorption kinetics (Figure 10.25e). The LiCl@rGO-SA adsorbent was expanded into a solar-powered fast-cycling continuous atmospheric water collector with synergistic heat and mass transfer enhancement. The water collector can achieve eight continuous water collection cycles per day with good cycling stability (Figure 10.25f). The LiCl@rGO-SA adsorbent has an ultra-high water yield in dry air, as high as 1050–2120 mL kg^{-1}·day^{-1} (Figure 10.25g), and does not consume any other energy.

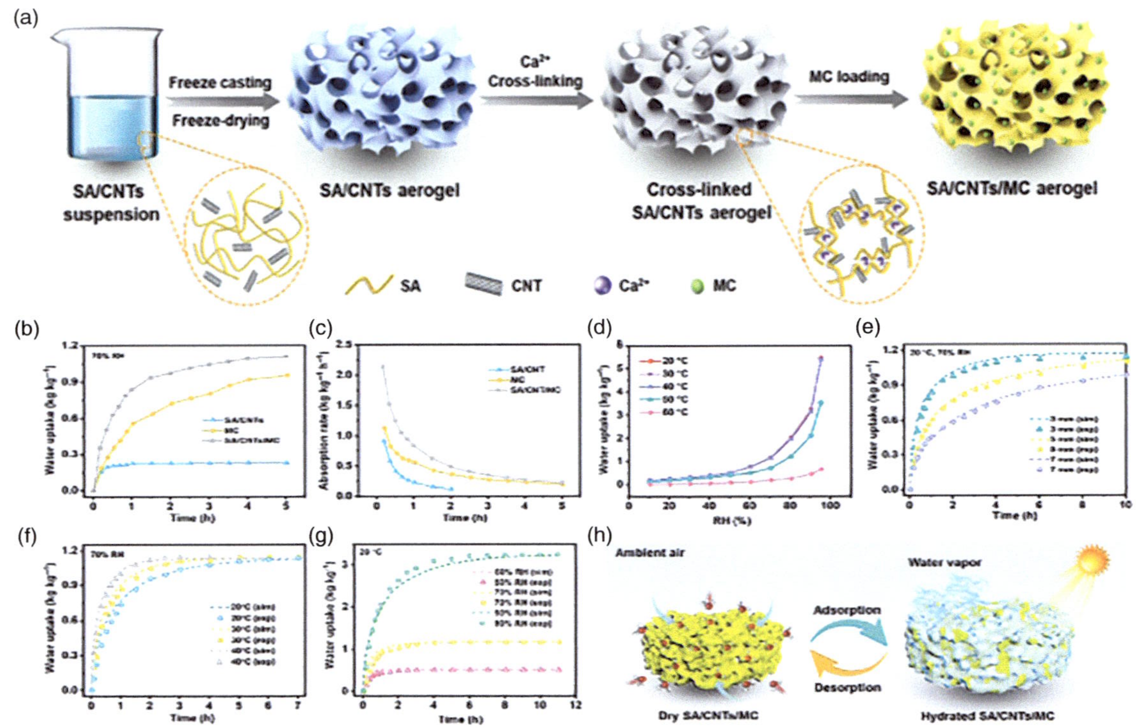

Figure 10.24 (a) Schematic illustration of the fabrication of SA/CNTs/MC aerogels. (b) Water absorption of SA/CNTs, MC, and SA/CNTs/MC versus time under environmental conditions (24 °C, 70% RH). The thickness of SA/CNTs and SA/CNTs/MC aerogels is 3 mm. (b-c) Water uptake and absorption rate of SA/CNTs/MC over time. (d) Water absorption isotherms of SA/CNTs/MC at different temperatures. (e) The dynamic water absorption of SA/CNTs/MC with different thicknesses under the same condition (20 °C, 70% RH). (f–g) The dynamic water absorption of SA/CNTs/MC under different temperatures and RHs, respectively. The thickness of SA/CNTs/MC is 3 mm. e, f, and g were conducted in a sorption analyzer (Aquadyne DVS) dynamic vapor sorption (DVS). sim and exp., simulated and experimental results. (h) Schematic illustration of the water vapor absorption/desorption process. Source: Reproduced from [12] / with permission of John Wiley & Sons.

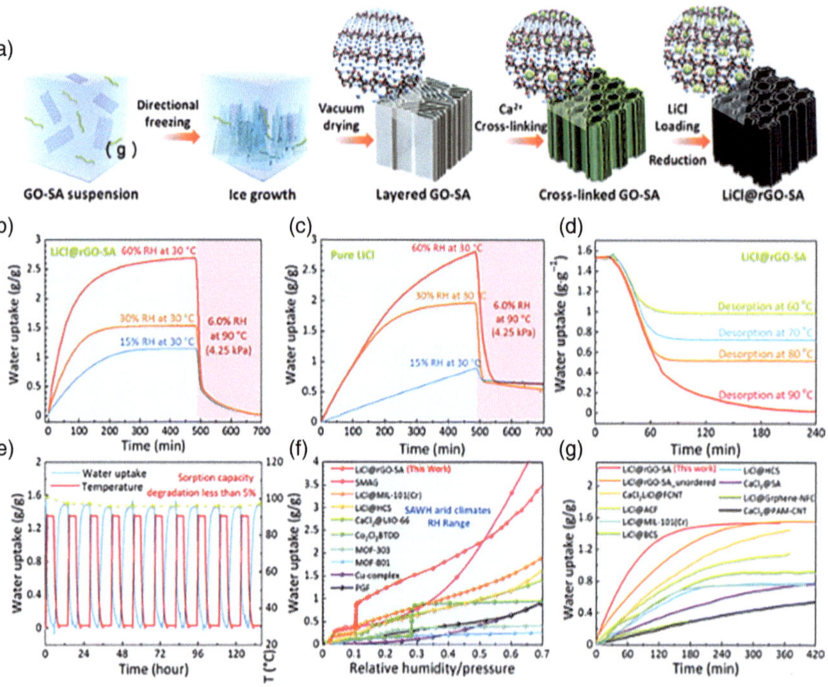

Figure 10.25 (a) Schematic of the synthesis route. The GO (black flakes) and SA (green chains) suspensions were mixed in optimal proportions in a designed mold, and then the directed self-assembly of GO flakes was achieved by ice-growth-induced directed freezing. A layered GO-SA matrix was obtained after vacuum drying. The SA chains were then crosslinked with Ca^{2+}, and LiCl crystals were loaded on the surface of the GO-SA matrix by salt solution immersion and drying. (b) dynamic water uptake process of LiCl@rGO-SA at different working conditions of 15%, 30%, and 60% RH at 30 °C and dynamic desorption process at 90 °C with RH = 6.0%; (c) curves of the dynamic water uptake process of pure LiCl under the same test conditions; (d) the dynamic water uptake process of LiCl@rGO-SA at different temperatures (60, 70, 80, and 90 °C) curves of the dynamic desorption process of water from LiCl@rGO-SA; (e) 10 water sorption–desorption cycling tests of the nanocomposite sorbent at 30 °C, 1.2 kPa (30% RH) for sorption and 90 °C, 4.2 kPa for desorption, showing satisfactory water sorption–desorption cycling stability. (f) The water sorption capacity comparison of recently reported state-of-the-art sorbents. (g) The water sorption kinetics comparison of recently reported state-of-the-art composite sorbents. Source: Reproduced from [93] / with permission of The Royal Society of Chemistry.

Additionally, Zhu et al. [94] constructed a hierarchically structured scaffold by 3D printing combined with freeze-drying, using cellulose nanofiber (CNF) carbon nanotubes and lithium chloride as building blocks in the Figure 10.26a–f. The multi-scale porous bottom layer derived from 3D printing and freeze-drying traps water vapor from the air with the help of hygroscopic lithium chloride. The top CNF/CNT layer converts the absorbed solar energy into heat and evaporates the absorbed water. The milli/μm multi-scale pores can effectively confine the lithium chloride while providing more exposed active areas for water adsorption and release, and the 3D printed structure accelerates the water absorption and evaporation

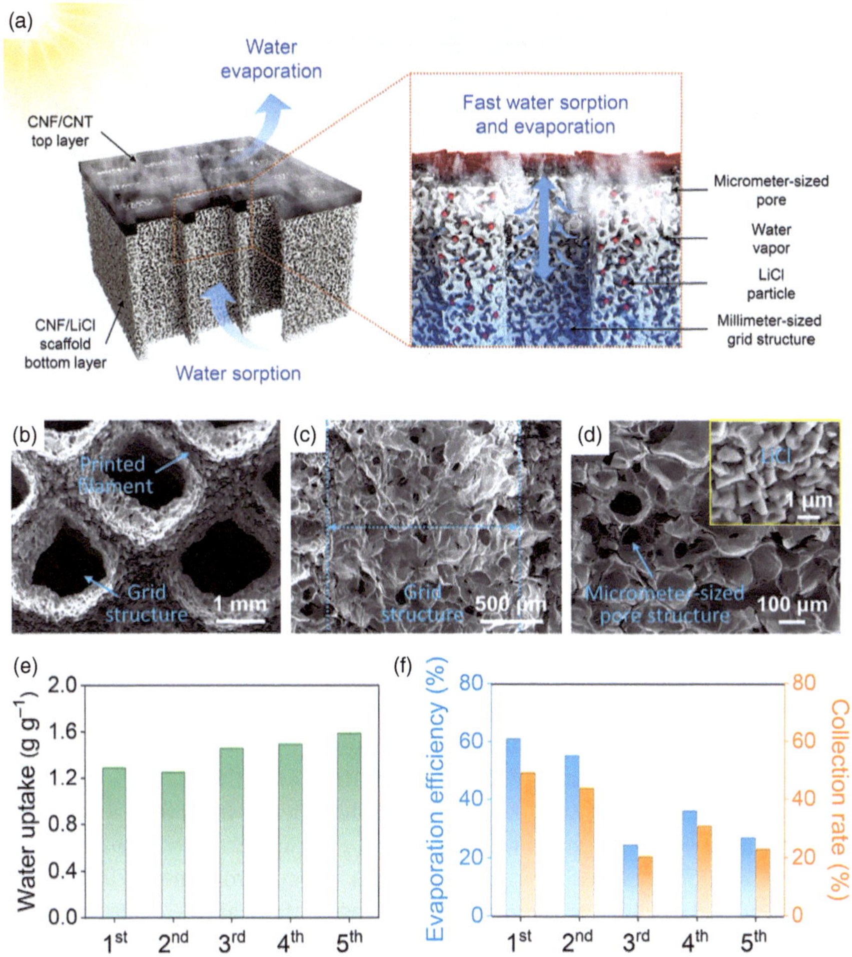

Figure 10.26 (a) Schematics illustrating the structure and working principle of the 3D printed bilayer scaffold sorbents. SEM images showing the (b) top-view and (c) vertical-section morphology of the printed grid structure of CNF/LiCl-2.0. (d) SEM image showing the micron-sized pore structure along the longitudinal direction of CNF/LiCl-2.0. The inset in (d) shows the morphology of LiCl particles loaded in the pores of CNF/LiCl-2.0. (e) Water uptake, (f) evaporation efficiency, and water collection rate of bilayer scaffold-1.5 array during five cycles. Source: Reproduced from [94] / with permission of John Wiley & Sons.

kinetics. Compared to conventional lyophilized aerogels, the 3D printed scaffolds have greater than 1.6-fold water absorption and more than 2.4-fold water release.

10.4.3 Other Types of Composite Adsorbents

During long-term desorption, some of the conventional moisture-absorbing hydrogels mentioned above may undergo excessive hydrolytic absorption, which in turn leads to severe mechanical deformation or even failure. By changing the solvent and replacing the hydrogel with an ionic gel or an organogel, the stability of the atmospheric hygroscopicity of the hydrogel can be effectively improved. The ionic gel is a solid mixture with ionic conductivity, usually made of a mixture of polymer organic polymers and salt electrolyte liquids with electrolyzable ions, where the polymer molecular chains are connected or entangled with each other to form a spatial mesh structure, and the structural voids are filled with anions and cations as a dispersion medium. The interactions between the molecular chains can be designed through the selection of anions and cations in the gel to achieve some special properties, such as excellent mechanics, non-volatility, thermal stability, and ionic conductivity. When used in atmospheric water harvesting, ionic gels can avoid material failure due to salinization effects, thus achieving high moisture absorption at low humidity [95]. Ni et al. [96] utilized the ionic liquid 1-ethyl-3-methylimidazole acetate ([EMIM][Ac]) with hygroscopic properties and dispersed it into a hydrophilic poly(acrylic-co-2-acrylamide-2-methylpropanesulfonic acid) (P(AA-co-AMPSA) network to prepare hygroscopic ionic gels, which were further applied to the evaporative refrigeration of thermoelectric generator (TEG), and to effectively enhance its thermoelectric properties. This hygroscopic ionogel showed excellent water absorption performance and achieved a moisture absorption efficiency of $0.96\,g\,g^{-1}$ for 12 h at 25 °C and RH = 90%.

In addition, organogels offer many excellent properties and a wide variety of functions. Typically, polymers and low molecular weight organogelators are polymerized to form organogels. Second, organogels can also be formed by simple soaking of a hydrogel, which contains water that can be displaced by a single or a mixture of solvents, such as glycerol, alditol, sorbitol, and the like [97]. There are also a few studies applying organogels to atmospheric water-harvesting techniques. Tillandsia is a typical epiphyte, which does not rely on rhizomes to absorb water from the soil but grows well by absorbing water directly from the air through its leaves. Under the action of osmotic pressure inside the leaf, the adsorbed water can realize the directional transport from the outermost tissue to the internal network and finally stored in the perfect tissue system inside the leaf to achieve continuous and rapid water absorption, as shown in Figure 10.27a [98]. Ni et al. [98] were inspired to investigate an integrated hygroscopic photothermal organogel (POG) for efficient AWH, as shown in Figure 10.27b. Due to the high polarity of the characteristic groups, such as hydroxyl and carboxyl groups, on the surface of the POG, water molecules in the air can be captured through hydrogen bonding interactions. Subsequently, the absorbed water molecules are guided by the osmotic pressure of the medium inside the glycerol to further diffuse into the interior of the POG, which can rapidly activate the adsorption binding sites on the surface. Moreover, these water molecules will take the place of the glycerol molecules on the polymer backbone by breaking the old hydrogen bonds to form new ones while releasing the initially bound glycerol molecules for new water adsorption and transportation, as shown in Figure 10.27c. Thus, these related properties give POG a sustainable high-capacity moisture adsorption capacity. The results showed that the ultra-high equilibrium moisture adsorption capacity of the composite POG was $16.01\,kg\,m^{-2}$ at 90% relative humidity, and the daily water production was as

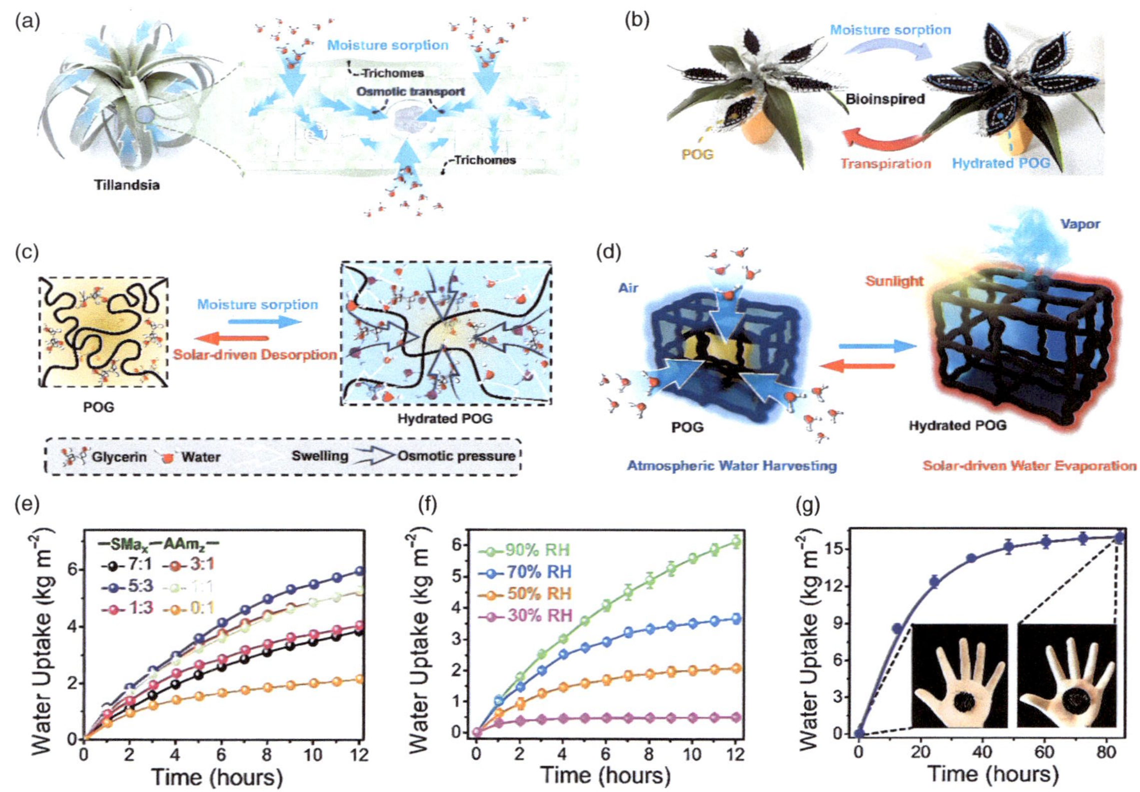

Figure 10.27 (a) Tillandsia absorbs water from the air for its vital activities; (b) hygroscopic POG mimicking the hygroscopic and transpirational behavior of Tillandsia; (c) Schematic representation of the hygroscopic effect of the POG, where water molecules are captured on the surface of the POG, then transported to its interior by osmotic pressure and stored inside the POG by swelling; (d) solar-powered POG based on realized atmospheric water harvesting; (e) moisture absorption of POG with different polymer networks showing that POG has the best moisture absorption when the ratio of sodium methacrylate (SMA) to AAm monomer is 5 : 3; (f) moisture absorption of optimized POG in the range of 30–90% RH; (g) long term moisture absorption of POG at RH = 90%, the inserted images are for the initial POG and the hydrated POG, respectively, and the hydrated POG had obvious solubilization. Source: Reproduced from [98] / with permission of John Wiley & Sons.

high as 2.43 kg m^{-2} in the actual outdoor test, as shown in Figure 10.27e–g. Such a bio-inspired integrated wicking material system could provide a new way to achieve continuous and high-capacity atmospheric wicking.

In summary, five elements, namely, the strength of interaction between water and adsorbent, porosity, pore structure, auxiliary facilities, and hydrophilicity, need to be considered in the study of polymer gels for atmospheric environmental water harvesting. Specifically, (i) for adsorbent materials, a stronger interaction between the material and the water molecules facilitates rapid moisture uptake; however, at the same time, this interaction should not be too strong so that the adsorbent can be regenerated quickly with less energy. (ii) High porosity is conducive to increasing the adsorption sites of the adsorbent; however, it will reduce the thermal conductivity of the adsorbent and increase the interfacial thermal resistance, thus hindering desorption. (iii) The larger the pore volume, the more water it can hold; however, due to the capillary coalescence phenomenon, desorption requires a higher temperature, i.e. "S" type hysteresis. In addition, the hydrophilicity of the pores of the porous adsorbent facilitates the penetration of water into the pores. Still, it may hinder the diffusion of the penetrating water within the pores, thus slowing down the overall rate of water absorption. (iv) In terms of device design, it is important to note that additional heating or cooling devices in conventional AWHs can accelerate desorption and condensation but, at the same time, greatly increase the weight of the device and reduce the mass capacity of the AWH. (v) Hydrophilic condensing surfaces can promote droplet nucleation during water vapor condensation due to their lower nucleation energy threshold, but difficulties in droplet separation then offset this gain. Superhydrophobic surfaces facilitate the shedding of condensed droplets but at the cost of lower nucleation density, which is another key limitation in water collection.

10.5 Reflections and Perspectives in SAWH

While humankind has made every effort to develop technology to make life better, it has neglected its impact on nature. Sewage is discharged freely, making the already scarce freshwater resources even more scarce. The rapid development of science and technology is based on the destruction of nature, and the greenhouse effect is becoming more and more serious. This has also led to many uncertain changes in the climate on a global scale, with the frequency of fog and the total number of days in many regions decreasing from year to year, which poses a great threat to the production and life of people who use non-phase-change fog collection as a source of fresh water. In the global water cycle system, the atmospheric water in the form of vapors and liquid droplets is a huge untapped freshwater resource equal to one-eighth of the total freshwater resources of rivers and lakes. Collecting these neglected but ubiquitous water resources in the atmosphere and rationally and fully utilizing them can greatly alleviate the difficulty of freshwater scarcity in arid regions.

In the global water cycle system, the abundant freshwater resources stored in the air are often neglected. If the atmospheric water resources can be fully utilized, the shortage of freshwater resources in arid areas can be greatly alleviated. Based on this, various atmospheric water collection methods have been proposed, which can be roughly categorized into passive cooling atmospheric water collection and adsorption-based atmospheric water collection. Passive cooling of atmospheric water collection requires external energy input to keep the temperature of the equipment surface below the dew point of atmospheric water so that the ambient water undergoes a phase change and is collected. The three most used methods of passive cooling are vapor compression cooling, thermoelectric cooling, and membrane-assisted cooling. The advantage of vapor compression cooling is that when the air is humid, the quality of water collected per unit of time is larger. Still, when the air is dry, the efficiency is low, the evaporation temperature is low, and the dew point temperature of the exit process air limits it. Thermoelectric cooling of the equipment structure is also very tightly arranged compared with the vapor compression cooling method. The mass of water collected per unit of time is low, energy consumption is high, and thermoelectric cooling requires power input. At the same time, the hot end of the need for effective heat dissipation, the cooling temperature is low. Membrane-assisted cooling requires effective heat dissipation at the hot end and low cooling temperature, but the mass transfer performance of the membrane needs to be improved. This passively cooled atmospheric water collection device should strive to concentrate the cooling energy provided by the radiatively cooled material, i.e. achieve concentrated high cooling power in a smaller area to achieve condensation at lower humidity. In addition, the thermal management of passively cooled atmospheric water-harvesting equipment should be adequately managed to reduce ambient heat input and further improve the overall cooling performance of the system.

Adsorption-based atmospheric ambient water harvesting has clear advantages in atmospheric water-harvesting technology, with high adaptability and less restriction to environmental conditions, as adsorbents have a high affinity for capturing water from the air over a wide humidity range. However, novel atmospheric water adsorbents with high water absorption, low energy demand, fast kinetics, and cyclic stability urgently need to be researched and developed to improve the atmospheric water-harvesting efficacy of adsorbents. By introducing hygroscopic salt crystals into MOFs, polymer gels, and other porous matrices to form molecularly modified composite atmospheric water-harvesting adsorbents. Unlike MOFs and polymer gels, which have a relatively limited moisture range for adsorption, molecularly modified composite atmospheric water-harvesting adsorbents can achieve excellent water absorption over a wide moisture range. The photothermal effect is intervened in the adsorption technology, and the release of collected water is driven by solar energy to realize the industrial requirement of low energy consumption and low cost.

Further, the state of water in the polymer gel determines the evaporation and transport behavior of water. Modulating the state of water by modulating the polymer network can further reduce the enthalpy of water evaporation and change the kinetics of water molecule adsorption and desorption in the bulk phase. However, most atmospheric water-harvesting technologies face difficulties with long cycle times and low water collection per unit mass due to small air mass flow rates and ambient vapor condensation temperatures. In addition, the mismatch between relatively low adsorption kinetics and desorption kinetics is considered, leading to a decrease in the coordination of collection efficacy and release efficacy. Due to the large difference in humidity and temperature between daytime and nighttime, the cyclicity of adsorption–desorption needs to be further improved by rationally designing the atmospheric water collection adsorbent to be collected and released alternately to maintain a high collection rate during the night and a high desorption rate during the day.

Based on the above, atmospheric water-harvesting technology can be innovated and developed in the future from the following two aspects: (i) Molecular and structural design of atmospheric water-harvesting materials. (ii) Most of the adsorbent materials suffer from a mismatch between the adsorption rate and the desorption rate. Among the currently reported adsorbents, the maximum adsorption rate is only one-fourth of the desorption rate, which makes it difficult to meet the rapid adsorption-desorption characteristics required for continuous systems. Thus, improving the adsorption-desorption kinetics becomes one of the key challenges in realizing a continuous system. In addition, the use of adsorption base materials with anisotropic thermal conductivity and oriented pore structure can bring additional mass and energy efficiency. Second, the adsorption-based atmospheric water collection equipment should be rationally optimized. In the future, the compactness and craftsmanship of the equipment should be improved, and convenient and intelligent equipment for water collection in arid environments should be designed. The mismatch between adsorption and desorption rates should be further addressed at the system level to achieve higher water release. Current AWH assessment strategies face the following challenges: (i) lack of reasonable comparability and fairness among different AWH technologies due to the lack of a unified consensus on the basic performance metrics; and (ii) the specific direction of the entire AWH research field is unclear due to the lack of comprehensive guidelines on the objectives. AWH needs to be further rationalized with high adsorption capacity, rapid heat and mass transfer performance, and low regeneration energy consumption, as well as long-term cyclic stability and low production process, costs to further realize the expectations of sustainable water production and large-scale use in arid water-scarce regions.

References

1 a) Wang, M., Sun, T., Wan, D. et al. (2021). *Nano Energy* 80: 105569. b) Qi, H., Wei, T., Zhao, W. et al. (2019). *Advanced Materials* 31: 1903378. c) Xia, M., Cai, D., Feng, J. et al. (2023). *Advanced Functional Materials* 33: 2214813.

2 Shi, W., Guan, W., Lei, C., and Yu, G. (2022). *Angewandte Chemie International Edition* 61: e202211267.

3 Wang, X., Li, X., Liu, G. et al. (2019). *Angewandte Chemie International Edition* 58: 12054.

4 a) Song, Y., Zeng, M., Wang, X. et al. (2024). *Advanced Materials* 36: 2209134. b) Graeber, G., Díaz-Marín, C.D., Gaugler, L.C. et al. (2023). *Advanced Materials* 2211783. c) Sun, C., Zhu, Y., Shao, P. et al. (2023). *Angewandte Chemie International Edition* 62: e202217103. d) Guan, W., Lei, C., Guo, Y. et al. (2024). *Advanced Materials* 36: 2207786.

5 Zhao, F., Guo, Y., Zhou, X. et al. (2020). *Nature Reviews Materials* 5: 388.

6 Zhang, Z., Fu, H., Li, Z. et al. (2022). *Chemical Engineering Journal* 439: 135756.

7 a) Wang, J., Li, Y., Deng, L. et al. (2017). *Advanced Materials* 29: 1603730. b) Liu, H., Chen, C., Wen, H. et al. (2018). *Journal of Materials Chefogry A* 6: 18839.

8 Gou, X. and Guo, Z. (2019). *Langmuir* 36: 64.

9 Guo, Y., Zhou, X., Zhao, F. et al. (2019). *ACS Nano* 13: 7913.

10 Maity, D., Teixeira, A.P., and Fussenegger, M. (2023). *Small* 19: 2301427.

11 Li, H., Zhang, W., Liu, J. et al. (2023). *Advanced Functional Materials* 33: 2308492.

12 Zhang, X., Qu, H., Li, X. et al. (2024). *Advanced Materials* 36: 2310219.

13 Zhang, L., Tang, B., Wu, J. et al. (2015). *Advanced Materials* 27: 4889.

14 Cui, S., Liu, H., Gan, L. et al. (2008). *Advanced Materials* 20: 2918.

15 a Xu, L., Cheng, L., Wang, C. et al. (2014). *Polymer Chefogry* 5: 1573. b Zhou, X., Zhao, F., Guo, Y. et al. (2018, 1985). *Energy and Environmental Science* 11.

16 Zhou, Z., Wang, G., Pei, X., and Zhou, L. (2023). *Chemical Engineering Journal* 474: 145605.

17 Zhang, P., Liu, F., Liao, Q. et al. (2018). *Angewandte Chemie International Edition* 57: 16343.

18 Ren, H., Tang, M., Guan, B. et al. (2017). *Advanced Materials* 38: 29.

19 Zhao, Q., Fan, T., Ding, J. et al. (2011). *Carbon* 49: 877.

20 Yang, Z.-P., Ci, L., Bur, J.A. et al. (2008). *Nano Letters* 8: 446.

21 a Jiang, Q., Tian, L., Liu, K.-K. et al. (2016). *Advanced Materials* 28: 9400. b Ghasemi, H., Ni, G., Marconnet, A.M. et al. (2014). *Nature Communications* 5: 4449.

22 Brassard, D., Alavitabari, S., Raphael, W. et al. (2024). *Advanced Sustainable Systems* 8: 2300309.

23 Anjali, C., Nidhisha, V., Amrutha, T.P. et al. (2024). *ACS Applied Nano Materials* https://doi.org/10.1021/acsanm.4c00110.

24 Song, Y., Xu, N., Liu, G. et al. (2022). *Nature Nanotechnology* 17: 857.

25 Kelly, K.L., Coronado, E., Zhao, L.L., and Schatz, G.C. (2003). *The Journal of Physical Chefogry B* 107: 668.

26 Boriskina, S.V., Ghasemi, H., and Chen, G. (2013). *Materials Today* 16: 375.

27 Liu, C., Cai, C., Ma, F. et al. (2020). *Journal of Colloid and Interface Science* 560: 103.

28 Liu, Z., Liu, L., Lu, H. et al. (2017). *Scientific Reports* 7: 43803.

29 He, Y., Fu, T., Wang, L. et al. (2023). *Chemical Engineering Journal* 472: 144786.

30 Hu, Y., Wang, Y., Fang, Z. et al. (2022). *Journal of Materials Chefogry A* 10: 15116.

31 Luo, F., Liao, T., Liang, X. et al. (2022). *Journal of Cleaner Production* 373: 133838.

32 a) Tu, Y., Wang, R., Zhang, Y., and Wang, J. (2018). *Joule* 2: 1452. b) Zhou, X., Lu, H., Zhao, F., and Yu, G. (2020). *ACS Materials Letters* 2: 671.

33 Bai, Z., Wang, P., Xu, J. et al. (2024). *Science Bulletin* 69: 671.

34 Lu, H., Shi, W., Guo, Y. et al. (2022). *Advanced Materials* 34: 2110079.

35 Yuan, Y., Zhang, H., Yang, F. et al. (2016). *Renewable and Sustainable Energy Reviews* 54: 761.

36 Aristov, Y.I., Restuccia, G., Cacciola, G., and Parmon, V.N. (2002). *Applied Thermal Engineering* 22: 191.

37 Wang, D., Zhang, J., Yang, Q. et al. (2014). *Applied Energy* 113: 734.

38 a)Eun, T.H., Song, H.K., Han, J.H. et al. (2000). *International Journal of Refrigeration* 23: 64. b) Jakubinek, M.B., Zhan, B.-Z., and White, M.A. (2007). *Microporous and Mesoporous Materials* 103: 108.

39 a) Li, R., Shi, Y., Wu, M. et al. (2020). *Nano Energy* 67: 104255. b) Li, R., Shi, Y., Shi, L. et al. (2018). *Environmental Science and Technology* 52: 5398. c) Yang, K., Shi, Y., Wu, M. et al. (2020). *Journal of Materials Chefogry A* 1887: 8.

40 Hanikel, N., Prévot, M.S., Fathieh, F. et al. (2019). *ACS Central Science* 5: 1699.

41 Hanikel, N., Pei, X., Chheda, S. et al. (2021). *Science* 374: 454.

42 a) Nguyen, H.L., Hanikel, N., Lyle, S.J. et al. (2020). *Journal of the American Chemical Society* 142: 2218. b) Fathieh, F., Kalmutzki, M.J., Kapustin, E.A. et al. *Science Advances* 4: eaat3198.

43 a) Zheng, Z., Hanikel, N., Lyu, H., and Yaghi, O.M. (2022). *Journal of the American Chemical Society* 144: 22669. b) Won, S., Jeong, S., Kim, D. et al. (2022). *Chemistry of Materials* 34: 273.

44 Hanikel, N., Prévot, M.S., and Yaghi, O.M. (2020). *Nature Nanotechnology* 15: 348.

45 Garibay, S.J. and Cohen, S.M. (2010). *Chemical Communications* 46: 7700.

46 a) Li, J.-X., Chang, G.-G., Tian, G. et al. (2021). *Advanced Functional Materials* 31: 2102868. b) Ameloot, R., Vermoortele, F., Hofkens, J. et al. (2013). *Angewandte Chemie International Edition* 52: 401. c) Jeremias, F., Lozan, V., Henninger, S.K., and Janiak, C. (2013). *Dalton Transactions* 42: 15967.

47 a) Hossain, M.I., Cunningham, J.D., Becker, T.M. et al. (2019). *Chemical Engineering Science* 203: 346. b) Fang, Z., Bueken, B., De Vos, D.E., and Fischer, R.A. (2015). *Angewandte Chemie International Edition* 54: 7234.

48 a) Hao, L., Li, X., Hurlock, M.J. et al. (2018). *Chemical Communications* 54: 11817. b) Cai, G. and Jiang, H.-L. (2017). *Angewandte Chemie International Edition* 56: 563. c) Schaate, A., Roy, P., Godt, A. et al. (2011). *Chemistry--A European Journal* 17: 6643.

49 a) Cavka, J.H., Jakobsen, S., Olsbye, U. et al. (2008). *Journal of the American Chemical Society* 130: 13850. b) Furukawa, H., Gándara, F., Zhang, Y.-B. et al. (2014). *Journal of the American Chemical Society* 136: 4369.

50 a) Wang, Y., Yang, Q., Yi, F. et al. (2021). *ACS Applied Materials and Interfaces* 13: 29916. b) Pakamorė, I., Rousseau, J., Rousseau, C. et al. (2018). *Green Chemistry* 20: 5292.

51 a) Kim, H., Yang, S., Rao, S.R. et al. (2017). *Science* 356: 430. b) Kim, H., Rao, S.R., Kapustin, E.A. et al. (2018). *Nature Communications* 9: 1191.

52 Zheng, Z., Nguyen, H.L., Hanikel, N. et al. (2023). *Nature Protocols* 18: 136.

53 Cong, S., Yuan, Y., Wang, J. et al. (2021). *Journal of the American Chemical Society* 143: 20055.

54 Lai, J.-Y., Wang, T.-Y., Zou, C. et al. (2022). *Journal of Membrane Science* 661: 120879.

55 Barjasteh, M., Vossoughi, M., Bagherzadeh, M., and Pooshang Bagheri, K. (2022). *International Journal of Pharmaceutics* 618: 121647.

56 Fang, Y., Yang, Z., Li, H., and Liu, X. (2020). *Environmental Science and Pollution Research* 27: 4703.

57 Yuan, B., Wang, X., Zhou, X. et al. (2019). *Chemical Engineering Journal* 355: 679.

58 a) Xian, S., Peng, J., Zhang, Z. et al. (2015). *Chemical Engineering Journal* 270: 385. b) Plaza, M.G., Ribeiro, A.M., Ferreira, A. et al. (2012). *Microporous and Mesoporous Materials* 153: 178.

59 Mahmoodi, N.M., Abdi, J., Oveisi, M. et al. (2018). *Materials Research Bulletin* 100: 357.

60 a) Horcajada, P., Serre, C., Vallet-Regí, M. et al. (2006). *Angewandte Chemie International Edition* 45: 5974. b) Barjasteh, M., Vossoughi, M., Bagherzadeh, M., and Bagheri, K.P. (2023). *Chemical Engineering Journal* 452: 138987.

61 a) Qiu, S., Wang, Y., Wan, J. et al. (2020). *Applied Surface Science* 525: 146511. b) Song, N., Sun, Y., Xie, X. et al. (2020). *Desalination* 492: 114601.

62 Férey, G., Mellot-Draznieks, C., Serre, C. et al. (2005). *Science* 309: 2040.

63 Zou, M., Dong, M., and Zhao, T. (2022). *International Journal of Molecular Sciences* 23: 9396.

64 Tan, B., Luo, Y., Liang, X. et al. (2019). *Industrial and Engineering Chefogry Research* 58: 2983.

65 a) Soltanolkottabi, F., Talaie, M.R., Aghamiri, S., and Tangestaninejad, S. (2019). *Journal of Environmental Management* 250: 109416. b) Celeste, A., Paolone, A., Itié, J.-P. et al. (2020). *Journal of the American Chemical Society* 142: 15012.

66 Taghizadeh, M. (2023). *S. Tahami* 39: 707.

67 Zhao, H., Li, Q., Wang, Z. et al. (2020). *Microporous and Mesoporous Materials* 297: 110044.

68 Feng, Y., Ge, T., Chen, B. et al. (2021). *Cell Reports Physical Science* 2: 100561.

69 Silva, M.P., Ribeiro, A.M., Silva, C.G. et al. (2021). *Adsorption* 27: 213.

70 Cadiau, A., Lee, J.S., Damasceno Borges, D. et al. (2015). *Advanced Materials* 27: 4775.

71 a) Lu, H., Shi, W., Zhang, J.H. et al. (2022). *Advanced Materials* 34: 2205344. b) Zhao, F., Zhou, X., Shi, Y. et al. (2018). *Nature Nanotechnology* 13: 489.

72 a) Shi, Y., Ilic, O., Atwater, H.A., and Greer, J.R. (2021). *Nature Communications* 12: 2797. b) Kallenberger, P.A. and Fröba, M. (2018). *Communications Chefogry* 1: 28.

73 a) Kou, S., Peters, L.M., and Mucalo, M.R. (2021). *International Journal of Biological Macromolecules* 169: 85. b) Vallejo-Domínguez, D., Rubio-Rosas, E., Aguila-Almanza, E. et al. (2021). *Ultrasonics Sonochefogry* 72: 105417.

74 a) Yang, J., Chen, X., Zhang, J. et al. (2021). *International Journal of Biological Macromolecules* 189: 53. b) Yu, F., Chen, Z., Guo, Z. et al. (2020). *ACS Sustainable Chefogry and Engineering* 8: 7139.

75 Wei, S., Ching, Y.C., and Chuah, C.H. (2020). *Carbohydrate Polymers* 231: 115744.

76 a) Sahariah, P. and Másson, M. (2017). *Biomacromolecules* 18: 3846. b) Pal, P., Pal, A., Nakashima, K., and Yadav, B.K. (2021). *Chemosphere* 266: 128934. c) Hamedi, H., Moradi, S., Hudson, S.M., and Tonelli, A.E. (2018). *Carbohydrate Polymers* 199: 445. d) Guo, Y., Dundas, C.M., Zhou, X. et al. (2021). *Advanced Materials* 33: 2102994.

77 a) Lee, K.Y. and Mooney, D.J. (2012). *Progress in Polymer Science* 37: 106. b) Goh, C.H., Heng, P.W.S., and Chan, L.W. (2012). *Carbohydrate Polymers* 88: 1.

78 a) Rahmanian, V., Pirzada, T., Wang, S., and Khan, S.A. (2021). *Advanced Materials* 33: 2102892. b) Zainal, S.H., Mohd, N.H., Suhaili, N. et al. (2021). *Journal of Materials Research and Technology* 10: 935. c) Zhang, W., Chen, X., Zhao, J. et al. (2023). *Nano Energy* 108: 108196.

79 a) Hou, Y., Sheng, Z., Fu, C. et al. (2022). *Nature Communications* 13: 1227. b) Sun, J., An, B., Zhang, K. et al. (2021). *Journal of Materials Chefogry A* 9: 24650. c) Lin, X., Wang, P., Hong, R. et al. (2022). *Advanced Functional Materials* 32: 2209262.

80 a) Choi, S.W., Guan, W., and Chung, K. (2021). *Cell* 184: 4115. b) Liu, X., Liu, J., Lin, S., and Zhao, X. (2020). *Materials Today* 36: 102. c) Guo, Y., Bae, J., Fang, Z. et al. (2020). *Chemical Reviews* 120: 7642.

81 Yang, M., Luo, H., Zou, W. et al. (2022). *ACS Applied Materials and Interfaces* 14: 24766.

82 Zhang, H., Zeng, H., Eklund, A. et al. (2022). *Nature Nanotechnology* 17: 1303.

83 a) Zhao, F., Zhou, X., Liu, Y. et al. (2019). *Advanced Materials* 31: 1806446. b) Yan, C., Kramer, P.L., Yuan, R., and Fayer, M.D. (2018). *Journal of the American Chemical Society* 140: 9466.

84 Xu, X., Ozden, S., Bizmark, N. et al. (2021). *Advanced Materials* 33: 2007833.

85 Zhang, Z., Wang, Y., Li, Z. et al. (2022). *ACS Applied Materials and Interfaces* 14: 55295.

86 An, H., Chen, Y., Wang, Y. et al. (2023). *Chemical Engineering Journal* 461: 141955.

87 Yang, C., Wu, H., Yun, J. et al. (2023). *Advanced Materials* 35: 2210235.

88 Xu, J., Li, T., Chao, J. et al. (2020). *Angewandte Chemie International Edition* 59: 5202.

89 Ge, L., Feng, Y., Xue, Y. et al. (2023). *Small Structures* 4: 2300055.

90 Hu, Q. and Zhao, H. (2019). *Journal of Molecular Structure* 1182: 191.

91 Aleid, S., Wu, M., Li, R. et al. (2022). *ACS Materials Letters* 4: 511.

92 a) Gao, Y., Zhang, W., Li, L. et al. (2023). *Chemical Engineering Journal* 452: 139248. b) Kamio, E., Yasui, T., Iida, Y. et al. (2017). *Advanced Materials* 29: 1704118.

93 Xu, J., Li, T., Yan, T. et al. (2021). *Energy and Environmental Science* 14: 5979.

94 Zhu, P., Yu, Z., Sun, H. et al. (2024). *Advanced Materials* 36: 2306653.

95 a) Yang, S., Tao, X., Chen, W. et al. (2022). *Advanced Materials* 34: 2200693. b) Ye, Y., Oguzlu, H., Zhu, J. et al. (2023). *Advanced Functional Materials* 33: 2209787. c) Wang, M., Zhang, P., Shamsi, M. et al. (2022). *Nature Materials* 21: 359.

96 Ni, F., Xiao, P., Zhang, C., and Chen, T. (2022). *Matter* 5: 2624.

97 a) Chen, F., Zhou, D., Wang, J. et al. (2018). *Angewandte Chemie International Edition* 57: 6568. bZhang, S., Shi, W., and Wang, X. (2022). *Science* 377: 100.

98 Ni, F., Qiu, N., Xiao, P. et al. (2020). *Angewandte Chemie International Edition* 59: 19237.

11

Applications of Fog Harvesting

Atmospheric fog collection is an effective method for harnessing natural resources by capturing water droplets suspended in the atmosphere to obtain freshwater [1]. Collectors designed according to various fog collection technologies can be integrated with systems that require a water supply [2]. In addition to harvesting water from the atmosphere, these systems can provide a stable and reliable source of water for electrical energy generation, chemical reactions, and other micro-reaction processes. However, it is important to note that the efficiency of fog collection is influenced by several factors, including natural conditions such as atmospheric humidity, temperature, and wind speed, as well as the design and operational methods of the collection equipment [3]. Therefore, proper planning and implementation of specific applications should be tailored to the local environment and needs. Below are several common applications of fog collection.

11.1 Collection of Water Resources

(i) **Domestic water.** In arid regions or areas lacking fresh water resources, fog collection can provide a reliable supply of potable water [4]. Fog collection devices condense and collect water droplets from the atmosphere, treating them to produce drinkable water. Natural organisms in water-scarce environments can survive by collecting airborne droplets through specialized structures [5]. To achieve maximize water collection in a single day, Jiang and coworkers. [6], inspired by the layered structure of the dogwood, designed a freshwater collection device that not only facilitates fog collection but also integrates solar-thermal energy conversion and humid-air solar steam generation (SSG) technologies. This innovative approach enhances access to freshwater throughout the day (Figure 11.1a). Hydrogel bionic S. viridis hydrogels with long-range vertically aligned lamellar structures (BSC-VAH), featuring cone spikes, grooves, and internal channels, was developed using 3D printing and the ice template method. Figure 11.1b illustrates the process of capturing and transporting droplets within the hydrogel BSC-VAH. The accelerated refreshment of water

Bioinspired Materials for Fog Harvesting: Principles, Design, Preparation, and Applications, First Edition. Zhiguang Guo and Weimin Liu.

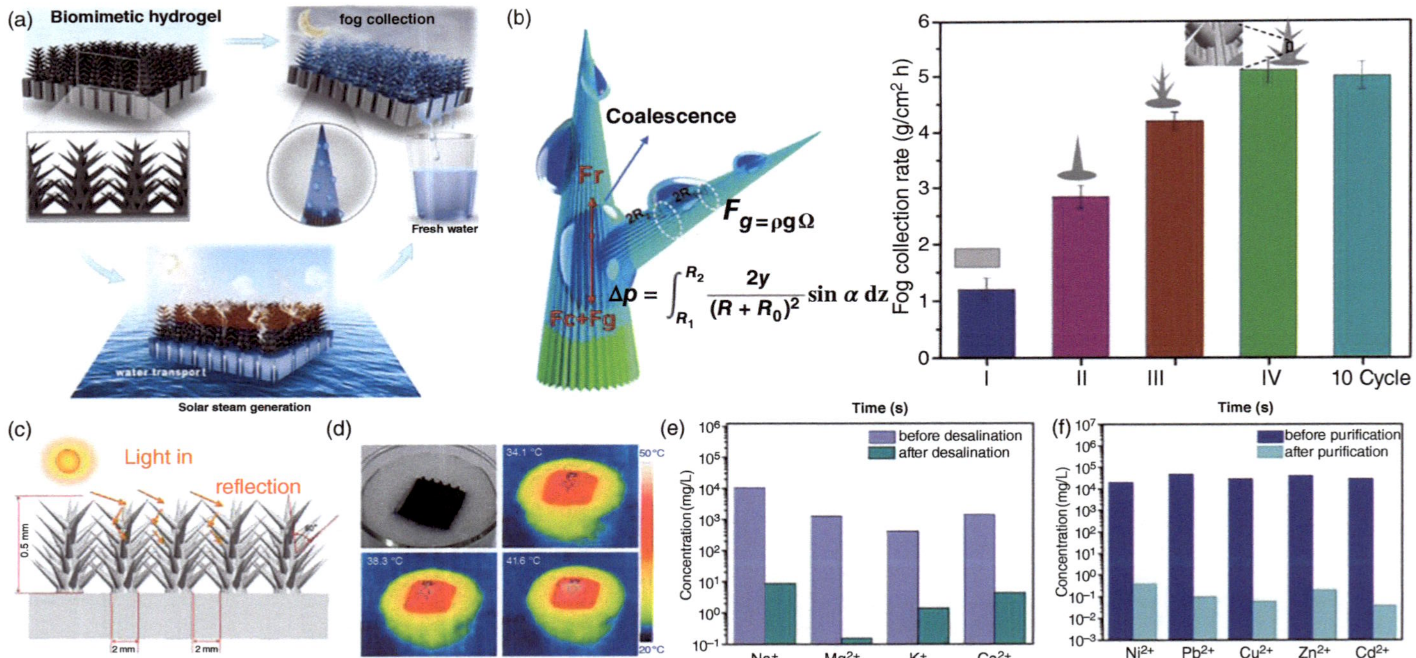

Figure 11.1 (a) Conceptual representation of the bionic hydrogel featuring multilevel structures capable of harvesting fresh water for 24 h. (b) Mechanism of fog collection on the BSC-VAH hydrogels. Comparison of fog collection for the hydrogels with different structures. (c) Schematic showing the absorption of sunlight and the excellent thermal localization of the bionic hydrogels. (d) The infrared images of the samples under one sun irradiation of the hydrogels. (e) Measured concentrations of the four primary ions in an actual seawater sample before and after desalination. Heavy ion concentration before and after purification. Source: Reproduced with permission from [6] / The Royal Society of Chemistry.

nucleation sites on the surface of the hydrogel cones is attributed to the synergistic transport effect of the Laplace pressure induced by the cone structure, along with the capillary action of the grooves. Figure 11.1b also presents the average water collection efficiency of hydrogels with various structures, including flat-structured hydrogels, cone-structured hydrogels, and dendritic and bionic groove-structured hydrogels. Among these, the hydrogels with dendritic and fluted structures exhibited the highest water collection efficiency, measuring $5.1\,g\,cm^{-2}\,h^{-1}$. These results underscore the significance of tree and groove structures in fog collection. The mechanism by which the bionic hydrogel absorbs and reflects sunlight is illustrated in Figure 11.1c. Its tapered spike array structure enhances absorption through multiple reflections of incident light, enabling it to demonstrate excellent photothermal conversion, as depicted in Figure 11.1d. Furthermore, as shown in Figure 11.1e, the concentrations of four ions (Na^+, Mg^{2+}, K^+, Ca^{2+}) in the collected water were significantly below the standards for drinking water. Additionally, the hydrogel BSC-VAH evaporator is capable of purifying water with high concentrations of heavy metals. At night, BSC-VAH, characterized by its oriented spines, grooved structure, and superhydrophilic surface, facilitates rapid water transport through the super-spreading of liquid films. During the day, long, aligned vertical channels transport water pumps to the surface of the enhanced evaporation zone. These richly structured, multi-layered systems achieve evaporation rates of $3.5\,kg\,m^{-2}\,h^{-1}$ and fog collection rates of $5\,g\,cm^{-2}\,h^{-1}$.

Unlike the Laplace-driven biomimetic multilayer structure mentioned above, Yao and coworkers [7] drew inspiration from the wettability gradient observed on the surfaces of desert beetles (Figure 11.2a), and successfully combined picosecond laser processing with chemical treatment to successfully create heterogeneous surfaces (HSs) on aluminum. These HSs are utilized for the capturing of droplets in multi-fog road sections and for desalinating seawater (Figure 11.2b). The resulting HS features both superhydrophilic and superhydrophobic regular periodic stripe structures. The spacing between the superhydrophilic and superhydrophobic regions can be adjusted based on the processing path. As illustrated in Figure 11.2c, the surface exhibits a 44% improvement in multi-fog section harvesting efficiency, achieving a rate of $0.53\,kg\,m^{-2}\,h^{-1}$ compared to the pristine aluminum plate. Figure 11.2d and e illustrates the seawater schematic. Two different methods were tested for seawater desalination. The first method involves positioning the sample at an angle and pumping seawater onto the surface of the sample using a superhydrophilic stripe. The second method consists of placing the sample horizontally on a sponge block, allowing it to float above the water surface and transport water to the HS surface through its small pores. The evaporation rate achieved was as high as $2.35\,kg\,m^{-2}\,h^{-1}$, with an efficiency of 52.39% under one solar irradiation. This functional surface can be utilized to obtain freshwater resources in coastal areas and arid regions where water vapor is relatively abundant, providing valuable reference and guidance for freshwater harvesting. This approach is promising for addressing the issue of water scarcity. In conclusion, the prepared aluminum-based evaporator demonstrates good stability.

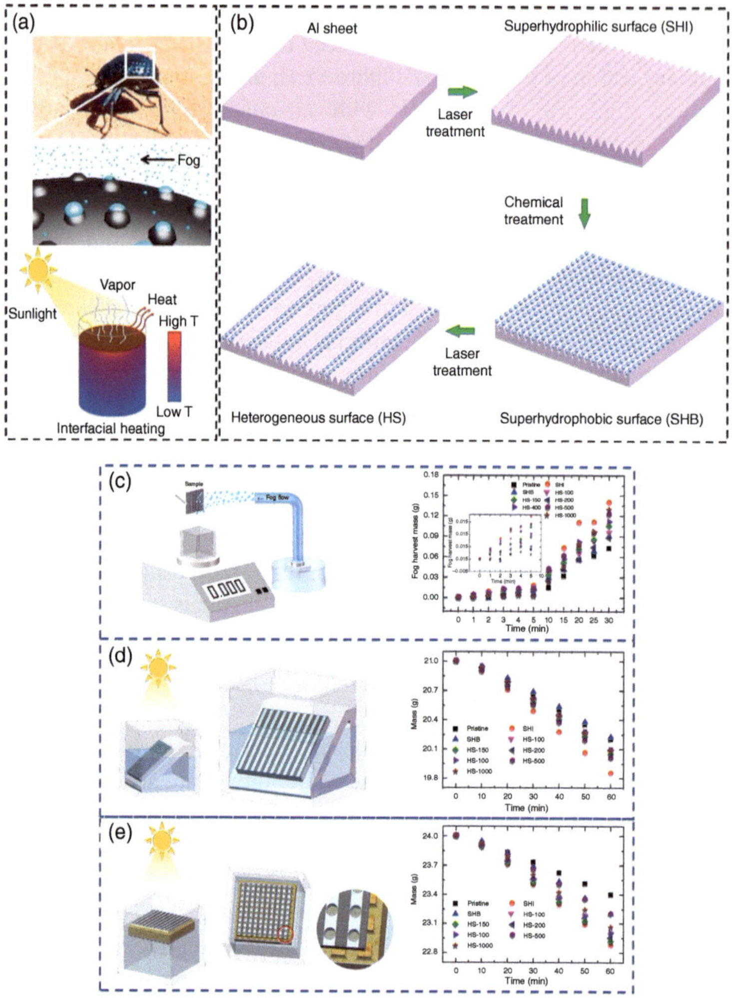

Figure 11.2 (a) Schematic diagram of fog harvesting by a desert beetle and interface evaporation. (b) Preparation process for HS. (c) Relationship between the fog-harvesting mass and time. (d–e) Change of the desalination quality of different samples over time in two different desalination modes. Source: Reproduced with permission from [7] / American Chemical Society.

In addition to considerations regarding collection volume, making fog-collected water potable presents another significant challenge. Thomas and coworkers [8] demonstrated a fog collection system that not only harvested water but also degraded organic pollutants simultaneously. At the core of this system design are TiO_2 nanoparticles embedded in a polymer matrix coating. The degradation of organic pollutants depends on the photosensitive properties of TiO_2, where photogenerated electrons and holes, produced by the photovoltaic effect, migrate to different locations on the TiO_2 surface in response to an electric field.

The photogenerated electron (e−) on the TiO_2 surface is readily captured by oxidizing substances in water, such as dissolved oxygen, resulting in the formation of superoxide radicals ($\cdot O_2^-$). Meanwhile, the hole (h+) oxidizes organic compounds that have adsorbed onto the TiO_2 surface or first removes hydroxide ions (OH−) from the surface of the TiO_2. OH− and H_2O molecules adsorbed on the surface of TiO_2 can be oxidized into hydroxyl radical (·OH). Both ·OH and $\cdot O^{2-}$ possess strong oxidizing capabilities, enabling them to break the chemical bonds of various organic compounds, Consequently, they can oxidize most of organic matter and inorganic pollutants, and mineralizing them into small inorganic molecules, CO_2, and H_2O and other substances [9]. As shown in Figure 11.3a, the TiO_2 nanocomposite coating, when activated by UV irradiation, remains reactive even in the absence of light, thereby purifying water without the need for continuous illumination from sources such as sunlight. Indoor experiments, as shown in Figure 11.3b–d, confirmed that the engineered hydrophilic surface outperformed the hydrophobic surface in both collection and treatment efficiencies. In outdoor tests, the device demonstrated

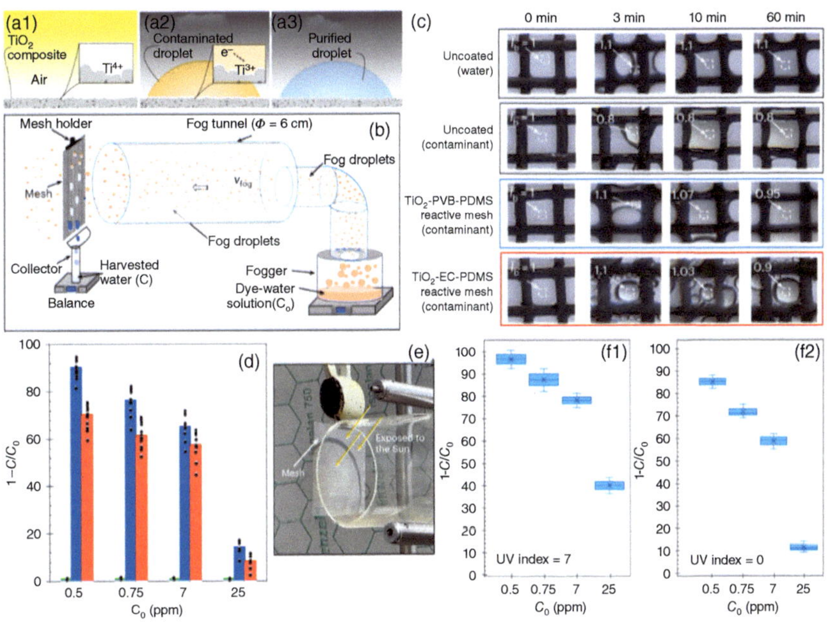

Figure 11.3 (a) The schematic diagram shows how a TiO_2 nanocomposite coating can be activated by UV irradiation and remain reactive even in the dark. (b) The schematic diagram shows the laboratory-scale fog-harvesting setup used to generate contaminated fog. (c) Color image sequence showing the mesh surfaces during fog-harvesting experiments in the presence ($C_0 = 25$ ppm) and absence ($C_0 = 0$ ppm) of contaminated fog. (d) Quantifying the effect of C_0 on the ability of the mesh coating to treat the intercepted fog: uncoated mesh (green bar), TiO_2-PVB-PDMS reactive mesh (blue bar) and TiO_2-EC-PDMS reactive mesh (red bar). (e–f) Outdoor demonstration of fog harvesting and treatment on sunny and cloudy days. Source: [8] / Springer Nature / CC BY 4.0.

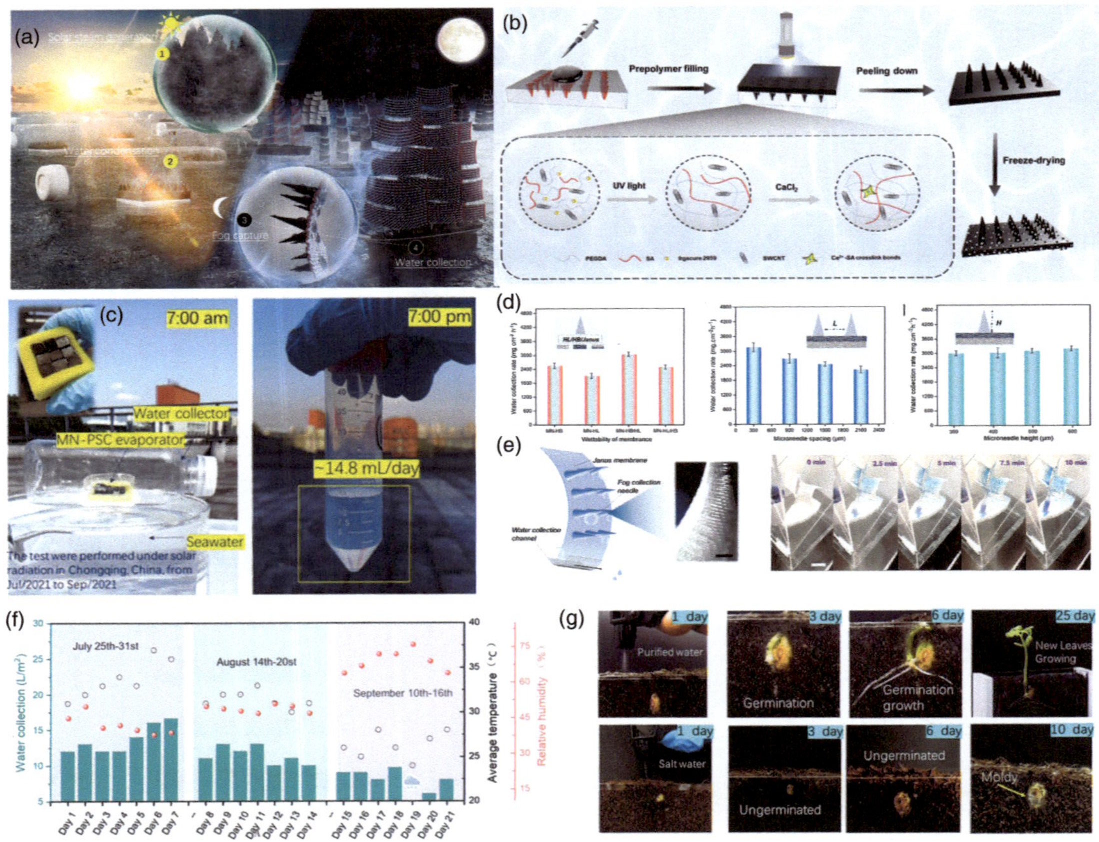
(a)
Water collection
Fog capture
(b)
Prepolymer filling
Peeling down
Freeze-drying
UV light
CaCl2
(c)
7:00 am
Water collector
MN-PSC evaporator
Seawater
The test were performed under solar radiation in Chongqing, China, from Jul/2021 to Sep/2021
7:00 pm
~14.8 mL/day
(d)
(e)
Janus membrane
Fog collection needle
Water collection channel
(f)
Water collection (L/m²)
Average temperature (°C)
Relative humidity (%)
July 25th-31st
August 14th-20st
September 10th-16th
(g)
1 day
Purified water
3 day
Germination
6 day
Germination growth
25 day
New Leaves Growing
1 day
Salt water
3 day
Ungerminated
6 day
Ungerminated
10 day
Moldy

Figure 11.4 (a) Schematic illustration of microneedle array architecture based all-day freshwater harvesting. (b) Schematic diagram of the evaporator preparation process of an MN-PSCx. (c) Photograph of a large-scale MN-PSC1% evaporator, water receiver based on a plastic bottle, and the actual volume of purified water from 7 a.m. to 7 p.m. (d) Wettability of membrane substrates, fixed the spacing at 1200 μm, height at 600 μm; spacing of microneedle arrays with MN-HB/HL membrane, fixed the height at 600 μm; and height of the microneedles with MN-HB/HL membrane, fixed the spacing at 300 μm. (e) Schematic diagram of the fog collection process of the arch-shaped MN-HB/HL (scale bar: 500 μm). Photograph of water collection process of arch-shaped MN-HB/HL on a small boat (scale bar: 3 cm). (f) Outdoor solar desalination, purified water volume during July, August, and September, in Chongqing, China. (g) Comparison of germination of soybeans treated with purified water and salt water. Source: Reproduced from [11] / with permission of John Wiley & Sons.

excellent fog collection performance, achieving water treatment efficiencies greater than 85% (Figure 11.3e,f). This study highlights that the continuous production of water through passive purification offers an energy-free solution to the challenge of water scarcity.

(ii) **Agricultural irrigation.** Fog collection can provide a vital water source for agriculture, particularly in arid regions. Atmospheric water droplets are captured using fog collection devices and utilized for crop irrigation, ensuring a reliable water supply [10].

In pursuit of maximizing daily freshwater collection, Tan and coworkers [11] combined fog collection at night with solar evaporation during the day to design a device that enables for all-weather freshwater collection (Figure 11.4a). The preparation process for the freshwater collection device, MN-PSC; PEGDA/SA/CNT, also as polyethylene (glycol) diacrylate/sodium alginate/carbon nano tube, is illustrated in Figure 11.4b. In which gel-forged microneedle arrays with Janus wettability are fabricated on a porous platform using a simple and controlled top-down microforming process. The porosity within the microneedles can be easily enhanced through an additional freeze-drying treatment. As shown in Figure 11.4d, optimization experiments were conducted to assess the effects of microneedles on the membrane surface, as well as the spacing and height of the microneedles in order to achieve optimal fog collection. The optimized membrane demonstrated excellent performance in terms of fog collection volume and recovery efficiency (Figure 11.4e). In addition, the collection device performed effectively in desalination. After one day of sunlight exposure (average solar intensity: 0.6 $KW\,m^{-2}$) with the MN-PSC designed evaporator, it produced approximately about 14.8 mL of freshwater (Figure 11.4c). Figure 11.4f also recorded and analyzed the temperature, heat flux, and relative humidity parameters of the outdoor system from July to September to confirm the impact of weather temperature on the desalination experiment. To evaluate the nontoxicity of MN-PSC1, soybean seed germination was studied using both MN-PSC1% pure water and 3.5 wt% NaCl solution. After 28 days of growth, soybeans germinated in the presence of pure water and developed green shoots (Figure 11.4g), whereas those exposed to the 3.5 wt% NaCl solution showed minimal germination and ultimately molded and died.

Cai and coworkers [12] developed a thermodynamically induced interfacial condensation-enhanced fog elimination fabric (AWF-6). AWF-6 is composed of three functional materials: Boron nitride nanosheets (BNNS) for creating a highly thermally conductive interface, $NiCo_2(OH)_6$ nano-needles to generate a Laplace pressure difference, and carbon cotton to establish a wettability gradient. As illustrated in Figure 11.5a, when fog is intercepted by AWF-6, the water vapor preferentially condenses and deposits on the surface of the BNNS, and then gradually infiltrating the interior of the material through the gaps. Due to its highly ordered lattice structure, the hexagonal BNNS on the surface efficiently transfers phonons during the condensation process, allowing latent heat to be rapidly dissipated through lattice vibrations. This mechanism significantly enhances the fabric's ability to trap fog effectively. As shown in Figure 11.5b, the effective heat and mass transfer related to fog collection efficiency has been analyzed and modeled through thermodynamics and surface dynamics to validate the improvements. In

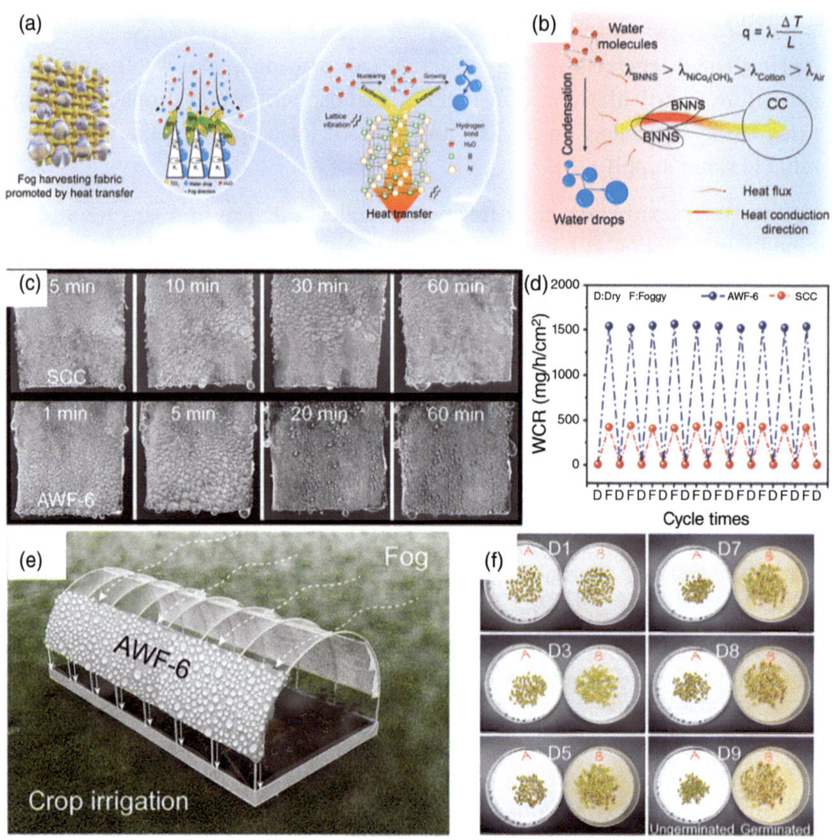

Figure 11.5 (a) Schematic diagram of fog collection for AWF-6. (b) Heat flow mechanism diagram of AWF-6 during fog harvesting. (c) Photos of SCC and AWF-6 during the cycle test. (d) Graph of cycle test for SCC and AWF-6. (e) Diagram of fog harvesting used for crop irrigation. (f) Photos of a mung bean seed germination experiment. Source: Reproduced with permission from [12] / John Wiley & Sons.

terms of practicality, AWF-6 exhibits faster fog collection efficiency and greater stability compared to superhydrophobic cotton (SCC) (Figure 11.5d). As shown in Figure 11.5e,f, fog-collecting fabrics have great potential in agricultural irrigation. Fabrics affixed to the surface of a greenhouse can effectively capture fog (Figure 11.5e). The harvested water then flows through a pipe to irrigate the soil for crops. In the soybean germination experiment, AWF-6 showed impressive results in fog-capture irrigation, with 98% of mung beans germinating on disk B after 9 days (Figure 11.5f). This study provides thermodynamic insights for the development of next-generation fiber fog-harvesting devices (FHDs).

In addition to water harvesting through a combination of bionic wettable surfaces and structural gradients, hygroscopic materials also excel in fog harvesting. Zhao and coworkers [13] reported an exceptional water-absorbing gel (SMAG), which can now be utilized to modify soil (SMAG-soil) for passive atmospheric water irrigation and sustainable solar environmental humidification in agricultural systems (Figure 11.6a). At 90% RH and 20 °C, SMAG-soil demonstrated a saturated water uptake of up to $1.1\,g\,g^{-1}$, which is nearly two orders of magnitude higher than that of sandy soil (Figure 11.6b). After milling, SMAG soils tend to aggregate into small clumps rather than remaining loose like sandy soils. Samples of fully hydrated SMAG soil and sandy soil were dried at 40% RH and 40 °C to evaluate their moisture retention capabilities. The SMAG soil retained approximately 40% of its water after four weeks, while the moisture content in the sandy soil decreased by approximately 80% water within just one week (Figure 11.6c). A simple atmospheric irrigation system, consisting of SMAG soil, transparent lids, and plastic trays was employed to demonstrate the cost-effectiveness of sustainable agriculture (Figure 11.6d). The system's water-trapping behavior facilitates uniform irrigation, while the water release behavior driven by solar radiation stabilizes the relative humidity in the growing room and prevents dehydration of the shoots (Figure 11.6e). At night when the environment is cool and humid, the system opens to the air, allowing for passive water inputs from the atmosphere and water storage in the SMAG soil. Realistically the tests were carried out on the roof of the building due to the high solar flux and temperature at this location (Figure 11.6f). Compared to radish seeds grown in sandy soil, seeds cultivated in SMAG solid germinated in just 2 days without any liquid watering on day 0 to ensure development, and the shoots grew rapidly, reaching 15 cm on day 14 without liquid irrigation (Figure 11.6g). In conclusion, Zhao and coworkers has introduced us an atmospheric water irrigation system that utilizes a superior moisture-absorbent gel (SMAG) developed to achieve sustainable irrigated agricultural production. Regardless of the availability of local liquid water resources, SMAG-soil collects water from the air and delivers it to the plants through solar heating. This solar-powered atmospheric water irrigation system enables agriculture to achieve geographical and hydrological independence. Thus, it frees crop cultivation in economically underdeveloped and arid regions from the limitations of long-distance water transportation.

Agriculture is a significant consumer of groundwater and surface water, accounting for 70% of the world's freshwater withdrawals. The increasing demand for agricultural water demand can lead to conflicts between urban, industrial, and agricultural sectors. This situation also poses threats to food security, human

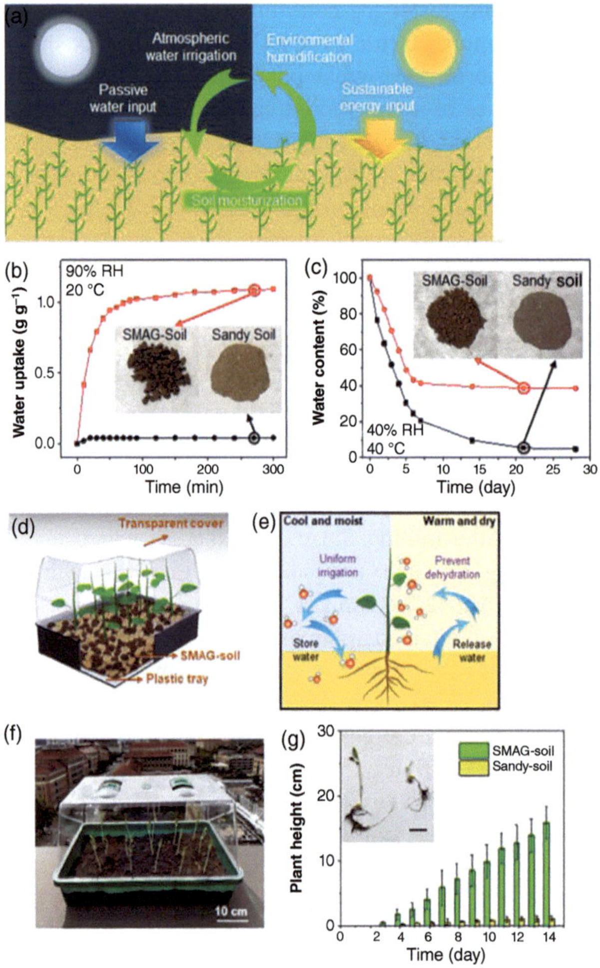

Figure 11.6 (a) Schematic of the atmospheric water irrigation based on super moisture absorbent gels (SMAG) for sustainable agriculture. (b) Water uptake of the SMAG-soil and the sandy soil at RH = 90% and 20 °C. (c) Water holding capacities of the SMAG-soil and the sandy soil at RH = 40% and 40 °C. (d) Schematic of atmospheric water irrigation system based on the SMAG-soil for plant growth. (e) Moisture capture and release cycles of the SMAG-soil. (f) Photograph of the device with growing radish plants. (g) Plant height with the SMAG-soil and the sandy soil without irrigation in 14 days. Source: Reproduced with permission from [13] / American Chemical Society.

health and the environment on a global scale. The urgent need for alternative water resources has spurred the exploration of atmospheric water as an abundant and untapped source of freshwater for irrigated agriculture. In this context, Xu and coworkers [14] developed self-contained kelp-derived hydrogel beads with various beneficial properties, including hygroscopicity, photothermal characteristics, temperature responsiveness, and durability, achieved through shape control and large-scale fabrication (Figure 11.7a). The maximum water absorption

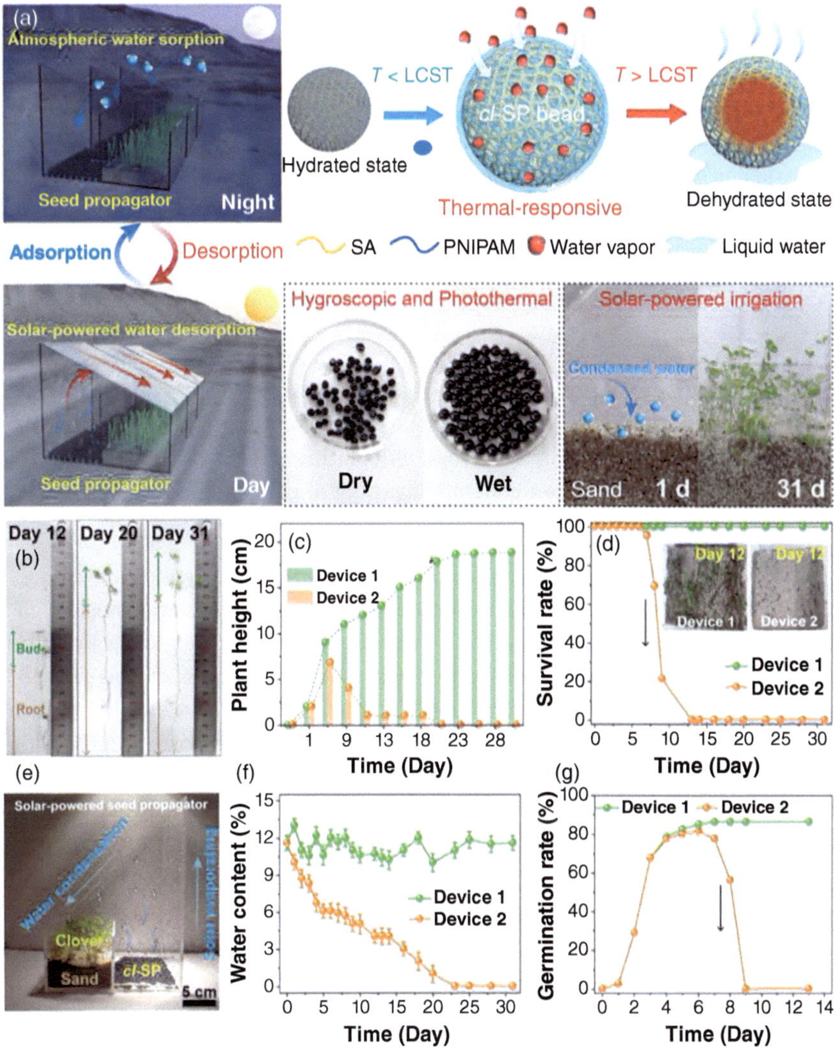

Figure 11.7 (a) Schematic illustration of hygroscopic, photothermal and temperature-responsive cl-SP beads and potential applications for atmospheric water harvesting and solar-powered irrigation. (b) Schematic illustration of the solar-powered seed propagator based on cl-SP beads. (c) Water content measurements of growing mediums in 31 days. (d) Germination rates of clover in 13 days. (e) Digital photographs of plant height in device 1 over time. (f) Summary of bud height in 31 days and (g) survival rates of clover in 31 days. Source: Reproduced with permission from [14] / The Royal Society of Chemistry.

capacity of these hygroscopic beads exceeded $5.0\,g\,g^{-1}$ at 90% relative humidity (RH). Synergistic photothermal heating and temperature-driven phase transitions facilitate multimodal hydrolysis uptake and efficient water release. More than 95% of the absorbed water can be rapidly released across a broad range of solar intensities from 0.6 to 1 solar and temperatures ranging from 40 to 60 °C. A solar and adsorption-based seed propagator was developed and demonstrated to support plant germination and growth by utilizing reversible water adsorption and desorption through hydrogel beads in a night day^{-1} $cycle^{-1}$. This system features high-quality irrigation water, zero energy consumption at the system level, and effective thermal management (Figure 11.7b–g). This research offers insights into the control and large-scale fabrication of hygroscopic hydrogels.

11.2 Other Energy Collection

In addition to being an essential resource for human survival, water is also a widely used raw material used in industrial production. The water collected by the fog collection system can serve as a reliable source for various reaction systems. In regions with abundant fog resources, integrating the fog collection system with other reaction systems can generate a variety of energy sources for the area, including electricity and hydrogen.

(i) **Electricity.** In fog collectors, aside from those simpler fog collection devices constructed from more durable materials, such as metal wire grid sheets combined with wettability and structural gradients for external fog collection, the majority of fog collection systems rely on hygroscopic materials (such as gels [15] or MOFs [16], etc.) supplemented by heating and filtration systems to produce potable water of acceptable quality. Admittedly, adsorption-based atmospheric water harvesting (AWH) is a promising technology for generating clean drinking water in arid regions where freshwater resources are limited. However, due to the slow water uptake kinetics, most adsorbents developed for this technology can only complete one cycle of water production per day. Furthermore, the heat generated during the process in current AWH systems is discarded and ultimately wasted. Han's team [17] designed and fabricated a hygroscopic aerogel material with high water absorption capacity, rapid adsorption kinetics, and excellent photothermal properties, enabling efficient solar-thermal-driven AWH across a wide range of relative humidities. In addition, this aerogel features a dual-functionality system that simultaneously generates electricity and extracts freshwater from the air. The water vapor adsorption isotherm (Figure 11.8b) shows that rGO exhibits a relatively low water adsorption capacity, while poly diallyldimethylammonium chloride (PDDA) effectively absorbs water across a broad spectrum of humidity levels. The reversible isotherm of PDDA is classified as type III, suggesting that the collected water molecules cluster around favorable sites and can be desorbed with fluctuations in relative humidity. The results presented in Figure 11.8c show that the water absorption capacity of G-PDDA is approximately 130 times greater than that of PDDA, which can be attributed to their structural differences. As shown in Figure 11.8a, the G-PDDA aerogel power generator equipped with a thermoelectric (TE) module, which harnesses the low-grade thermal energy involved in the AWH process (i.e., adsorbed heat and excess solar thermal energy) to generate electricity.

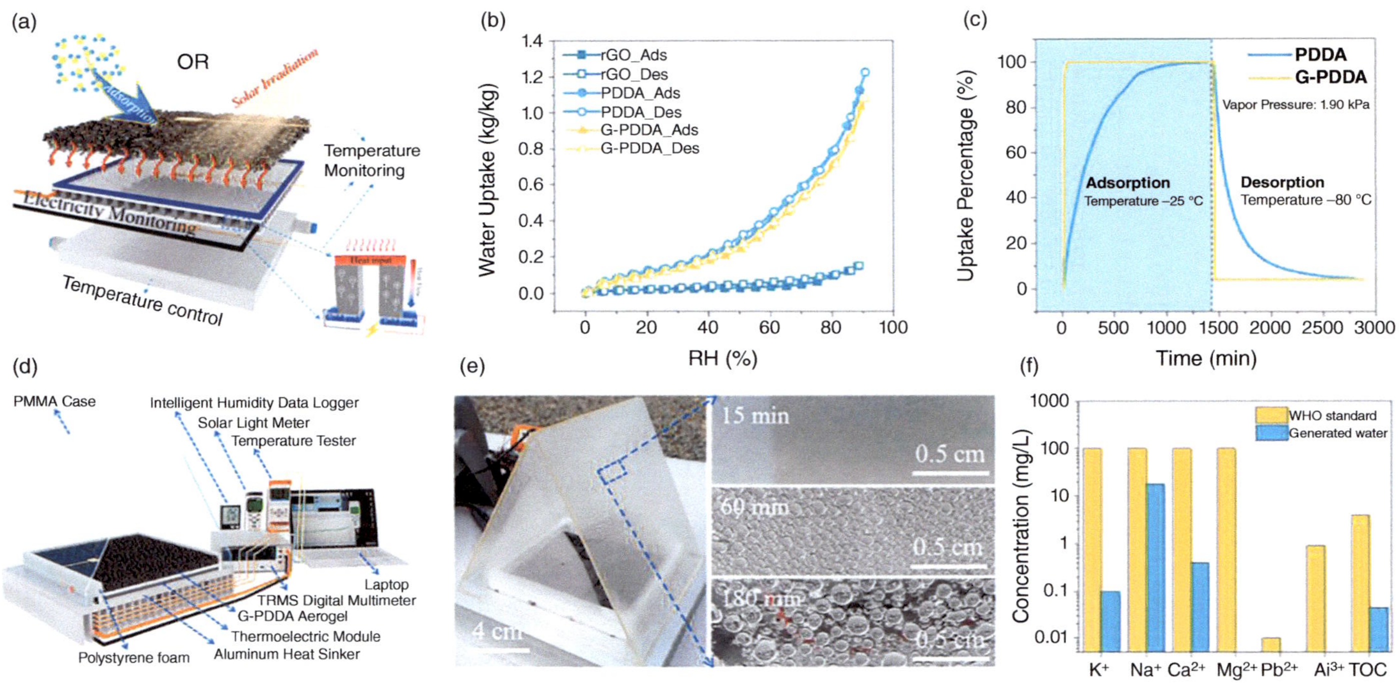

Figure 11.8 (a) Schematic diagram of the device that combines a layer of AWH sorbent and a thermoelectric module. (b) Water sorption isotherms of rGO, PDDA, and G-PDDA. (c) Water sorption kinetics of PDDA and G-PDDA, measured with 1.90 kPa vapor pressure at 25 °C (60% RH) for uptake and 80 °C (4% RH) for desorption. (d) Schematic illustration of the apparatus used for outdoor experiments. (e) the accumulation of condensed water droplets on the PMMA wall. (f) Quality assessment of the collected water. Source: Reproduced with permission from [17] / Elsevier.

Figure 11.8d shows the bifunctional system used to further verify its feasibility under outdoor, real-world weather conditions. In this device, vapor generated from the G-PDDA aerogel through solar thermal heating diffuses upward and condenses on the PMMA walls (Figure 11.8e). The small droplets coalesce into larger droplets, which eventually flow into the collection tank, where approximately 1.5 mL liquid water is collected during this cycle. The collected water was analyzed and the results showed that the ionic concentrations (K^+, Na^+, Ca^{2+}, Mg^{2+}, Pb^{2+}, Al^{3+}) and total organic carbon content (TOC) fully comply with the World Health Organization (WHO) drinking water standards (Figure 11.8f). In conclusion, this device demonstrates high practicality.

With the advancement of flexible electronics, there is an increasing demand for a power source that offers high flexibility, mobility, good stability and sustainability. Converting thermal energy into electrical energy through water evaporation presents a compelling and promising approach. Zhang and coworkers {[18] demonstrated a sustainable, flexible hydrokinetic power generator (HPG) powered by water evaporation (Figure 11.9a). It mainly consists of a superabsorbent hydrogel, a PVA@FCB@3DS film and a polyethylene terephthalate (PET)/PDMS shell. The commercial superabsorbent hydrogel is encapsulated within a PET/PDMS shell, serving as a water reservoir. Its main component is composed of branched chains of polyacrylic acid, which can absorb more than 150 times its mass and gradually release water under ambient conditions (Figure 11.9b). The path of water flow within the HPG device is shown in Figure 11.9c. Driven by the pressure drop induced by capillary action, water in the superabsorbent hydrogel seeps out and spontaneously flows through the nanochannels between the functionalized carbon nanoparticles. Simultaneously, the water molecules on the surface absorb ambient heat and evaporate, thereby driving the water in the porous PVA@FCB@3DS film to replenish by continuously flowing along the nanochannels. Among them, Figure 11.9d and e show the schematic diagrams of the initial and steady-state phases of power generation, respectively. Due to gravity, liquid water is distributed in a gradient from the bottom to the top of the PVA@FCB@3DS film, which plays a crucial role in power generation (Figure 11.9f). Under ambient conditions (20.4 °C, 55% RH), a sustained short-circuit current of up to 63 μA and an optimized output power of 8.1 μW were achieved, with a lifetime of exceeding 150 h. In addition, the flexible hydrovolt nanogenerator can endure a wide range of bending deformations (up to 120°) without significant changes in power generation. This capability allows the evaporation-induced HPG the ability to function as a flexible, portable power source for wearable electronics.

Efforts have been made to utilize nanogenerators based on carbon nanomaterials to harvest electricity from ubiquitous water for renewable and clean energy and to combat fossil fuel depletion and pollution. At the same time, although many living organisms can generate bioelectricity from water, producing bio-nanogenerators based on bio-nanomaterials, which are produced by nature in billions of tons $year^{-1}$, remains challenging. Bio-nanogenic fibers (cellulose, chitin, filament proteins, and amyloid proteins) are produced by liquid stripping of biomass or by supramolecular assembly of biomacromolecules. By virtue of their inherent

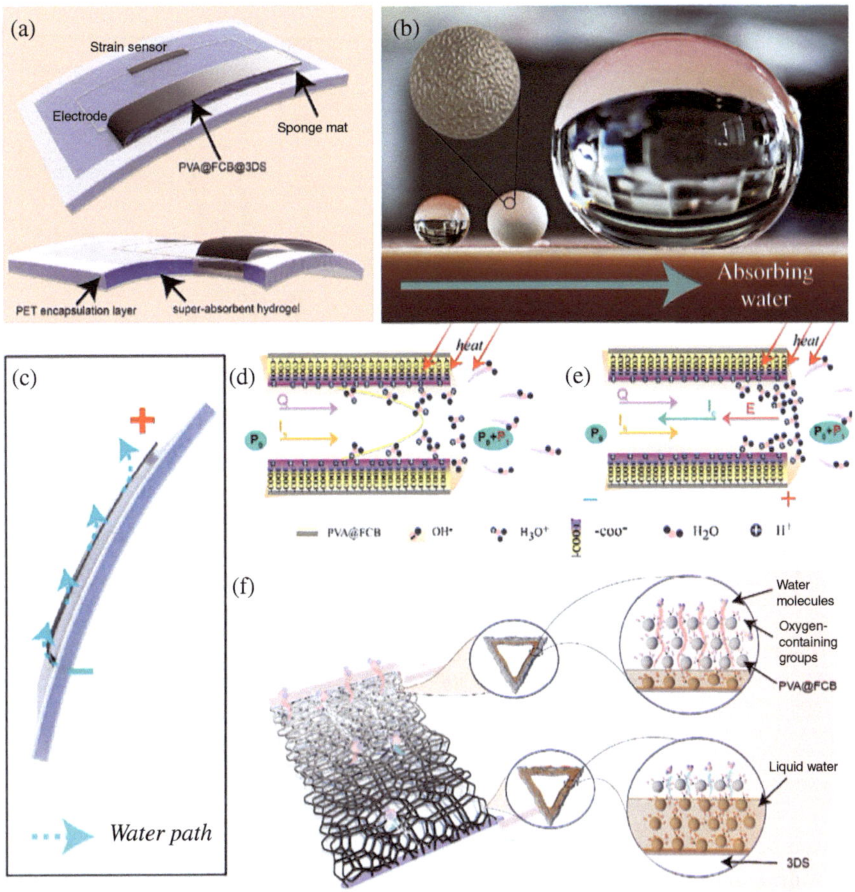

Figure 11.9 (a) Schematic diagram of the hydrovoltaic nanogenerator. (b) Typical photographs showing the water absorption states of the superabsorbent hydrogel. (c) Schematic diagram of water flowing path in HPGs. (d–e) Working principle of the water evaporation-driven HPG. (f) PVA@FCB Schematic diagram of surface liquid water distribution and flow path. Source: Reproduced with permission from [18] / Elsevier.

hydrophilicity and charged state, these fibers can capture moisture from the air and form hydrated nano-channels, akin to the ion channels found in cell membranes. When Li et al. [19] exposed the aerogels to moist air streams, the absorption and evaporation of water reached an equilibrium, resulting in the generation of flow potentials and open-circuit voltages across the aerogels (Figure 11.10a–c). These bio-nanogenerators are flexible, sustainable, biocompatible, and biodegradable, and can harvest energy from humid air streams found in nature (e.g. wind, breath, and sweat) as well as in industrial applications, making them suitable for environmentally friendly, low-cost, high-efficiency, wearable, and compact power devices. Mechanical energy is abundant in nature, yet its potential remains largely untapped. Despite extensive research, triboelectric nanogenerators (TENGs) have primarily

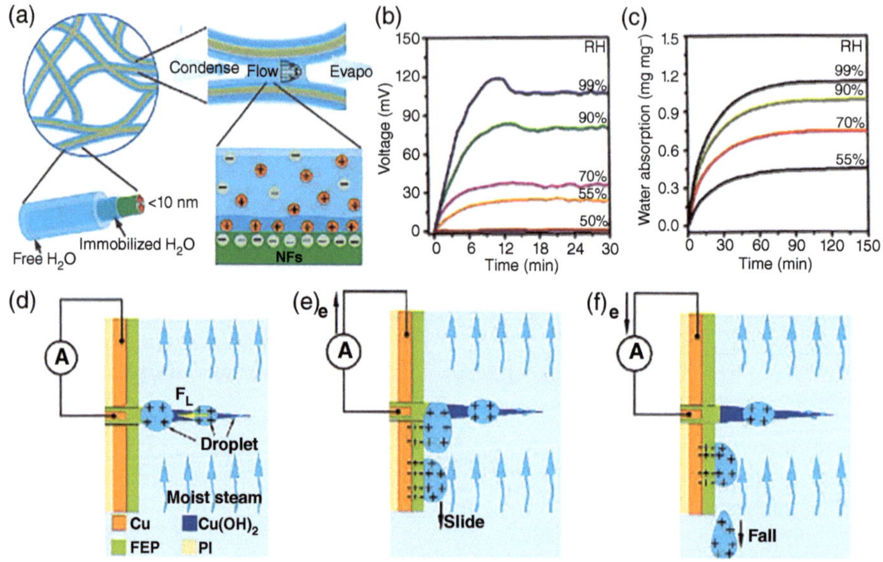

Figure 11.10 (a) Schematic illustration of hydrated channels around and between NFs. (b) Voc variation upon exposing to air flow with different RHs. Flow velocity: 15 cm·s^{-1}. (c) Water absorption curves upon exposing to air flow with different RHs. Source: Reproduced from [19] / with permission of John Wiley & Sons. (d) Water fog condensed on the tapered electrode surface forming small water droplets in a humid environment. (e) The droplet oscillated slightly after moving to the hydrophobic junction. (f) The accumulated electrons flowed back after the droplet had completely left the dielectric layer. Source: Reproduced from [20] / with permission of Elsevier.

focused on harvesting energy from bulk water, with limited studies on capturing environmental mechanical energy from small-scale water sources, such as water fog. In this study, inspired by the ability of tapered micropillars to collect water from the air, Wong and coworkers [20] proposed a solid–liquid TENG to harvest mechanical energy from water fog. The investigation into the contact area between the water droplets and the dielectric layer revealed that the current generated in the TENG is a critical factor influencing its performance. It was found that higher currents and voltages can be achieved by utilizing large-diameter conical electrodes or by configuring multiple conical electrodes in parallel. Figure 11.10d–f shows the working principle of TENG. In a humid environment, water fog condenses on the surface of the conical electrode, forming small droplets. These small droplets converge to create a larger droplet, which is then transported from the tip of the electrode to its base, driven by the Laplace force. As the droplet moves toward the hydrophobic connection point, it oscillates slightly. When the droplet grows large enough to strike the dielectric layer, it generates an induced potential and current are. The droplet subsequently leaves the tapered electrode due to gravity and slides along the dielectric layer. Once the droplet has completely exited the dielectric layer, the accumulated electrons flow back. With optimized parameters, the TENG can simultaneously generate a maximum open circuit voltage of 9.5 V and a short circuit current of 250 nA. The TENG can easily light up to seven commercial light emitting diodes (LEDs) and their brightness is very rate sensitive to varying fog currents. These findings can inform the design and fabrication of a new self-powered humidity sensor and TENG for harvesting energy from the air.

Generally, the dissociation of water requires an external stimulus, such as light energy or electricity. Vanaraj's team [21] has proposed a stimulus-free method for water dissociation using a mesoporous SnO_2-based hydrodynamic cell, which can be directly utilized to generate electricity for portable applications. The device is configured almost identically to a metal-air battery, but follows a completely different reaction pathway. The mesoporous SnO_2-based hydronic cell dissociates water molecules into hydroxide ions (OH−) and hydronium hydrate ions without any external stimulus, transporting the hydronium hydrate ions to the opposite end while functioning as a separator. The OH− reacts with the Al electrode to release electrons, while the hydronium hydrate ions are reduced at the Ag electrode, generating a potential difference between the electrodes of approximately 1000 ± 20 mV, which can be stabilized for which can be stabilized for over 3500 h (**Figure 11.11a**). The device also demonstrates the potential for generating electricity from atmospheric moisture through a copper oxide layer that act as a moisture pump (**Figure 11.11b**).

(ii) **Hydrogen energy.** Alternatives to fossil fuels, such as solar, wind and wave energy, have been extensively studied in response to the global energy crisis. However, the potential of atmospheric moisture as an alternative energy source has not been thoroughly investigated. With nearly 13 trillion kiloliters of water, atmospheric moisture can serve as a sustainable energy source and a plentiful source of hydrogen. Tan's team [22] utilized moisture in the air to absorb and

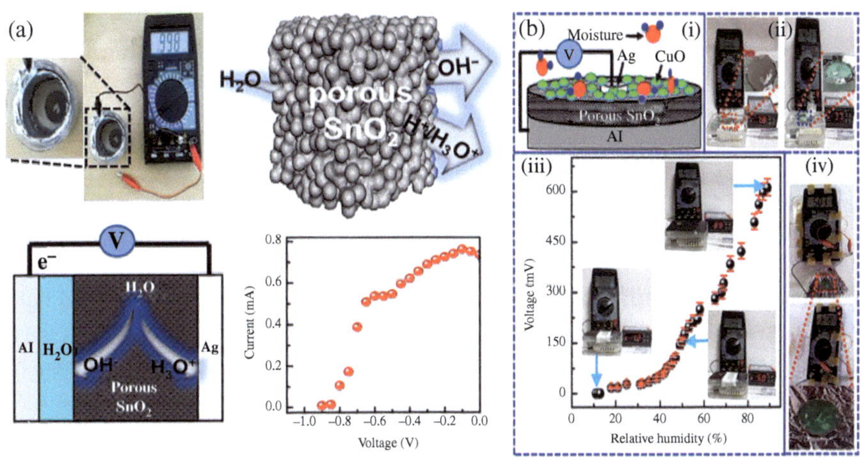

Figure 11.11 (a) Static hydroelectric power generation and tunable power generation characteristics of mesoporous SnO_2. (b) Schematic of the hydroelectric power generator through atmospheric moisture absorption. (ii) An effect of moisture pumping through CuO-based absorbing layer on voltage production. (iii) Response of fabricated hydroelectric power generator in varying humid atmosphere. (iv) Voltage production ability of the device at costal area of Arabian Sea with a relative humidity of ~91%. Source: Reproduced with permission from [21] / John Wiley & Sons.

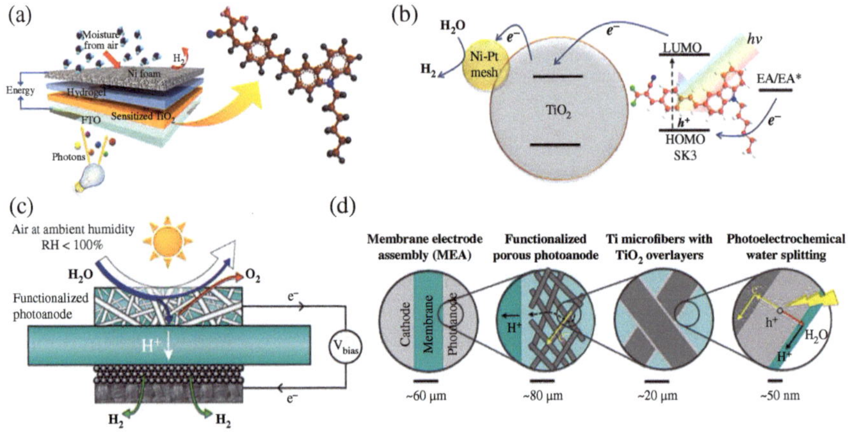

Figure 11.12 (a) Schematic diagram showing the construction of the hybrid device for simultaneous dehumidification and energy/hydrogen generation. (b) Working principle of the device. Source: Reproduced from [22] / with permission of Elsevier. (c) Schematic illustration of functionalized photoanodes at RH < 100 and (d) close view on the interfaces developed upon functionalization and pathways for proton conduction. Source: Reproduced from [23] / with permission of American Chemical Society.

subsequently convert it into electrical energy. To facilitate moisture absorption and energy conversion, they constructed a hybrid solar device by integrating customized hygroscopic materials with a metal-free organic dye-excited photosystem (Figure 11.12a). The hygroscopic material absorbs water from humid air, providing a neutral source of water for the photoelectrochemical reaction. This process directly converts atmospheric humidity into electricity and hydrogen in the presence of light, as shown in Figure 11.12b. In addition, the humidity level in the confined space is significantly reduced from 80% to 40%. The hybrid system operates with simultaneous dehumidification and generates a photocurrent of 240 μA cm^{-2} under ambient indoor light.

Solar hydrogen production is a promising sustainable energy carrier, and steady progress has been made in the development of photoelectrochemical (PEC) cells. Most of the research in this field has concentrated on ion transfer using acidic or alkaline liquid electrolytes. However, this performance is constrained by (i) light scattering and bubble obstruction of the catalytic sites, and (ii) limitations in mass transfer. An attractive alternative to liquid water feedstocks is the use of water vapor as a humidity presence in the surrounding air, which has been shown to alleviate the aforementioned issues and to expand the geographic range in which these devices can be used. Here, Tsampas and coworkers [23] demonstrates that porous TiO_2 and WO_3 functionalized photoanodes, combined solid electrolytes such as the proton-conducting Aquivion and Nafion ionomers, are capable of capturing water from the ambient air and thereby facilitating subsequent PEC hydrogen production (Figure 11.12c,d). Optimization strategies for photoanode functionalization were investigated by testing the effects of ionomer loading and ionomer composition. The optimized functionalized photoanodes, operating at 60% relative humidity (RH) and T-cells when water is introduced into the liquid phase (i.e. conventional PEC operation), recovered up to 90% of their performance with respect to reverse hydrogen electrodes (RHEs) at temperatures ranging from 30 to 70 °C. Full performance recovery can be attained at higher application potentials. In addition, long-term experiments demonstrated excellent stability, with 64 h of cycling at 60% RH, indicating that this concept is suitable for outdoor applications.

Unlike traditional water decomposition in aqueous media, the direct decomposition of atmospheric water presents a promising method for simultaneously dehumidifying living spaces and generating electricity. Bosman and coworkers [24] integrated a customized super-absorbent hydrogel, a catalyst, and a solar cell into a humidity digester that breaks down ambient moisture into hydrogen and oxygen, resulting in a highly efficient electrochemical cell (Figure 11.13). The hydrogel's function is to collect water from the surrounding humidity and transfer the collected water to the catalyst. Barium titanate and vertical 2D MoS_2 nanosheets are integrated as catalysts. The negatively polarized cathode enhances electron transfer and attracts H^+ to the MoS_2 surface, where water reduction occurs, while water oxidation takes place at the positively polarized anode. By employing this mechanism, the relative humidity of a moderately sized room can be maintained at less than 60% without any additional energy input. Furthermore, the humidity ablator generates a steady current of 12.5 mA cm^{-2} when exposed to ambient light.

(iii) **Methane.** Solid adsorbent-based carbon dioxide air capture facilities (DAC) exhibit significant water co-sorption [25]. Integrating this process into a power-to-gas (PtG) framework may mitigate this limitation by utilizing water extracted from the air as feedstock for electrolytic hydrogen production [26]. Agar's study of the entire integrated system, conducted through detailed simulations, demonstrates the potential for achieving a high degree of synergy

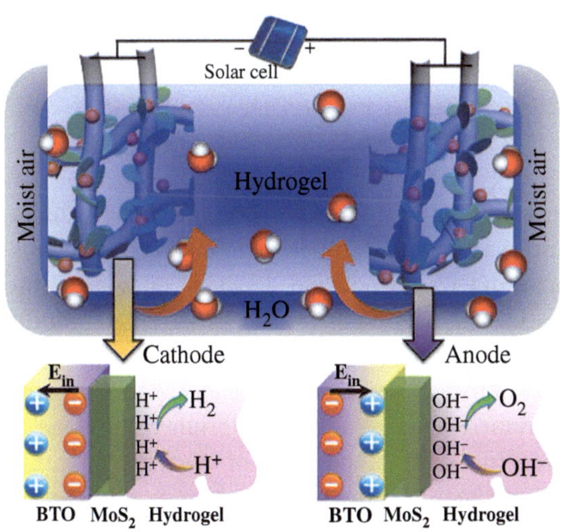

Figure 11.13 Schematic diagram of humidity splitting mechanism. Source: Reproduced from [24] / with permission of John Wiley & Sons.

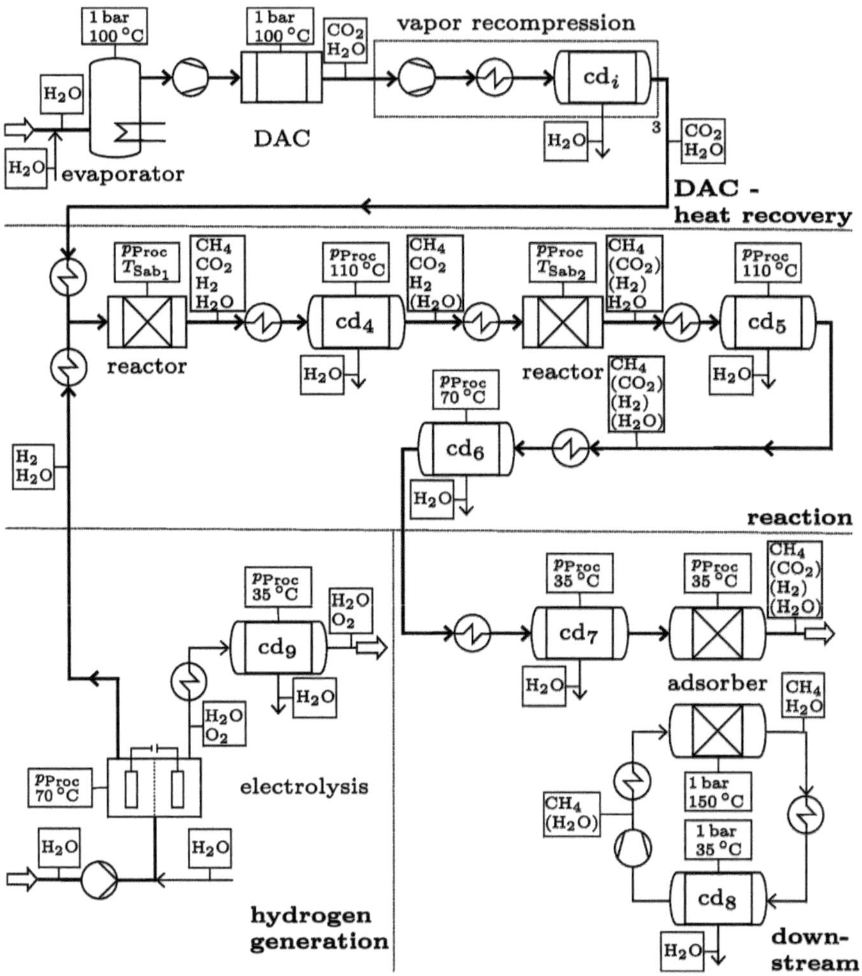

Figure 11.14 Basic flow sheet of the intensified heat and mass integrated DAC-PtG concept. Source: Reproduced from [27] / with permission of Elsevier.

in a strongly coupled DAC-PtG process [27]. Efficient water recycling reduces the process water requirement by a factor of 4. A detailed pinch point analysis of the integrated DAC-PtG concept was conducted, focusing on the overall energy requirements (Figure 11.14). High-temperature water condensation effectively recovers heat, yielding approximately 1.5 times the reaction heat released during the methanation process. The point of the electrolysis process was found to be critically important. Operating at an equilibrium cell voltage of 1.5 results in an excess heat 2 capture of 475 kJ per mole of CO, whereas at this equilibrium cell voltage a process heat demand of 16 kJ mol^{-1} was observed. Overall, this study demonstrates the feasibility of designing an autothermally operated process for methane production using carbon dioxide and water from ambient air, The synergies identified between the process components are particularly relevant for general power-to-X (PtX) applications.

In summary, fog collection technology, as an innovative method for obtaining water resources, has broad application prospects. In addition to the classic application scenarios mentioned above, this technology can be particularly valuable in natural disasters or other emergency situations, such as droughts and earthquakes, where water resources become extremely scarce. In these instances, fog collection technology can serve as a rapid and effective emergency measure to provide urgently needed water for disaster-stricken areas. Furthermore, in ecological restoration projects, fog collection technology can be utilized to supply essential water to support the reconstruction and rehabilitation of ecosystems. However, it is important to note that the efficiency and effectiveness of fog collection technology are influenced by various factors, including climate conditions, geographical location, equipment design, and operational methods. Therefore, when implementing fog collection projects, it is crucial to conduct thorough planning and evaluation based on the local environment and needs to ensure the feasibility and effectiveness of the project.

References

1 (a) Gao, S. and Wang, Z. (2023). *Nature Sustainability* 6: 1514. (b) Yang, K., Duan, C., Meng, Z. et al. (2024). *Carbohydrate Polymers* 335: 122069.
2 (a) Feng, J., Zhong, L., and Guo, Z. (2020). *Chemical Engineering Journal* 388: 124283. (b) Yue, H., Zeng, Q., Huang, J. et al. (2022). *Advances in Colloid and Interface Science* 300: 102583.
3 Xu, T., Lin, Y., Zhang, M. et al. (2016). *ACS Nano* 10: 10681.
4 Zhan, D. and Guo, Z. (2023). *Materials Horizons* 10: 4827.
5 Li, J., Ran, R., Wang, H. et al. (2021). *Nature Communications* 12: 5484.
6 Su, X., Hao, D., Li, P. et al. (2023). *Journal of Materials Chemistry A* 11: 7702.
7 Liang, Z., Wang, Q., Chu, D. et al. (2023). *ACS Applied Materials and Interfaces* 15: 46195.
8 Ghosh, R., Butt, H.-J., Baut, A. et al. (2023). *Nature Sustainability* 6: 1663.
9 (a) Yu, L., Xi, J., Li, M.-D. et al. (2012). *Physical Chefogry Chemical Physics* 14: 3589. (b) Chen, Y., Xu, M., Wen, J. et al. (2021). *Nature Sustainability* 4: 618. (c) Keerthana, S.P., Yuvakkumar, R., Ravi, G. et al. (2022). *Chemosphere* 293: 133540.
10 Yang, J., Zhang, X., Qu, H. et al. (2020). *Advanced Materials* 32: e2002936.
11 Zhou, W., Zhou, C., Deng, C. et al. (2022). *Advanced Functional Materials* 32: 2113264.
12 Ji, Y., Yang, W., Li, X. et al. (2023). *Small* 19: e2304037.
13 Zhou, X., Zhang, P., Zhao, F., and Yu, G. (2020). *ACS Materials Letters* 2: 1419.
14 Chang, X., Li, S., Li, N. et al. (2022). *Journal of Materials Chefogry A* 10: 18170.

15 Hu, J., Wei, J., Li, J. et al. (2024). *Energy and Environmental Science* 17: 1664.
16 Hanikel, N., Kurandina, D., Chheda, S. et al. (2023). *ACS Central Science* 9: 551.
17 Yang, K., Pan, T., Pinnau, I. et al. (2020). *Nano Energy* 78: 105326.
18 Wei, C., Lin, L., Zhao, Y. et al. (2020). *ACS Applied Materials and Interfaces* 12: 19130.
19 Li, M., Zong, L., Yang, W. et al. (2019). *Advanced Functional Materials* 29: 1901798.
20 Chen, Y., Kuang, Y., Shi, D. et al. (2020). *Nano Energy* 73: 104765.
21 Solanki, V., Krupanidhi, S.B., and Nanda, K.K. (2020). *International Journal of Energy Research* 44: 1276.
22 Nandakumar, D.K., Vaghasiya, J.V., Yang, L. et al. (2020). *Nano Energy* 68: 104263.
23 Zafeiropoulos, G., Johnson, H., Kinge, S. et al. (2019). *ACS Applied Materials and Interfaces* 11: 41267.
24 Yang, L., Loh, L., Nandakumar, D.K. et al. (2020). *Advanced Materials* 32: e2000971.
25 Thema, M., Bauer, F., and Sterner, M. (2019). *Renewable and Sustainable Energy Reviews* 112: 775.
26 Peters, R., Baltruweit, M., Grube, T. et al. (2019). *Journal of CO_2 Utilization* 34: 616.
27 Drechsler, C. and Agar, D.W. (2020). *Applied Energy* 273: 115076.

12

Challenges of Fog Harvesting

Fog harvesting is a technology that captures water vapor from the air and condense it into water droplets, providing access freshwater resources. While this technology has promising applications in arid regions or areas with limited traditional water sources, it faces a number of challenges. The following are some of the major challenges faced by fog collection.

12.1 Fog Volume and Humidity

The efficiency of fog collection is significantly influenced by the volume of fog and the humidity levels in the area. In regions with low fog density or insufficient humidity, the quantity of water collected will be minimal. Consequently, thorough assessments of humidity and fog volume should be conducted when selecting a site for the implementation of a fog collection program.

12.2 Collector Surface Characterization

Effective fog collection necessitates the use of specialized collector surface materials that can effectively adsorb and coalesce water droplets. However, the performance of these materials can be affected by factors such as particle deposition, contaminants, and biological growth. Consequently, maintaining and cleaning collector surfaces presents a significant challenge.

12.3 Cost-Effectiveness

The cost of fog collection technology is an important factor in its adoption and implementation. Currently, the construction and operational expenses associated with fog collection systems are relatively high, encompassing collector installation, maintenance, and post-treatment processes. Consequently, a significant challenge lies in finding ways to reduce costs while enhancing effectiveness.

Bioinspired Materials for Fog Harvesting: Principles, Design, Preparation, and Applications,
First Edition. Zhiguang Guo and Weimin Liu.

12.4 Water Quality Treatment

As fog condenses from the air, it may contain atmospheric pollutants and microorganisms. Therefore, the proper treatment and filtration of the collected fog to ensure that the water quality meets acceptable standards is also a significant challenge.

12.5 Sustainability and Environmental Impacts

Fog collection technologies have a minimal environmental impact; however, their sustainability and effects on ecosystems must be carefully evaluated. For instance, the collection of significant amounts of fog could disrupt local humidity and temperature distributions, potentially affecting the surrounding ecosystem.

In summary, despite the potential applications of fog collection technologies, several challenges persist including fog volume and humidity, collector surface properties, cost-effectiveness, water quality and treatment, and sustainability and environmental impacts. Further research and technological improvements are necessary to address these challenges and facilitate broader applications.

13

Challenges and Prospects

Water is an essential resource for human survival and socioeconomic development, serving as the fundamental basis for all living organisms on Earth. Despite the fact that approximately 70% of the Earth's surface is covered by water, the availability of freshwater resources remains critically limited, posing challenges for sustaining human populations. Moreover, climate change is anticipated to induce significant alterations to water supply across various regions [1]. Over the past century, global warming has profoundly influenced the distribution of water resources and regional hydroclimatic cycles, resulting in unpredictable shifts in rainfall patterns. Concurrently, water demand continues to rise, exacerbating the already pressing issue of diminishing water availability [2].

The uneven distribution of freshwater resources, compounded by demographic and geographic variations across countries and regions, has heightened the potential for conflicts. This disparity can lead to tensions between regions and may even escalate into violent confrontations over water access [3]. Consequently, the exploration of innovative methods for harvesting freshwater from fog and atmospheric moisture has garnered considerable attention among researchers. Advances in biomimetic materials – engineered to mimic the natural water-collecting abilities of specific plants and animals – show significant promise in addressing water scarcity challenges. These materials exhibit high theoretical fog-trapping efficiency, low production costs, and versatility in application, making them a compelling solution for mitigating issues related to water supply.

In this book, Chapter 1 examines the current state of global water resources [4], highlights the necessity of fog harvesting, and discusses the development of innovative fog harvesting materials. It emphasizes the bionic design principles inspired by natural organisms and outlines methods for evaluating the performance of these engineering materials. Chapter 2 identifies natural prototypes for fog collection, detailing the collection strategies employed by these biomimetic models [4]. Notable examples include cacti, iron orchids, pitcher plants, desert beetles, spider silk, and desert lizards. Chapter 3 introduces the fundamental principles of fog harvesting, including the nucleation mechanism of water molecules, droplet growth theory, and droplet transport mechanisms. A theoretical model for atmospheric water collection based on adsorption is also presented.

Bioinspired Materials for Fog Harvesting: Principles, Design, Preparation, and Applications, First Edition. Zhiguang Guo and Weimin Liu.

Chapter 4 provides a comprehensive summary of the factors influencing the fog collection process. Each factor's impact on its efficiency is discussed, establishing a theoretical basis for designing highly effective fog collection devices. Chapter 5 reviews a range of microscopic materials utilized in fog collection. It emphasizes how variations in microstructure lead to differing wettability, which significantly affects the efficiency of fog collection. Chapter 6 organizes the fog collection materials introduced in the previous chapter into distinct categories based on their microstructural characteristics. Chapter 7 explores how the mechanical properties of fog collection materials influence their efficiency. The harshness of real-world applications necessitates a focus on the durability of materials, making an understanding of how to enhance their mechanical properties crucial.

Chapter 8 outlines various surface preparation methods, highlighting how they yield different micro- and nano-scale surface characteristics. These variations in surface structure affect wettability and, ultimately, fog collection efficiency. Chapters 9 and 10 present two types of fog harvesting systems: non-phase change

Figure 13.1 Engineering biomimetic materials. for fog collecting. Source: Reproduced from [5] / with permission of Elsevier.

fog collection systems and phase change atmospheric water collection systems. Detailed discussions cover the common base materials, structural designs, and surface wettability adaptations used in non-phase change systems. Additionally, the theory behind both types of phase change systems is elaborated upon, alongside recent advancements in materials and their collection processes, along with developmental insights. Chapter 11 investigates the applications of fog collection, assessing its feasibility and potential in current and future contexts. Chapters 12 and 13 summarize the challenges faced in the field of fog harvesting and propose future research directions (Figure 13.1).

13.1 Research Status and Challenges

Freshwater resources are essential for the survival of all living organisms on Earth and play a crucial role in daily human activities and production processes. According to data from the China Geological Survey, freshwater constitutes approximately 2.53% of the Earth's total water supply, with nearly 70% of this freshwater stored in solid glaciers. Consequently, the availability of consumable freshwater for human use is severely limited [4, 6]. Climate change is exacerbating existing water stress. Projections indicate that an increase in global average temperature by 0.4 °C by 2030 will significantly diminish average runoff and overall water resources in river basins [7]. If current trends persist, it is estimated that 5 billion people worldwide will experience relative water scarcity, with 450 million facing severe water stress [8]. Furthermore, the excessive use of chemical products in various human activities has led to the pollution of freshwater sources, further exacerbating the scarcity of potable water. Given these challenges, finding effective methods to maximize freshwater availability has become an urgent priority.

At the technical level, seawater desalination is a well-established method for procuring freshwater resources [9]. As of 2018, there were approximately 15,906 desalination facilities globally, producing over 95 million cubic meters of desalinated water per day [10]. However, the high energy consumption, significant costs, and geographical limitations associated with desalination technologies render them inadequate for meeting the freshwater needs of inland arid regions. Consequently, there is a pressing necessity for more cost-effective approaches to freshwater acquisition.

Currently, large mesh fog collectors are widely utilized in practical applications for fog harvesting. These collectors operate by trapping droplets on a grid as fog passes through [11]. The collected small droplets coalesce with other droplets during the fog harvesting process. Once the volume of the droplets reaches a critical threshold, they are able to overcome the adhesive forces at the surface of the fog collector, thereby sliding off and being collected into a storage tank. The mesh used in water collectors can be fabricated using various techniques, typically taking the form of woven or knitted structures made from textile materials [12]. However, there are three notable challenges associated with the mesh in fog harvesting. First, droplet convection loss may occur. As the volume of water droplets increases and approaches

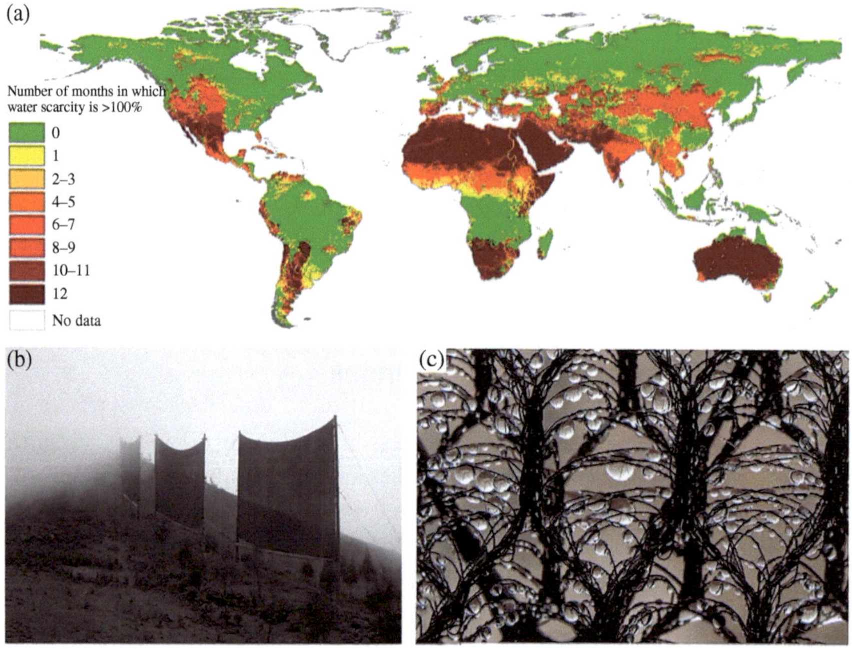

Figure 13.2 (a) Distribution of water scarcity on a global scale, Source: [14] / John Wiley & Sons / CC BY 4.0; (b) Fog nets in the mountains; (c) Details of fog nets. Source: Reproduced with permission from [15, 16] / John Wiley & Sons.

the critical volume, aerodynamic forces can cause the droplets to detach from the collector surface and return to the air, overcoming the adhesion forces. Second, the phenomenon of gridlock can lead to a reduction in the efficiency of fog collection over time [13]. Lastly, large mesh fog collectors tend to entail higher costs, which may yield lower economic returns. Addressing these challenges is essential for improving the efficiency and viability of fog harvesting as an alternative freshwater resource (Figure 13.2).

Fog collection technology has been developed and has demonstrated its potential benefits; however, several challenges persist [17]. These challenges include a lack of supportive policies, limited local institutions, and an absence of expert communities. Effectively addressing these challenges is essential for mitigating conflicts in regions where competition for clean water resources is becoming increasingly intense.

The author asserts that water scarcity represents a persistent and long-term challenge, particularly in remote areas that lack adequate water resources. This reality serves as a critical backdrop for addressing local water scarcity issues. Therefore, the authors propose the following recommendations: (i) Review and Evaluate National Water Management Action Plans: A comprehensive review of existing national water management action plans is essential, with a specific focus on evaluating the value and potential of fog harvesting as a sustainable water source. (ii) Analyze Local Conditions: A thorough analysis of the local context is necessary to understand the trade-offs between fog harvesting initiatives and the socioeconomic dynamics of the community. This understanding will ensure that such projects are not only environmentally viable but also socially and economically beneficial. (iii) Establish Supportive Water Policies and Budgets: It is crucial to develop appropriate policies and allocate budgets that facilitate the integration of fog harvesting technologies into broader water management strategies. (iv) Educate and Inform Investors: Providing detailed explanations of fog harvesting technologies and their benefits to potential investors is vital. This information will empower decision-makers with the necessary insights to formulate supportive policies and make informed investment choices.

In conclusion, addressing these issues will not only enhance the feasibility of fog harvesting as a viable solution to water scarcity but will also contribute to the development of sustainable water management practices in vulnerable regions.

13.2 Future Expectations

Over the course of 3.7 billion years of evolution, numerous dryland flora and fauna have developed remarkable adaptations for survival in harsh and fluctuating environments. Inspired by the fog collection strategies observed in various organisms, such as desert beetles [18], spider silk [19], cacti [20], and pitcher plants [21], there has been a proliferation of fog harvesting devices designed using biomimetic materials. In comparison to conventional seawater desalination technologies, these fog collection systems are more cost-effective and practical. They are not limited by geographical constraints, do not require specialized maintenance personnel, and can create job opportunities for women in arid regions, thereby fostering economic development in those areas [21c]. Future advancements in biomimetic fog harvesting materials can be approached from several perspectives:

(1) Exploration of biological diversity is essential. Nature hosts a vast array of undiscovered strategies for water harvesting, and many organisms with potentially superior water management capabilities remain unidentified. It is imperative for researchers to continue exploring these biological mechanisms to enhance biomimetic fog harvesting technologies. Besides, the dynamic nature of evolution is also very important. The evolutionary landscape is continually shifting, with organisms constantly adapting and developing novel fog-collecting mechanisms. As new fog-collecting species emerge, their unique strategies can serve as valuable inspiration for the design of improved biomimetic fog harvesting systems. In conclusion, the ongoing search for lesser-known organisms with exceptional fog-trapping abilities is essential, paving the way for innovative developments in this field.
(2) The theory and mechanisms underlying fog collection warrant further investigation. While many fog-collecting devices have been developed, comprehensive explanations of their operational mechanisms remain scarce. Notably, research has thoroughly examined the role of wettability in fog

collection, notably the interactions between hydrophilic and hydrophobic surfaces, as well as the influence of the distribution of these surface characteristics on fog collection efficiency.

(3) Currently, fog collection technologies are primarily at the laboratory stage, making it essential to address how to transition these systems for practical outdoor applications. In particular, optimizing the stability and durability of fog collection devices in harsh environmental conditions should be prioritized in future research (Figure 13.3).

(4) The fog collection process is inherently lengthy, and environmental factors often lead to the evaporation of small droplets, which can diminish the total

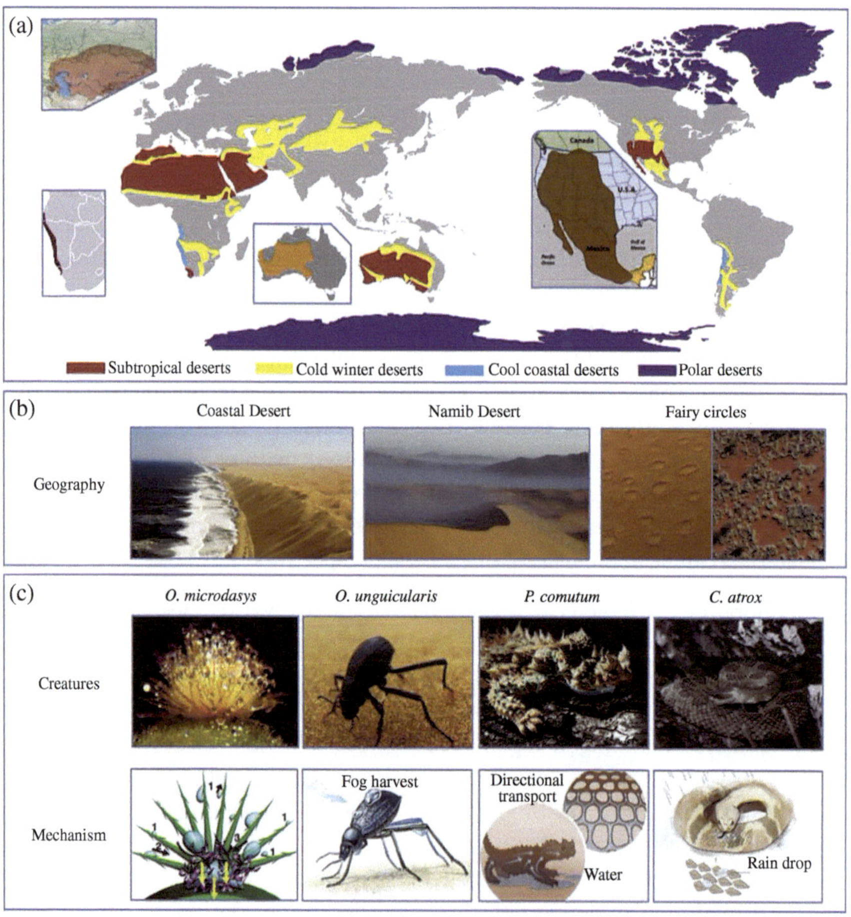

Figure 13.3 (a) Geographical distribution of the two extreme conditions, including the deserts and the rainforests. (b) Potential to incorporate attributes of creatures from desert and rainforest to increase efficiency and enable new application. (c) Examples that utilizes different characteristic including hetero-wettability bumps in Namib Desert beetles, asymmetric spine structures in cactus, slipperiness of Nepenthes pitcher plants, pumping strategy of emergent aquatic plants, and capillary channel of honeycomb network. Source: Reproduced with permission from [22] / Elsevier / CC BY 4.0.

volume of collected fog. Consequently, it is crucial to investigate the thermal properties of materials used in fog collection to enhance their efficiency.

(5) During fog collection, the direction of fog flow may frequently vary. Therefore, it is important to examine the impacts of external environmental factors, such as the stability of fog flow and ambient pressure, on the performance of fog collection systems.

(6) In the initial design phases of fog collection materials and devices, researchers should aim to integrate biological inspirations to identify species with exceptional water-harvesting capabilities. By harnessing the strengths of various organisms within an integrated system, the performance of fog collection materials and devices can be significantly enhanced. For instance, certain desert lizards efficiently transport water using their skin to draw moisture from multiple sources. By learning from such adaptations and combining traits from other species, innovative applications could emerge, including anti-gravity water transfer systems and spontaneous irrigation solutions.

(7) To optimize land resource utilization, it is essential to develop multifunctional integrated fog collecting devices for arid regions. These designs should emphasize sustainable materials, environmentally friendly preparation processes, and cost-effective solutions.

(8) Given that fog harvesting aims to alleviate the scarcity of freshwater resources, ensuring the safety of collected water is critical. Previous research has often overlooked water safety considerations, which can include contamination from microbial growth on wet surfaces and the potential leaching of functional coatings from droplets over time. Moreover, fog collection systems, such as nets, plates, and desiccant collectors, may be susceptible to contamination from algae, bacteria, and animal droppings, necessitating appropriate water treatment before potable use.

(9) Many existing bionic material fog collectors occupy extensive areas, which may inadvertently capture low-flying wildlife, disrupt ecosystems, and compromise the safety of the collected water. Therefore, developing fog collectors with reduced surface area while maintaining or enhancing efficiency would be advantageous.

(10) Finally, it is imperative to establish a standardized evaluation criterion for measuring the efficiency of fog collection systems. The variability in experimental conditions and parameters documented across studies makes it challenging to compare the overall performance of different fog-collecting devices (Figure 13.4).

Greatly alleviate the problem of freshwater scarcity in arid regions. Based on this, a variety of atmospheric water collection methods have been proposed, which can be roughly divided into passive cooling atmospheric environmental water collection and adsorption-based atmospheric environmental water collection.

Currently, the rapid advancement of science and technology is often accompanied by significant environmental degradation, exacerbating the greenhouse effect and

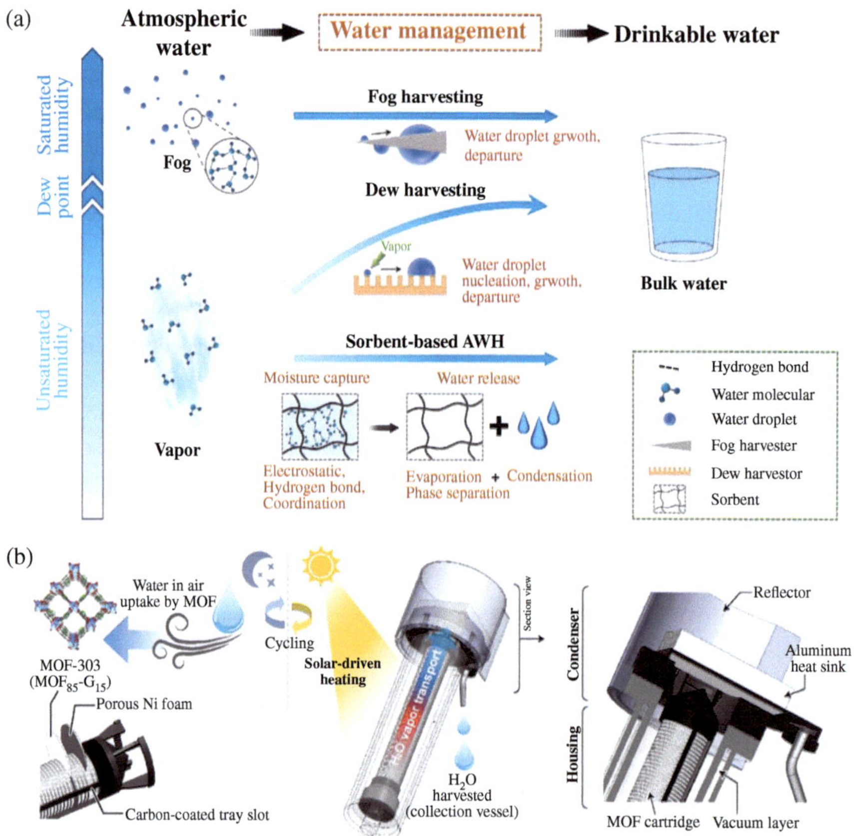

Figure 13.4 (a) Three different ways of collecting water. Source: Reproduced from [23] / with permission of John Wiley & Sons; (b) MOF-based device for atmospheric water adsorption. Source: Reproduced from [24] / with permission of Springer Nature.

contributing to increasingly unpredictable changes in global climate patterns [25]. Notably, many regions are experiencing a yearly decline in both the frequency and duration of fog events, which poses a serious threat to communities that rely on non-phase change fog collection for their freshwater supply [26]. In the context of the global water cycle, water exists in the atmosphere in the form of vapor and droplets, representing a substantial yet largely untapped reservoir of freshwater – equivalent to approximately one-eighth of the total freshwater resources found in rivers and lakes [27]. Harnessing these abundant, albeit often overlooked, atmospheric water resources not only presents an opportunity to enhance water security but could also play a critical role in mitigating freshwater scarcity, particularly in arid regions [28]. To this end, a variety of atmospheric water collection methods have been developed. These methods can be broadly categorized into two principal types: passive cooling atmospheric water collection and adsorption-based atmospheric water collection. Each of these approaches offers unique advantages and can be strategically deployed to maximize the efficiency and effectiveness of freshwater harvesting.

Passive cooling atmospheric water collection involves the utilization of external energy to lower the temperature of the collection surface below the ambient dew point. This temperature differential facilitates the phase transition of water vapor from the atmosphere, allowing for its collection. However, the high energy consumption associated with this process presents significant challenges that limit the broader application of passive cooling technologies. In contrast, adsorption-based atmospheric water harvesting (AWH) technologies offer several advantages, including high adaptability and fewer restrictions on environmental conditions. Adsorbents exhibit strong adsorption capabilities over a wide humidity range, effectively capturing moisture from the air. Nonetheless, there is a pressing need for the development of new atmospheric water adsorbents that possess high water absorption capacity, low energy requirements, rapid kinetics, and cycle stability. These improvements are essential for enhancing the efficiency of atmospheric water collection. For example, the future innovations in atmospheric water collection technology can be pursued along two primary avenues: the first one is the molecular and structural design of adsorbent materials. There exists a notable mismatch between the adsorption and desorption rates of many existing adsorbent materials. Among the reported adsorbents, the maximum adsorption rate rarely exceeds 25% of the desorption rate. This imbalance hinders the ability to meet the rapid adsorption and desorption demands characteristic of continuous systems. Therefore, enhancing the kinetics of adsorption and desorption should be prioritized to facilitate continuous operation. Additionally, the incorporation of anisotropic thermal conductivity and engineered pore structures in adsorbent materials can contribute to improved mass and energy efficiency. The second one is the optimization of adsorption-based water collection equipment. Future efforts should focus on enhancing the compactness and manufacturability of water collection devices, particularly for deployment in arid environments. There is a need for the design of smart and user-friendly equipment that addresses the mismatch between adsorption and desorption rates at the system level, thereby enabling more efficient water release and collection [29].

Current strategies for assessing atmospheric water harvesting technologies face several challenges that must be addressed to advance the field. For instance, the absence of a unified consensus on fundamental performance indicators hinders the comparability and fairness of various atmospheric water collection technologies. In addition, the lack of comprehensive guidelines for target metrics results in ambiguity regarding the developmental trajectory of atmospheric water collection research as a whole. To achieve the desired outcomes of sustainable water production and large-scale implementation in arid and water-scarce regions, AWH systems must be strategically designed. Key considerations should include enhancing adsorption capacity, optimizing heat and mass transfer efficiency, minimizing renewable energy consumption, and ensuring long-term stability while reducing production costs.

This book begins by exploring the natural phenomenon of fog collection, offering a thorough analysis of the underlying mechanisms. It provides an extensive overview of current biomimetic materials utilized for fog harvesting and presents unique insights into the design of contemporary fog collectors. Furthermore, it discusses the potential applications of biomimetic fog-collection functional materials based on practical considerations. The author posits that drawing inspiration from nature serves as a foundational principle in the development of fog harvesting engineering materials. Bridging the gap between basic research and applied technology remains a crucial direction for further study in fog collection materials. Ultimately, enhancing material design and optimizing equipment to improve efficiency is the foremost objective. The authors firmly advocate that the development of innovative, efficient, and cost-effective engineering materials for fog harvesting is essential for fostering sustainable human development.

References

1 a) Mekonnen, M.M. and Hoekstra, A.Y. *Science Advances* 2: e1500323. b) Bai, H., Zhang, C., Long, Z. et al. (2018). *Journal of Materials Chefogry A* 6: 20966.

2 Orlowsky, B., Hoekstra, A.Y., Gudmundsson, L., and Seneviratne, S.I. (2014). *Environmental Research Letters* 9: 074007.

3 Bernauer, T. and Böhmelt, T. (2020). *Nature Sustainability* 3: 350.

4 Vörösmarty, C.J., Green, P., Salisbury, J., and Lammers, R.B. (2000). *Science* 289: 284.

5 Zhang, Y., Cai, Y., Shi, J. et al. (2022). *Desalination* 540: 115975.

6 Oki, T. and Kanae, S. (2006). *Science* 313: 1068.

7 Jenkins, S., Povey, A., Gettelman, A. et al. (2022). *Journal of Climate* 35: 4273.

8 Cooley, S.W., Ryan, J.C., and Smith, L.C. (2021). *Nature* 591: 78.

9 a) Lee, K. and Jepson, W. (2021). *Desalination* 509: 115066. b) Oviroh, P.O., Jen, T.-C., Ren, J., and van Duin, A. (2023). *npj Clean Water* 6: 14. c) Wang, Q., Li, N., Bolto, B. et al. (2016). *Desalination* 387: 46. d) Ahmed, F.E., Hashaikeh, R., and Hilal, N. (2020). *Desalination* 495: 114659.

10 Jones, E., Qadir, M., van Vliet, M.T.H. et al. (2019). *Science of the Total Environment* 657: 1343.

11 a) Gude, V.G. (2016). *Water Research* 89: 87. b) Pinto, F.S. and Marques, R.C. (2017). *Renewable and Sustainable Energy Reviews* 78: 904.

12 a) Algarni, S. (2018). *Water and Environment Journal* 32: 301. b) Noman, M.T., Militky, J., Wiener, J. et al. (2018). *Ultrasonics* 83: 203.

13 Park, K.-C., Chhatre, S.S., Srinivasan, S. et al. (2013). *Langmuir* 29: 13269.

14 Jiang, Y., Machado, C., and Park, K.-C.K. (2023). *Droplet* 2: e55.

15 Davtalab, R., Salamat, A., and Oji, R. (2013). *Irrigation and Drainage* 62: 281.

16 Kennedy, B.S. and Boreyko, J.B. (2023). *Advanced Functional Materials* 2306162.

17 Pan, Z., Pitt, W.G., Zhang, Y. et al. (2016). *Nature Plants* 2: 16076.

18 a) Gido, B., Friedler, E., and Broday, D.M. (2016). *Atmospheric Research* 182: 156. b) Wang, X., Zeng, J., Yu, X. et al. (2020). *RSC Advances* 10: 282. c) Wu, J., Yan, Z., Yan, Y. et al. (2022). *ACS Applied Materials and Interfaces* 14: 49338.

19 a) Wen, C., Guo, H., Bai, H. et al. (2019). *ACS Applied Materials and Interfaces* 11: 34330. b) Chen, W. and Guo, Z. (2019). *Nanoscale* 11: 15448.

20 a) Zhang, M., Zheng, Z., Zhu, Y. et al. (2022). *Chemical Engineering Journal* 433: 134495. b) Heng, X., Xiang, M., Lu, Z., and Luo, C. (2014). *ACS Applied Materials and Interfaces* 6: 8032.

21 a) Wang, J., Yi, S., Yang, Z. et al. (2020). *ACS Applied Materials and Interfaces* 12: 21080. b) Zhou, H., Jing, X., Li, S., and Guo, Z. (2021). *Chemical Engineering Journal* 417: 129215. c) Feng, R., Song, F., Xu, C. et al. (2021). *Chemical Engineering Journal* 422: 130119.

22 He, G., Zhang, C., and Dong, Z. (2023). *iScience* 26: 105819.

23 Abdulbakee Muhammed, H., Shahadat, M., Tweib, S.A. et al. (2023). *ChemBioEng Reviews* 197.

24 Song, W., Zheng, Z., Alawadhi, A.H., and Yaghi, O.M. (2023). *Nature Water* 1: 626.

25 Wang, Y., Gao, S., Xu, W., and Wang, Z. (2020). *Advanced Functional Materials* 30: 1908252.

26 Loeb, S. and Sourirajan, S. (1963). *Saline Water Conversion—II* Ch. 9, vol. 38, 117. American Chemical Society.

27 Cess, R.D., Zhang, M.H., Ingram, W.J. et al. (1996). *Journal of Geophysical Research: Atmospheres* 101: 12791.

28 Katata, G. (2014). *Journal of Geophysical Research: Atmospheres* 119: 8137.

29 a) Poredoš, P., Shan, H., Wang, C. et al. (2022). *Energy and Environmental Science* 15: 3223. b) Deng, F., Chen, Z., Wang, C. et al. (2022). *Advanced Science* 9: 2204724.

Index

Bioinspired Materials for Fog Harvesting: Principles, Design, Preparation, and Applications, First Edition. Zhiguang Guo and Weimin Liu.